Contents

Annotated Instructor's Edition

MATHEMATICS IN ACTION

An Introduction to Algebraic, Graphical, and Numerical Problem Solving

FOURTH EDITION

The Consortium for Foundation Mathematics

Ralph Bertelle	*Columbia-Greene Community College*
Judith Bloch	*University of Rochester*
Roy Cameron	*SUNY Cobleskill*
Carolyn Curley	*Erie Community College—South Campus*
Ernie Danforth	*Corning Community College*
Brian Gray	*Howard Community College*
Arlene Kleinstein	*SUNY Farmingdale*
Kathleen Milligan	*Monroe Community College*
Patricia Pacitti	*SUNY Oswego*
Rick Patrick	*Adirondack Community College*
Renan Sezer	*LaGuardia Community College*
Patricia Shuart	*Polk State College—Winter Haven, Florida*
Sylvia Svitak	*Queensborough Community College*
Assad J. Thompson	*LaGuardia Community College*

Addison-Wesley

Boston   Columbus   Indianapolis   New York   San Francisco   Upper Saddle River
Amsterdam   Cape Town   Dubai   London   Madrid   Milan   Munich   Paris   Montréal   Toronto
Delhi   Mexico City   São Paulo   Sydney   Hong Kong   Seoul   Singapore   Taipei   Tokyo

Editorial Director, Mathematics: Christine Hoag
Editor in Chief: Maureen O'Connor
Content Editor: Courtney Slade
Assistant Editor: Mary St. Thomas
Senior Managing Editor: Karen Wernholm
Production Project Manager: Beth Houston
Senior Designer/Cover Designer: Barbara Atkinson
Interior Designer: Studio Montage
Digital Assets Manager: Marianne Groth
Production Coordinator: Katherine Roz
Associate Producer: Christina Maestri
Associate Marketing Manager: Tracy Rabinowitz
Marketing Coordinator: Alicia Frankel
Senior Author Support/Technology Specialist: Joe Vetere
Rights and Permissions Advisor: Michael Joyce
Senior Manufacturing Buyer: Carol Melville
Production Management/Composition: PreMediaGlobal
Cover photo: Eric Michaud/iStockphoto

Many of the designations used by manufacturers and sellers to distinguish their products are claimed as trademarks. Where those designations appear in this book, and Addison-Wesley was aware of a trademark claim, the designations have been printed in initial caps or all caps.

Library of Congress Cataloging-in-Publication Data
Mathematics in action: an introduction to algebraic, graphical, and numerical problem solving / the Consortium for Foundation Mathematics. — 4th ed.
   p. cm.
  ISBN-13: 978-0-321-69860-5 (student ed.)
  ISBN-10: 0-321-69860-6 (student ed.)
  ISBN-13: 978-0-321-69273-3 (instructor ed.)
  ISBN-10: 0-321-69273-X (instructor ed.)
   1. Mathematics. I. Consortium for Foundation Mathematics.

QA39.3.M42 2012
510—dc22

2009052323

Addison-Wesley
is an imprint of

www.pearsonhighered.com

1 2 3 4 5 6 7 8 9 10—EB—14 13 12 11 10

Objectives:
1. Identify signed numbers.

2. Use signed numbers to represent quantities in real-world situations.

3. Compare signed numbers.

4. Calculate the absolute value of numbers.

5. Identify and use properties of addition and subtraction of signed numbers.

6. Add and subtract signed numbers using absolute value.

Objective:
1. Multiply and divide signed numbers.

Objectives:
1. Use the order of operations convention to evaluate expressions involving signed numbers.

2. Evaluate expressions that involve negative exponents.

3. Distinguish between such expressions as -5^4 and $(-5)^4$.

4. Write very small numbers in scientific notation.

Objectives:
1. Identify input and output in situations involving two variable quantities.

2. Determine the replacement values for a variable within a given situation.

3. Use a table to numerically represent a relationship between two variables.

4. Represent a relationship between two variables graphically.

5. Identify trends in data pairs that are represented numerically and graphically.

APPENDIXES

Preface

Our Vision

Mathematics in Action: An Introduction to Algebraic, Graphical, and Numerical Problem Solving, Fourth Edition, is intended to help college mathematics students gain mathematical literacy in the real world and simultaneously help them build a solid foundation for future study in mathematics and other disciplines.

Our team of fourteen faculty, primarily from the State University of New York and the City University of New York systems, used the AMATYC *Crossroads* standards to develop this three-book series to serve a very large population of college students who, for whatever reason, have not yet succeeded in learning mathematics. It became apparent to us that teaching the same content in the same way to students who have not previously comprehended it is not effective, and this realization motivated us to develop a new approach.

Mathematics in Action is based on the principle that students learn mathematics best by doing mathematics within a meaningful context. In keeping with this premise, students solve problems in a series of realistic situations from which the crucial need for mathematics arises. *Mathematics in Action* guides students toward developing a sense of independence and taking responsibility for their own learning. Students are encouraged to construct, reflect on, apply, and describe their own mathematical models, which they use to solve meaningful problems. We see this as the key to bridging the gap between abstraction and application, and as the basis for transfer learning. Appropriate technology is integrated throughout the books, allowing students to interpret real-life data verbally, numerically, symbolically, and graphically.

We expect that by using the *Mathematics in Action* series, all students will be able to achieve the following goals:

- Develop mathematical intuition and a relevant base of mathematical knowledge.

- Gain experiences that connect classroom learning with real-world applications.

- Prepare effectively for further college work in mathematics and related disciplines.

- Learn to work in groups as well as independently.

- Increase knowledge of mathematics through explorations with appropriate technology.

- Develop a positive attitude about learning and using mathematics.

- Build techniques of reasoning for effective problem solving.

- Learn to apply and display knowledge through alternative means of assessment, such as mathematical portfolios and journal writing.

Our vision for you is to join the growing number of students using our approaches who discover that mathematics is an essential and learnable survival skill for the 21st century.

Pedagogical Features

The pedagogical core of *Mathematics in Action* is a series of guided-discovery activities in which students work in groups to discover mathematical principles embedded in realistic situations. The key principles of each activity are highlighted and summarized at the activity's conclusion. Each activity is followed by exercises that reinforce the concepts and skills revealed in the activity.

The activities are clustered within each chapter. Each cluster contains regular activities along with project and lab activities that relate to particular topics. The lab activities require more than just paper, pencil, and calculator; they also require measurements and data collection and are ideal for in-class group work. The project activities are designed to allow students to explore specific topics in greater depth, either individually or in groups. These activities are usually self-contained and have no accompanying exercises. For specific suggestions on how to use the three types of activities, we strongly encourage instructors to refer to the *Instructor's Resource Manual with Tests* that accompanies this text.

Each cluster concludes with two sections: What Have I Learned? and How Can I Practice? The What Have I Learned? exercises are designed to help students pull together the key concepts of the cluster. The How Can I Practice? exercises are designed primarily to provide additional work with the numeric and algebraic skills of the cluster. Taken as a whole, these exercises give students the tools they need to bridge the gaps between abstraction, skills, and application.

In Chapter 1, two sets of Skills Check exercises follow Clusters 3 and 4 to provide students with more opportunities to practice basic numerical skills. Additionally, each chapter ends with a Summary containing a brief description of the concepts and skills discussed in the chapter, plus examples illustrating these concepts and skills. The concepts and skills are also referenced to the activity in which they appear, making the format easier to follow for those students who are unfamiliar with our approach. Each chapter also ends with a Gateway Review, providing students with an opportunity to check their understanding of the chapter's concepts and skills.

Changes from the Third Edition

The Fourth Edition retains all the features of the previous edition, with the following content changes.

- All data-based activities and exercises have been updated to reflect the most recent information and/or replaced with more relevant topics.

- The language in many activities is now clearer and easier to understand.

- Activity 2.9, "Are They the Same?" (now Activity 2.10) is now covered in Cluster 3, "Problem Solving Using Algebra."

- Three new activities have been added to Chapters 2, 3, and 4.

  - Activity 2.9, "Four out of Five Dentists Prefer Crest"
  - Lab Activity 3.11, "Body Parts"
  - Activity 4.9, "A Thunderstorm"

- Activity 1.6, "Everything Is Relative," and Activity 2.4, "Symbolizing Arithmetic," have been rewritten extensively.

- An additional objective on determining the replacement values for a variable within a given situation has been added to Activity 2.1, "Blood-Alcohol Levels."

- Activities 2.14 and 2.15 from the previous edition have been combined into one activity, Activity 2.11, "Do It Two Ways."

- Several activities have moved to MyMathLab to streamline the course without loss of content.

Supplements

Instructor Supplements
Annotated Instructor's Edition

ISBN-13 978-0-321-69273-3
ISBN-10 0-321-69273-X

This special version of the student text provides answers to all exercises directly beneath each problem.

Instructor's Resource Manual with Tests

ISBN-13 978-0-321-69274-0
ISBN-10 0-321-69274-8

This valuable teaching resource includes the following materials:

- Sample syllabi suggesting ways to structure the course around core and supplemental activities and within different credit-hour options.

- Sample course outlines containing timelines for covering topics.

- Teaching notes for each chapter. These notes are ideal for those using the *Mathematics in Action* approach for the first time.

- Extra practice worksheets for topics with which students typically have difficulty.

- Sample chapter tests and final exams for in-class and take-home use by individual students and groups.

- Information about incorporating technology in the classroom, including sample graphing calculator assignments.

TestGen®

ISBN-13 978-0-321-69275-7
ISBN-10 0-321-69275-6

TestGen enables instructors to build, edit, print, and administer tests using a computerized bank of questions developed to cover all the objectives of the text. TestGen is algorithmically based, allowing instructors to create multiple but equivalent versions of the same question or test with the click of a button. Instructors can also modify test bank questions or add new questions. The software and testbank are available for download from Pearson Education's online catalog.

Instructor's Training Video

ISBN-13 978-0-321-69279-5
ISBN-10 0-321-69279-9

This innovative video discusses effective ways to implement the teaching pedagogy of the *Mathematics in Action* series, focusing on how to make collaborative learning, discovery learning, and alternative means of assessment work in the classroom.

Student Supplements
Worksheets for Classroom or Lab Practice

ISBN-13 978-0-321-73836-3
ISBN-10 0-321-73836-5

- Extra practice exercises for every section of the text with ample space for students to show their work.

- These lab- and classroom-friendly workbooks also list the learning objectives and key vocabulary terms for every text section, along with vocabulary practice problems.

- Concept Connection exercises, similar to the What Have I Learned? exercises found in the text, assess students' conceptual understanding of the skills required to complete each worksheet.

MathXL® Tutorials on CD

ISBN-13 978-0-321-69276-4
ISBN-10 0-321-69276-4

This interactive tutorial CD-ROM provides algorithmically generated practice exercises that are correlated at the objective level to the exercises in the textbook. Every practice exercise is accompanied by an example and a guided solution designed to involve students in the solution process. The software provides helpful feedback for incorrect answers and can generate printed summaries of students' progress.

InterAct Math Tutorial Web Site www.interactmath.com

Get practice and tutorial help online! This interactive tutorial Web site provides algorithmically generated practice exercises that correlate directly to the exercises in the textbook. Students can retry an exercise as many times as they like with new values each time for unlimited practice and mastery. Every exercise is accompanied by an interactive guided solution that provides helpful feedback for incorrect answers, and students can also view a worked-out sample problem that steps them through an exercise similar to the one they're working on.

Pearson Math Adjunct Support Center

The **Pearson Math Adjunct Support Center** (http://www.pearsontutorservices.com/math-adjunct.html) is staffed by qualified instructors with more than 100 years of combined experience at both the community college and university levels. Assistance is provided for faculty in the following areas:

- Suggested syllabus consultation

- Tips on using materials packed with your book

- Book-specific content assistance

- Teaching suggestions, including advice on classroom strategies

Supplements for Instructors and Students
MathXL® Online Course (access code required)

MathXL® is a powerful online homework, tutorial, and assessment system that accompanies Pearson Education's textbooks in mathematics or statistics. With MathXL, instructors can:

- Create, edit, and assign online homework and tests using algorithmically generated exercises correlated at the objective level to the textbook.

- Create and assign their own online exercises and import TestGen tests for added flexibility.

- Maintain records of all student work tracked in MathXL's online gradebook.

With MathXL, students can:

- Take chapter tests in MathXL and receive personalized study plans and/or personalized homework assignments based on their test results.

- Use the study plan and/or the homework to link directly to tutorial exercises for the objectives they need to study.

- Access supplemental animations and video clips directly from selected exercises.

MathXL is available to qualified adopters. For more information, visit our Web site at www.mathxl.com, or contact your Pearson representative.

MyMathLab® Online Course (access code required)

MyMathLab® is a text-specific, easily customizable online course that integrates interactive multimedia instruction with textbook content. MyMathLab gives you the tools you need to deliver all or a portion of your course online, whether your students are in a lab setting or working from home.

- **Interactive homework exercises,** correlated to your textbook at the objective level, are algorithmically generated for unlimited practice and mastery. Most exercises are free-response and provide guided solutions, sample problems, and tutorial learning aids for extra help.

- **Personalized homework** assignments that you can design to meet the needs of your class. MyMathLab tailors the assignment for each student based on their test or quiz scores. Each student receives a homework assignment that contains only the problems they still need to master.

- **Personalized Study Plan,** generated when students complete a test or quiz or homework, indicates which topics have been mastered and links to tutorial exercises for topics students have not mastered. You can customize the Study Plan so that the topics available match your course content.

- **Multimedia learning aids,** such as video lectures and podcasts, animations, and a complete multimedia textbook, help students independently improve their understanding and performance. You can assign these multimedia learning aids as homework to help your students grasp the concepts.

- **Homework and Test Manager** lets you assign homework, quizzes, and tests that are automatically graded. Select just the right mix of questions from the MyMathLab exercise bank, instructor-created custom exercises, and/or TestGen® test items.

- **Gradebook,** designed specifically for mathematics and statistics, automatically tracks students' results, lets you stay on top of student performance, and gives you control over how to calculate final grades. You can also add offline (paper-and-pencil) grades to the gradebook.

- **MathXL Exercise Builder** allows you to create static and algorithmic exercises for your online assignments. You can use the library of sample exercises as an easy starting point, or you can edit any course-related exercise.

- **Pearson Tutor Center** (www.pearsontutorservices.com) access is automatically included with MyMathLab. The Tutor Center is staffed by qualified math instructors who provide textbook-specific tutoring for students via toll-free phone, fax, e-mail, and interactive Web sessions.

Students do their assignments in the Flash®-based MathXL Player, which is compatible with almost any browser (Firefox®, Safari™, or Internet Explorer®) on almost any platform (Macintosh® or Windows®). MyMathLab is powered by CourseCompass™, Pearson Education's online teaching and learning environment, and by MathXL®, our online homework, tutorial, and assessment system. MyMathLab is available to qualified adopters. For more information, visit www.mymathlab.com or contact your Pearson representative.

Acknowledgments

The Consortium would like to acknowledge and thank the following people for their invaluable assistance in reviewing and testing material for this text in the past and current editions:

Mark Alexander, *Kapi'olani Community College*

Kathleen Bavelas, *Manchester Community College*

Shirley J. Beil, *Normandale Community College*

Carol Bellisio, *Monmouth University*

Barbara Burke, *Hawai'i Pacific University*

San Dong Chung, *Kapi'olani Community College*

Marjorie Deutsch, *Queensboro Community College*

Jennifer Dollar, *Grand Rapids Community College*

Irene Duranczyk, *University of Minnesota*

Brian J. Garant, *Morton College*

Maryann Justinger, *Erie Community College—South Campus*

Brian Karasek, *South Mountain Community College*

Miriam Long, *Madonna University*

Kathy Potter, *St. Ambrose University*

Ellen Musen, *Brookdale Community College*

Robbie Ray, *Sul Ross State University*

Janice Roy, *Montcalm Community College*

Andrew S. Russell, *Queensborough Community Collge*

Amy C. Salvati, *Adirondack Community College*

Philomena Sawyer, *Manchester Community College*

Kurt Verderber, *SUNY Cobleskill*

We would also like to thank our accuracy checkers, Shannon d'Hemecourt, Diane E. Cook, Jon Weerts, and James Lapp.

Finally, a special thank you to our families for their unwavering support and sacrifice, which enabled us to make this text a reality.

The Consortium for Foundation Mathematics

To the Student

The book in your hands is most likely very different from any mathematics textbook you have seen before. In this book, you will take an active role in developing the important ideas of arithmetic and beginning algebra. You will be expected to add your own words to the text. This will be part of your daily work, both in and out of class. It is the belief of the authors that students learn mathematics best when they are actively involved in solving problems that are meaningful to them.

The text is primarily a collection of situations drawn from real life. Each situation leads to one or more problems. By answering a series of questions and solving each part of the problem, you will be led to use one or more ideas of introductory college mathematics. Sometimes, these will be basic skills that build on your knowledge of arithmetic. Other times, they will be new concepts that are more general and far reaching. The important point is that you won't be asked to master a skill until you see a real need for that skill as part of solving a realistic application.

Another important aspect of this text and the course you are taking is the benefit gained by collaborating with your classmates. Much of your work in class will result from being a member of a team. Working in small groups, you will help each other work through a problem situation. While you may feel uncomfortable working this way at first, there are several reasons we believe it is appropriate in this course. First, it is part of the learning-by-doing philosophy. You will be talking about mathematics, needing to express your thoughts in words. This is a key to learning. Secondly, you will be developing skills that will be very valuable when you leave the classroom. Currently, many jobs and careers require the ability to collaborate within a team environment. Your instructor will provide you with more specific information about this collaboration.

One more fundamental part of this course is that you will have access to appropriate technology at all times. You will have access to calculators and some form of graphics tool—either a calculator or computer. Technology is a part of our modern world, and learning to use technology goes hand in hand with learning mathematics. Your work in this course will help prepare you for whatever you pursue in your working life.

This course will help you develop both the mathematical and general skills necessary in today's workplace, such as organization, problem solving, communication, and collaborative skills. By keeping up with your work and following the suggested organization of the text, you will gain a valuable resource that will serve you well in the future. With hard work and dedication you will be ready for the next step.

The Consortium for Foundation Mathematics

Number Sense

Your goal in this chapter is to use the numerical mathematical skills you already have—and those you will learn or relearn—to solve problems. Chapter activities are based on practical, real-world situations that you may encounter in your daily life and work. Before you begin the activities in Chapter 1, we ask you to think about your previous encounters with mathematics and choose one word to describe those experiences.   (Answers will vary.)

Cluster 1   Introduction to Problem Solving

Activity 1.1

The Bookstore

Objectives

1. Practice communication skills.

2. Organize information.

3. Write a solution in sentences.

4. Develop problem-solving skills.

By 11:00 A.M., a line has formed outside the crowded bookstore. You ask the guard at the gate how long you can expect to wait. She provides you with the following information: She is permitted to let 6 people into the bookstore only after 6 people have left; students are leaving at the rate of 2 students per minute; and she has just let 6 new students in. Also, each student spends an average of 15 minutes gathering books and supplies and 10 minutes waiting in line to check out.

Currently 38 people are ahead of you in line. You know that it is a 10-minute walk to your noon class. Can you buy your books and still expect to make it to your noon class on time? Use the following questions to guide you in solving this problem.

1. What was your initial reaction after reading the problem?

   (Answers will vary.) "I can't do word problems." "There is way too much information here. My brain can't handle it." "I should be able to figure this one out. Let me think."

2. Have you ever worked a problem such as this before?

   (Answers will vary.) "Not successfully." "Sometimes, but maybe not with so much information to use."

3. Organizing the information will help you solve the problem.

   a. How many students must leave the bookstore before the guard allows more to enter?

      Six students must leave before more may enter.

   b. How many students per minute leave the bookstore?

      Two students per minute leave the bookstore.

   c. How many minutes are there between groups of students entering the bookstore?

      There are 3 minutes between groups of students entering the bookstore.

   d. How long will you stand in line outside the bookstore?

      $6\overline{)38}$ remainder 2

      Six groups of students will enter before me. I will enter with the seventh group. I will stand in line $7 \cdot 3 = 21$ minutes.

    e. Now finish solving the problem and answer the question: How early or late for class will you be?

I enter the bookstore at 11:21 and spend 15 minutes looking for books and 10 minutes in the checkout line. So, it is 11:21 + 15 + 10 = 11:46 when I leave the bookstore. It takes 10 minutes to walk to class, so I arrive at 11:56 and am 4 minutes early.

   4. In complete sentences, write what you did to solve this problem. Then, explain your solution to a classmate.

(Answers will vary.)

SUMMARY: ACTIVITY 1.1

Steps in Problem Solving

1. Sort out the relevant information and organize it.

2. Discuss the problem with others to increase your understanding of the problem.

3. Write your solution in complete sentences to review your steps and check your answer.

EXERCISES: ACTIVITY 1.1

1. Think about the various approaches you and your classmates used to solve Activity 1.1, The Bookstore. Choose the approach that is best for you, and describe it in complete sentences.

(Answers will vary.)

2. What mathematical operations and skills did you use?

(Answers will vary.) I used division, multiplication, addition, subtraction, logical reasoning, and problem solving.

The Classroom

Objectives

1. Organize information.

2. Develop problem-solving strategies.

 • Draw a picture.

 • Recognize a pattern.

 • Do a simpler problem.

3. Communicate problem-solving ideas.

The Handshake

This algebra course involves working with other students in the class, so form a group of 3, 4, or 5 students. Introduce yourself to every other student in your group with a firm handshake. Share some information about yourself with the other members of your group.

1. How many people are in your group?

(Answers will vary.) "There are 3 (or 4 or 5) people in my group."

2. How many handshakes in all were there in your group?

If there are 3 members, then there are 3 handshakes.

If there are 4 members, then there are 6 handshakes.

If there are 5 members, then there are 10 handshakes.

3. Discuss how your group determined the number of handshakes. Be sure everyone understands and agrees with the method and the answer. Write the explanation of the method here.

(Answers will vary.) "We each counted a number aloud as we alternated shaking hands."

4. Share your findings with the other groups, and fill in the table.

 Shaking Hands

NUMBER OF STUDENTS IN GROUP	NUMBER OF HANDSHAKES
2	1
3	3
4	6
5	10
6	15
7	21

5. a. Describe a rule for determining the number of handshakes in a group of seven students.

(Answers will vary.) Student A shakes hands with the 6 other students (B, C, D, E, F, G) for a total of 6 handshakes. Student A has now shaken hands with everyone in his group. Now Student B shakes hands with the remaining 5 students (C, D, E, F, G). The pattern continues. The total number of handshakes is $6 + 5 + 4 + 3 + 2 + 1 = 21$.

Or, each of the 7 students shakes hands with each of the other 6 students, yielding 42 handshakes. Because handshakes occur in pairs, divide 42 by 2 to get a total of 21 handshakes. The total number of handshakes is $\frac{7 \cdot 6}{2} = 21$.

b. Describe a rule for determining the number of handshakes in a class of n students.

Subtract 1 from the number of students in the group. Add that number to the sum of all the whole numbers that come before it.

Or, n students shake hands with $n - 1$ other students; multiply n by $n - 1$, and then divide the product by 2.

6. If each student shakes hands with each other student, how many handshakes will be needed in your algebra class?

    (Answers will vary.) If there are 20 students in the class, there will be $\frac{20 \cdot 19}{2} = 190$ handshakes.

7. Is shaking hands during class time a practical way for students to introduce themselves? Explain.

    Probably not. In a class of 24 students, $\frac{24 \cdot 23}{2} = 276$ handshakes would be needed for each student to shake hands with every other student exactly once.

George Polya's book *How to Solve It* **outlines a four-step process for solving problems.**

   i. Understand the problem (determine what is involved).

   ii. Devise a plan (look for connections to obtain the idea of a solution).

   iii. Carry out the plan.

   iv. Look back at the completed solution (review and discuss it).

---

8. Describe how your experiences with the handshake problem correspond with Polya's suggestions.

    (Answers will vary.)

The Classroom

Suppose the tables in your classroom have square tops. Four students can comfortably sit at each table with ample working space. Putting tables together in clusters as shown will allow students to work in larger groups.

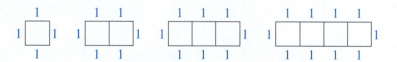

9. Construct a table of values for the number of tables and the corresponding total number of students.

NUMBER OF SQUARE TABLES IN EACH CLUSTER	TOTAL NUMBER OF STUDENTS
1	4
2	6
3	8
4	10
5	12

10. How many students can sit around a cluster of seven square tables?

    Sixteen students can sit around a cluster of seven square tables.

11. Describe the pattern that connects the number of square tables in a cluster and the total number of students that can be seated. Write a rule in a sentence that will determine the total number of students who can sit in a cluster of a given number of square tables.

    The number of students seated is equal to two times the number of tables in the cluster plus the two students sitting at the ends of the tables.

12. There are 24 students in a math course at your college.

    a. How many tables must be put together to seat a group of 6 students?

    Two tables must be placed together to seat a group of 6 students.

    b. How many clusters of tables are needed?

    Four clusters of two tables with six seats are needed.

13. Discuss the best way to arrange the square tables into clusters given the number of students in your class.

    (Answers will vary.)

SUMMARY: ACTIVITY 1.2

1. Problem-solving strategies include:

   - discussing the problem
   - organizing information
   - drawing a picture
   - recognizing patterns
   - doing a simpler problem

2. George Polya's book *How to Solve It* outlines **a four-step process for solving problems**.

   i. Understand the problem (determine what is involved).

   ii. Devise a plan (look for connections to obtain the idea of a solution).

   iii. Carry out the plan.

   iv. Look back at the completed solution (review and discuss it).

EXERCISES: ACTIVITY 1.2

1. At the opening session of the United States Supreme Court, each justice shakes hands with all the others.

   a. How many justices are there?

   Nine justices sit on the Supreme Court.

   b. How many handshakes do they make?

   The justices make 36 handshakes.

Exercise numbers appearing in color are answered in the Selected Answers appendix.

2. Identify how the numbers are generated in this triangular arrangement, known as Pascal's triangle. Fill in the missing numbers.

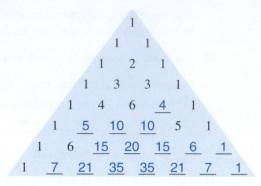

3. An **arithmetic sequence** is a list of numbers in which consecutive numbers share a common difference. Each number after the first is calculated by adding the common difference to the preceding number. For example, the arithmetic sequence $1, 4, 7, 10, \ldots$ has 3 as its common difference. Identify the common difference in each arithmetic sequence that follows.

   a. $2, 4, 6, 8, 10, \ldots$ The common difference is 2.

   b. $1, 3, 5, 7, 9, 11, \ldots$ The common difference is 2.

   c. $26, 31, 36, 41, 46, \ldots$ The common difference is 5.

4. A **geometric sequence** is a list of numbers in which consecutive numbers share a common ratio. Each number after the first is calculated by multiplying the preceding number by the common ratio. For example, $1, 3, 9, 27, \ldots$ has 3 as its common ratio. Identify the common ratio in each geometric sequence that follows.

   a. $2, 4, 8, 16, 32, \ldots$ The common ratio is 2.

   b. $1, 5, 25, 125, 625, \ldots$ The common ratio is 5.

5. The operations needed to get from one number to the next in a sequence can be more complex. Describe a relationship shared by consecutive numbers in the following sequences.

   a. $2, 4, 16, 256, \ldots$ Each number is generated by squaring the preceding number.

   b. $2, 5, 11, 23, 47, \ldots$ Each number is generated by multiplying the preceding number by 2 and then adding 1.

   c. $1, 2, 5, 14, 41, 122, \ldots$ Each number is generated by multiplying the preceding number by 3 and then subtracting 1.

6. In biology lab, you conduct the following experiment. You put two rabbits in a large caged area. In the first month, the pair produces no offspring (rabbits need a month to reach adulthood). At the end of the second month the pair produces exactly one new pair of rabbits (one male and one female). The result makes you wonder how many male/female pairs you might have if you continue the experiment and each existing pair of rabbits produces a new pair each month, starting after their first month. The numbers for the first 4 months are calculated and recorded for you in the following table. The arrows in the table illustrate that the number of pairs produced in a given month equals the number of pairs that existed at the beginning of the preceding month. Continue the pattern and fill in the rest of the table.

Reproducing Rabbits

MONTH	NUMBER OF PAIRS AT THE BEGINNING OF THE MONTH	NUMBER OF NEW PAIRS PRODUCED	TOTAL NUMBER OF PAIRS AT THE END OF THE MONTH
1	1	0	1
2	1	1	2
3	2	1	3
4	3	2	5
5	5	3	8
6	8	5	13
7	13	8	21
8	21	13	34

The list of numbers in the second column is called the **Fibonacci sequence**. This problem on the reproduction of rabbits first appeared in 1202 in the mathematics text *Liber Abaci*, written by Leonardo of Pisa (nicknamed Fibonacci). Using the first two numbers, 1 and 1, as a starting point, describe how the next number is generated. Your rule should generate the rest of the numbers shown in the sequence in column 2.

Each number is generated by adding the two numbers that precede it.

7. If you shift all the numbers in Pascal's triangle so that all the 1s are in the same column, you get the following triangle.

    a. Add the numbers crossed by each arrow. Put the sums at the tip of the arrow.

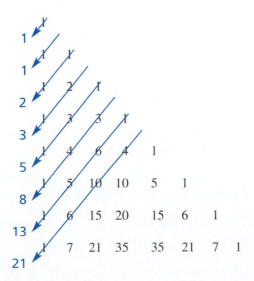

    b. What is the name of the sequence formed by these sums?

The sequence formed is the Fibonacci sequence.

8. There are some interesting patterns within the Fibonacci sequence itself. Take any number in the sequence and multiply it by itself, and then subtract the product of the number immediately before it and the number immediately after it. What is the result? Pick another number and follow the same procedure. What result do you obtain? Try two more numbers in the sequence.

For example, choose 5.

$5 \cdot 5 = 25$

$3 \cdot 8 = 24$

$25 - 24 = 1$

For example, choose 3.

$3 \cdot 3 = 9$

$2 \cdot 5 = 10$

$9 - 10 = -1$

The result is 1 or −1. Similar results occur if I choose 8 and 13. From the four examples, it appears that the result will always be 1 or −1.

Activity 1.3

Properties of Arithmetic

Objectives

1. Identify and use the commutative property in calculations.

2. Use the distributive property to evaluate arithmetic expressions.

3. Use the order of operations convention to evaluate arithmetic expressions.

4. Identify and use the properties of exponents in calculations.

5. Convert numbers to and from scientific notation.

6. Identify, understand, and use formulas.

7. Use the order of operations convention in formulas involving whole numbers.

A calculator is a powerful tool in the problem-solving process. A quick look at the desks of those around you should show you that calculators come in many sizes and shapes and with varying capabilities. Some calculators perform only basic operations such as addition, subtraction, multiplication, division, and square roots. Others also handle operations with exponents, perform operations with fractions, and do trigonometry and statistics. There are calculators that graph equations and generate tables of values. Some even manipulate algebraic symbols.

Unlike people, however, calculators do not think for themselves and can perform tasks only in the way that you instruct them (or program them). Therefore, you need to understand the properties of algebra and become familiar with the way your calculator operates with numbers. In particular, you will learn how your calculator deals with very large numbers and the order in which it performs the operations you request.

Note: It is assumed that you will have access to at least a scientific calculator. If you have a TI-83/84 Plus graphing calculator, you can find detailed information on many of its features in Appendix D.

Commutative Property

1. Use your calculator to determine the sum $146 + 875$.

   1021

2. **a.** Now, input $875 + 146$ into your calculator and evaluate the answer. How does this sum compare to the sum in Problem 1?

   $875 + 146 = 1021$ The sum is the same as in Problem 1.

   b. If you use numbers other than 146 and 875, does the order in which you add the numbers change the result? Explain by giving examples.

   Changing the order of the numbers to be added does not change the result.

   Examples:

   $345 + 678 = 1023$ and $678 + 345 = 1023$; $13 + 7 = 20$ and $7 + 13 = 20$

If the order in which two numbers are added is reversed, the sum remains the same. This property is called the **commutative property** of addition and can be written symbolically as

$$a + b = b + a.$$

3. Is the commutative property true for the operation of subtraction? Multiplication? Division? Explain by giving examples for each operation.

SUBTRACTION	MULTIPLICATION	DIVISION
$13 - 7 = 6$; $7 - 13 = -6$	$12 \cdot 5 = 60$; $5 \cdot 12 = 60$	$6 \div 2 = 3$; $2 \div 6 = \frac{2}{6} = \frac{1}{3}$
The result is different if the order is reversed, so subtraction is *not* commutative. Reversing the order in which two numbers are subtracted changes the sign of the result.	The result is the same if the order is reversed. This demonstrates that multiplication is commutative.	The result is different if the order is reversed, so division is *not* commutative. Reversing the order in which two numbers are divided produces the reciprocal of the original quotient.

Distributive Property

It is sometimes convenient to do mental arithmetic (i.e., without the aid of your calculator).

Example 1   *Evaluate* $4 \cdot 27$ *without the aid of your calculator.*

SOLUTION

Think about the multiplication as follows: 27 can be written as $20 + 7$. Therefore, $4 \cdot 27$ can be written as $4 \cdot (20 + 7)$, which can be evaluated as $4 \cdot 20 + 4 \cdot 7$.

The product $4 \cdot 27$ can now be thought of as $80 + 28$ or 108.

To summarize, $4 \cdot 27 = \underline{4 \cdot (20 + 7) = 4 \cdot 20 + 4 \cdot 7} = 80 + 28 = 108$.

The underlined step in Example 1 demonstrates a very important property called the **distributive property**. In particular, multiplication is distributed over addition or subtraction.

The **distributive property** is written symbolically as:

$$c \cdot (a + b) = c \cdot a + c \cdot b \text{ for addition,}$$

or

$$c \cdot (a - b) = c \cdot a - c \cdot b \text{ for subtraction.}$$

Note that $c \cdot (a + b)$ can also be written as $c(a + b)$, and $c \cdot (a - b)$ can also be written as $c(a - b)$. When a number is immediately followed by parentheses, the operation of multiplication is implied.

4. Another way to express 27 is as $25 + 2$ or $30 - 3$.

   a. Express 27 as $25 + 2$, and use the distributive property to multiply $4 \cdot 27$.
   $4(27) = 4(25 + 2) = 4 \cdot 25 + 4 \cdot 2 = 100 + 8 = 108$

   b. Express 27 as $30 - 3$, and use the distributive property to multiply $4 \cdot 27$.
   $4(27) = 4(30 - 3) = 4 \cdot 30 - 4 \cdot 3 = 120 - 12 = 108$

5. Use the distributive property to mentally evaluate the following multiplication problems. Verify your answer on your calculator.

   a. $7 \cdot 82$
   "Think" $7 \cdot 82 = 7 \cdot (80 + 2) = 7 \cdot 80 + 7 \cdot 2 = 560 + 14 = 574$.

   b. $5 \cdot 108$
   "Think" $5 \cdot (108) = 5 \cdot (100 + 8) = 5 \cdot 100 + 5 \cdot 8 = 500 + 40 = 540$.

Order of Operations

6. **a.** Evaluate $5 + 6 \cdot 2$ in your head and record the result. Use your calculator to verify your answer.
   $5 + 12 = 17$

   b. What operations are involved in this problem?
   This problem involves addition and multiplication.

   c. In what order do you and your calculator perform the operations to obtain the answer?
   First, I multiply $6 \cdot 2$ to obtain 12, and then I add $5 + 12$ to obtain 17.

Scientific and graphing calculators are programmed to perform operations in a universally accepted order, as illustrated by the previous problems. *Part* of the **order of operations convention** is as follows:

1. Do all multiplications and divisions in the order in which they occur, working from left to right.

2. Next, do all additions and subtractions in the order in which they occur, working from left to right.

Example 2   *Evaluate* $10 - 2 \cdot 4 + 7$ *without a calculator. Then use your calculator to verify your result.*

SOLUTION

$10 - 2 \cdot 4 + 7$ **Do multiplication first.**

$= 10 - 8 + 7$ **Subtract 8 from 10 because you encounter subtraction first as you read from left to right.**

$= 2 + 7$ **Add.**

$= 9$

7. Perform the following calculations without a calculator. Then use your calculator to verify your result.

   a. $42 \div 3 + 4$
     $= 14 + 4 = 18$

   b. $42 + 8 \div 4$
     $= 42 + 2 = 44$

   c. $24 \div 6 - 2 \cdot 2$
     $= 4 - 2 \cdot 2$
     $= 4 - 4$
     $= 0$

   d. $4 + 8 \div 2 \cdot 3 - 9$
     $= 4 + 4(3) - 9$
     $= 4 + 12 - 9$
     $= 16 - 9$
     $= 7$

8. **a.** Write the calculation $\dfrac{24}{2 + 6}$ in words.

     Twenty-four is divided by the sum of 2 and 6.

   b. Perform the calculation $\dfrac{24}{2 + 6}$ without your calculator.

     $24 \div 8 = 3$

   c. If you enter the expression $\dfrac{24}{2 + 6}$ as $24 \div 2 + 6$ in your calculator, would you obtain the correct answer? Explain.

     No. The calculator would give the result 18.

Note that $\frac{24}{2 + 6}$ is the same as the quotient $\frac{24}{8}$. Therefore, the addition in the denominator is done first, followed by the division. To write $\frac{24}{2 + 6}$ in a horizontal format, you must use parentheses to group the expression in the denominator to indicate that the addition is performed first. That is, write $\frac{24}{2 + 6}$ as $24 \div (2 + 6)$.

9. Use the parentheses key to enter $24 \div (2 + 6)$ into your calculator and verify that the result is 3.

   The result is 3.

Parentheses are grouping symbols that are used to override the standard order of operations. Operations contained in parentheses are performed first.

Example 3 *Evaluate* $2(3 + 4 \cdot 5)$.

SOLUTION

$2(3 + 4 \cdot 5)$ First evaluate the arithmetic expression in parentheses using the order of operations indicated. That is, multiply $4 \cdot 5$ and then add 3.

$= 2(3 + 20)$

$= 2 \cdot 23$ Finally, multiply 23 by 2.

$= 46$

10. Mentally evaluate the following, and verify your answers on your calculator.

a. $\dfrac{6}{3 + 3}$

$\dfrac{6}{6} = 1$

b. $\dfrac{2 + 8}{4 - 2}$

$\dfrac{10}{2} = 5$

c. $5 + 2(4 \div 2 + 3)$

$= 5 + 2 \cdot (2 + 3)$
$= 5 + 2 \cdot 5$
$= 5 + 10$
$= 15$

d. $10 - (12 - 3 \cdot 2) \div 3$

$= 10 - (12 - 6) \div 3$
$= 10 - 6 \div 3$
$= 10 - 2$
$= 8$

Exponentiation

Calculators can easily perform repeated multiplication. Recall that $5 \cdot 5$ can be written as 5^2—that is, 5 squared. There are two ways to square a number on most calculators.

11. a. One method to evaluate 5^2 is to use the x^2 key. Input 5 and then press the x^2 key. Do this now and record your answer.

$5^2 = 25$

b. Another way you can evaluate 5^2 is to use the exponent key. Depending on your calculator, the exponent key may be x^y, y^x, or $\wedge$. To calculate 5^2, input 5, press the exponent key, then enter the exponent as 2 and press enter. Do this now and record your answer.

$5^2 = 25$

c. The exponent key can be used with any exponent. For example, $5 \cdot 5 \cdot 5$ can be written as 5^3. Evaluate $5 \cdot 5 \cdot 5$ as written, and then use the exponent key to evaluate 5^3.

$5 \cdot 5 \cdot 5 = 125$
$5^3 = 125$

An expression such as 5^3 is called a **power** of 5. The base is 5 and the exponent is 3. Note that the exponent indicates how many times the base is written as a factor. Note also that 5^3 is read as "five raised to the third power." When a power (also known as an *exponential expression*) is contained in an expression, it is evaluated *before* any multiplication or division.

Procedure

Order of operations convention for evaluating arithmetic expressions containing parentheses, addition, subtraction, multiplication, division, and exponentiation:

Before performing any operations outside parentheses, first perform all operations contained within parentheses. Perform all operations in the following order.

1. Evaluate all exponential expressions as you read the expression from left to right.

2. Perform all multiplication and division as you read the expression from left to right.

3. Perform all addition and subtraction as you read the expression from left to right.

Example 4

Evaluate the expression $20 - 2 \cdot 3^2$.

SOLUTION

$20 - 2 \cdot 3^2$ Evaluate all exponential expressions as you read the arithmetic expression from left to right.

$= 20 - 2 \cdot 9$ Perform all multiplication and division as you read the expression from left to right.

$= 20 - 18$ Perform all addition and subtraction as you read the expression from left to right.

$= 2$

12. Enter the expression $20 - 2 \cdot 3^2$ into your calculator, and verify the result.

    The result is 2.

13. Evaluate the following.

    a. $6 + 3 \cdot 4^3$

    $= 6 + 3 \cdot 64$
    $= 6 + 192$
    $= 198$

    b. $2 \cdot 3^4 - 5^3$

    $= 2 \cdot 81 - 125$
    $= 162 - 125$
    $= 37$

    c. $12 - 2(8 - 2 \cdot 3) + 3^2$

    $= 12 - 2 \cdot (8 - 6) + 3^2$
    $= 12 - 2 \cdot 2 + 9$
    $= 12 - 4 + 9$
    $= 17$

    d. $\dfrac{128}{16 - 2^3}$

    $= \dfrac{128}{16 - 8}$
    $= \dfrac{128}{8}$
    $= 16$

14. **a.** Use the exponent key to evaluate the following powers of 10: 10^2, 10^3, 10^4, and 10^5. What relationship do you notice between the exponent and the number of zeros in the result?

    $10^2 = 100$ $10^3 = 1000$ $10^4 = 10,000$ $10^5 = 100,000$

    The exponent of 10 equals the number of zeros in the result.

    b. Evaluate 10^6. Is the result what you expected? How many zeros are in the result?

    $10^6 = 1,000,000$ The result is as expected. There are 6 zeros in the result.

c. Complete the following table.

EXPONENTIAL FORM	EXPANDED FORM	VALUE
10^5	$10 \times 10 \times 10 \times 10 \times 10$	100,000
10^4	$10 \times 10 \times 10 \times 10$	10,000
10^3	$10 \times 10 \times 10$	1000
10^2	10×10	100
10^1	10	10

15. a. Follow the patterns in the preceding table and determine a reasonable value for 10^0.

1

b. Use your calculator to evaluate 10^0. Is the result what you expected? How many zeros are in the result?

$10^0 = 1$ The result is as expected. There are no zeros in the result.

c. Use the exponent key to evaluate the following powers: 3^0, 8^0, 23^0, and 526^0.

1, 1, 1, 1

d. Evaluate other nonzero numbers with a zero exponent.

(Answers will vary.) $6^0 = 1$ $1478^0 = 1$ $12.3^0 = 1$

e. Write a sentence describing the result of raising a nonzero number to a zero exponent.

Any nonzero number raised to the zero power equals 1.

Scientific Notation

Very large numbers appear in a variety of situations. For example, in the year 2007, the total personal income of all Americans was approximately $11,700,000,000,000. Because such numbers contain so many zeros, they are difficult to read and to use in computations. Scientific notation is a convenient way to write these numbers.

16. a. The number 5000 is said to be written in **standard notation**. Because 5000 is 5×1000, the number 5000 can also be written as 5×10^3. Use your calculator to verify this by evaluating 5×10^3.

$5 \times 10^3 = 5000$

b. On your calculator there is another way to evaluate a number multiplied by a power of 10, such as 5×10^3. Find the key labeled (EE) or (E) or (EXP); sometimes it is a second function. This key takes the place of the (×) (1) (0) and (^) keystroke sequence. To evaluate 5×10^3, enter 5, press the (EE) key, and then enter 3. Try this now and verify the result.

$5\text{E}3 = 5000$

A number written as the product of a decimal number between 1 and 10 and a power of 10 is said to be written in **scientific notation**.

> **Example 5** *Write 9,420,000 in scientific notation.*
>
> **SOLUTION**
>
> 9,420,000 can be written in scientific notation as follows:
>
> Place a decimal point immediately to the right of the first digit of the number: 9.420000.
>
> Count the number of digits to the right of the decimal point. This number will be the exponent. In this example, these digits are 420000, that is, 6 digits.
>
> Multiply the decimal number by 10^6. Drop all the trailing zeros in the decimal number and then multiply by 10^6.
>
> Therefore, $9,420,000 = 9.42 \times 10^6$.

The number 9,420,000 can be entered into your calculator in scientific notation using the ⌷EE⌷ or ⌷E⌷ key as follows: 9.42 ⌷EE⌷ 6. Try it.

17. Write each of the following numbers in scientific notation.

   a. 5120   **b.** 2,600,000

   5.12×10^3 2.6×10^6

   c. \$11,700,000,000,000 (personal income of Americans in 2007)

   $\$1.17 \times 10^{13}$

18. a. Write each of the following numbers in standard notation.

   i. 4.72×10^5 **ii.** 3.5×10^{11}

   472,000   350,000,000,000

   b. Describe the process you used to convert a number from scientific notation to standard notation.

   (Answers will vary.)

   Note the value of the exponent in the number's scientific notation. Move the decimal point that many places to the right, inserting trailing zeros as placeholders if needed.

19. Input 4.23 ⌷EE⌷ 2 into your calculator, press the ⌷ENTER⌷ or ⌷=⌷ key, and see if you obtain the result that you expect.

   4.23ᴇ2 = 423, as expected.

Note: Scientific notation can also be used in writing very small numbers such as 0.00000027. This is discussed in Activity 1.14.

Computations Using Scientific Notation

One advantage of using scientific notation becomes evident when you have to perform computations involving very large numbers.

20. a. The average distance from Earth to the Sun is 93,000,000 miles. Write this number in words and in scientific notation. Then use the ⌷EE⌷ key to enter it into your calculator.

   ninety-three million = 9.3×10^7

   b. Estimate 5 times the average distance from Earth to the Sun.

   (Answers will vary.) Five times the average distance from Earth to the Sun is a little more than 450,000,000 miles, as an estimate.

c. Use your calculator to verify your estimate from part b. Write the answer in standard notation, in scientific notation, and in words.

$5(93,000,000) = 465,000,000 = 4.65 \times 10^8$

four hundred sixty-five million miles

21. There are heavenly bodies that are thousands of times farther away from Earth than is the Sun. Multiply 93,000,000 miles by 1000. Write the result in scientific notation, in standard notation, and in words.

$9.3 \times 10^{10} = 93,000,000,000$

ninety-three billion miles

22. Use your calculator to perform each of the following calculations. Then express each result in standard notation.

a. $(3.26 \times 10^4)(5.87 \times 10^3)$

Enter the numbers in scientific notation and calculate to obtain 1.91362×10^8 or 191,362,000.

b. $\dfrac{750 \times 10^{17}}{25 \times 10^{10}}$

3×10^8 or 300,000,000

c. $(25 \times 10^3) + (750 \times 10^2)$

1×10^5 or 100,000

Formulas

In almost every field of study, you are likely to encounter formulas that relate two or more quantities that are represented by letters.

Definition

A **formula** shows the arithmetic relationship between two or more quantities. Each quantity, called a **variable**, is represented by a letter or symbol. You evaluate a formula by replacing the variables with their appropriate value and then performing the indicated operations.

---

23. The distance traveled by an object moving at a constant speed is given by the formula

$$d = rt,$$

where d represents the distance traveled,

r represents the rate or speed of the object, and

t represents the time the object is in motion.

Light travels at a rate of approximately 1.86×10^5 miles per second. It takes light 500 seconds to travel from the Sun to Earth. Use the formula $d = rt$ to determine the distance between Earth and the Sun.

$(1.86 \times 10^5)(500) = (1.86 \times 10^5)(5 \times 10^2) = 9.3 \times 10^7$ or 93,000,000 miles

24. The formula $d = rt$ in Problem 23 can be rewritten as

$$t = \frac{d}{r}.$$

The star Merak (contained in the Big Dipper constellation) is about 4.65×10^{14} miles from Earth. Use the formula $t = \frac{d}{r}$ to determine how long it takes light from Merak to reach Earth. Light travels at a rate of approximately 5.87×10^{12} miles per year.

$$\frac{4.65 \times 10^{14} \text{ mi}}{5.87 \times 10^{12} \text{ mi/yr}} \approx 79.216 \text{ or about } 79 \text{ yr}$$

It takes approximately 79 years for light to travel from Merak to Earth.

The use of variables to help solve problems will be studied in Chapter 2 (Variable Sense).

SUMMARY: ACTIVITY 1.3

1. The **commutative property** states that changing the order in which you add or multiply two numbers does not change the result. That is, the commutative property for addition is written symbolically as

$$a + b = b + a,$$

and the commutative property for multiplication is written symbolically as

$$a \cdot b = b \cdot a.$$

The commutative property does not hold for subtraction or division.

2. The **distributive property** of multiplication over addition or subtraction is written symbolically as

$$c(a + b) = c \cdot a + c \cdot b \text{ for addition}$$

or

$$c(a - b) = c \cdot a - c \cdot b \text{ for subtraction.}$$

3. Following are the **order of operations conventions** for evaluating arithmetic expressions containing parentheses, addition, subtraction, multiplication, division, and exponentiation.

   First, before performing any operations outside parentheses, perform all operations contained within parentheses. Perform all operations in the following order.

   a. Evaluate all exponential expressions as you read the expression from left to right.

   b. Perform all multiplication and division as you read the expression from left to right.

   c. Perform all addition and subtraction as you read the expression from left to right.

4. Any number, except zero, raised to the zero power equals 1. 0^0 has no numerical value.

5. A number is expressed in **scientific notation** when it is written as the product of a decimal number between 1 and 10 and a power of 10.

6. A **formula** shows the arithmetic relationship between two or more quantities; each quantity is represented by a letter or symbol called a **variable**. You evaluate a formula by replacing the variables with their appropriate value and then performing the operations.

EXERCISES: ACTIVITY 1.3

1. Evaluate each expression, and use your calculator to check your answers.

   a. $7(20 + 5)$

   175

   b. $7 \cdot 20 + 5$

   145

   c. $7 \cdot 25$

   175

   d. $20 + 7 \cdot 5$

   55

   e. Which two arithmetic expressions above yield the same answer?

   a and c yield the same result.

2. Evaluate each expression, and use your calculator to check your answers.

   a. $20(100 - 2)$

   1960

   b. $20 \cdot 100 - 2$

   1998

   c. $20 \cdot 98$

   1960

   d. $100 - 20 \cdot 2$

   60

   e. Which two arithmetic expressions above yield the same answer?

   a and c yield the same result, 1960.

3. Use the order of operations convention to evaluate each expression.

   a. $17(50 - 2)$

   $17 \cdot 48 = 816$

   b. $(90 - 7)5$

   $83 \cdot 5 = 415$

4. Use the distributive property to evaluate each expression.

   a. $17(50 - 2)$

   $850 - 34 = 816$

   b. $(90 - 7)5$

   $450 - 35 = 415$

5. Perform the following calculations without a calculator. Then use your calculator to verify your result.

   a. $45 \div 3 + 12$

   $15 + 12 = 27$

   b. $54 \div 9 - 2 \cdot 3$

   $6 - 6 = 0$

   c. $12 + 30 \div 2 \cdot 3 - 4$

   $12 + 15 \cdot 3 - 4 = 12 + 45 - 4 = 53$

   d. $26 + 2 \cdot 7 - 12 \div 4$

   $26 + 14 - 3 = 40 - 3 = 37$

6. **a.** Explain why the result of $72 \div 8 + 4$ is 13.

   Divide 72 by 8 first to get 9. Then add 4 to obtain 13.

   b. Explain why the result of $72 \div (8 + 4)$ is 6.

   Add 8 and 4 first to get 12. Then divide 72 by 12 to obtain 6.

7. Evaluate the following and verify on your calculator.

   a. $48 \div (4 + 4)$

   $48 \div 8 = 6$

   b. $\dfrac{8 + 12}{6 - 2}$

   $\frac{20}{4} = 5$

c. $120 \div (6 + 4)$

  $120 \div 10 = 12$

d. $64 \div (6 - 2) \cdot 2$

  $64 \div 4 \cdot 2 = 16 \cdot 2 = 32$

e. $(16 + 84) \div (4 \cdot 3 - 2)$

  $100 \div 10 = 10$

f. $(6 + 2) \cdot 20 - 12 \div 3$

  $8 \cdot 20 - 4 = 160 - 4 = 156$

g. $39 + 3 \cdot (8 \div 2 + 3)$

  $39 + 3 \cdot (7) = 39 + 21 = 60$

h. $100 - (81 - 27 \cdot 3) \div 3$

  $100 - (0) \div 3 = 100 - 0 = 100$

8. Evaluate the following. Use a calculator to verify your answers.

a. $15 + 2 \cdot 5^3$

  $15 + 250 = 265$

b. $5 \cdot 2^4 - 3^3$

  $80 - 27 = 53$

c. $5^2 \cdot 2^3 \div 10 - 6$

  $200 \div 10 - 6 = 20 - 6 = 14$

d. $5^2 \cdot 2 - 5 \cdot 2^3$

  $50 - 40 = 10$

9. Evaluate each of the following arithmetic expressions by performing the operations in the appropriate order. Use a calculator to check your results.

a. $37 - 2(18 - 2 \cdot 5) + 1^2$

  $37 - 16 + 1 = 22$

b. $3^5 + 2 \cdot 10^2$

  $243 + 200 = 443$

c. $\dfrac{243}{36 - 3^3}$

  $\frac{243}{9} = 27$

d. $(75 - 2 \cdot 15) \div 9$

  $45 \div 9 = 5$

e. $7 \cdot 2^3 - 9 \cdot 2 + 5$

  $56 - 18 + 5 = 43$

f. $2^5 \cdot 5^2$

  $32 \cdot 25 = 800$

g. $2^3 \cdot 2^2$

  $8 \cdot 4 = 32$

h. $6^2 + 2^6$

  $36 + 64 = 100$

i. $1350 \div 75 \cdot 5 - 15 \cdot 2$

  $18 \cdot 5 - 15 \cdot 2 = 90 - 30 = 60$

j. $(3^2 - 4)^2$

  $5^2 = 25$

10. The following numbers are written in standard notation. Convert each number to scientific notation.

a. 213,040,000,000

  2.1304×10^{11}

b. 555,140,500,000,000

  5.551405×10^{14}

11. The following numbers are written in scientific notation. Convert each number to standard notation.

a. 4.532×10^{11}

  453,200,000,000

b. 4.532×10^7

  45,320,000

12. To solve the following problem, change the numbers to scientific notation and then perform the appropriate operations.

  The distance that light travels in 1 second is 186,000 miles. How far will light travel in 1 year? (There are approximately 31,500,000 seconds in 1 year.)

  $(1.86 \times 10^5)(3.15 \times 10^7) = 5.859 \times 10^{12}$

  Light travels 5.859×10^{12} (or 5,859,000,000,000) miles in 1 year.

13. Evaluate each formula for the given values.

   a. $d = r \cdot t$, for $r = 35$ and $t = 6$

   $d = 35 \cdot 6 = 210$

   b. $F = ma$, for $m = 120$ and $a = 25$

   $F = 120 \cdot 25 = 3000$

   c. $V = lwh$, for $l = 100$, $w = 5$, and $h = 25$

   $V = 100 \cdot 5 \cdot 25 = 12{,}500$

   d. $F = \dfrac{mv^2}{r}$, for $m = 200$, $v = 25$, and $r = 125$

   $F = 200 \cdot 25^2 \div 125 = 200 \cdot 625 \div 125 = 125{,}000 \div 125 = 1000$

   e. $A = \dfrac{t_1 + t_2 + t_3}{3}$, for $t_1 = 76$, $t_2 = 83$, and $t_3 = 81$

   $A = \dfrac{76 + 83 + 81}{3} = \dfrac{240}{3} = 80$

Cluster 1 | **What Have I Learned?**

Activities 1.1 and 1.2 gave you an opportunity to develop some problem-solving strategies. Apply the skills you used in this cluster to solve the following problems.

1. Your last class for the day is over! As you grab a water bottle and settle down to read a chapter of text for tomorrow's class, you notice a group of students forming a circle outside and beginning to randomly kick an odd-looking ball from person to person. You notice that there are 12 students in the circle and that they are able to keep the object in the air as they kick it. Sometimes they kick it to the person next to them; other times they kick it to someone across the circle.

   a. Suppose each student kicks the Hacky-Sack (you've discovered the odd-looking ball has a name) exactly once to each of the others in the circle. How many total kicks would that take?

   The total number of kicks in this situation would be $12 \times 11 = 132$ kicks.

   b. One student in the circle invites you and another student to join them for a total of 14. How many kicks would it now take if each student kicks the Hacky-Sack exactly once to each of the others? How do you arrive at your answer?

   Each of the 14 students kicks the ball to the 13 other students exactly once for a total of $14 \cdot 13 = 182$ kicks.

   c. George Polya's book *How to Solve It* outlines a four-step process for solving problems.

      i. Understand the problem (determine what is involved).

      ii. Devise a plan (look for connections to obtain the idea of a solution).

      iii. Carry out the plan.

      iv. Look back at the completed solution (review and discuss it).

   Describe how your procedures in parts a and b correspond with Polya's suggestions.

   (Answers will vary.)

      i. I know how many persons are playing the game (12) and I figured out that each one kicks the Hacky-Sack exactly once to each of the other players.

      ii. Each player would kick the Hacky-Sack 11 times (once to each of the other players) so I would multiply 12 and 11 to get the total number of kicks.

      iii. The total is $12 \times 11 = 132$ kicks.

      iv. My answer makes sense. I tested my reasoning by noting that if two persons played, the ball would be kicked twice, once by each player. Similarly for three players, the ball would be kicked twice by each player for a total of $3 \times 2 = 6$ times. The same pattern holds for 4 players, 5 players, and so forth.

2. You are assigned to read *War and Peace* for your literature class. The edition you have contains 1232 pages. You time yourself and estimate that you can read 12 pages in 1 hour. You have 5 days before your exam on this book. Will you be able to finish reading it before the exam?

   (Answers will vary.) If I can devote 10 hours per day reading the book, at the end of the fifth day, I will have read only $12 \cdot 10 \cdot 5 = 600$ pages. So I would not be able to finish the book before the exam.

Cluster 1   How Can I Practice?

1. Describe the relationship shared by consecutive numbers in the following sequences.

a. $1, 4, 7, 10, \ldots$

Each number is generated by adding 3 to the preceding number.

b. $1, 2, 4, 8, \ldots$

Each number is generated by multiplying the preceding number by 2.

c. $1, 3, 7, 15, 31, 63, \ldots$

Each number is generated by multiplying the preceding number by 2, and then adding 1.

2. Evaluate each of the following arithmetic expressions by performing the operations in the appropriate order. Use your calculator to check your results.

a. $4(6 + 3) - 9 \cdot 2$

$= 4 \cdot 9 - 9 \cdot 2$
$= 36 - 18$
$= 18$

b. $5 \cdot 9 \div 3 - 3 \cdot 4 \div 6$

$= 45 \div 3 - 3 \cdot 4 \div 6$
$= 15 - 12 \div 6$
$= 15 - 2$
$= 13$

c. $2 + 3 \cdot 4^3$

$= 2 + 3 \cdot 64$
$= 2 + 192$
$= 194$

d. $\dfrac{256}{28 + 6^2}$

$= \dfrac{256}{28 + 36}$

$= \dfrac{256}{64}$

$= 4$

e. $7 - 3(8 - 2 \cdot 3) + 2^2$

$7 - 3(2) + 4 = 7 - 6 + 4 = 5$

f. $3 \cdot 2^5 + 2 \cdot 5^2$

$3 \cdot 32 + 2 \cdot 25 = 96 + 50 = 146$

g. $144 \div (24 - 2^3)$

$144 \div (24 - 8) = 144 \div 16 = 9$

h. $\dfrac{36 - 2 \cdot 9}{6}$

$\dfrac{36 - 18}{6} = \dfrac{18}{6} = 3$

i. $9 \cdot 5 - 5 \cdot 2^3 + 5$

$9 \cdot 5 - 5 \cdot 8 + 5 = 45 - 40 + 5 = 10$

j. $1^5 \cdot 5^1$

$1 \cdot 5 = 5$

k. $2^3 \cdot 2^0$

$8 \cdot 1 = 8$

l. $7^2 + 7^2$

$49 + 49 = 98$

m. $(3^2 - 4 \cdot 0)^2$

$(9 - 0)^2 = 9^2 = 81$

3. Write each number in standard notation.

a. 8^0

1

b. 12^2

144

c. 3^4

81

d. 2^6

64

4. a. Write 214,000,000,000 in scientific notation.

2.14×10^{11}

b. Write 7.83×10^4 in standard notation.

78,300

5. A newly discovered binary star, Shuart 1, is located 185 light-years from Earth. One light-year is 9,460,000,000,000 kilometers. Express the distance to Shuart 1 in kilometers. Write the answer in scientific notation.

$185(9.46 \times 10^{12}) = 1.7501 \times 10^{15}$ km

The distance from Earth to Shuart 1 is 1.7501×10^{15} kilometers.

6. The human brain contains about 10,000,000,000 nerve cells. If the average cell contains about 200,000,000,000,000 molecules, determine the number of molecules in the nerve cells of the human brain.

$10,000,000,000(200,000,000,000,000) = 1 \times 10^{10} \cdot 2 \times 10^{14} = 2 \times 10^{24}$ molecules

7. Identify the arithmetic property expressed by each numerical statement.

a. $25 \cdot 30 = 30 \cdot 25$

commutative property of multiplication

b. $15 \cdot 9 = 15(10 - 1) = 15 \cdot 10 - 15 \cdot 1$

distributive property of multiplication over subtraction

8. Determine if each of the following numerical statements is true or false. In each case, justify your answer.

a. $7(20 + 2) = 7 \cdot 22$

true; order of operations

b. $25 - 10 - 4 = 25 - (10 - 4)$

false; left-hand side = 11 and right-hand side = 19

c. $\frac{1}{0} = 1$

false; cannot divide by 0

9. Evaluate each formula for the given values.

a. $A = lw$, for $l = 5$ and $w = 3$

$A = (5)(3) = 15$

b. $A = s^2$, for $s = 7$

$A = 7^2 = 49$

c. $A = \dfrac{a + b}{2}$, for $a = 65$ and $b = 85$

$A = \frac{65 + 85}{2} = \frac{150}{2} = 75$

Cluster 2

Problem Solving with Fractions and Decimals (Rational Numbers)

Activity 1.4

Delicious Recipes

Objectives

1. Add and subtract fractions.

2. Multiply and divide fractions.

Problem-solving situations frequently require the use of fractions and decimals in their solutions. If you need some help working with fractions and decimals, Appendixes A and B contain detailed explanations, examples, and practice exercises with worked-out solutions.

As part of the final exam in your culinary arts class, you are asked the following questions. Remember to reduce each fraction and to use mixed numbers when appropriate. Good luck!

Crab Supreme

4 small (6 oz.) cans crabmeat

1 egg, hard-boiled and mashed

$\frac{1}{2}$ cup mayonnaise

$2\frac{1}{2}$ tbsp. chopped onion

$3\frac{2}{3}$ tbsp. plain yogurt

2 dashes of Tabasco sauce

$3\frac{1}{2}$ tbsp. chopped fresh chives

$\frac{1}{4}$ tsp. salt

$\frac{1}{2}$ tsp. garlic powder

1 tsp. lemon juice

Drain and rinse crab in cold water. Mash crab and egg together. Add all remaining ingredients except chives. Stir well. Chill and serve with chips or crackers. SERVES 6.

---

1. Determine the ingredients for one-half of the crab recipe. Fill in the blanks below.

__2__ small (6 oz.) cans crabmeat	__1__ dash Tabasco
__$\frac{1}{2}$__ egg, hard-boiled and mashed	__$1\frac{3}{4}$__ tbsp. chives
__$\frac{1}{4}$__ cup(s) mayonnaise	__$\frac{1}{8}$__ tsp. salt
__$1\frac{1}{4}$__ tbsp. chopped onion	__$\frac{1}{4}$__ tsp. garlic powder
__$1\frac{5}{6}$__ tbsp. plain yogurt	__$\frac{1}{2}$__ tsp. lemon juice

2. List the ingredients needed for the crab recipe if 18 people attend the party.

Use 3 times the amounts in the original recipe.

__12__ small (6 oz.) cans crabmeat	__6__ dashes Tabasco
__3__ eggs, hard-boiled and mashed	__$10\frac{1}{2}$__ tbsp. chives
__$1\frac{1}{2}$__ cup(s) mayonnaise	__$\frac{3}{4}$__ tsp. salt
__$7\frac{1}{2}$__ tbsp. chopped onion	__$1\frac{1}{2}$__ tsp. garlic powder
__11__ tbsp. plain yogurt	__3__ tsp. lemon juice

3. If a container of yogurt holds 1 cup, how many full batches of crab appetizer can you make with one container (1 cup = 16 tbsp.)?

$16 \div 3\frac{2}{3} = 16 \div \frac{11}{3} = \frac{16}{1} \cdot \frac{3}{11} = \frac{48}{11} = 4\frac{4}{11}$

I can make 4 whole batches.

4. If each person drinks $2\frac{2}{3}$ cups of soda, how many cups of soda will be needed for 18 people?

$2\frac{2}{3} \cdot 18 = \frac{8}{3} \cdot \frac{\overset{6}{\cancel{18}}}{1} = 48$ cups of soda will be needed.

Apple Crisp

4 cups tart apples	$\frac{1}{3}$ cup softened butter
peeled, cored, and sliced	$\frac{1}{2}$ tsp. salt
$\frac{2}{3}$ cup packed brown sugar	$\frac{3}{4}$ tsp. cinnamon
$\frac{1}{4}$ cup rolled oats	$\frac{1}{8}$ tsp. allspice or nutmeg
$\frac{1}{2}$ cup flour	

Preheat oven to 375°F. Place apples in a greased 8-inch square pan. Blend remaining ingredients until crumbly, and spread over the apples. Bake approximately 30 minutes uncovered, until the topping is golden and the apples are tender. SERVES 4.

---

5. List the ingredients needed for the apple crisp recipe if 18 people attend the party.

Since each recipe serves 4, I need $\frac{18}{4} = \frac{9}{2}$ times each ingredient to serve 18 people.

apples: $\frac{9}{2} \cdot 4 = 18$ cups   brown sugar: $\frac{9}{2} \cdot \frac{2}{3} = 3$ cups

oats: $\frac{9}{2} \cdot \frac{1}{4} = \frac{9}{8} = 1\frac{1}{8}$ cups   flour: $\frac{9}{2} \cdot \frac{1}{2} = \frac{9}{4} = 2\frac{1}{4}$ cups

butter: $\frac{9}{2} \cdot \frac{1}{3} = \frac{9}{6} = 1\frac{1}{2}$ cups   salt: $\frac{9}{2} \cdot \frac{1}{2} = \frac{9}{4} = 2\frac{1}{4}$ tsp.

cinn.: $\frac{9}{2} \cdot \frac{3}{4} = \frac{27}{8} = 3\frac{3}{8}$ tsp.   nutmeg: $\frac{9}{2} \cdot \frac{1}{8} = \frac{9}{16}$ tsp.

6. How many times would you need to fill a $\frac{2}{3}$-cup container to measure 4 cups of apples?

$4 \div \frac{2}{3} = 4 \cdot \frac{3}{2} = 6$ I need to fill it 6 times.

7. If it takes $\frac{3}{4}$ teaspoon of cinnamon to make one batch of apple crisp and you have only 6 teaspoons of cinnamon left in the cupboard, how many batches can you make?

$6 \div \frac{3}{4} = \frac{\overset{2}{\cancel{6}}}{1} \cdot \frac{4}{\underset{1}{\cancel{3}}} = 8$ I can make 8 batches with the cinnamon I have left.

Potato Pancakes

6 cups potato	$\frac{1}{3}$ cup flour
(pared and grated)	$3\frac{3}{8}$ tsp. salt
9 eggs	$2\frac{1}{4}$ tbsp. grated onion

Drain the potatoes well. Beat eggs and stir into the potatoes. Combine and sift the flour and salt, then stir in the onions. Add to the potato mixture. Shape into patties and sauté in hot fat. Best served hot with applesauce. MAKES 36 three-inch pancakes.

8. If you were to make one batch each of the crab supreme, apple crisp, and potato pancake recipes, how much salt would you need? How much flour? How much onion?

salt: $\frac{1}{4} + \frac{1}{2} + 3\frac{3}{8} = \frac{2}{8} + \frac{4}{8} + \frac{27}{8} = \frac{33}{8} = 4\frac{1}{8}$ tsp.

flour: $0 + \frac{1}{2} + \frac{1}{3} = \frac{3}{6} + \frac{2}{6} = \frac{5}{6}$ cup

onion: $2\frac{1}{2} + 0 + 2\frac{1}{4} = 2\frac{2}{4} + 2\frac{1}{4} = 4\frac{3}{4}$ tbsp.

To make one batch of each recipe, I would need $4\frac{1}{8}$ teaspoon salt, $\frac{5}{6}$ cups flour, and $4\frac{3}{4}$ tablespoon onion.

9. If you have 2 cups of flour in the cupboard before you start cooking for the party and you make one batch of each recipe, how much flour will you have left?

$2 - \frac{5}{6} = \frac{12}{6} - \frac{5}{6} = \frac{7}{6} = 1\frac{1}{6}$

I will have $1\frac{1}{6}$ cups of flour left.

EXERCISES: ACTIVITY 1.4

1. The year that you enter college, your freshman class consists of 760 students. According to statistical studies, about $\frac{4}{7}$ of these students will actually graduate. Approximately how many of your classmates will receive their degrees?

$\frac{4}{7}(760) = \frac{3040}{7} \approx 434$ classmates will receive their degrees.

2. You rent an apartment for the academic year (two semesters) with three of your college friends. The rent for the entire academic year is $10,000. Each semester you receive a bill for your share of the rent. If you and your friends divide the rent equally, how much must you pay each semester?

$10{,}000\left(\frac{1}{4}\right) = \$2500$ each per year

$\frac{1}{2}(2500) = \$1250$ each per semester

We each pay $1250 rent per semester.

3. Your residence hall has been designated a quiet building. This means that there is a no-noise rule from 10:00 P.M. every night to noon the next day. During what fraction of a 24-hour period is one allowed to make noise?

14 hours without noise

10 hours with noise $\frac{10}{24} = \frac{5}{12}$

One is allowed to make noise $\frac{5}{12}$ of the day.

4. You would like to learn to play the harp but are concerned with time constraints. A friend of yours plays, and for three consecutive days before a recital, she practices $1\frac{1}{4}$ hours, $2\frac{1}{2}$ hours, and $3\frac{2}{3}$ hours. What is her total practice time before a recital?

$1\frac{1}{4} + 2\frac{1}{2} + 3\frac{2}{3} = 1\frac{3}{12} + 2\frac{6}{12} + 3\frac{8}{12} = 6\frac{17}{12} = 7\frac{5}{12}$ hr

Her total practice time before a recital is $7\frac{5}{12}$ hours.

5. You are planning a summer cookout and decide to serve quarter-pound hamburgers. If you buy $5\frac{1}{2}$ pounds of hamburger meat, how many burgers can you make?

$5\frac{1}{2} \div \frac{1}{4} = \frac{11}{2} \div \frac{1}{4} = \frac{11}{2} \cdot \frac{4}{1} = \frac{44}{2} = 22$ hamburgers

I can make 22 hamburgers with $5\frac{1}{2}$ pounds of meat.

6. Your favorite muffin recipe calls for $2\frac{2}{3}$ cups of flour, 1 cup of sugar, $\frac{1}{2}$ cup of crushed cashews, and $\frac{5}{8}$ cup of milk, plus assorted spices. How many cups of mixture do you have?

$2\frac{2}{3} + 1 + \frac{1}{2} + \frac{5}{8} = 2\frac{16}{24} + 1 + \frac{12}{24} + \frac{15}{24} = 3\frac{43}{24} = 4\frac{19}{24}$

I have $4\frac{19}{24}$ cups of the muffin recipe mixture.

Exercise numbers appearing in color are answered in the Selected Answers appendix.

7. You must take medicine in four equal doses each day. Each day's medicine comes in a single container and measures $3\frac{1}{5}$ tablespoons. How much medicine is in each dose?

$3\frac{1}{5} \div 4 = \frac{16}{5} \div \frac{4}{1} = \frac{16}{5} \cdot \frac{1}{4} = \frac{16}{20} = \frac{4}{5}$ tbsp. per dose

Each dose contains $\frac{4}{5}$ tablespoons of medicine.

8. Perform the indicated operations.

a. $4\frac{2}{3} - 1\frac{6}{7}$

$= 4\frac{14}{21} - 1\frac{18}{21}$

$= 3\frac{35}{21} - 1\frac{18}{21}$

$= 2\frac{17}{21}$

b. $5\frac{1}{2} + 2\frac{1}{3}$

$= 5\frac{3}{6} + 2\frac{2}{6}$

$= 7\frac{5}{6}$

c. $2\frac{1}{6} \cdot 4\frac{1}{2}$

$= \frac{13}{6} \cdot \frac{\overset{3}{\cancel{9}}}{\underset{2}{\cancel{2}}}$

$= \frac{39}{4} = 9\frac{3}{4}$

d. $2\frac{3}{7} + \frac{14}{5}$

$= 2\frac{3}{7} + 2\frac{4}{5}$

$= 2\frac{15}{35} + 2\frac{28}{35}$

$= 4\frac{43}{35}$

$= 5\frac{8}{35}$

e. $\dfrac{4}{5} \div \dfrac{8}{3}$

$= \frac{4}{5} \cdot \frac{3}{\underset{2}{\cancel{8}}}$

$= \frac{3}{10}$

f. $4\frac{1}{5} \div \frac{10}{3}$

$= \frac{21}{5} \div \frac{10}{3}$

$= \frac{21}{5} \cdot \frac{3}{10}$

$= \frac{63}{50}$

$= 1\frac{13}{50}$

Activity 1.5

Course Grades and Your GPA

Objective

1. Recognize and calculate a weighted average.

You are a college freshman and the end of the semester is approaching. You are concerned about keeping a B− (80 through 83) average in your English literature class. Your grade will be determined by computing the **simple average**, or mean, of your exam scores. To calculate a simple average, you add all your scores and divide the sum by the number of exams. So far, you have scores of 82, 75, 85, and 93 on four exams. Each exam has a maximum score of 100.

1. What is your current average for the four exams?

$$\frac{82 + 75 + 85 + 93}{4} = \frac{335}{4} = 83.75; \quad \text{My current average is approximately 84.}$$

2. There is another way you can view simple averages that will lead to the important concept of weighted average. Note that the 4 in the denominator of the fraction $\frac{82 + 75 + 85 + 93}{4}$ divides each term of the numerator. Therefore, you can write

$$\frac{82 + 75 + 85 + 93}{4} = \frac{82}{4} + \frac{75}{4} + \frac{85}{4} + \frac{93}{4}.$$

a. The sum $\frac{82}{4} + \frac{75}{4} + \frac{85}{4} + \frac{93}{4}$ can be calculated by performing the four divisions and then adding the results to obtain the average score 83.75. Do this calculation.

$$\frac{82}{4} + \frac{75}{4} + \frac{85}{4} + \frac{93}{4} = 20.5 + 18.75 + 21.25 + 23.25 = 83.75 \text{ or } 84$$

b. Part a shows that the average 83.75 is the sum of 4 values, each one of which is $\frac{1}{4}$ of a test. This means that each test contributed $\frac{1}{4}$ of its value to the average and you will note that each test contributed equally to the average. For example, the first test score 82 contributed 20.5 to the average. How much did each of the other tests contribute?

The test score 75 added 18.75 to the average; the test scores 85 and 93 added 21.25 and 23.25, respectively.

3. This calculation of averages can also be viewed in a way that allows you to understand a weighted average. Recall that dividing by 4 is the same as multiplying by its reciprocal $\frac{1}{4}$. So you can rewrite the sum from Problem 2a as follows:

$$\frac{82}{4} + \frac{75}{4} + \frac{85}{4} + \frac{93}{4} = \frac{1}{4} \cdot 82 + \frac{1}{4} \cdot 75 + \frac{1}{4} \cdot 85 + \frac{1}{4} \cdot 93$$

Calculate the expression $\frac{1}{4} \cdot 82 + \frac{1}{4} \cdot 75 + \frac{1}{4} \cdot 85 + \frac{1}{4} \cdot 93$ by first multiplying each test score by $\frac{1}{4}$ and then adding the terms to get the average 83.75.

$$\frac{1}{4} \cdot 82 + \frac{1}{4} \cdot 75 + \frac{1}{4} \cdot 85 + \frac{1}{4} \cdot 93$$

$$= 20.5 + 18.75 + 21.25 + 23.25 = 83.75 \text{ or } 84$$

From the point of view in Problems 2 and 3, the ratio $\frac{1}{4}$ is called the **weight** of a test score. It means that a test with a weight of $\frac{1}{4}$ contributes $\frac{1}{4}$ (or 0.25 or 25%) of its score to the overall average. In the preceding problems, each test had the same weight. However there are many other situations where not every score contributes the same weight. One such situation occurs when your college determines an average for your semester's work. The average is called a **grade point average**, or **GPA**.

Weighted Averages

The semester finally ended and your transcript has just arrived in the mail. As a part-time student this semester you took 8 credits—a 3-credit English literature course and a 5-credit biology course. You open the transcript and discover that you earned an A (numerically equivalent to 4.0) in biology and a B (numerically 3.0) in English.

4. Do you think that your biology grade should contribute more to your semester average (GPA) than your English literature grade? Discuss this with your classmates and give a reason for your answer.

Yes, I think my biology grade should count for more in my average because my biology class met 5 times a week compared to 3 times for my literature class. It also meant that I put more time and work into the biology course, which also contained more material.

Perhaps you recognized and wrote in Problem 4 that a course that has more credit usually means more time and effort spent on more material. So, it seems fair that you should earn more points towards your GPA from a higher credit course.

5. a. One way to give your 5-credit biology course more weight in your GPA than your literature course is to determine what part of your 8 total credits does your biology course represent. Write that ratio.

$$\frac{5}{8}$$

b. Write the ratio that represents the weight your 3-credit literature grade contributes to your GPA.

$$\frac{3}{8}$$

c. Calculate the sum of the two weights from part a and part b.

$$\frac{5}{8} + \frac{3}{8} = \frac{8}{8} = 1$$

As you continue working with weighted averages, you will notice that the sum of the weights for a set of scores will always total 1, as it did in Problem 5c.

6. In Problem 3, you multiplied each test score by its weight and then summed the products to obtain your test average. Do the same kind of calculation here to obtain your GPA; that is, multiply each grade (4.0 for biology and 3.0 for literature) by its respective weight (from Problem 5) and then sum the products. Write your GPA to the nearest hundredth.

GPA $= \frac{5}{8} \cdot 4.0 + \frac{3}{8} \cdot 3.0 = \frac{29}{8} = 3.625$. Therefore my GPA, to the nearest hundredth, is 3.63

Problem 6 illustrates the procedure for computing a **weighted average**. The general procedure is given in the following.

Procedure

Computing the Weighted Average of Several Data Values

1. Multiply each data value by its respective weight, and then

2. Sum these weighted data values.

Example 1   *You will have a 12-credit load next semester as a full-time student. You will have a 2-credit course, two 3-credit courses and a 4-credit course. Determine the weights of each course.*

SOLUTION

The 2-credit course will have a weight of $\frac{2}{12} = \frac{1}{6}$; the 3-credit courses will each have a weight of $\frac{3}{12} = \frac{1}{4}$. Finally, the 4-credit course has a weight of $\frac{4}{12} = \frac{1}{3}$.

7. Do the weights in Example 1 sum to 1? Explain.

Yes, because $\frac{2}{12} + (2)\frac{3}{12} + \frac{4}{12} = \frac{2 + 6 + 4}{12} = \frac{12}{12} = 1.$

Your letter grade for a course translates into a numerical equivalent according to the following table:

LETTER GRADE	A	A−	B+	B	B−	C+	C	C−	D+	D	D−	F
NUMERICAL EQUIVALENT	4.00	3.67	3.33	3.00	2.67	2.33	2.00	1.67	1.33	1.00	0.67	0.00

Suppose you took 17 credit-hours this past semester, your third semester in college. You earned an A− in psychology (3 hours), a C+ in economics (3 hours), a B+ in chemistry (4 hours), a B in English (3 hours), and a B− in mathematics (4 hours).

8. a. Use the first four columns of the following table to record the information regarding the courses you took. As a guide, the information for your psychology course has been recorded for you.

1 COURSE	2 LETTER GRADE	3 NUMERICAL EQUIVALENT	4 CREDIT HOURS	5 WEIGHT	6 CONTRIBUTION TO GPA
Psychology	A−	3.67	3	$\frac{3}{17}$	$\frac{3}{17} \cdot 3.67 \approx 0.648$
Economics	C+	2.33	3	$\frac{3}{17}$	0.411
Chemistry	B+	3.33	4	$\frac{4}{17}$	0.784
English	B	3.00	3	$\frac{3}{17}$	0.529
Mathematics	B−	2.67	4	$\frac{4}{17}$	0.628

b. Calculate the weight for each course and enter it in column 5.

c. For each course, multiply your numerical grade (column 3) by the course's weight (column 5). Round to three decimal places and enter this product in column 6, the course's contribution to your GPA.

d. You can now calculate your semester's GPA by summing the contributions of all your courses. What is your semester GPA?

My semester GPA is 3.000.

9. ESR Manufacturing Corporation of Tampa, Florida, makes brass desk lamps that require three levels of labor to make and finish. The table shows the number of hours each level of labor requires to make each lamp and how much each level of labor costs per hour.

LEVEL OF LABOR	LABOR HOURS REQUIRED	HOURLY WAGE ($)
Skilled	6	10.00
Semiskilled	3	8.00
Unskilled	1	6.00
Total:	10	

a. What are the weights for each level of labor?

Skilled: $\frac{6}{10} = 0.6$; Semiskilled: $\frac{3}{10} = 0.3$; Unskilled: $\frac{1}{10} = 0.1$

b. Determine the average hourly wage.

Average hourly wage is $0.6(10) + 0.3(8) + 0.1(6) = \$9$ per hour.

SUMMARY: ACTIVITY 1.5

1. To calculate a **simple average** (also called a **mean**), add all the values and divide the sum by the number of values.

2. To compute a **weighted average** of several data values:

   i. Multiply each data value by its respective weight, and then

   ii. Sum the weighted data values.

3. The sum of the weights used to compute a weighted average will always be equal to 1.

EXERCISES: ACTIVITY 1.5

1. A grade of W is given if you withdraw from a course before a certain date. The W appears on your transcript but is not included in your grade point average. Suppose that instead of a C+ in economics in Problem 8, you receive a W. Use this new grade to recalculate your GPA.

Withdrawing from economics leaves 14 credits.

$\frac{3}{14} \approx 0.214$, $\frac{4}{14} \approx 0.286$

psychology $\frac{3}{14}(3.67) = 0.786$;   English $\frac{3}{14}(3.00) = 0.643$;

chemistry $\frac{4}{14}(3.33) = 0.951$; and mathematics $\frac{4}{14}(2.67) = 0.763$

My GPA is 3.143 if I withdraw from economics.

2. Now suppose that you earn an F in economics. The F is included in your grade point average. Recalculate your GPA from Problem 8.

$3.000 - 0.411 = 2.589$

My GPA is 2.589 if I earn an F in economics.

3. In your first semester in college, you took 13 credit hours and earned a GPA of 2.13. In your second semester, your GPA of 2.34 was based on 12 credit hours. You calculated the third semester's GPA in Problem 8.

a. Explain why the calculation of your overall GPA for the three semesters requires a weighted average.

(Answers will vary.) The larger the number of credits taken, the greater the effect of that semester's GPA.

b. Calculate your overall GPA for the three semesters.

SEMESTER	GPA	CREDITS	WEIGHT	NUMERICAL EQUIVALENT
1	2.13	13	$\frac{13}{42}$	$2.13\left(\frac{13}{42}\right) \approx 0.659$
2	2.34	12	$\frac{12}{42}$	$2.34\left(\frac{12}{42}\right) \approx 0.669$
3	3.00	17	$\frac{17}{42}$	$3.00\left(\frac{17}{42}\right) \approx 1.214$
				Total: 2.542

My overall GPA for the three semesters is 2.542 ≈ 2.54.

4. You are concerned about passing your economics class with a C– (70) average. Your grade is determined by averaging your exam scores. So far, you have scores of 78, 66, 87, and 59 on four exams. Each exam is based on 100 points. Your economics teacher uses the simple average method to determine your average.

a. What is your current average for the four exams?

78 + 66 + 87 + 59 = 290

$\frac{290}{4} = 72.5$ My current average is 72.5.

b. What is the lowest score you could achieve on the fifth exam to have at least a 70 average?

For a 70 average on five exams, I need 70 · 5 = 350 points. I already have 290 points total for the first four exams, so I need 350 − 290 = 60 more points to achieve at least a 70 average.

5. Suppose you took 15 credit hours last semester. You earned an A– in English (3 hours), a B in mathematics (4 hours), a C+ in chemistry (3 hours), a B+ in health (2 hours), and a B– in history (3 hours). Calculate your GPA for the semester.

COURSE	LETTER GRADE	NUMERICAL EQUIVALENT	CREDIT HOURS	WEIGHT	WEIGHT × NUM. EQUIV.
English	A–	3.67	3	$\frac{3}{15}$	0.734
Math	B	3.00	4	$\frac{4}{15}$	0.800
Chemistry	C+	2.33	3	$\frac{3}{15}$	0.466
Health	B+	3.33	2	$\frac{2}{15}$	0.444
History	B–	2.67	3	$\frac{3}{15}$	0.534
					Sum: 2.978

My GPA for the semester is 2.978.

6. Your history professor discovers an error in his calculation of your grade from last semester. Your newly computed history grade is a B+. Use this new grade and the information in Exercise 5 to recalculate your GPA.

B+ = 3.33   0.2(3.33) = 0.666

Use 0.666 in place of 0.534 in Exercise 5.

The revised sum is 0.734 + 0.800 + 0.466 + 0.444 + 0.666 = 3.11.

The corrected GPA is 3.11.

7. In baseball, weighted averages may lead to surprising results. For example, this happened in 1995 and 1996 in the comparison of batting averages for Derek Jeter of the New York Yankees and David Justice of the Atlanta Braves. In 1995, Derek Jeter made 12 hits and each hit that he made was weighted $\frac{1}{48}$ because he went up to bat 48 times in 1995. Therefore, his batting average was $\frac{1}{48} \cdot 12 = 12 \div 48 = .250$ to the nearest thousandth.

a. In 1995, David Justice of the Atlanta Braves went up to bat 411 times, so each of his 1995 hits was weighted $\frac{1}{411}$. He made 104 hits. Calculate his batting average and record your answer to the nearest thousandth in the table in part c.

$\frac{1}{411} \cdot 104 = 104 \div 411 = .253$

b. In the 1996 baseball season, Jeter made 183 hits in 582 times at bat. Justice made 45 hits in 140 at bats. Calculate each of their batting averages for 1996 and record in the table.

Jeter: $\frac{1}{582} \cdot 183 = .314$;   Justice: $\frac{1}{140} \cdot 45 = .321$

c. Other statistics that could be of interest to ballplayers, managers, team owners, and fans would be batting averages over two years, over three years, over entire careers. For example, over the 2-year period, 1995 and 1996, Justice had 104 + 45 = 149 hits in 411 + 140 = 551 at bats. Therefore, his batting average combining the two years is $\frac{1}{551} \cdot 149 = .270$. Similarly, calculate Jeter's combined batting average combining the 2 years. Record your result in the table.

$\frac{1}{630} \cdot 195 = .310$

	BATTING AVERAGE		
	1995	**1996**	**1995 & 1996 COMBINED**
DEREK JETER	.250	.314	.310
DAVID JUSTICE	.253	.321	.270

d. According to the statistics in the preceding table, who was the better hitter in 1995? In 1996? Give a reason for each answer.

David Justice had higher batting averages in 1995 (.253 > .250) and in 1996 (.321 > .314), which indicates that he could be a better hitter than Derek Jeter.

e. Is David Justice a better hitter according to the combined 1995–1996 baseball seasons?

The combined 2-year averages show that Derek Jeter's batting average is significantly higher than David Justice's average (.310 > .270). This indicates Jeter could be the better hitter.

f. As it turned out, these contradictory results continued for Jeter and Justice into the 1997 baseball season. In 1997, Jeter had 190 hits in 654 at bats; Justice hit 163 times in 495 at bats. Calculate their batting averages to the nearest thousandth and record in the following table.

BATTING AVERAGES			
	1995	**1996**	**1997**
DEREK JETER	.250	.314	.291
DAVID JUSTICE	.253	.321	.329

g. Use the appropriate data from parts a, b, and c to determine the total number of hits and at bats for each player. Then use those totals to determine the batting averages over the three-year period. Record your results here.

BATTING AVERAGE FOR COMBINED DATA FOR 1995 THROUGH 1997		
TOTAL HITS	**TOTAL AT BATS**	**BATTING AVERAGE**
DEREK JETER 12 + 183 + 190 = 385	48 + 582 + 654 = 1284	385/1284 = .300
DAVID JUSTICE 104 + 45 + 163 = 312	411 + 140 + 495 = 1046	312/1046 = .298

h. What conclusions would you draw from the results you recorded in part g?

(Answers will vary). One possible conclusion may be: On a yearly basis, Justice appears to be a slightly better hitter. It appears that Jeter may be the better hitter in the long run. Other data, statistics, and deeper analysis are most likely needed to make more definite conclusions.

You may want to research these curious results further to find out what else baseball followers and statisticians have to say about this result (which is known as the Simpson-Yule Paradox in statistics).

Cluster 2 · What Have I Learned?

1. To add or subtract fractions, you must write them in equivalent form with common denominators. However, to multiply or divide fractions, you do not need a common denominator. Why is this reasonable?

(Answers will vary.)

When combining fractional parts, you need to count portions of the same size. To do this, you write fractions equivalently with a common denominator. When multiplying or dividing fractions, you do not need a common denominator because you are not combining or comparing.

2. The operation of division can be viewed from several different points of view. For example, $24 \div 3$ has at least two meanings:

- Write 24 as the sum of some number of 3s.

- Divide 24 into three equal-sized parts, whose size you must determine.

These interpretations can be applied to fractions as well as to whole numbers.

a. Calculate $2 \div \frac{1}{2}$ by answering this question: 2 can be written as the sum of how many $\frac{1}{2}$s?

$\frac{1}{2} + \frac{1}{2} + \frac{1}{2} + \frac{1}{2} = 2$

2 can be written as the sum of four $\frac{1}{2}$s, so, $2 \div \frac{1}{2} = 4$.

b. Calculate $\frac{1}{5} \div 2$ by answering this question: If you divide $\frac{1}{5}$ into two equal parts, how large is each part?

Each part is $\frac{1}{10}$ unit because $\frac{1}{10} + \frac{1}{10} = \frac{1}{5}$.

c. Do your answers to parts a and b agree with the results you would obtain by using the procedures for dividing fractions reviewed in this cluster? Explain.

Yes, the answers agree. (Explanations will vary.)

Cluster 2   How Can I Practice?

1. You are in a golf tournament. There is a prize for the person who drives the ball closest to the green on the sixth hole. You drive the ball to within 4 feet $2\frac{3}{8}$ inches of the hole and your nearest competitor is 4 feet $5\frac{1}{4}$ inches from the hole. By how many inches do you win?

4 ft. $5\frac{1}{4}$ in. $-$ 4 ft. $2\frac{3}{8}$ in. $= 5\frac{1}{4} - 2\frac{3}{8}$ in. $= 5\frac{2}{8} - 2\frac{3}{8}$ in. $= 4\frac{10}{8} - 2\frac{3}{8}$ in. $= 2\frac{7}{8}$ in.

I won by $2\frac{7}{8}$ inches.

2. One of your jobs as the assistant to a weather forecaster is to determine the average thickness of the ice in a bay on the St. Lawrence River. Ice fishermen use this report to determine if the ice is safe for fishing. You must chop holes in five different areas, measure the thickness of the ice, and take the average. During the first week in January, you record the following measurements: $2\frac{3}{8}, 5\frac{1}{2}, 6\frac{3}{4}, 4,$ and $5\frac{7}{8}$ inches. What do you report as the average thickness of the ice in this area?

$2\frac{3}{8} + 5\frac{1}{2} + 6\frac{3}{4} + 4 + 5\frac{7}{8} = 2\frac{3}{8} + 5\frac{4}{8} + 6\frac{6}{8} + 4 + 5\frac{7}{8} = 22\frac{20}{8} = 24\frac{4}{8} = 24\frac{1}{2}$

$24\frac{1}{2} \div 5 = \frac{49}{2} \div \frac{5}{1} = \frac{49}{2} \cdot \frac{1}{5} = \frac{49}{10} = 4\frac{9}{10}$ in.

The average thickness of the ice is $4\frac{9}{10}$ inches.

3. You and two others in your family will divide 120 shares of a computer stock left by a relative who died. The stock is worth $\$10\frac{9}{16}$ per share. If you decide to sell your portion of the stock, how much money will you receive?

$\frac{120}{3} = 40$ shares each

$40\left(10\frac{9}{16}\right) = \frac{\overset{10}{\cancel{40}}}{1} \cdot \frac{169}{\underset{4}{\cancel{16}}} = \frac{1690}{4} = \$422.50$ is what I will receive for my portion.

4. You are about to purchase a rug for your college dorm room. The rug's length is perfect for your room. The width of the rug you want to purchase is $6\frac{1}{2}$ feet. If you center the rug in the middle of your room, which is 10 feet wide, how much of the floor will show on each side of the rug?

$10 - 6\frac{1}{2} = 9\frac{2}{2} - 6\frac{1}{2} = 3\frac{1}{2}$ ft. short

$3\frac{1}{2} \div 2 = \frac{7}{2} \div \frac{2}{1} = \frac{7}{2} \cdot \frac{1}{2} = \frac{7}{4} = 1\frac{3}{4}$ feet of the floor will show on each side.

5. A plumber has $12\frac{1}{2}$ feet of plastic pipe. She uses $3\frac{2}{3}$ feet for the sink line and $5\frac{3}{4}$ feet for the washing machine. She needs approximately $3\frac{1}{2}$ feet for a disposal. Does she have enough pipe left for a disposal?

$3\frac{2}{3} + 5\frac{3}{4} = 3\frac{8}{12} + 5\frac{9}{12} = 8\frac{17}{12} = 9\frac{5}{12}$ feet of pipe is used.

$12\frac{1}{2} - 9\frac{5}{12} = 12\frac{6}{12} - 9\frac{5}{12} = 3\frac{1}{12}$ feet are left.

There is not enough pipe left for a disposal.

6. Perform the following operations. Write the result in simplest terms or as a mixed number.

a. $\dfrac{5}{7} + \dfrac{2}{7}$

$\dfrac{7}{7} = 1$

b. $\dfrac{3}{4} + \dfrac{3}{8}$

$\dfrac{6}{8} + \dfrac{3}{8} = \dfrac{9}{8} = 1\frac{1}{8}$

c. $\dfrac{3}{8} + \dfrac{1}{12}$

$\dfrac{9}{24} + \dfrac{2}{24} = \dfrac{11}{24}$

d. $\dfrac{4}{5} + \dfrac{5}{6}$

$\dfrac{24}{30} + \dfrac{25}{30} = \dfrac{49}{30} = 1\frac{19}{30}$

Exercise numbers appearing in color are answered in the Selected Answers appendix.

e. $\dfrac{1}{2} + \dfrac{3}{5} + \dfrac{4}{15}$

$\dfrac{15}{30} + \dfrac{18}{30} + \dfrac{8}{30} = \dfrac{41}{30} = 1\frac{11}{30}$

f. $\dfrac{11}{12} - \dfrac{5}{12}$

$\dfrac{6}{12} = \dfrac{1}{2}$

g. $\dfrac{7}{9} - \dfrac{5}{12}$

$\dfrac{28}{36} - \dfrac{15}{36} = \dfrac{13}{36}$

h. $\dfrac{2}{3} - \dfrac{1}{4}$

$\dfrac{8}{12} - \dfrac{3}{12} = \dfrac{5}{12}$

i. $\dfrac{7}{30} - \dfrac{3}{20}$

$\dfrac{14}{60} - \dfrac{9}{60} = \dfrac{5}{60} = \dfrac{1}{12}$

j. $\dfrac{4}{5} - \dfrac{3}{4} + \dfrac{1}{2}$

$\dfrac{16}{20} - \dfrac{15}{20} + \dfrac{10}{20} = \dfrac{11}{20}$

k. $\dfrac{3}{5} \cdot \dfrac{1}{2}$

$\dfrac{3}{10}$

l. $\dfrac{2}{3} \cdot \dfrac{7}{8}$

$\dfrac{14}{24} = \dfrac{7}{12}$

m. $\dfrac{15}{8} \cdot \dfrac{24}{5}$

$\dfrac{15}{8} \cdot \dfrac{24}{5} = 9$

n. $5 \cdot \dfrac{3}{10}$

$\dfrac{15}{10} = \dfrac{3}{2} = 1\frac{1}{2}$

o. $\dfrac{3}{8} \div \dfrac{3}{4}$

$\dfrac{3}{8} \cdot \dfrac{4}{3} = \dfrac{1}{2}$

p. $8 \div \dfrac{1}{2}$

$\dfrac{8}{1} \cdot \dfrac{2}{1} = 16$

q. $\dfrac{5}{7} \div \dfrac{20}{21}$

$\dfrac{5}{7} \cdot \dfrac{21}{20} = \dfrac{3}{4}$

r. $4\frac{5}{6} + 3\frac{2}{9}$

$4\frac{15}{18} + 3\frac{4}{18} = 7\frac{19}{18} = 8\frac{1}{18}$

s. $12\frac{5}{12} - 4\frac{1}{6}$

$12\frac{5}{12} - 4\frac{2}{12} = 8\frac{3}{12} = 8\frac{1}{4}$

t. $6\frac{2}{13} - 4\frac{7}{26}$

$6\frac{4}{26} - 4\frac{7}{26} = 5\frac{30}{26} - 4\frac{7}{26} = 1\frac{23}{26}$

u. $2\frac{1}{4} \cdot 5\frac{2}{3}$

$\dfrac{9}{4} \cdot \dfrac{17}{3} = \dfrac{51}{4} = 12\frac{3}{4}$

v. $6\frac{3}{4} \div 1\frac{2}{7}$

$\dfrac{27}{4} \div \dfrac{9}{7} = \dfrac{27}{4} \cdot \dfrac{7}{9} = \dfrac{21}{4} = 5\frac{1}{4}$

7. At the end of the semester, the bookstore will buy back books that will be used again in courses the next semester. Usually, the store will give you one-sixth of the original cost of the book. If you spend \$243 on books this semester and the bookstore will buy back all your books, how much money can you expect to receive?

$243\left(\frac{1}{6}\right) = \$40.50$ is what I can expect to receive if I sell my books back to the bookstore.

8. a. The driving distance between Buffalo, New York, and Orlando, Florida, is approximately 1178 miles. If your average speed is 58.5 miles per hour, calculate the total driving time.

$$t = \frac{d}{r} = \frac{1178}{58.5} = 20.13675 \approx 20 \text{ hr.}$$

The driving time would be about 20 hours.

b. The driving distance between Erie, Pennsylvania, and Daytona Beach, Florida, is approximately 1048 miles. If your average speed is 68.2 miles per hour, calculate the total driving time. (Round to the nearest tenths place.)

$$t = \frac{d}{r} = \frac{1048}{68.2} = 15.36657 \approx 15.4 \text{ hours is the total driving time.}$$

c. If you need to make the trip in part b in 14 hours, calculate the average speed needed. Use $r = \frac{d}{t}$.

$$r = \frac{d}{t} = \frac{1048}{14} = 74.85714 \approx 75 \text{ miles per hour is the average speed required}$$

to make the trip in 14 hours.

Cluster 3   Comparisons and Proportional Reasoning

Activity 1.6

Everything Is Relative

Objectives

1. Distinguish between absolute and relative measure.

2. Write ratios in fraction, decimal, and percent formats.

3. Determine equivalence of ratios.

During the 2008–2009 National Basketball Association (NBA) season, LeBron James of the Cleveland Cavaliers, Kobe Bryant of the Los Angeles Lakers, Dwayne Wade of the Miami Heat, and Carmelo Anthony of the Denver Nuggets were among the NBA's highest scorers. The statistics in the table represent each player's 3-point field goal totals for the entire season.

PLAYER	NUMBER OF 3-POINT FIELD GOALS MADE	NUMBER OF 3-POINT FIELD GOALS ATTEMPTED
James	132	384
Bryant	118	336
Wade	88	278
Anthony	63	170

1. Using only the data in column 2, Number of 3-Point Field Goals Made, rank the players from best to worst according to their field goal performance.

   James, Bryant, Wade, Anthony

2. You can also rank the players by using the data from both column 2 and column 3. For example, LeBron James made 132 3-point field goals out of the 384 he attempted. The 132 successful baskets can be compared to the 384 attempts by dividing 132 by 384. You can represent that comparison numerically as the fraction $\frac{132}{384}$ or, equivalently, as the decimal 0.344 (rounded to thousandths). Complete the following table, and use your results to determine another ranking of the four players from best to worst performance.

 Nothing but Net

PLAYER	NUMBER OF 3-PT. FIELD GOALS MADE	NUMBER OF 3-PT. FIELD GOALS ATTEMPTED	RELATIVE PERFORMANCE		
			VERBAL	FRACTIONAL	DECIMAL
James	132	384	132 out of 384	$\frac{132}{384}$	0.344
Bryant	118	336	118 out of 336	$\frac{118}{336}$	0.351
Wade	88	278	88 out of 278	$\frac{88}{278}$	0.317
Anthony	63	170	63 out of 170	$\frac{63}{170}$	0.371

The ranking from best to worst is Anthony, Bryant, James, Wade.

The two sets of rankings are based on two different points of view. The first ranking (in Problem 1) takes an **absolute** viewpoint in which you just count the actual number of 3-pt. field goals made. The second ranking (in Problem 2) takes a **relative** perspective in which you take into account the number of successes *relative* to the number of attempts.

3. Does one measure, either absolute or relative, better describe field goal performance than the other? Explain.

Relative measure is a better measure of performance because it can compare the performance of players who each have a different number of scoring opportunities.

4. Identify which statements refer to an absolute measure and which refer to a relative measure. Explain your answers.

a. I got seven answers wrong.

It is an *absolute* measure because no comparison to another number is made.

b. I guessed on four answers.

The measure is *absolute* for the same reason as in part a.

c. Two-thirds of the class failed.

The measure is *relative* because the number who failed is being compared to the entire class. The use of a fraction is often a clue that the measure is relative.

d. I saved $10.

The measure is *absolute* for the same reason as in part a.

e. I saved 40%.

The measure is *relative* because a comparison to an original price is implied.

f. Four out of five students work to help pay tuition.

The measure is relative because a number of students who work is being compared to a number of students being considered.

g. Johnny Damon's batting average in 2005, his final year with the Red Sox, was .316.

The clue that the measure is relative is that a decimal point is used. Damon's performance was equivalent to having 316 hits in 1000 at bats.

h. Barry Bonds hit 73 home runs, a new major league record, in 2001.

The measure is *absolute* for the same reason as in part a.

i. In 2007, the U.S. Census Bureau estimated that 45,504,000 members of the U.S. population were Hispanic (of any race).

The measure is *absolute* for the same reason as in part a.

j. In 2007, the U.S. Census Bureau indicated that 15.1% of the nation's population was Hispanic.

This is a relative measure because the Hispanic population is being compared to the entire U.S. population.

5. What mathematical notation or verbal phrases in Problem 4 indicate a relative measure?

The terms *fraction, a percent, a decimal*, and *out of* indicate relative measure.

Definition

Relative measure is a quotient that compares two similar quantities, often a "part" and a "total." The part is *divided* by the total.

Ratio is the term used to describe the relative measure quotient. Ratios can be expressed in several forms—words (verbal form), fractions, decimals, or percents.

6. Use the free-throw statistics from the NBA 2008–2009 regular season to express each player's relative performance as a ratio in verbal, fractional, and decimal form. The data for LeBron James is completed for you. Which player had the best relative performance?

Kobe Bryant, who made 85.6% of his free throws.

 **Net Results**

PLAYER	NUMBER OF FREE THROWS MADE	NUMBER OF FREE THROWS ATTEMPTED	RELATIVE PERFORMANCE		
			VERBAL	FRACTIONAL	DECIMAL
James	594	762	594 out of 762	$\frac{594}{762}$	0.780
Bryant	483	564	483 out of 564	$\frac{483}{564}$	0.856
Wade	590	771	590 out of 771	$\frac{590}{771}$	0.765
Anthony	371	468	371 out of 468	$\frac{371}{468}$	0.793

7. a. Three friends, shooting baskets in the schoolyard, kept track of their performance. Andy made 9 out of 15 shots, Pat made 28 out of 40, and Val made 15 out of 24. Rank their relative performance.

1. Pat, $\frac{28}{40} = 0.700$;   2. Val, $\frac{15}{24} = 0.625$;   3. Andy, $\frac{9}{15} = 0.600$

b. Which ratio form (fractional, decimal, or other) did you use to determine the ranking?

I used decimals.

8. How would you determine whether two ratios, such as "12 out of 20" and "21 out of 35," were equivalent?

Write the ratios in decimal form and compare the decimals.

9. a. Match each ratio from column A with the equivalent ratio in column B.

COLUMN A	COLUMN B
15 out of 25	84 out of 100
42 out of 60	65 out of 100
63 out of 75	60 out of 100
52 out of 80	70 out of 100

15 out of 25 is equivalent to 60 out of 100;

42 out of 60 is equivalent to 70 out of 100;

63 out of 75 is equivalent to 84 out of 100; and

52 out of 80 is equivalent to 65 out of 100.

b. Which ratio in each matched pair is more useful in comparing and ranking the four pairs? Why?

The ratios in column B are more useful because they are all relative to the same amount (100).

Percents

Relative measure based on 100 is familiar and seems natural. There are 100 cents in a dollar, 100 points on a test. You have probably been using a ranking scale from 0 to 100 since childhood. You most likely possess an instinctive understanding of ratios relative to 100. A ratio such as 40 out of 100 can be expressed as 40 **per** 100, or, more commonly, as 40 **percent**, written as 40%. Percent always indicates a ratio, number of parts out of 100.

10. Express each ratio in column B of Problem 9 as a percent, using the symbol %.

84 out of 100 = 84%, 65 out of 100 = 65%, 60 out of 100 = 60%, 70 out of 100 = 70%

Each ratio in Problem 10 is already a ratio "out of 100," so you just need to replace the phrase "out of 100" with the % symbol. However, suppose you need to write a ratio such as 21 out of 25 in percent format. You may recognize that the denominator, 25, is a factor of 100 ($25 \cdot 4 = 100$). Then the fraction $\frac{21}{25}$ can be written equivalently as $\frac{21 \cdot 4}{25 \cdot 4} = \frac{84}{100}$, which is precisely 84%.

A more general method to convert a ratio such as 21 out of 25 into percent format is described next.

Procedure

Converting a Fraction to a Percent

1. Obtain the equivalent decimal form by dividing the numerator of the fraction by the denominator.

2. Move the decimal point in the quotient from step 1 two places to the right, inserting placeholder zeros, if necessary.

3. Attach the % symbol to the right of the number.

---

11. Write the following ratios in percent format.

a. 35 out of 100

$$\frac{35}{100} = 0.35 = 35\%$$

b. 16 out of 50

$$\frac{16}{50} = 0.32 = 32\%$$

c. 8 out of 20

$$\frac{8}{20} = 0.40 = 40\%$$

d. 7 out of 8

$$\frac{7}{8} = 0.875 = 87.5\%$$

In many applications, you will also need to convert a percent into decimal form. This can be done by replacing the % symbol with its equivalent meaning "out of 100," $\frac{1}{100}$.

Procedure

Converting a Percent to a Decimal

1. Locate the decimal point in the number preceding the % symbol.

2. Move the decimal point two places to the left, inserting placeholding zeros if needed. (Note that moving the decimal point two places to the left is the same as dividing by 100.)

3. Delete the % symbol.

12. Write the following percents in decimal format.

a. 75%   **b.** 3.5%   **c.** 200%   **d.** 0.75%

0.75   0.035   2.00   0.0075

13. Use the field goal statistics from the regular 2008–2009 NBA season to express each player's relative performance as a ratio in verbal, fractional, decimal, and percent form. Round the decimal to the nearest hundredth.

 Get to the Points

PLAYER	NUMBER OF FIELD GOALS MADE	NUMBER OF FIELD GOALS ATTEMPTED	RELATIVE PERFORMANCE			
			VERBAL	FRACTION	DECIMAL	PERCENT
D. Wade (Miami)	854	1739	854 out of 1739	$\frac{854}{1739}$	0.49	49%
K. Bryant (LA)	800	1712	800 out of 1712	$\frac{800}{1712}$	0.47	47%
L. James (Cleveland)	789	1613	789 out of 1613	$\frac{789}{1613}$	0.49	49%
C. Anthony (Denver)	535	1207	535 out of 1207	$\frac{535}{1207}$	0.44	44%

SUMMARY: ACTIVITY 1.6

1. **Relative measure** is a quotient that compares two similar quantities, often a "part" and a "total." The part is divided by the total.

2. **Ratio** is the term used to describe a relative measure.

3. Ratios can be expressed in several forms: **verbal** (4 out of 5), **fractional** $\left(\frac{4}{5}\right)$, **decimal** (0.8), or as a **percent** (80%).

4. **Percent** always indicates a ratio out of 100.

5. To convert a fraction or decimal to a percent:

   i. Convert the fraction to decimal form by dividing the numerator by the denominator.

   ii. Move the decimal point two places to the right and then attach the % symbol.

6. To convert a percent to a decimal:

   i. Locate the decimal point in the number preceding the % symbol.

   ii. Move the decimal point two places to the left and then drop the % symbol.

7. Two ratios are **equivalent** if their decimal or reduced-fraction forms are equal.

EXERCISES: ACTIVITY 1.6

1. Complete the table below by representing each ratio in all four formats. Round decimals to thousandths and percents to tenths.

 Relatively Speaking...

VERBAL FORM	REDUCED-FRACTION FORM	DECIMAL FORM	PERCENT FORM
1 out of 3	$\frac{1}{3}$	$0.\overline{3} \approx 0.333$	$33\frac{1}{3}\% \approx 33.3\%$
2 out of 5	$\frac{2}{5}$	0.4	40%
18 out of 25	$\frac{18}{25}$	0.72	72%
8 out of 9	$\frac{8}{9}$	$0.\overline{8} \approx 0.889$	$88\frac{8}{9}\% \approx 88.9\%$
3 out of 8	$\frac{3}{8}$	0.375	37.5%
25 out of 45	$\frac{5}{9}$	$0.\overline{5} \approx 0.556$	$55\frac{5}{9}\% \approx 55.6\%$
120 out of 40	3	3.0	300%
3 out of 4	$\frac{3}{4}$	0.75	75%
27 out of 40	$\frac{27}{40}$	0.675	67.5%
3 out of 5	$\frac{3}{5}$	0.6	60%
2 out of 3	$\frac{2}{3}$	$0.\overline{6} \approx 0.667$	$66\frac{2}{3}\% \approx 66.7\%$
4 out of 5	$\frac{4}{5}$	0.8	80%
1 out of 200	$\frac{1}{200}$	0.005	0.50%
2 out of 1	2	2.0	200%

2. **a.** Match each ratio from column I with the equivalent ratio in column II.

COLUMN I	COLUMN II
1. 12 out of 27	A. 60 out of 75
2. 28 out of 36	B. 25 out of 40
3. 45 out of 75	C. 21 out of 27
4. 64 out of 80	D. 42 out of 70
5. 35 out of 56	E. 20 out of 45

See answers to part b for the matching.

b. Write each matched pair of equivalent ratios as a percent.

1. and E, $\frac{12}{27} = \frac{20}{45} = 0.44\overline{4} = 44.\overline{4}\%$

2. and C, $\frac{28}{36} = \frac{21}{27} = 0.77\overline{7} = 77.\overline{7}\%$

3. and D, $\frac{45}{75} = \frac{42}{70} = 0.6 = 60\%$

4. and A, $\frac{64}{80} = \frac{60}{75} = 0.8 = 80\%$

5. and B, $\frac{35}{56} = \frac{25}{40} = 0.625 = 62.5\%$

3. Your biology instructor returned three quizzes today. On which quiz did you perform best? Explain how you determined the best score.

Quiz 1: 18 out of 25   Quiz 2: 32 out of 40   Quiz 3: 14 out of 20

$\frac{18}{25} = 72\%$ $\frac{32}{40} = 80\%$ $\frac{14}{20} = 70\%$

I performed best on quiz 2.

4. Baseball batting averages are the ratios of hits to 'at bats.' They are reported as three-digit decimals. Determine the batting averages and ranking of three players with the following records.

a. 16 hits out of 54 at bats

0.296

b. 25 hits out of 80 at bats

0.313

c. 32 hits out of 98 at bats

0.327

From highest to lowest, the ranking is c, b, a.

5. There are 1720 women among the 3200 students at the local community college. Express this ratio in each of the following forms:

a. fraction $\frac{1720}{3200}$

b. reduced fraction $\frac{43}{80}$

c. decimal   0.5375

d. percent   53.75%

6. At the state university campus, there are 2304 women and 2196 men enrolled. In which school, the community college in Exercise 5 or the state university, is the relative number of women greater? Explain your reasoning.

The total number of students at the university is 2304 + 2196 = 4500.

The relative number of women at the university is $\frac{2304}{4500} = 0.512 = 51.2\%$.

The relative number of women is greater at the community college.

7. In the 2008 Major League Baseball season, the world champion Philadelphia Phillies ended their regular season with a 92–70 win-loss record. During the playoff season, their win-loss record was 10–4. Did they play better in the regular season or in the postseason? Justify your answer mathematically.

number of games	ratio of wins to games played
regular season: 92 + 70 = 162	$\frac{92}{162} \approx 0.568 = 56.8\%$
playoff season: 10 + 4 = 14	$\frac{10}{14} \approx 0.714 = 71.4\%$

Comparing the percentages shows that the Phillies played better in the playoff season.

8. A random check of 150 Southwest Airlines flights last month identified that 113 of them arrived on time. What "on-time" percent does this represent?

$\frac{113}{150} \approx 0.753 = 75.3\%$

Southwest had a 75.3% 'on-time' percent last month.

9. A consumer magazine reported that of the 13,350 subscribers who owned brand A dishwasher, 2940 required a service call. Only 730 of the 1860 subscribers who owned brand B needed dishwasher repairs. Which brand dishwasher has the better repair record?

brand A: $\frac{2940}{13350} \approx 0.22 = 22\%$

brand B: $\frac{730}{1860} \approx 0.392 = 39.2\%$

Relatively speaking, brand A dishwashers have a better repair record.

10. In the following table, the admissions office has organized data describing the number of men and women who are currently enrolled in your college in order to formulate recruitment and retention efforts for the coming academic year. A student is matriculated if he or she is enrolled in a program that leads to a degree (such as associate or baccalaureate).

MATRICULATION BY GENDER		
FULL-TIME MATRICULATED ($\geq$ 12 CREDITS)	PART-TIME MATRICULATED (< 12 CREDITS)	PART-TIME NONMATRICULATED (< 12 CREDITS)
Men	214	174	65
Women	262	87	29

a. How many men attend your college?

214 + 174 + 65 = 453 men attend my college.

b. What percent of the men are full-time students?

$\frac{214}{453} \approx 0.472 \approx 47.2\%$ of the full-time students are men.

c. How many women attend your college?

262 + 87 + 29 = 378 women attend my college.

d. What percent of the women are full-time students?

$\frac{262}{378} \approx 0.693 \approx 69.3\%$ of the women are full-time.

e. How many students are enrolled full-time?

214 + 262 = 476 students are enrolled full-time.

f. How many students are enrolled part-time?

174 + 87 + 65 + 29 = 355 students are enrolled part-time.

g. Women constitute what percent of the full-time students?

$\frac{262}{476} \approx 0.55 \approx 55\%$ of the full-time students are women.

h. Women constitute what percent of the part-time students?

$\frac{87 + 29}{355} = \frac{116}{355} \approx 0.327 \approx 32.7\%$ of the part-time students are women.

i. How many students are nonmatriculated?

65 + 29 = 94 students are nonmatriculated.

j. What percent of the student body is nonmatriculated?

The total number of students
= 214 + 174 + 65 + 262 + 87 + 29 = 831.

$\frac{94}{831} \approx 0.113 \approx 11.3\%$ of the student body is nonmatriculated.

Activity 1.7

The Devastation of AIDS in Africa

Objective

1. Use proportional reasoning to apply a known ratio to a given piece of information.

International public health experts are desperately seeking to stem the spread of the AIDS epidemic in sub-Saharan Africa, especially in the small nation of Botswana. According to a recent United Nations report, approximately 23.9% of Botswana's 1,000,000 adults (ages 15–49) were infected with the AIDS virus.

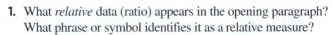

The data in the paragraph above is typical of the kind of numerical information you will find in reading virtually any printed document, report, or article.

1. What *relative* data (ratio) appears in the opening paragraph? What phrase or symbol identifies it as a relative measure?

   23.9%; the % symbol identifies it as a relative measure.

2. Express the ratio in Problem 1 in fractional, decimal, and verbal form.

   $23.9\% = \frac{23.9}{100} = \frac{239}{1000} = 0.239$, 23.9 out of 100.

Recall from Activity 1.6 that a ratio represents a relative measure, the quotient of a part divided by a total.

3. The opening paragraph contains a second piece of information, namely, 1,000,000. Does this number represent a part or a total in this situation?

   1,000,000 represents the total adult population (ages 15–49) of Botswana.

Once you know two of the three values in the basic relationship

$$ratio = \frac{part}{total},$$

the third unknown value can be determined using proportional reasoning.

Definition

Proportional reasoning is the thought process by which you apply a known ratio to one piece of information (part or total) to determine a related, but yet unknown, piece of information (total or part).

The following describes the direct method of applying a known ratio to a given piece of information.

Procedure

Applying the Known Ratio Directly by Multiplication/Division

Step 1: Identify the known ratio.

Step 2: Identify whether the given piece of information represents a total or a part.

Step 3: Rewrite the basic relationship, $ratio = \frac{part}{total}$, in one of two equivalent forms:

- If the given information represents a *total*, then the part is unknown. You can determine the unknown part by *multiplying* the known total by the known ratio as follows:

  $$unknown\ part = total \cdot ratio.$$

- If the given information represents a *part*, then the total is the unknown

  $$unknown\ total = part \div ratio.$$

  In this case, you *divide* the known part by the known ratio to determine the total.

Examples 1 and 2 demonstrate this procedure.

Example 1 *How many adults in Botswana are infected with AIDS (refer to the introductory paragraph)?*

SOLUTION

Proportional reasoning is needed to determine the actual number of adults in Botswana who are infected with AIDS.

Step 1: The known ratio is 23.9%, or 0.239 in decimal form—that is, 239 out of every 1000 adults.

Step 2: The given piece of information is Botswana's total adult population, 1,000,000.

Step 3: The unknown piece of information is the actual number, or part, of Botswana's adults infected with AIDS. Therefore,

$$\text{unknown part} = \text{total} \cdot \text{ratio},$$

$$\text{unknown part} = 1{,}000{,}000 \cdot 0.239 = 239{,}000.$$

The number of adults in Botswana infected with AIDS is 239,000.

Example 2 *You do further research about this disturbing health problem and discover that, in 2007, there were approximately 22,000,000 adults and children living with HIV/AIDS in sub-Saharan Africa. This represented 5% of the total population there. Use proportional reasoning to determine the total population of sub-Saharan Africa in 2007.*

SOLUTION

The known ratio is 5% or 0.05.

The given information, 22,000,000 adults and children with HIV/AIDS, is a part of the total population.

$$\text{unknown total} = \text{part} \div \text{ratio},$$

$$\text{unknown total} = \frac{22{,}000{,}000}{0.05} = 440{,}000{,}000.$$

The total population of sub-Saharan Africa is approximately 440 million people.

4. In 2007, sub-Saharan Africa was home to approximately 22,000,000 people living with HIV, which is 67% of all people living with HIV. From this, estimate the number of people in the world living with HIV in 2007.

22 million is part of the total number of people with HIV, so the total is 22,000,000 ÷ 0.67 = 32,835,821. There were approximately 32.8 million people in the world living with HIV in 2007.

5. In 2008 the U.S. Census Bureau estimated the U.S. population at approximately 304 million. Approximately 27% of this population was under the age of 20 years. How many people were under the age of 20 in the United States in 2008?

27% of 304,000,000 = 0.27 · 304,000,000 = 82,080,000

There were approximately 82 million people under 20 years of age.

SUMMARY: ACTIVITY 1.7

1. **Proportional reasoning** is the process by which you apply a known ratio to one piece of information (part or total) to determine a related, but yet unknown, piece of information (total or part).

2. Given the total, *multiply* the total by the ratio to obtain the unknown part:

$$\text{total} \cdot \text{ratio} = \text{unknown part}$$

3. Given the part, *divide* the part by the ratio to obtain the unknown total:

$$\text{part} \div \text{ratio} = \text{unknown total}$$

EXERCISES: ACTIVITY 1.7

Determine the value of each expression.

1. a. $12 \cdot \dfrac{2}{3}$

   8

   b. $18 \div \dfrac{2}{3}$

   27

   c. $\dfrac{3}{5}$ of 25

   15

   d. $24 \div \dfrac{3}{4}$

   32

   e. 30% of 60

   18

   f. $150 \div 0.60$

   250

   g. $45 \cdot \dfrac{4}{9}$

   20

   h. 25% of 960

   240

   i. $5280 \div 75\%$

   7040

   j. $2000 \div \dfrac{4}{5}$

   2500

   k. 80% of 225

   180

   l. $30,000 \div 50\%$

   60,000

2. Children (under age 15) accounted for 18.4% of the 3,100,000 people who died of AIDS in the year 2005. How many children died of AIDS that year?

   a. Identify the known ratio.

   18.4% = 0.184

   b. Identify the given piece of information. Does this number represent a "part" or a "total"?

   3,100,000; The number represents a total.

   c. How many children died of AIDS in 2005?

   18.4% of 3,100,000 = 0.184 · 3,100,000 = 570,400 children died of AIDS in 2005.

3. The customary tip on waiter service in New York City restaurants is approximately 20% of the food and beverage cost. This means that your waiter or waitress is counting on a tip of $0.20 for each dollar you have spent on your meal. What is the expected tip on a dinner for two costing $65?

   20% of 65 = 0.2 · 65 = 13

   The tip should be $13.00.

Exercise numbers appearing in color are answered in the Selected Answers appendix.

4. Your new car costs $22,500. The state sales tax rate is 8%. How much sales tax will you pay on the car?

8% of 22,500 = 0.08 · 22,500 = 1800

The sales tax on my car is $1800.

5. You happened to notice that the 8% sales tax that your uncle paid on his new luxury car came to $3800. How much did the car cost him?

cost = $\frac{3800}{0.08}$ = 47,500

The car cost my uncle $47,500.

6. In your recent school board elections, only 45% of the registered voters went to the polls. Approximately how many voters are registered in your school district if 22,000 votes were cast?

$\frac{22,000}{0.45}$ ≈ 48,889   There are almost 49,000 registered voters in my district.

7. In 2008, approximately 13,200,000 new cars and trucks were sold in the United States. The market share breakdown among the "big three" car manufacturers was:

General Motors   22.6%

Ford   15.1%

Chrysler   11.0%

a. Approximately how many new vehicles did each company sell that year?

General Motors:

22.6% of 13,200,000 = 0.226 · 13,200,000 = 2,983,200 cars

Ford: 15.1% of 13,200,000 = 0.151 · 13,200,000 = 1,993,200 cars

Chrysler: 11.0% of 13,200,000 = 0.11 · 13,200,000 = 1,452,000 cars

b. Honda sold 1,428,765 vehicles in 2008. What was its market share?

1,428,765 out of 13,200,000 = $\frac{1,428,765}{13,200,000}$ = 0.108 = 10.8%

Honda's market share was 10.8%.

8. In a very disappointing season your softball team won only 40% of the games it played this year. If your team won 8 games, how many games did it play? How many did the team lose?

Number of games played is $\frac{8}{0.4}$ = 20.

Number of games lost is 20 − 8 = 12.

9. In a typical telemarketing campaign, 5% of the people contacted agree to purchase the product. If your quota is to sign up 50 people, approximately how many phone calls do you anticipate making?

$\frac{50}{0.05}$ = 1000

I anticipate making a total of 1000 phone calls.

10. Approximately three-fourths of the teacher-education students in your college pass the licensing exam on their first attempt. This year, 92 students will sit for the exam. How many are expected to pass?

$\frac{3}{4}$ · 92 = 69

Approximately 69 students are expected to pass.

11. You paid $90 for your economics textbook. At the end of the semester, the bookstore will buy back your book for 20% of the purchase price. The book sells used for $65. How much money does the bookstore net on the resale?

20% of 90 = 0.2 · 90 = 18 and 65 − 18 = 47

The bookstore pays me $18 and nets $47.

Activity 1.8

Who Really Did Better?

Objectives

1. Define actual and relative change.

2. Distinguish between actual and relative change.

3. Calculate relative change as a percent increase or percent decrease.

After researching several companies, you and your cousin decided to purchase your first shares of stock. You bought $200 worth of stock in one company; your cousin invested $400 in another. At the end of a year, your stock was worth $280, whereas your cousin's stock was valued at $500.

1. How much money did you earn from your purchase?

   $280 − $200 = $80   The amount I earned is $80.

2. How much did your cousin earn?

   $500 − $400 = $100   The amount my cousin earned is $100.

3. Who earned more money?

   My cousin earned more money than I did.

Your answers to Problems 1 and 2 represent **actual change**, the actual numerical value by which a quantity has changed. When a quantity increases in value, such as your stock, the actual change is positive. When a quantity decreases in value, the actual change is negative.

4. Keeping in mind the amount each of you originally invested, whose investment do you believe did better? Explain your answer.

   My investment increased by $\frac{80}{200} = 0.4 = 40\%$.

   My cousin's investment increased by $\frac{100}{400} = 0.25 = 25\%$.

   My investment did better relative to my cousin's investment.

Relative Change

In answering Problem 4, you considered the actual change relative to the original amount invested.

Definition

The ratio formed by comparing the **actual change** to the **original value** is called the **relative change**. Relative change is written in fraction form, with the actual change placed in the numerator and the original (or earlier) amount placed in the denominator.

$$\text{relative change} = \frac{\text{actual change}}{\text{original value}}$$

---

5. Determine the relative change in your $200 investment. Write your answer as a fraction and as a percent.

   The relative change is $\dfrac{\text{actual change}}{\text{original value}} = \dfrac{80}{200} = \dfrac{2}{5} = 40\%$.

6. Determine the relative change in your cousin's $400 investment. Write your answer as a fraction and as a percent.

   The relative change is $\dfrac{\text{actual change}}{\text{original value}} = \dfrac{100}{400} = 25\%$.

7. In relative terms, whose stock performed better?

   My investment did better.

Percent Change

Because *relative change* (increase or decrease) is frequently reported as a percent, it is often called *percent change*. The two terms are interchangeable.

8. Determine the actual change and the percent change of the following quantities.

a. You paid $20 for a rare baseball card that is now worth $30.

The actual change in worth is: $30 − $20 = $10.

The percent change in worth is: $\frac{10}{20} = \frac{1}{2} = 50\%$.

b. Last year's graduating class had 180 students; this year's class size is 225.

225 − 180 = 45   The actual change is 45 students.

The percent change is $\frac{45}{180} = \frac{1}{4} = 25\%$.

The size of the graduating class increased 25%.

c. Ann invested $300 in an energy stock that is now worth $540.

actual change: $540 − $300 = $240

percent change: $\frac{240}{300} = \frac{4}{5} = 80\%$

Ann's investment increased 80%.

d. Bob invested $300 in a technology stock that is now worth only $225.

The stock's actual change is $225 − $300 = −$75, a decrease of $75.

The percent change in the stock is $\frac{-75}{300} = -\frac{1}{4} = -25\%$.

Bob lost 25% of his investment.

e. Pat invested $300 in a risky dot-com venture, and his stock is now worth $60.

The actual change in the stock's value is $60 − $300 = −$240, a decrease of $240.

The percent change is $\frac{-240}{300} = -\frac{4}{5} = -80\%$.

Pat lost 80% of his investment in the risky venture.

Notice that when you speak about percent change, a percent greater than 100 makes sense. For example, suppose your $50 stock investment tripled to $150. The actual increase in your investment is $150 − $50 = $100.

Therefore, the relative change,

$$\frac{\text{actual increase}}{\text{original value}}$$

is calculated as $\dfrac{\$100}{\$50} = 2.00$. Converting from decimal to percent, your investment has increased by 200%.

9. Determine the actual change and the percent change of the following quantities.

a. In 1970, your parents purchased their home for $40,000. Recently, they sold it for $200,000.

The actual change is $200,000 − $40,000 = $160,000.

The percent change is $\frac{\$200,000 - \$40,000}{\$40,000} = \frac{\$160,000}{\$40,000} = 400\%$.

The house is now worth 400% more than when it was purchased.

b. After a new convention center was built, the number of hotel rooms in the city increased from 1500 to 6000.

The actual change in number of hotel rooms is 6000 − 1500 = 4500.

The percent change is $\frac{6000 - 1500}{1500} = \frac{4500}{1500} = 300\%$.

There are 300% more hotel rooms.

Actual change and relative (percent) change provide two perspectives for understanding change. Actual change looks only at the size of the change (ignoring the original value), whereas relative change indicates the significance or importance of the change by comparing it to the original value.

The following problems will help to clarify these two perspectives.

10. You and your colleague are bragging about your respective stock performances. Each of your stocks earned $1000 last year. Then it slipped out that your original investment was $2000, whereas your colleague had invested $20,000 in her stock.

a. What was the actual increase in each investment?

The actual increase in each investment was $1000.

b. Calculate the relative increase (in percent format) for each investment.

My investment increased by $\frac{1000}{2000} = \frac{1}{2} = 50\%$.

My colleague's investment increased by $\frac{1000}{20,000} = \frac{1}{20} = 5\%$.

c. Whose investment performance was more impressive? Why?

My investment increased by 50%, whereas my colleague's increased by 5%; therefore my investment performed better.

11. During the first week of classes, campus clubs actively recruit members. The Ultimate Frisbee Club signed up 8 new members, reaching an all-time high membership of 48 students. The Yoga Club also attracted 8 new members and now boasts 18 members.

a. What was the membership of the Ultimate Frisbee Club before recruitment week?

48 − 8 = 40

The Ultimate Frisbee Club had 40 members before recruitment week.

b. By what percent did the Frisbee Club increase its membership?

The percent increase in club membership was $\frac{8}{40} = \frac{1}{5} = 20\%$.

c. What was the membership of the Yoga Club before recruitment week?

18 − 8 = 10

The Yoga Club had 10 members before recruitment week.

d. By what percent did the Yoga Club increase its membership?

$\frac{8}{10} = \frac{4}{5} = 80\%$

There was an 80% increase in Yoga Club membership.

EXERCISES: ACTIVITY 1.8

1. Last year, the full-time enrollment in your college was 3200 students. This year, there are 3560 full-time students on campus.

  a. Determine the actual increase in full-time enrollment.

  The actual increase in enrollment was 3560 − 3200 = 360 students.

  b. Determine the relative increase (as a percent) in full-time enrollment.

  $\frac{360}{3200} = \frac{9}{80} \approx 11.3\%$ The relative increase was approximately 11.3%.

2. In April 2005, the average price of a gallon of unleaded gasoline in the United States was $2.238. One year later, the average price had jumped to $2.923.

  a. Determine the actual increase in the price of gasoline.

  The actual increase in the price of 1 gallon of gasoline was $2.923 − $2.238 = $0.685.

  b. Determine the relative increase (as a percent) in the price of gasoline.

  $\frac{0.685}{2.238} = 0.306 = 30.6\%$ The relative increase was 30.6%.

3. This year, your hourly wage from your part-time job is $6.50, up from last year's $6.25. What percent increase does this represent?

  actual increase: 6.50 − 6.25 = $0.25

  percent increase: $\frac{0.25}{6.25} = 0.04 = 4\%$

  My hourly wage went up $0.25, which was a 4% increase.

4. Last month, Acme and Arco corporations were forced to lay off 500 workers each. Acme's workforce is now 1000. Arco currently has 4500 employees. What percent of its workforce did each company lay off?

  Acme: $\frac{500}{1500} = \frac{1}{3} \approx 33.3\%$ Arco: $\frac{500}{5000} = \frac{1}{10} = 10\%$

  Last month Acme lost 33.3% of its workforce and Arco lost 10%.

5. You are now totally committed to dieting. You have decided to cut back your daily caloric intake from 2400 to 1800 calories. By what percent are you cutting back on calories?

  2400 − 1800 = 600

  $\frac{600}{2400} = 0.25 = 25\%$

  I am cutting back 600 calories, which is a decrease of 25%.

6. a. You paid $10 per share for a promising technology stock. Within a year, the stock price sky-rocketed to $50 per share. By what percent did your stock value increase?

  50 − 10 = $40

  $\frac{40}{10} = 4 = 400\%$

  The stock value increased 400%.

  b. You held on to the stock for a bit too long. The price has just fallen from its $50 high back to $10. By what percent did your stock decrease?

  amount of decrease: 50 − 10 = $40

  percent decrease: $\frac{40}{50} = 0.8 = 80\%$

  My stock's value decreased 80%.

Exercise numbers appearing in color are answered in the Selected Answers appendix.

7. Perform the following calculations.

   a. Start with a value of 10, double it, and then calculate the percent increase.

      $10 \cdot 2 = 20$; actual increase: $20 - 10 = 10$

      percent increase: $\frac{10}{10} = 1 = 100\%$

   b. Start with a value of 25, double it, and then calculate the percent increase.

      $25 \cdot 2 = 50$; actual increase: $50 - 25 = 25$

      percent increase: $\frac{25}{25} = 1 = 100\%$

   c. Start with a value of 40, double it, and then calculate the percent increase.

      $40 \cdot 2 = 80$; actual increase: $80 - 40 = 40$

      percent increase: $\frac{40}{40} = 1 = 100\%$

   d. When any quantity doubles in size, by what percent has it increased?

      A quantity increases 100% when it doubles in size.

8. Perform a set of calculations similar to the ones you did in Exercise 7 to help determine the percent increase of any quantity that triples in size.

   $10 \cdot 3 = 30$; actual increase: $30 - 10 = 20$; percent increase: $\frac{20}{10} = 2 = 200\%$

   For a quantity that triples in size, the percent increase is 200%.

Activity 1.9

Going Shopping

Objectives

1. Define growth factor.

2. Determine growth factors from percent increases.

3. Apply growth factors to problems involving percent increases.

4. Define decay factor.

5. Determine decay factors from percent decreases.

6. Apply decay factors to problems involving percent decreases.

To earn revenue (income), many state and local governments require merchants to collect sales tax on the items they sell. In several localities, the sales tax is assessed at as much as 8% of the selling price and is passed on directly to the purchaser.

1. Determine the total cost (including the 8% sales tax) to the customer of the following items. Include each step of your calculation in your answer.

   a. A greeting card selling for $1.50

   sales tax: $0.08 \cdot 1.50 = \$0.12$

   total cost: $1.50 + 0.12 = \$1.62$

   The total cost of the greeting card is $1.62.

   b. A DVD player priced at $300

   sales tax: $0.08 \cdot 300 = \$24$

   total cost: $300 + 24 = \$324$

   The total cost of the DVD player is $324.

Many people correctly determine the total costs for Problem 1 in two steps:

- They first compute the sales tax on the item.

- Then they add the sales tax to the selling price to obtain the total cost.

It is also possible to compute the total cost in one step by using the idea of a **growth factor**.

Growth Factors

2. **a.** If a quantity increases by 50%, how does its new value compare to its original value? That is, what is the ratio of the new value to the original value? The first row of the following table is completed for you. Complete the remaining rows to confirm or discover your answer.

 A Matter of Values

ORIGINAL VALUE	NEW VALUE (INCREASED BY 50%)	RATIO OF NEW VALUE TO ORIGINAL VALUE		
		FRACTIONAL FORM	DECIMAL FORM	PERCENT FORM
20	$10 + 20 = 30$	$\frac{30}{20} = \frac{3}{2} = 1\frac{1}{2}$	1.50	150%
50	75	$\frac{75}{50} = \frac{3}{2} = 1\frac{1}{2}$	1.50	150%
100	150	$\frac{150}{100} = \frac{3}{2} = 1\frac{1}{2}$	1.50	150%
Your choice of an original value 200	300	$\frac{300}{200} = \frac{3}{2} = 1\frac{1}{2}$	1.50	150%

b. Use the results from the preceding table to answer the question, "What is the ratio of the new value to the original value of any quantity that increases by 50%?" Express this ratio as a reduced fraction, a mixed number, a decimal, and a percent.

The ratio of the new value to the original value can be represented by the equivalent forms $\frac{3}{2} = 1\frac{1}{2} = 1.5 = 150\%$.

Definition

For any specified percent *increase*, no matter what the original value may be, the ratio of the new value to the original value is always the same. This ratio of the new value to the original value is called the **growth factor** associated with the specified percent increase.

The growth factor is most often written in decimal form. As you saw previously, the growth factor associated with a 50% increase is 1.50.

Procedure

The growth factor is formed by adding the specified percent increase to 100% and then changing this percent into its decimal form.

Example 1 *The growth factor corresponding to a 50% increase is*

$$100\% + 50\% = 150\%.$$

If you change 150% to a decimal, the growth factor is 1.50.

3. Determine the growth factor of any quantity that increases by the given percent.

a. 30%   100% + 30% = 130% = 1.30

b. 75%   100% + 75% = 175% = 1.75

c. 15%   100% + 15% = 115% = 1.15

d. 7.5%   100% + 7.5% = 107.5% = 1.075

4. A growth factor can always be written in decimal form. Explain why this decimal will always be greater than 1.

The growth factor is obtained by adding the percent increase to 100%. Because 100% = 1, the growth factor is greater than 1.

Procedure

When a quantity increases by a specified percent, its new value can be obtained by multiplying the original value by the corresponding growth factor. That is,

original value · growth factor = new value.

5. a. Use the growth factor 1.50 to determine the new value of a stock portfolio that increased 50% over its original value of $400.

400 · 1.50 = $600

The new value of the stock portfolio is $600.

b. Use the growth factor 1.50 to determine the population of a town that has grown 50% over its previous size of 120,000 residents.

120,000 · 1.50 = 180,000

The new population is 180,000 residents.

6. a. Determine the growth factor for any quantity that increases 20%.

The growth factor associated with a 20% increase is
100% + 20% = 120% = 1.20.

b. Use the growth factor to determine this year's budget, which has increased 20% over last year's budget of $75,000.

This year's budget is 75,000 · 1.20 = $90,000.

7. a. Determine the growth factor of any quantity that increases 8%.

The growth factor associated with an 8% increase is
100% + 8% = 108% = 1.08.

b. Use the growth factor to determine the total cost of each item in Problem 1.

1a: 1.50 · 1.08 = $1.62

1b: 300 · 1.08 = $324

In the previous problems, you were given an original (earlier) value and a percent increase and were asked to determine the new value. Suppose you are asked a "reverse" question: Sales in your small retail business have increased 20% over last year. This year you had gross sales receipts of $75,000. How much did you gross last year?

Using the growth factor 1.20, you can write

$$\text{original value} \cdot 1.20 = 75,000.$$

To determine last year's sales receipts, you divide both sides of the equation by 1.20:

$$\text{original value} = 75,000 \div 1.20.$$

Therefore, last year's sales receipts totaled $62,500.

> **Procedure**
>
> When a quantity has *already increased* by a specified percent, its original value is obtained by dividing the new value by the corresponding growth factor. That is,
>
> $$\text{new value} \div \text{growth factor} = \text{original value}.$$

8. You determined a growth factor for an 8% increase in Problem 7. Use it to answer the following questions.

a. The cash register receipt for your new coat, which includes the 8% sales tax, totals $243. What was the ticketed price of the coat?

The ticketed price of the coat was 243 ÷ 1.08 = $225.

b. The credit-card receipt for your new sofa, which includes the 8% sales tax, totals $1620. What was the ticketed price of the sofa?

1620 ÷ 1.08 = $1500

The ticketed price of the sofa was $1500.

Decay Factors

It's the sale you have been waiting for all season. Everything in the store is marked 40% off the original ticket price.

9. Determine the discounted price of a pair of sunglasses originally selling for $25.

The amount of discount is 25 · 0.40 = $10.00.

The discounted price is 25 − 10 = $15.00.

Many people correctly determine the discounted prices for Problem 9 in two steps:

- They first compute the actual dollar discount on the item.

- Then they subtract the dollar discount from the original price to obtain the sale price.

It is also possible to compute the sale price in one step by using the idea of a **decay factor**.

10. a. If a quantity decreases 20%, how does its new value compare to its original value? That is, what is the ratio of the new value to the original value? Complete the rows in the following table. The first row has been done for you. Notice that the new values are less than the originals, so the ratios of new values to original values will be less than 1.

Up and Down

ORIGINAL VALUE	NEW VALUE (DECREASED BY 20%)	RATIO OF NEW VALUE TO ORIGINAL VALUE		
		FRACTIONAL FORM	DECIMAL FORM	PERCENT FORM
20	16	$\frac{16}{20} = \frac{4}{5}$	0.80	80%
50	$50 \cdot 0.20 = 10$ $50 - 10 = 40$	$\frac{40}{50} = \frac{4}{5}$	0.80	80%
100	$100 \cdot 0.20 = 20$ $100 - 20 = 80$	$\frac{80}{100} = \frac{4}{5}$	0.80	80%
Choose any value 200	$200 \cdot 0.20 = 40$ $200 - 40 = 160$	$\frac{160}{200} = \frac{4}{5}$	0.80	80%

b. Use the results from the preceding table to answer the question, "What is the ratio of the new value to the original value of any quantity that decreases 20%?" Express this ratio in reduced-fraction, decimal, and percent forms.

The ratio can be expressed in the equivalent forms $\frac{4}{5} = 0.80 = 80\%$.

Definition

For a specified percent *decrease*, no matter what the original value may be, the ratio of the new value to the original value is always the same. This ratio of the new to the original value is called the **decay factor** associated with the specified percent decrease.

As you have seen, the decay factor associated with a 20% decrease is 80%. For calculation purposes, a decay factor is usually written in decimal form, so the decay factor 80% is expressed as 0.80.

Procedure

Note that a percent decrease describes the percent that has been *removed* (such as the discount taken off an original price), but the corresponding decay factor always represents the percent remaining (the portion that you still must pay). Therefore, a decay factor is formed by *subtracting* the specified percent decrease from 100% and then changing this percent into its decimal form.

Example 2 *The decay factor corresponding to a 10% decrease is*

$$100\% - 10\% = 90\%.$$

*If you change 90% to a decimal, the decay factor is **0.90**.*

11. Determine the decay factor of any quantity that decreases by the following given percents.

a. 30%

100% − 30% = 70% = 0.70

b. 75%

100% − 75% = 25% = 0.25

c. 15%

100% − 15% = 85% = 0.85

d. 7.5%

100% − 7.5% = 92.5% = 0.925

Remember that multiplying the original value by the decay factor will always result in the amount remaining, not the amount that has been removed.

12. Explain why every decay factor will have a decimal format that is less than 1.

The decay factor is determined by subtracting the percent decrease from 100%. Because 100% = 1, the decay factor will be less than 1.

Procedure

When a quantity *decreases* by a specified percent, its new value is obtained by multiplying the original value by the corresponding decay factor. That is,

$$\text{original value} \cdot \text{decay factor} = \text{new value.}$$

13. Use the decay factor 0.80 to determine the new value of a stock portfolio that has lost 20% of its original value of $400.

The stock's new value is 400 · 0.8 = $320.

In the previous questions, you were given an original value and a percent decrease and were asked to determine the new (*smaller*) value. Suppose you are asked a "reverse" question: Voter turnout in the local school board elections declined 40% from last year. This year only 3600 eligible voters went to the polls. How many people voted in last year's school board elections?

Using the decay factor 0.60 you can write

$$\text{original value} \cdot 0.60 = 3600.$$

To determine last year's voter turnout, divide both sides of the equation by 0.60:

$$\text{original value} = 3600 \div 0.60.$$

Therefore, last year 6000 people voted in the school board elections.

Procedure

When a quantity has *already decreased* by a specified percent, its original value is obtained by dividing the new value by the corresponding decay factor. That is,

$$\text{new value} \div \text{decay factor} = \text{original value.}$$

14. **a.** You purchased a two-line answering machine and telephone on sale for $150. The discount was 40%. What was the original price?

decay factor: $1 - 0.40 = 0.60$

The original price was $150 \div 0.60 = \$250$.

b. Determine the original price of a Nordic Track treadmill that is on sale for $1140. Assume that the store has reduced all merchandise by 25%.

decay factor: 0.75

The original price of the treadmill was $1140 \div 0.75 = \$1520$.

SUMMARY: ACTIVITY 1.9

1. When a quantity increases by a specified percent, the ratio of its new value to its original value is called the **growth factor**. The growth factor is formed by adding the specified percent increase to 100% and then changing this percent into its decimal form.

2. **a.** When a quantity increases by a specified percent, its new value can be obtained by multiplying the original value by the corresponding growth factor. So,

$$\text{original value} \cdot \text{growth factor} = \text{new value}.$$

b. When a quantity has already increased by a specified percent, its original value can be obtained by dividing the new value by the corresponding growth factor. So,

$$\text{new value} \div \text{growth factor} = \text{original value}.$$

3. When a quantity decreases by a specified percent, the ratio of its new value to its original value is called the **decay factor** associated with the specified percent decrease. The decay factor is formed by subtracting the specified percent decrease from 100% and then changing this percent into its decimal form.

a. When a quantity decreases by a specified percent, its new value can be obtained by multiplying the original value by the corresponding decay factor. So,

$$\text{original value} \cdot \text{decay factor} = \text{new value}.$$

b. When a quantity has already decreased by a specified percent, its original value can be obtained by dividing the new value by the corresponding decay factor. So,

$$\text{new value} \div \text{decay factor} = \text{original value}.$$

EXERCISES: ACTIVITY 1.9

1. Complete the following table.

Percent Increase	5%	35%	15%	4.5%	100%	300%	1000%
Growth Factor	1.05	1.35	1.15	1.045	2	4	11.00

2. You are purchasing a new car. The price you negotiated with a car dealer in Staten Island, New York, is $19,744, excluding sales tax. The sales tax rate is 8.375%.

    a. What growth factor is associated with the sales tax rate?

    The growth factor is 1.08375.

    b. Use the growth factor to determine the total cost of the car.

    $19,744 \cdot 1.08375 = \$21,397.56$ The total cost of the car is $21,397.56.

3. The cost of first-class postage in the United States increased approximately 4.8% in May 2009 to 44¢. What was the pre-May cost of the first-class stamp? Round your answer to the nearest penny.

    $0.44 \div 1.048 = \$0.42$ The pre-May cost was 42 cents.

4. A state university in the western United States reported that its enrollment increased from 11,952 students in 2008–2009 to 12,144 students in 2009–2010.

    a. What is the ratio of the number of students in 2009–2010 to the number of students in 2008–2009?

    The ratio is $\frac{12,144}{11,952} \approx 1.016$.

    b. From your result in part a, determine the growth factor. Write it in decimal form to the nearest thousandth.

    The growth factor is 1.016.

    c. By what percent did the enrollment increase?

    The enrollment increased by 1.6%.

5. The 2000 United States Census showed that from 1990 to 2000, Florida's population increased by 23.5%. Florida reported its population in 2000 to be 15,982,378. What was Florida's population in 1990?

    The 1990 population was $15,982,378 \div 1.235 = 12,941,197$.

6. In 1967, your grandparents bought a house for $26,000. The United States Bureau of Labor Statistics reports that, due to inflation, the cost of the same house in 2005 was $152,030.

    a. What is the inflation growth factor of housing from 1967 to 2005?

    The growth factor from 1967 to 2005 is $152,030 \div 26,000 \approx 5.85$.

    b. What is the inflation rate (percent increase to the nearest whole number percent) for housing from 1967 to 2005?

    The inflation rate is 485%.

    c. In 2005, your grandparents sold their house for $245,000. What profit did they make in terms of 2005 dollars?

    My grandparents' profit was $245,000 - 152,030 = \$92,970$.

7. Your friend plans to move from Newark, New Jersey, to Anchorage, Alaska. She earns $50,000 a year in Newark. How much must she earn in Anchorage to maintain the same standard of living if the cost of living in Anchorage is 35% higher?

    She must earn $50,000 \cdot 1.35 = \$67,500$.

8. You decide to invest $3000 in a 1-year certificate of deposit that earns an annual percentage yield of 5.05%. How much will your investment be worth in a year?

    The investment will be worth $3000 \cdot 1.0505 = \$3151.50$.

9. Your company is moving you from Albany, New York, to one of the other cities listed in the following table. For each city, the table lists the cost of goods and services that would cost $100 in Albany, New York.

A World of Difference

CITY	COST OF GOODS AND SERVICES ($)	GROWTH FACTOR IN COST OF LIVING	PERCENT INCREASE IN COST OF LIVING
Albany, NY	100	1.00	0%
Buenos Aires, Argentina	131	1.31	31%
Sydney, Australia	107	1.07	7%
Brussels, Belgium	103	1.03	3%
Toronto, Canada	113	1.13	13%
Riyadh, Saudi Arabia	106	1.06	6%
Singapore, Singapore	114	1.14	14%

a. Determine the growth factor for each cost in the table, and list it in the third column.

b. Determine the percent increase in your salary that would allow you the same standard of living as you have in Albany, New York. List the percent increase in the fourth column of the table.

c. You earn a salary of $45,000 in Albany, New York. What would your salary have to be in Buenos Aires, Argentina, for you to maintain the same standard of living?

My salary would have to be 45,000 · 1.31 = $58,950.

10. Complete the following table.

Percent Decrease	5%	55%	15%	6%	12.5%
Decay Factor	0.95	0.45	0.85	0.94	0.875

11. You wrote an 8-page article that will be published in a journal. The editor has asked you to revise the article and reduce the number of pages to 6. By what percent must you reduce the length of your article?

The length of the article must be reduced by 2 pages: $\frac{2}{8} = \frac{1}{4} = 0.25 = 25\%$.

12. In 1995, the state of California recorded 551,226 live births. In 2000, the number of live births dropped to 531,285. For the 5-year period from 1995 to 2000, what is the decay factor for live births? What is the percent decrease (also called the decay rate)?

The decay factor is 531,285 ÷ 551,226 = 0.964.

percent decrease: 1 − 0.964 − 0.036 = 3.6%

The decay rate for live births in the 5-year period was 3.6%.

13. A car dealer will sell you a car you want for $18,194, which is just $200 over the dealer invoice price (the price the dealer pays the manufacturer for the car). You tell him that you will think about it. The dealer is anxious to meet his monthly quota of sales, so he calls the next day to offer you the car for $17,994 if you agree to buy it tomorrow. You decide to accept the deal.

a. What is the decay factor associated with the decrease in the price to you? Write your answer in decimal form to the nearest thousandth.

The decay factor is 17,994 ÷ 18,194 = 0.989 to the nearest thousandth.

b. What is the percent decrease of the price, to the nearest tenth of a percent?

The percent decrease is 1 − 0.989 = 0.011 = 1.1% to the nearest tenth of a percent.

c. The sales tax is 6.5%. How much did you save on sales tax by taking the dealer's second offer?

200 · 0.065 = $13

I saved $13 in sales tax.

14. Your company is moving you from Boston, Massachusetts, to one of the other cities listed in the following table. For each city, the table lists the cost of goods and services that would cost $100 in Boston, Massachusetts.

 Of Boston and Budapest

CITY	COST OF GOODS AND SERVICES ($)	DECAY FACTOR IN COST OF LIVING	PERCENT DECREASE IN COST OF LIVING
Boston, MA	100	1.00	0%
Santa Fe De Bogota, Colombia	81	0.81	19%
Budapest, Hungary	94	0.94	6%
Calcutta, India	88	0.88	12%
Medan, Indonesia	68	0.68	32%
Krakow, Poland	63	0.63	37%
Cape Town, South Africa	79	0.79	21%

a. Determine the decay factor for each cost in the table, and list it in the third column.

b. Determine the percent decrease in the cost of living for each city compared to living in Boston, Massachusetts. List the percent decrease in the fourth column of the table.

c. You earn a salary of $45,000 in Boston, Massachusetts. If you move to Cape Town, South Africa, what salary will you need to maintain your standard of living there?

I will need 45,000 · 0.79 = $35,550 to live in Cape Town, South Africa.

d. How much of your salary is available to save if you move to Budapest, Hungary, and your salary remains at $45,000?

45,000 · 0.06 = $2700

I can save as much as $2700 if I move to Budapest, Hungary.

15. You need a fax machine but plan to wait for a sale. Yesterday the model you want was reduced by 30%. It originally cost $129.95. Use the decay factor to determine the sale price of the fax machine.

Decay factor: 100% − 30% = 70% = 0.70

The sale price is 129.95 · 0.70 = $90.97.

16. You are on a weight-reducing diet. Your doctor advises you that losing 10% of your body weight will significantly improve your health.

a. You weigh 175 pounds. Determine and use the decay factor to calculate your goal weight to the nearest pound.

Decay factor: 0.90

My goal weight is 175 · 0.90 = 158 pounds.

b. After 6 months, you reach your goal weight. However, you still have more weight to lose to reach your ideal weight range, which is 125–150 pounds. If you lose 10% of your new weight, will you be in your ideal weight range? Explain.

158 · 0.9 = 142 pounds

Yes, I will be in my ideal weight range because 142 is between 125 and 150.

17. Aspirin is typically absorbed into the bloodstream from the duodenum (the beginning portion of the small intestine). In general, 75% of a dose of aspirin is eliminated from the bloodstream in an hour. A patient is given a dose of 650 milligrams. How much aspirin remains in the patient after 1 hour?

650 mg (.25) = 162.5 mg

After 1 hour, 162.5 milligrams of aspirin remains in the bloodstream.

18. Airlines often encourage their customers to book online by offering a 5% discount on their ticket. You are traveling from Kansas City to Denver. A fully refundable fare is $558.60. A restricted, nonrefundable fare is $273.60. Determine and use the decay factor to calculate the cost of each fare if you book online.

Decay factor: 100% − 5% = 95% = 0.95

558.60 · 0.95 = $530.67 is the discounted online fare for the fully refundable ticket.

273.60 · 0.95 = $259.92 is the discounted online fare for the restricted ticket.

19. The annual inflation rate is calculated as the percent increase of the Consumer Price Index (CPI). According to the Department of Labor Statistics, the CPI for 2004 was 188.9, and for 2005 it was 195.3. Calculate the annual inflation rate for 2005.

$$\frac{195.3 - 188.9}{188.9} \approx 0.03388 \approx 3.4\%$$

The annual inflation rate for 2005 was 3.4%.

Activity 1.10

Take an Additional 20% Off

Objectives

1. Define consecutive growth and decay factors.

2. Determine a consecutive growth or decay factor from two or more consecutive percent changes.

3. Apply consecutive growth and/or decay factors to solve problems involving percent changes.

Cumulative Discounts

Your friend arrives at your house in a foul mood. Today's newspaper contains a 20%-off coupon at Old Navy. The $100 jacket she had been eyeing all season was already reduced by 40%. She clipped the coupon, drove to the store, selected her jacket, and walked up to the register. The cashier brought up a price of $48; your friend insisted that the price should have been only $40. The store manager arrived and reentered the transaction, and, again, the register displayed $48. Your friend left the store without purchasing the jacket and drove to your house.

1. How do you think your friend calculated a price of $40?

   My friend may have added 40% and 20% to figure she would get a 60% discount.

2. After your friend calms down, you grab a pencil and start your own calculation. First, you determine the ticketed price that reflects the 40% reduction. At what price is Old Navy selling the jacket? Explain how you calculated this price.

   The decay factor is 100% − 40% = 60% = 0.60.

   sale price: 100 · 0.60 = $60

   The ticketed sale price is $60.

3. To what price does the 20%-off coupon apply?

   The 20%-off coupon applies to the $60 sale price.

4. Apply the 20% discount to determine the final price of the jacket.

   The decay factor is 100% − 20% = 80% = 0.80.

   final price: 60 · 0.80 = $48

   The final sale price of the jacket is $48.

You are now curious: Could you justify a better price by applying the discounts in the reverse order? That is, applying the 20%-off coupon, followed by a 40% reduction. You start a new set of calculations.

5. Starting with the list price, determine the sale price after taking the 20% reduction.

   sale price: 100 · 0.80 = $80

6. Apply the 40% discount to the intermediate sale price.

   final price: 80 · 0.60 = $48

7. Which sequence of discounts gives a better sale price?

   Both sequences of discounts give the same final price.

The important point to remember here is that when multiple discounts are given, they are always applied sequentially, one after the other—never all at once.

In the following example, you will see how to use decay factors to simplify calculations such as the ones above.

Example 1 *A stunning $2000 gold and diamond necklace you saw was far too expensive to even consider purchasing. However, over several weeks you tracked the following successive discounts: 20% off list; 30% off marked price; and an additional 40% off every item. Determine the selling price after each of the discounts is taken.*

SOLUTION

To calculate the first reduced price, apply the decay factor corresponding to a 20% discount. Recall that you form a decay factor by subtracting the 20% discount from 100% to obtain

80%, or 0.80 as a decimal. Apply the 20% reduction by multiplying the original price by the decay factor.

The first sale price = $2000 · 0.80; the selling price is now $1600.

In a similar manner, determine the decay factor corresponding to a 30% discount by subtracting 30% from 100% to obtain 70%, or 0.70 as a decimal. Apply the 30% reduction by multiplying the already discounted price by the decay factor.

The second sale price = $1600 · 0.70; the selling price is now $1120.

Finally, determine the decay factor corresponding to a 40% discount by subtracting 40% from 100% to obtain 60%, or 0.60 as a decimal. Apply the 40% reduction by multiplying the most current discounted price by the decay factor.

$1120 · 0.60; the final selling price is $672.

You may have noticed that it is possible to *calculate the final sale price from the original price using a single chain of multiplications:*

$2000 · 0.80 · 0.70 · 0.60 = $672

The final sale price is $672.

8. Use a chain of multiplications as in Example 1 to determine the final sale price of the necklace if the discounts are taken in the reverse order (40%, 30%, and 20%).

$2000 · 0.60 · 0.70 · 0.80 = $672

The final sale price is $672.

You can form a single decay factor that represents the cumulative effect of applying the three consecutive percent decreases; the single decay factor is the *product* of the three decay factors.

In Example 1, the effective decay factor is given by the product 0.80 · 0.70 · 0.60, which equals 0.336 (or 33.6%). The effective discount is calculated by subtracting the decay factor (in percent form) from 100%, to obtain 66.4%. Therefore, the effect of applying 20%, 30%, and 40% consecutive discounts is identical to a single discount of 66.4%.

9. a. Determine a single decay factor that represents the cumulative effect of consecutively applying Old Navy's 40% and 20% discounts.

The single decay factor is 0.60 · 0.80 = 0.48.

b. Use this decay factor to determine the effective discount on your friend's jacket.

The effective discount is 1 − 0.48 = 0.52 = 52%.

10. a. Determine a single decay factor that represents the cumulative effect of consecutively applying discounts of 40% and 50%.

The cumulative decay factor is 0.60 · 0.50 = 0.30.

b. Use this decay factor to determine the effective discount.

The effective discount is 1 − 0.30 = 0.70 = 70%.

Cumulative Increases

Your fast-food franchise is growing faster than you had ever imagined. Last year you had 100 employees statewide. This year you opened several additional locations and increased the number of workers by 30%. With demand so high, next year you will open new stores nationwide and plan to increase your employee roll by an additional 50%.

To determine the projected number of employees next year, you can use growth factors to simplify the calculations.

11. a. Determine the growth factor corresponding to a 30% increase.

The growth factor is 100% + 30% = 130% = 1.30.

b. Apply this growth factor to calculate your current workforce.

The current workforce is 100 · 1.30 = 130 employees.

12. a. Determine the growth factor corresponding to a 50% increase.

The growth factor is 100% + 50% = 150% = 1.50.

b. Apply this growth factor to your current workforce to determine the projected number of employees next year.

Next year's workforce is expected to be 130 · 1.50 = 195 employees.

13. Starting from last year's workforce of 100, write a single chain of multiplications to calculate the projected number of employees.

100 · 1.30 · 1.50 = 195 employees are projected for my franchise.

> You can form a single growth factor that represents the cumulative effect of applying the two consecutive percent increases; the single growth factor is the *product* of the two growth factors.

In Problem 13, the effective growth factor is given by the product 1.30 · 1.50, which equals 1.95 (an increase of 95%), nearly double the number of employees last year.

What Goes Up Often Comes Down

You purchased $1000 of a recommended stock last year and watched gleefully as it rose quickly by 30%. Unfortunately, the economy turned downward, and your stock recently fell 30% from last year's high. Have you made or lost money on your investment? The answer might surprise you. Find out by solving the following sequence of problems.

14. a. Determine the growth factor corresponding to a 30% increase.

The growth factor is 100% + 30% = 130% = 1.30.

b. Determine the decay factor corresponding to a 30% decrease.

The decay factor is 100% − 30% = 70% = 0.70.

You can form a single factor that represents the cumulative effect of applying the consecutive percent increase and decrease—the single factor is the *product* of the growth and decay factors. In this example, the effective factor is given by the product 1.30 · 0.70, which equals 0.91.

15. Does 0.91 represent a growth factor or a decay factor? How can you tell?

0.91 is a decay factor because it is less than 1.

16. a. What is the current value of your stock?

The current value is 1000 · 1.30 · 0.70 = 1000 · 0.91 = $910.

b. What is the cumulative effect (as a percent change) of applying a 30% increase followed by a 30% decrease?

The cumulative effect is a 1.00 − 0.91 = .09, or 9% decrease.

17. What is the cumulative effect if the 30% decrease had been applied first, followed by the 30% increase?

The current value would be 1000 · 0.70 · 1.30 = $910; therefore, the cumulative effect is the same.

SUMMARY: ACTIVITY 1.10

1. The cumulative effect of a sequence of percent changes is the *product* of the associated growth or decay factors. For example,

a. To calculate the effect of consecutively applying 20% and 50% increases, form the respective growth factors and multiply. The effective growth factor is

$$1.20 \cdot 1.50 = 1.80,$$

which represents an effective increase of 80%.

b. To calculate the effect of applying a 25% *increase* followed by a 20% *decrease*, form the respective growth and decay factors, and then multiply them. The effective factor is $1.25 \cdot 0.80 = 1.00$, which indicates neither growth nor decay. That is, the quantity has returned to its original value.

2. The cumulative effect of a sequence of percent changes is the same, regardless of the order in which the changes are applied. For example, the cumulative effect of applying a 20% *increase* followed by a 50% *decrease* is equivalent to having first applied the 50% *decrease*, followed by the 20% *increase*.

EXERCISES: ACTIVITY 1.10

1. A $300 suit is on sale for 30% off. You present a coupon at the cash register for an additional 20% off.

a. Determine the decay factor corresponding to each percent decrease.

30% off: 0.70 is the decay factor.

20% off: 0.80 is the decay factor.

b. Use these decay factors to determine the price you paid for the suit.

The sale price of the suit was 300 · 0.70 · 0.80 = $168.

2. Your union has just negotiated a 3-year contract containing annual raises of 3%, 4%, and 5% during the term of contract. Your current salary is $42,000. What will you be earning in 3 years?

In 3 years my salary will be 42,000 · 1.03 · 1.04 · 1.05 = $47,239.92.

3. You anticipated a large demand for a popular toy and increased your inventory of 1600 by 25%. You sold 75% of your inventory. How many toys remain?

1600(1.25)(0.25) = 500

There are 500 units of this toy in inventory.

Exercise numbers appearing in color are answered in the Selected Answers appendix.

4. You deposit $2000 in a 5-year certificate of deposit that pays 4% interest compounded annually. Determine to the nearest dollar your account balance when your certificate comes due.

$$2000 \cdot 1.04 \cdot 1.04 \cdot 1.04 \cdot 1.04 \cdot 1.04 = 2000 \cdot 1.04^5 = \$2433.31$$

The account balance will be approximately $2433.

5. Budget cuts have severely crippled your department over the last few years. Your operating budget of $600,000 has decreased 5% in each of the last 3 years. What is your current operating budget?

The current budget is

$$600,000 \cdot 0.95 \cdot 0.95 \cdot 0.95 = 600,000 \cdot 0.95^3 = \$514,425.$$

6. When you became a manager, your $60,000 annual salary increased 25%. You found the new job too stressful and requested a return to your original job. You resumed your former duties at a 20% reduction in salary. How much more are you making at your old job after the transfer and return?

new salary: $60,000 \cdot 1.25 \cdot 0.80 = \$60,000$

My new salary is the same as my original salary.

Activity 1.11

Fuel Economy

Objectives

1. Apply rates directly to solve problems.

2. Use unit or dimensional analysis to solve problems that involve consecutive rates.

You are excited about purchasing a new car for your commute to college and a part-time job. Concerned about the cost of driving, you do some research on the Internet and come across a Web site that lists fuel efficiency, in miles per gallon (mpg), for five cars that you are considering. You record the mpg for city and highway driving in the table below.

1. a. For each of the cars listed in the following table, how many city miles can you travel per week on 5 gallons of gasoline? Explain the calculation you will do to obtain the answers. Record your answers in the third column of the table.

I multiply each mpg by 5 to obtain miles traveled on 5 gallons.

Every Drop Counts

FUEL ECONOMY GUIDE MODEL YEAR 2009					
MAKE/MODEL	CITY MPG	CITY MILES ON 5 GAL OF GAS	HIGHWAY MPG	GALLONS NEEDED TO DRIVE 304 MILES	FUEL TANK CAPACITY IN GAL
Chevrolet Cobalt	25	125	37	8.2	13.0
Ford Focus	24	120	35	8.7	13.5
Honda Civic	25	125	36	8.4	12.3
Hyundai Accent	27	135	33	9.2	11.9
Toyota Corolla	26	130	35	8.7	13.2

b. The round-trip drive to your college, part-time job, and back home is 32 city miles, which you do 4 days a week. Which of the cars would get you to school and back each week on 5 gallons of gas?

I drive 32 · 4 = 128 city miles to school each week. Therefore, the Hyundai and Toyota will get me back and forth on about 5 gallons of gasoline per week.

2. Suppose your round-trip commute to a summer job would be 304 highway miles each week. How many gallons of gas would each of the cars require? Explain the calculation you do to obtain the answers. Record your answers to the nearest tenth in the fifth column of the above table.

I divide 304 miles by the mpg to obtain the number of gallons of gas for each trip.

Rates and Unit Analysis

Miles per gallon (mpg) is an example of a rate. Mathematically, a **rate** is a comparison of two quantities that have different units of measurement. Numerically, you calculate with rates as you would with ratios. Paying attention to what happens to the units of measurement during the calculation is critical to determining the unit of the result.

Definition

A **rate** is a comparison, expressed as a quotient, of two quantities that have different units of measurement.

Unit analysis (sometimes called *dimensional analysis*) uses units of measurement as a guide in setting up a calculation or writing an equation involving one or more rates. When the calculation or equation is set up properly, the result is an answer with the appropriate units of measurement.

The most common method for solving problems involving rates is to apply the known rate directly by multiplication or division (direct method). You use the units of measurement as a guide in setting up the calculation. Example 1 demonstrates this method.

Example 1

Direct Method: Apply a known rate directly by multiplication or division to solve a problem.

In Problem 1, the known rate is miles per gallon. You can write miles per gallon in fraction form, $\dfrac{\text{number of miles}}{1 \text{ gallon}}$. The city fuel economy rating for the Chevrolet Cobalt is $\dfrac{25 \text{ mi.}}{1 \text{ gal.}}$. To determine how many city miles you can travel on 5 gallons of gasoline, you multiply the known rate by 5 gallons as follows:

$$\frac{25 \text{ mi.}}{1 \text{ gal.}} \cdot 5 \text{ gal.} = 125 \text{ mi.}$$

Notice that the unit of measurement, gallon, occurs in both the numerator and the denominator. You divide out common units of measurement in the same way that you divide out common numerical factors.

Therefore, in the Chevrolet Cobalt, you can drive 125 miles on 5 gallons of gas.

3. a. The gas tank of a Ford Focus holds 13.5 gallons. How many highway miles can you travel on a full tank of gas?

$\dfrac{35 \text{ mi.}}{1 \text{ gal.}} \cdot 13.5 \text{ gal.} = 472.5 \text{ mi.}$

I can travel about 472.5 highway miles on a full tank of gas in the Ford Focus.

b. The Chevrolet Cobalt's gas tank holds 13.0 gallons. Is it possible to travel as far on the highway in this car as in the Ford Focus?

$\dfrac{37 \text{ mi.}}{1 \text{ gal.}} \cdot 13.0 \text{ gal.} = 481 \text{ mi.}$

Actually, I can travel about 8.5 miles farther.

In Problem 2, you were asked to determine how many gallons of gas were required to drive 304 highway miles. Using the direct method, you may have known to divide the total miles by the miles per gallon. For example, in the case of the Honda Civic, $\dfrac{304}{36} = 8.4$ gallons.

One way to be sure the problem is set up correctly is to use the units of measurement. In Problem 2, you are determining the number of gallons. Set up the calculation so that miles will divide out and the remaining measurement unit will be gallons.

$$304 \, \text{mi.} \cdot \frac{1 \, \text{gal.}}{36 \, \text{mi.}} = \frac{304}{36} \, \text{gal.} = 8.4 \, \text{gal.}$$

This example shows that a rate may be expressed in two ways. For miles and gallons, the rate can be expressed as $\frac{\text{miles}}{\text{gallon}}$ or as $\frac{\text{gallon}}{\text{miles}}$. To solve a problem, you choose the form so that all measurement units can be divided out except the measurement unit of the answer.

4. After you purchase your new car, you would like to take a trip to see a good friend in another state. The highway distance is approximately 560 miles. Solve the following problems by the direct method.

a. If you bought the Hyundai Accent, how many gallons of gas would you need to make the round-trip?

$$1120 \, \text{mi.} \cdot \frac{1 \, \text{gal.}}{33 \, \text{mi.}} = \frac{1120}{33} \, \text{gal.} \approx 33.9 \, \text{gal.}$$

I would need about 33.9 gallons of gas to make the round-trip.

b. How many tanks of gas would you need for the trip?

$$33.9 \, \text{gal.} \cdot \frac{1 \, \text{tank}}{11.9 \, \text{gal.}} \approx 2.8 \, \text{tanks}$$

I would need about 2.8, or 3, tanks of gas to make the trip.

Procedure

Method for Solving Problems Involving Rates

Direct Method: Multiply or Divide by the Known Rate

1. Identify the unit of the result.

2. Set up the calculation so the appropriate units will divide out, leaving the unit of the result.

3. Multiply or divide the numbers as indicated to obtain the numerical part of the result.

4. Divide out the common units to obtain the unit of the answer.

Using Unit Analysis to Solve a Problem

5. You are on a part of a 1500-mile trip where gas stations are far apart. Your car is a hybrid model that is averaging 50 miles per gallon and you are traveling at 60 miles per hour (mph). The fuel tank holds 12 gallons of gas, and you just filled the tank. How long is it before you have to fill the tank again?

You can solve this problem in two parts, as follows.

a. Determine how many miles you can travel on one tank of gas.

$$\frac{50 \, \text{mi.}}{1 \, \text{gal.}} \cdot \frac{12 \, \text{gal.}}{1 \, \text{tank}} = 600 \, \text{mi. per tank}$$

I can travel 600 miles on one tank of gas.

b. Use the result of part a to determine how many hours you can drive before you have to fill the tank again.

$$\frac{600 \, \text{mi.}}{1 \, \text{tank}} \div \frac{60 \, \text{mi.}}{1 \, \text{hr.}} = 10 \, \text{hr. per tank}$$

I can drive for 10 hours before I have to refill the tank.

Problem 5 is an example of applying consecutive rates to solve a problem. Part a was the first step and part b the second step. Alternatively, you can also determine the answer in a *single calculation* by considering what the measurement unit of the answer should be. In this case, it is hours per tank. Therefore, you will need to set up the calculation so that you can divide out miles and gallons.

$$\frac{50 \text{ mi.}}{1 \text{ gal.}} \cdot \frac{12 \text{ gal.}}{1 \text{ tank}} \cdot \frac{1 \text{ hr.}}{60 \text{ mi.}} = \frac{10 \text{ hr.}}{1 \text{ tank}}$$

Notice that miles and gallons divide out to leave hours per tank as the measurement unit of the answer.

Procedure

Using Unit Analysis to Solve Problems

To apply consecutive rates:

1. Identify the measurement unit of the result.

2. Set up the sequence of multiplications so that the appropriate units divide out, leaving the appropriate measurement unit of the result.

3. Multiply and divide the numbers as indicated to obtain the numerical part of the result.

4. Check that the appropriate measurement units divide out, leaving the expected unit for the result.

---

6. You have been driving for several hours and notice that your car's 13.2-gallon fuel tank registers half empty. How many more miles can you travel if your car is averaging 45 miles per gallon?

$$\frac{1}{2} \text{ tank} \cdot \frac{13.2 \text{ gal.}}{1 \text{ tank}} \cdot \frac{45 \text{ mi.}}{1 \text{ gal.}} = 297 \text{ mi.}$$

I will be able to travel about 297 miles more.

Unit Conversion

In many countries, distance is measured in kilometers (km) and gasoline in liters (L). A mile is equivalent to 1.609 kilometers, and a liter is equivalent to 0.264 gallon. Each of these equivalences can be treated as a rate and written in fraction form. Thus, the fact that a mile is equivalent to 1.609 kilometers is written as $\frac{1.609 \text{ km}}{1 \text{ mi.}}$ or $\frac{1 \text{ mi.}}{1.609 \text{ km}}$. Using the fraction form, you can convert one measurement unit to another by applying multiplication or division directly.

7. Your friend joins you on a trip through Canada, where gasoline is measured in liters and distance, in kilometers.

   a. Write the equivalence of liters and gallons in fraction form.

   The fraction form is $\frac{0.264 \text{ gal.}}{1 \text{ L}}$ or $\frac{1 \text{ L}}{0.264 \text{ gal.}}$.

   b. If you bought 20 liters of gas, how many gallons did you buy?

   $\frac{0.264 \text{ gal.}}{1 \text{ L}} \cdot 20 \text{ L} = 5.28 \text{ gal.}$ I purchased about 5.3 gallons of gas.

8. To keep track of mileage and fuel needs in Canada, your friend suggests that you convert your car's miles per gallon into kilometers per liter (kpL). Your car's highway fuel efficiency is 45 miles per gallon. What is its fuel efficiency in kilometers per liter?

$$\frac{45 \text{ mi.}}{1 \text{ gal.}} \cdot \frac{1.609 \text{ km}}{1 \text{ mi.}} \cdot \frac{0.264 \text{ gal.}}{1 \text{ L}} \approx 19.1 \text{ kpL}$$

The car's fuel efficiency is 19.1 kilometers per liter.

SUMMARY: ACTIVITY 1.11

1. A **rate** is a comparison, expressed as a quotient, of two quantities that have different units of measurement.

2. **Unit analysis** (sometimes called *dimensional analysis*) uses units of measurement as a guide in setting up a calculation or writing an equation involving one or more rates. When the calculation or equation is set up properly, the result is an answer with the appropriate units of measurement.

3. **Direct Method** for Solving Problems Involving Rates: Multiply or divide directly by rates to solve problems.

  i. Identify the measurement unit of the result.

  ii. Set up the calculation so that the appropriate measurement units will divide out, leaving the unit of the result.

  iii. Multiply or divide the numbers as indicated to obtain the numerical part of the result.

  iv. Divide out the common measurement units to obtain the unit of the answer.

4. Applying several rates consecutively:

  i. Identify the measurement unit of the result.

  ii. Set up the sequence of multiplications so the appropriate measurement units divide out, leaving the unit of the result.

  iii. Multiply and divide the numbers as usual to obtain the numerical part of the result.

  iv. Check that the appropriate measurement units divide out, leaving the expected unit of the result.

EXERCISES: ACTIVITY 1.11

Use the conversion tables on the inside front cover of the textbook for conversion equivalencies.

1. The length of a football playing field is 100 yards between the opposing goal lines. What is the length of the football field in feet?

$$100 \text{ yd.} \cdot \frac{3 \text{ ft.}}{1 \text{ yd.}} = 300 \text{ ft.}$$

The length of the playing field is 300 feet.

Exercise numbers appearing in color are answered in the Selected Answers appendix.

2. The distance between New York City, New York, and Los Angeles, California, is approximately 4485 kilometers. What is the distance between these two major cities in miles?

$$4485 \text{ km} \cdot \frac{1 \text{ mi.}}{1.609 \text{ km}} \approx 2787 \text{ mi.}$$

The distance between New York City and Los Angeles is approximately 2787 miles.

3. The aorta is the largest artery in the human body. In the average adult, the aorta attains a maximum diameter of about 1.18 inches where it adjoins the heart. What is the maximum diameter of the aorta in centimeters? In millimeters?

The maximum diameter of the aorta is $1.18 \text{ inches} \cdot \dfrac{2.54 \text{ cm}}{1 \text{ in.}} = 2.9972 \text{ cm}$

and in millimeters, it is

$$2.9972 \text{ cm} \cdot \frac{10 \text{ mm}}{1 \text{ cm}} = 29.972 \text{ mm.}$$

4. In the Himalayan Mountains along the border of Tibet and Nepal, Mt. Everest reaches a record height of 29,035 feet. How high is Mt. Everest in miles? In kilometers? In meters?

$$29{,}035 \text{ ft.} \cdot \frac{1 \text{ mi.}}{5280 \text{ ft.}} \approx 5.4991 \approx 5.5 \text{ mi.}$$

$$5.4991 \text{ mi.} \cdot \frac{1.609 \text{ km}}{1 \text{ mi.}} \approx 8.8481 \text{ km}$$

$$8.8481 \text{ km} \cdot \frac{1000 \text{ m}}{1 \text{ km}} \approx 8848 \text{ m}$$

Mt. Everest is approximately 5.5 miles high, which is 8.8481 kilometers or 8848 meters.

5. The average weight of a mature human brain is approximately 1400 grams. What is the equivalent weight in kilograms? In pounds?

$$1400 \text{ g} \cdot \frac{1 \text{ kg}}{1000 \text{ g}} = 1.4 \text{ kg}$$

$$1.4 \text{ kg} \cdot \frac{2.20 \text{ lb.}}{1 \text{ kg}} = 3.08 \text{ lb.}$$

The average weight of the human brain is 1.4 kilograms or 3.08 pounds.

6. Approximately 4.5 liters of blood circulates in the body of the average human adult. How many quarts of blood does the average person have? How many pints?

$$4.5 \text{ L} \cdot \frac{1 \text{ gal.}}{3.785 \text{ L}} \cdot \frac{4 \text{ qt.}}{1 \text{ gal.}} \approx 4.76 \text{ qt.}$$

$$4.5 \text{ L} \cdot \frac{1 \text{ gal.}}{3.785 \text{ L}} \cdot \frac{4 \text{ qt.}}{1 \text{ gal.}} \cdot \frac{2 \text{ pt.}}{1 \text{ qt.}} \approx 9.51 \text{ pt. or}$$

$$4.76 \text{ qt.} \cdot \frac{2 \text{ pt.}}{1 \text{ qt.}} = 9.52 \text{ pt.}$$

An average person has approximately 4.8 quarts or 9.5 pints of blood.

7. How many seconds are in a day? In a week? In a year?

$$1 \text{ day} \cdot \frac{24 \text{ hr.}}{1 \text{ day}} \cdot \frac{60 \text{ min.}}{1 \text{ hr.}} \cdot \frac{60 \text{ sec.}}{1 \text{ min.}} = 86{,}400 \text{ sec. in 1 day}$$

$$1 \text{ week} \cdot \frac{7 \text{ days}}{1 \text{ week}} \cdot \frac{86{,}400 \text{ sec.}}{1 \text{ day}} = 604{,}800 \text{ sec. in 1 week}$$

$$1 \text{ yr.} \cdot \frac{365 \text{ days}}{1 \text{ yr.}} \cdot \frac{86{,}400 \text{ sec.}}{1 \text{ day}} = 31{,}536{,}000 \text{ sec. in 1 year}$$

8. The following places are three of the wettest locations on Earth. For each site, determine the rainfall in centimeters per year. Record your results in the third column of the table.

 Water, Water Everywhere

LOCATIONS	ANNUAL RAINFALL (IN INCHES)	ANNUAL RAINFALL (IN CM)
Mawsynram, Meghalaya, India	467	1186
Tutenendo, Colombia	463.5	1177
Mt. Waialeale, Kauai, Hawaii	410	1041

$$467 \text{ in.} \cdot \frac{2.54 \text{ cm}}{1 \text{ in.}} = 1186.18 \text{ cm}$$

$$463.5 \text{ in.} \cdot \frac{2.54 \text{ cm}}{1 \text{ in.}} = 1177.29 \text{ cm}$$

$$410 \text{ in.} \cdot \frac{2.54 \text{ cm}}{1 \text{ in.}} = 1041.4 \text{ cm}$$

9. Jewelry is commonly weighed in carats. Five carats are equivalent to 1 gram. What is the weight of a 24-carat diamond in grams? In ounces?

$$24 \text{ carats} \cdot \frac{1 \text{ g}}{5 \text{ carats}} = 4.8 \text{ g} \qquad 4.8 \text{ g} \cdot \frac{0.035 \text{ oz.}}{1 \text{ g}} = 0.168 \text{ oz.}$$

A 24-carat diamond weighs 4.8 grams or 0.168 ounces.

10. Tissues of living organisms consist primarily of organic compounds; that is, they contain carbon molecules known as proteins, carbohydrates, and fats. A healthy human body is approximately 18% carbon by weight. Determine how many pounds of carbon your own body contains. How many kilograms?

(Answers will vary); for example, if a human body weighs 200 pounds, then it contains $200 \cdot 0.18 = 36$ pounds of carbon.

$$36 \text{ lb.} \cdot \frac{0.454 \text{ kg}}{1 \text{ lb.}} = 16.344 \text{ kg}$$

A 200-pound body contains about 16 kilograms of carbon.

Cluster 3   What Have I Learned?

1. On a 40-question practice test for this course, you answered 32 questions correctly. On the test itself, you correctly answered 16 out of 20 questions. Does this mean that you did better on the practice test than you did on the test itself? Explain your answer using the concepts of actual and relative comparison.

   I answered more questions correctly on the practice exam (32) than on the actual exam (16). This does not mean I did better on the practice exam because if I compare my scores to the number of questions, my percentage of correct answers is the same in both cases, 80%.

   I scored $\frac{32}{40} = \frac{4}{5} = 0.8 = 80\%$ on the practice exam.

   I scored $\frac{16}{20} = \frac{4}{5} = 0.8 = 80\%$ on the actual exam.

   So I did the same on the actual exam as I did on the practice exam.

2. **a.** Ratios are often written as a fraction in the form $\frac{a}{b}$. List the other ways you can express a ratio.

   (Answers will vary.)

   Verbally, you can say or write a out of b, for example, 95 out of 100 dentists.

   A ratio can also be expressed as a decimal and as a percent, for example, 0.95 and 95%.

   b. Thirty-three of the 108 colleges and universities in Michigan are two-year institutions. Write this ratio in each of the ways you listed in part a.

   fraction: $\frac{33}{108} = \frac{11}{36}$ words: 33 out of 108 or 11 out of 36

   decimal: $33 \div 108 \approx 0.31$ percent: 31%

3. In 2000, Florida's total population was 15,982,378 persons and 183 out of every 1000 residents were 65 years and older. Show how you would determine the actual number of Florida residents who were 65 years or older in 2000.

   $15{,}982{,}378 \text{ Florida residents} \cdot \dfrac{183 \text{ Florida residents over } 65}{1000 \text{ Florida residents}} \approx 2{,}924{,}775$

   Approximately 2,924,800 Florida residents were 65 years or older in 2000.

4. Explain how you would determine the percent increase in the cost of a cup of coffee at your local diner if the price increased from \$1.69 to \$1.89 last week.

   1. Find the amount of the increase: \$1.89 − \$1.69 = \$0.20.

   2. Divide the amount of the increase by the original amount: $0.20 \div 1.69 \approx 0.118$.

   3. Write the result as a percent: $0.118 = 11.8\%$.

   Therefore, the cost of a cup of coffee increased 11.8%.

Exercise numbers appearing in color are answered in the Selected Answers appendix.

5. a. Think of an example, or find one in a newspaper or magazine or on the Internet, and use it to show how to determine the growth factor associated with a percent increase. For instance, you might think of or find growth factors associated with yearly interest rates for a savings account.

(Answers will vary.)

The growth factor is determined by adding the percent increase to 100% and converting the result to a decimal.

The growth factor for a savings account that yields a 3% annual increase is 100% + 3% = 103% = 1.03.

b. Show how you would use the growth factor to apply a percent increase twice, then three times.

(Answers will vary and correspond to part a.)

If $100 is deposited, the balance in 1 year is 100 · 1.03 = $103.00.

The balance in 2 years is 100 · 1.03 · 1.03 = 100 · 1.03^2 = $106.09.

The balance in 3 years is

100 · 1.03 · 1.03 · 1.03 = 100 · 1.03^3 = $109.27.

6. Current health research shows that losing just 10% of body weight produces significant health benefits, including a reduced risk for a heart attack. Suppose a relative who weighs 199 pounds begins a diet to reach a goal weight of 145 pounds.

a. Use the idea of a decay factor to show your relative how much he will weigh after losing the first 10% of his body weight.

decay factor: 100% − 10% = 90% = 0.90

199 · 0.90 = 179.1 lb.

My relative will weigh 179 pounds after he loses 10% of his body weight.

b. Explain to your relative how he can use the decay factor to determine how many times he has to lose 10% of his body weight to reach his goal weight of 145 pounds.

179.1 · 0.90 = 161.2 lb.

161.2 · 0.90 = 145 lb.

He must lose 10% of his body weight 3 times to reach 145 pounds.

Cluster 3   How Can I Practice?

1. Write the following percents in decimal format.

   a. 25%   0.25   **b.** 87.5%   0.875

   c. 6%   0.06   **d.** 3.5%   0.035

   e. 250%   2.50   **f.** 0.3%   0.003

2. An error in a measurement is the difference between the measured value and the true value. The relative error in measurement is the ratio of the error to the true value. That is,

$$\text{relative error} = \frac{\text{error}}{\text{true value}}$$

   Determine the actual and relative errors of the measurements in the following table.

 Measure for Measure

MEASUREMENT	TRUE VALUE	ACTUAL ERROR	RELATIVE ERROR (AS A PERCENT)
107 in.	100 in.	7 in.	7 ÷ 100 = 0.07 = 7%
5.7 oz.	5.0 oz.	0.7 oz.	0.7 ÷ 5.0 = 0.14 = 14%
4.3 g	5.0 g	0.7 g	0.7 ÷ 5.0 = 0.14 = 14%
11.5 cm	12.5 cm	1 cm	1 ÷ 12.5 = 0.08 = 8%

3. A study compared three different treatment programs for cocaine addiction. Subjects in the study were divided into three groups. One group was given the standard therapy, the second group was treated with an antidepressant, and the third group was given a placebo. The groups were tracked for 3 years. Data from the study is given in the following table.

TREATMENT THERAPY	NO. OF SUBJECTS RELAPSED	NO. OF SUBJECTS STILL SUBSTANCE FREE
Standard therapy	36	20
Antidepressant therapy	27	18
Placebo	30	9

   a. Interpret this data to determine which, if any, of the three therapies was most effective. Explain the calculations you would do and how you use them to interpret the study.

   For each treatment, I would determine the ratio of substance-free subjects to the total number receiving the treatment. The therapy with the highest percentage is the most effective.

b. Complete the following table. Which therapy appears most effective?

TREATMENT THERAPY	TOTAL NUMBER OF SUBJECTS	PERCENT OF SUBJECTS SUBSTANCE FREE
Standard therapy	$36 + 20 = 56$	$20 \div 56 = 0.357 = 35.7\%$
Antidepressant therapy	$27 + 18 = 45$	$18 \div 45 = 0.40 = 40\%$
Placebo	$30 + 9 = 39$	$9 \div 39 = 0.231 = 23.1\%$

From comparing the percentages of subjects substance free after the experiment, the antidepressant therapy appears most effective.

4. The 2619 women students on a campus constitute 54% of the entire student body. What is the total enrollment of the college?

total enrollment: $2619 \div 0.54 = 4850$ students

5. The New York City Board of Education recently announced plans to eliminate 1500 administrative staff jobs over the next several years, reducing the number of employees at its central headquarters by 40%. What is the size of its administrative staff prior to any layoffs?

current staff: $1500 \div 0.40 = 3750$ employees

6. According to the 2000 U.S. Census, the Cajun community in Louisiana has dwindled from 407,000 to 44,000 in the preceding 10 years. What percent decrease does this represent?

amount of decrease: $407{,}000 - 44{,}000 = 363{,}000$

percent decrease: $363{,}000 \div 407{,}000 \approx 0.892 = 89.2\%$

7. Hourly workers at a fiber-optics cable plant saw their workweek cut back from 60 hours to 36. What percent decrease in income does this represent?

amount of decrease: $60 - 36 = 24$ hr

percent decrease: $24 \div 60 = 0.40 = 40\%$

8. A suit with an original price of $300 was marked down 30%. You have a coupon good for an additional 20% off.

a. What is the suit's final cost?

final cost: $300 \cdot 0.70 \cdot 0.80 = \$168$

b. What is the equivalent percent discount?

decay factor: $0.70 \cdot 0.80 = 0.56$

percent discount: $1 - 0.56 = 0.44 = 44\%$

9. In 2006 contract negotiations between the Transit Workers Union and the New York City Metropolitan Transit Authority, the workers rejected a 3-year contract, with 3%, 4%, and 3.5% wage increases respectively over each year of the contract. At the end of this 3-year period, what would have been the salary of a motorman who had been earning $48,000 at the start of the contract?

salary at end of the 3-year contract:
$48{,}000 \cdot 1.03 \cdot 1.04 \cdot 1.035 = \$53{,}217.22$

10. You are moving into a rent-controlled apartment, and the landlord is raising the rent by the maximum allowable 20%. You will be paying $900. What was the former tenant's rent?

former rent: $900 \div 1.20 = \$750$

11. Gasoline prices increased from $2.29 per gallon to a price of $2.57. What percent increase does this represent?

amount increase: 2.57 − 2.29 = $0.28

percent increase: 0.28 ÷ 2.29 = 0.122 = 12.2%

12. The Ford Motor Company released its Focus sales figures for the month of April in 2009. Dealers sold a total of 11,691 new Focus automobiles, a decrease of 51.0% from the same month in 2008. Approximately how many new Focus cars were sold in April 2008?

Decay factor: 100% − 51.0% = 49.0% = 0.49

Number of 2008 Focuses: 11,691 ÷ 0.49 = 23,859 cars

13. The Mariana Trench in the South Pacific Ocean near the Mariana Islands extends down to a maximum depth of 35,827 feet. How deep is this deepest point in miles? In kilometers? In meters?

$35,827 \text{ ft.} \cdot \frac{1 \text{ mi.}}{5280 \text{ ft.}} \approx 6.79 \text{ mi.}$

$6.79 \text{ mi.} \cdot \frac{1.609 \text{ km}}{1 \text{ mi.}} \approx 10.93 \text{ km}$

$10.93 \text{ km} \cdot \frac{1000 \text{ m}}{1 \text{ km}} = 10,930 \text{ m}$

The Mariana Trench is about 6.8 miles, 10.9 kilometers, or 10,930 meters deep.

14. In the middle of the 2008–2009 NBA playing season, the three offensive leaders were Kobe Bryant (LA), Dwayne Wade (Miami), and LeBron James (Cleveland). Their field goal statistics were:

Bryant: 489 field goals out of 1082 attempts $\frac{489}{1082} \approx 45.2\%$

Wade: 470 field goals out of 1038 attempts $\frac{470}{1038} \approx 45.3\%$

James: 465 field goals out of 980 attempts $\frac{465}{980} \approx 47.4\%$

Rank them according to their relative (%) performance.

1. James   2. Wade   3. Bryant

Skills Check 1

1. Subtract 187 from 406.   219

2. Multiply 68 by 79.   5372

3. Write 16.0709 in words.

   Sixteen and seven hundred nine ten-thousandths

4. Write three thousand four hundred two and twenty-nine thousandths in standard form.

   3402.029

5. Round 567.0468 to the nearest hundredth.

   567.05

6. Round 2.59945 to the nearest thousandth.

   2.599

7. Add 48.2 + 36 + 2.97 + 0.743.

   87.913

8. Subtract 0.48 from 29.3.

   28.82

9. Multiply 2.003 by 0.36.

   0.72108

10. Divide 28.71 by 0.3.

    95.7

11. Change 0.12 to a percent.

    12%

12. Change 3 to a percent.

    300%

13. The sales tax rate in some states is 6.5%. Write this percent as a decimal.

    0.065

14. Write 360% in decimal form.

    3.6

15. Write $\dfrac{2}{15}$ as a decimal.

    $0.1\overline{3}$

16. Write $\dfrac{3}{7}$ as a percent. Round the percent to the nearest tenth.

    $\dfrac{3}{7} \approx 0.4286 \approx 42.86\% \approx 42.9\%$

17. The membership in the Nautilus Club increased by 12.5%. Write this percent as a fraction.

    $\dfrac{1}{8}$

Answers to all Skills Check exercises are included in the Selected Answers appendix.

18. Write the following numbers in order from smallest to largest:

   3.027   3.27   3.0027   3.0207

   3.0027   3.0207   3.027   3.27

19. Find the average of 43, 25, 37, and 58.

   $\frac{163}{4} = 40.75$ The average is 40.75.

20. In a recent survey, 14 of the 27 people questioned preferred Coke to Pepsi. What percent of the people preferred Coke?

   $\frac{14}{27} \approx 52\%$ preferred Coke.

21. If the sales tax is 7%, what is the cost, including tax, of a new baseball cap that is priced at $8.10?

   $\$8.10(1.07) \approx \$8.67$ is the total cost of the cap.

22. If you spend $63 a week for food and you earn $800 per month, what percent of your monthly income is spent on food?

   $63(4) = \$252$ a month for food $\frac{252}{800} \approx 0.32$

   Approximately 32% of monthly income is spent on food.

23. The enrollment at a local college has increased 5.5%. Last year's enrollment was 9500 students. How many students are expected this year?

   $9500(1.055) \approx 10{,}023$ About 10,023 students are expected this year.

24. You decide to decrease the number of calories in your diet from 2400 to 1500. Determine the percent decrease.

   $\frac{900}{2400} = 37.5\%$ decrease

25. The total area of the world's land masses is about 57,000,000 square miles. If the area of 1 acre is about 0.00156 square mile, determine the number of acres of land in the world.

   $57{,}000{,}000 \ \cancel{\text{sq.mi.}} \cdot \frac{1 \ \text{acre}}{0.00156 \ \cancel{\text{sq.mi.}}} \approx 3.654 \cdot 10^{10}$ acres $= 36{,}540{,}000{,}000$ acres

   The total area of the world's land masses is 36.54 billion acres.

26. You and a friend have been waiting for the price of a winter coat to come down sufficiently from the manufacturer's suggested retail price (MSRP) of $500 so you each can afford to buy one. The retailer always discounts the MSRP by 10%. Between Thanksgiving and New Year's, the price is further reduced by 40%, and in January, it is reduced again by 50%.

   a. Your friend remarks that it looks like you can get the coat for free in January. Is that true? Explain.

   No, the percent reductions are based on the price at the time of the specific reduction, so it is not correct to sum up the percent reductions.

   b. What does the coat cost before Thanksgiving, in December, and in January?

   The decay factors are 0.9, 0.6, and 0.5, respectively.

   Therefore, the coat will cost $0.9(500) = \$450$ before Thanksgiving, $0.6(450) = \$270$ in December, and $0.5(270) = \$135$ in January.

c. How would the final price differ if the discounts had been taken in the reverse order (50% off, 40% off, and 10% off)? How would the intermediate prices be affected?

The final price in January would be $135, the same as before, because $0.9(0.6)(0.5)(500) = 0.5(0.6)(0.9)(500) = 135$.

However, the intermediate prices would be different.

Before Thanksgiving, the price would be $0.5(500) = \$250$.

In December, the cost would be $0.6(250) = \$150$.

27. A local education official proudly declares that although math scores had fallen by nearly 55% since 1995, by 2005 they had rebounded over 65%. This sounds like great news for your district, doesn't it? Determine how the 2005 math scores actually compare with the 1995 scores.

The initial decay factor was 0.45.

The subsequent growth factor was 1.65.

Therefore, the 2005 math scores were $(0.45)(1.65) = 0.7425$, or approximately 74% of the 1995 scores. This is not good news.

28. The MAI Corporation noted that its January sales were up 10.8% over December's sales. Sales took a dip of 4.7% in February and then increased 12.4% in March. If December's sales were approximately $6 million, what was the sales figure for March?

The January growth factor was 1.108.

The February decay factor was $1.0 - 0.047 = 0.953$.

The March growth factor was 1.124.

$(1.108)(0.953)(1.124)(6{,}000{,}000) = 7{,}121{,}151.5$

Sales for March were a little over $7 million.

29. Each year, Social Security payments are adjusted for cost of living. Your grandmother relies on Social Security for a significant part of her retirement income. In 2003, her Social Security income was approximately $1250 per month. If cost-of-living increases were 2.1% for 2004, 2.7% for 2005, and 4.1% for 2006, what was the amount of your grandmother's monthly Social Security checks in 2006?

The growth factors for 2003–2006 are 1.021, 1.027, and 1.041, respectively: $1.021(1.027)(1.041)(1250) \approx 1364.45$.

Therefore, my grandmother received monthly checks of approximately $1364 in 2006.

Cluster 4   **Problem Solving with Signed Numbers**

Activity 1.12

Celsius Thermometers

Objectives

1. Identify signed numbers.

2. Use signed numbers to represent quantities in real-world situations.

3. Compare signed numbers.

4. Calculate the absolute value of numbers.

5. Identify and use properties of addition and subtraction of signed numbers.

6. Add and subtract signed numbers using absolute value.

1. What are some situations in which you have encountered negative numbers?

   (Answers will vary.) Some examples are temperatures, changes in stock prices, checkbook balances, and elevations below sea level.

--- **Definition** ---

The collection of positive counting numbers, negative counting numbers, and zero is basic to our number system. For easy referral, this collection is called the set of **integers.**

A good technique for visualizing the relationship between positive and negative numbers is to use a number line, scaled with integers, much like an outdoor thermometer's scale.

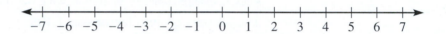

On a number line, 0 separates the positive and negative numbers. Recall that numbers increase in value as you move to the right. Thus, -1 has a value greater than -4, written as the inequality $-1 > -4$. This relationship between -1 and -4 makes sense in the thermometer model, because $-1°$ represents a (higher) warmer temperature than $-4°$.

2. Insert the appropriate inequality symbol between the two numbers in each of the following pairs.

   a. $2 < 6$ **b.** $-2 > -6$ **c.** $2 > -3$

   d. $-5 < -1$ **e.** $-3 > -8$

Adding Signed Numbers

In this activity, a thermometer model illustrates addition and subtraction of signed numbers. On a Celsius thermometer, $0°$ represents the temperature at which water freezes, and $100°$ represents the temperature at which water boils. The thermometers in the following problems show temperatures from $-10°C$ to $+10°C$.

3. **a.** What is the practical meaning of positive numbers in the Celsius thermometer model?

   Positive numbers in the Celsius thermometer model represent temperatures above water's freezing point.

   b. What is the practical meaning of negative numbers in the Celsius thermometer model?

   Negative numbers in the Celsius thermometer model represent temperatures below water's freezing point.

You can use signed numbers to represent a change in temperature. A rise in temperature is indicated by a positive number, and a drop in temperature is indicated by a negative number. As shown on the thermometer at the right, a rise of 6° from −2° results in a temperature of 4°, symbolically written as −2° + 6° = 4°.

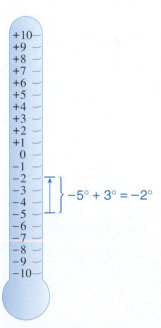

4. Answer the following questions, using the thermometer models.

a. What is the result if a temperature starts at 5° and rises 3°? Symbolically, you are calculating 5° + 3°. Use the thermometer below to demonstrate your calculation.

b. If a temperature starts at −5° and rises 3°, what is the result? Write the calculation symbolically. Use the thermometer below to demonstrate your calculation.

c. If a temperature starts at 5° and drops 3°, the resulting temperature is 2°. Symbolically, you are calculating $5° + (-3°)$. Use the thermometer below to demonstrate your calculation.

d. What is the result when a temperature drops 3° from $-5°$? Write the calculation symbolically. Use the thermometer below to demonstrate your calculation.

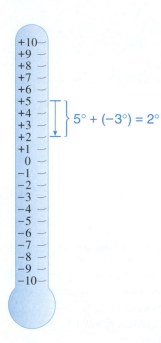

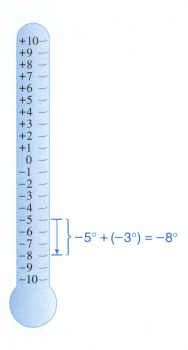

5. a. In what direction do you move on the thermometer when you add positive degrees to a starting temperature in Problem 4?

I move up on the thermometer when I add positive degrees to a starting temperature.

b. In each case, is the result greater or less than the starting number?

The result is a greater number.

6. a. When you add negative degrees to a starting temperature in Problem 4, in what direction do you move on the thermometer?

I move down on the thermometer when I add negative degrees to a starting temperature.

b. In each case, is the result greater or less than the starting number?

The result is a lesser number.

7. a. Evaluate each of the following mentally. Use your calculator to check your result.

i. $4 + 6 = 10$ **ii.** $6 + 8 = 14$ **iii.** $-7 + (-2) = -9$

iv. $-16 + (-10)$ **v.** $(-0.5) + (-1.4)$ **vi.** $-\frac{5}{2} + \left(-\frac{3}{2}\right)$

$= -26$ $= -1.9$ $= -\frac{8}{2} = -4$

b. In each calculation in part a, what do you notice about the signs of the two numbers being added?

In each case, the signs of the numbers are the same.

c. How do the signs of the numbers being added determine the sign of the result?

If the numbers have the same sign, the result also has that sign.

d. How do you calculate the numerical part of the result from the numerical parts of the numbers being added?

The numerical parts of the original values are added to obtain the numerical part of the result.

8. a. Evaluate each of the following mentally. Use your calculator to check your result.

i. $4 + (-6) = -2$ **ii.** $-6 + 8 = 2$ **iii.** $7 + (-2) = 5$

iv. $-16 + (10) = -6$ **v.** $-\dfrac{6}{2} + \dfrac{7}{2} = \dfrac{1}{2}$ **vi.** $0.5 + (-2.8) = -2.3$

b. In each calculation in part a, what do you notice about the signs of the numbers being added?

The signs are different.

c. How do the signs of the numbers being added determine the sign of the result?

If two numbers being added have different signs, then the sign of the result is the same as the sign of the number with the larger numerical part.

d. How do you calculate the numerical part of the result from the numerical parts of the numbers being added?

The numerical parts of the original values are subtracted to obtain the numerical part of the result.

9. Evaluate each of the following. Use your calculator to check your result.

a. $-8 + (-6)$
$= -14$

b. $9 + (-12)$
$= -3$

c. $6 + (-8)$
$= -2$

d. $-7 + (-9)$
$= -16$

e. $16 + 10$
$= 26$

f. $16 + (-10)$
$= 6$

g. $-9 + 8$
$= -1$

h. $-2 + (-3)$
$= -5$

i. $9 + (-5)$
$= 4$

j. $3 + (-6)$
$= -3$

k. $-\dfrac{5}{8} + \dfrac{1}{8}$
$= -\dfrac{4}{8} = -\dfrac{1}{2}$

l. $-\dfrac{5}{6} + \left(-\dfrac{1}{6}\right)$
$= -\dfrac{6}{6} = -1$

m. $\dfrac{3}{4} + \left(-\dfrac{1}{4}\right)$
$= \dfrac{2}{4} = \dfrac{1}{2}$

n. $\dfrac{1}{2} + \dfrac{1}{3}$
$= \dfrac{3}{6} + \dfrac{2}{6} = \dfrac{5}{6}$

o. $-5.9 + (-4.7)$
$= -10.6$

p. $0.50 + 0.06$
$= 0.56$

q. $(-5.75) + 1.25$
$= -4.50$

r. $-12.1 + 8.3$
$= -3.8$

s. $-6 + \left(-\dfrac{2}{3}\right)$
$= -6\dfrac{2}{3}$

t. $5 + \left(-1\dfrac{3}{4}\right)$
$= 3\dfrac{1}{4}$

Subtracting Signed Numbers

Suppose you know the starting and ending temperatures for a certain period of time and that you are interested in the change in temperature over that period.

> **Procedure**
>
> **Calculating Change in Value**
>
> **Change in value** is calculated by subtracting the initial (original) value from the final value. That is,
>
> final number − initial number = change in value (difference).

Example 1   *The change in value from 2° to 9° is 9° − 2° = 7°. Here, 9° is the final temperature, 2° is the original, or initial, temperature, and 7° is the change in temperature. Notice that the temperature has risen; therefore, the change in temperature of 7° is positive, as shown on the thermometer on the left.*

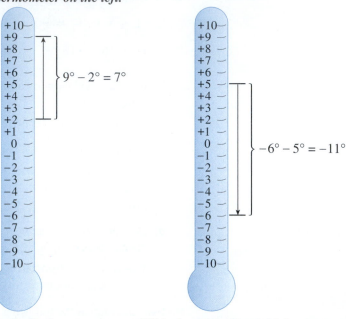

Suppose that −6° is the final temperature and 5° is the original, or initial, temperature. Symbolically, this is written −6° − 5°, which produces a result of −11°, as indicated on the thermometer on the right. The significance of a negative change is that the temperature has fallen.

10. a. What is the change in temperature from 3° to 5°? That is, what is the difference between the final temperature, 5°, and the initial temperature, 3°? Symbolically, you are calculating +5° − (+3°). Use the thermometer below to demonstrate this calculation. Has the temperature risen or fallen?

b. The change in temperature from 3° to −5° is −8°. Symbolically, you write −5° − (+3°) = −8°. Use the thermometer below to demonstrate this calculation. Has the temperature risen or fallen?

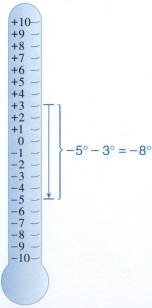

The temperature has risen 2°.   The temperature has fallen 8°.

c. A temperature in Montreal last winter rose from $-3°$ to $5°$. What was the change in temperature? Write the calculation symbolically, and determine the result. Use the thermometer below to demonstrate your calculation.

d. The temperature on a March day in Montana was $-3°$ at noon and $-5°$ at 5 P.M. What was the change in temperature? Write the calculation symbolically, and determine the result. Use the thermometer below to demonstrate your calculation. Has the temperature risen or fallen?

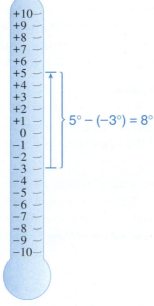

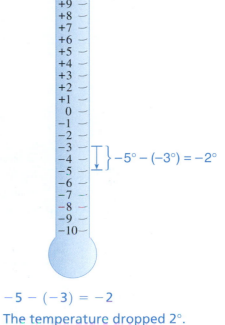

$5 - (-3) = 8$

The change in temperature was $8°$.

$-5 - (-3) = -2$

The temperature dropped $2°$.

The change in temperature was $-2°$.

11. When calculating a change in value, in which position in the subtraction calculation must the initial value be placed—to the left or to the right of the subtraction symbol?

The initial value must be placed to the right of the subtraction symbol.

Your calculations in Problem 10 demonstrated that subtraction is used to calculate the *change* or *difference* in values.

Opposites

You can use the concept of opposites to relate the operations of addition and subtraction of integers.

Opposites are numbers that when added together give a sum of zero.

Example 2 *The numbers 10 and -10 are opposites because $10 + (-10) = 0$.*

12. a. What is the opposite of 5?   The opposite of 5 is -5.

b. What is the opposite of -8?   The opposite of -8 is $+8$.

c. $0 - 5 = -5$

$0 + (-5) = -5$

d. $0 - 12 = -12$

$0 + (-12) = -12$

e. $0 - (-8) = 8$

$0 + 8 = 8$

f. $0 - (-6) = 6$

$0 + 6 = 6$

g. From your experience in parts c–f, what can you conclude about the result of subtracting a number from zero? What about adding a number to zero?

When I subtract a number from zero, I obtain its opposite. When I add a number to zero, the result is the number itself.

h. From your experience with parts c–f, is it reasonable to believe that subtracting a number gives the same result as adding its opposite? Explain.

From the results of part g, it seems that subtraction can be considered addition of the opposite.

Subtracting a signed number is equivalent to *adding its opposite*. Therefore, when subtracting a signed number, *two* changes must be made. The subtraction symbol is changed to an addition symbol, *and* the number following the subtraction symbol is replaced by its opposite. The transformed problem is evaluated as an addition problem.

Example 3 *The expression* $(-4) - 6$ *becomes* $(-4) + (-6)$*, which equals* -10 *by the addition rule of signed numbers.*

opposite

$$-4 - 6 = -4 + (-6) = -10$$

change
subtraction
to addition

13. Convert each of the following to an equivalent addition problem and evaluate. Use a calculator to check your results.

a. $4 - 11$

$= 4 + (-11) = -7$

b. $-8 - (-6)$

$= -8 + 6 = -2$

c. $9 - (-2)$

$= 9 + 2 = 11$

d. $-7 - 1$

$= -7 + (-1) = -8$

e. $10 - 15$

$= 10 + (-15) = -5$

f. $6 - (-4)$

$= 6 + 4 = 10$

g. $-8 - 10$

$= -8 + (-10) = -18$

h. $-2 - (-5)$

$= -2 + 5 = 3$

i. $-\dfrac{5}{8} - \dfrac{1}{8}$

$= -\frac{5}{8} + \left(-\frac{1}{8}\right)$

$= -\frac{6}{8} = -\frac{3}{4}$

j. $\dfrac{5}{6} - \left(-\dfrac{1}{6}\right)$

$= \frac{5}{6} + \frac{1}{6} = 1$

k. $-\dfrac{3}{4} - \left(-\dfrac{1}{4}\right)$

$= -\frac{3}{4} + \frac{1}{4}$

$= -\frac{2}{4} = -\frac{1}{2}$

l. $\dfrac{1}{2} - \dfrac{1}{3}$

$= \frac{1}{2} + \left(-\frac{1}{3}\right)$

$= \frac{3}{6} + \left(-\frac{2}{6}\right) = \frac{1}{6}$

m. $5.9 - (-4.7)$

$= 5.9 + 4.7 = 10.6$

n. $-3.75 - 1.25$

$= -3.75 + (-1.25) = -5$

o. $-6 - \left(-\dfrac{2}{3}\right)$

$= -6 + \frac{2}{3} = -5\frac{1}{3}$

p. $5 - \left(-\dfrac{3}{4}\right)$

$= 5 + \frac{3}{4} = 5\frac{3}{4}$

Number lines and models such as thermometers provide a good visual approach to adding and subtracting signed numbers. However, you will find it more efficient to use a general method when dealing with signed numbers in applications. A convenient way to add and subtract signed numbers is to use the concept of absolute value.

Absolute Value

The **absolute value** of a number represents the distance that the number is from zero on the number line. For example, $+9$ and -9 are both 9 units from 0 but in opposite directions, so they have the same absolute value, 9. Symbolically, you write $|9| = 9$ and $|-9| = 9$ and $|0| = 0$. Notice that opposites have the same absolute value. More generally, the absolute value of a number represents the size or magnitude of the number. *The absolute value of a nonzero number is always positive.*

14. Evaluate each of the following.

   a. $|17|$ **b.** $\left|-12\frac{1}{4}\right|$ **c.** $|0|$ **d.** $|-3.75|$

   17 $12\frac{1}{4}$ 0 3.75

15. a. What number has the same absolute value as -57?

   57 has the same absolute value as -57.

   b. What number has the same absolute value as 38.5?

   -38.5 has the same absolute value as 38.5.

   c. What number has an absolute value of zero?

   The absolute value of zero is zero.

   d. Can the absolute value of a number be negative? Explain.

   The absolute value of a number represents a distance and, therefore, cannot be negative.

The procedures you have developed in this activity for adding and subtracting signed numbers can be restated using absolute value.

Procedure

Addition Rule for Signed Numbers When *adding* two numbers with the *same* sign, add the absolute values of the numbers. The sign of the sum is the same as the sign of the original numbers.

When *adding* two numbers with *opposite* signs, determine their absolute values and then subtract the smaller from the larger. The sign of the sum is the same as the sign of the number with the larger absolute value.

Example 4   *Add the following signed numbers.*

    a. $-6 + (-13)$ **b.** $-29 + 14$

SOLUTION

  a. The numbers being added have the same sign, so first add their absolute values.

$$|-6| + |-13| = 6 + 13 = 19$$

Since the signs of given numbers are the same, the answer will have that sign. Therefore, $-6 + (-13) = -19$.

  b. The numbers being added have different signs, so determine their absolute values and subtract the smaller from the larger.

$$|-29| - |14| = 29 - 14 = 15$$

The sign of the answer is the sign of the number with the greater absolute value. Therefore, $-29 + 14 = -15$

16. Add the signed numbers using absolute value.

  a. $-31 + (-23)$
$$= -(|-31| + |-23|)$$
$$= -(31 + 23)$$
$$= -54$$

  b. $-31 + 23$
$$= -31 + 23 = -(|-31| - |23|)$$
$$= -(8)$$
$$= -8$$

  c. $-52 + (-12)$
$$= -(|-52| + |-12|)$$
$$= -(52 + 12)$$
$$= -64$$

  d. $-14 + (-23)$
$$= -(|-14| + |-23|)$$
$$= -(14 + 23)$$
$$= -37$$

  e. $0 + (-23)$
$$= -23$$

  f. $23 + (-15)$
$$= 8$$

SUMMARY: ACTIVITY 1.12

1. The collection of positive counting numbers, negative counting numbers, and zero is known as the set of **integers**.

2. The following is a number line scaled with the integers.

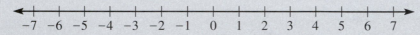

3. Change in value is calculated by subtracting the initial (original) value from the final value.

4. Two numbers whose sum is zero are called **opposites**. On a number line, opposites are equidistant from 0, one lying to the left of 0 and the other to the right of 0.

5. Subtracting a signed number is equivalent to *adding the opposite* of the signed number.

6. The **absolute value** of a number, also called its *magnitude*, represents the distance that the number is from zero on the number line.

7. The absolute value of a nonzero number is always positive.

8. When *adding* **two numbers with the** *same* **sign**, add the absolute values of the numbers. The sign of the sum is the same as the sign of the numbers being added.

9. When *adding* **two numbers with** *opposite* **signs**, determine their absolute values and then subtract the smaller from the larger. The sign of the sum is the same as the sign of the number with the larger absolute value.

EXERCISES: ACTIVITY 1.12

1. Another illustration of adding signed numbers can be found on the gridiron—that is, on the football field. On the number line, a gain is represented by a move to the right, while a loss results in a move to the left. For example, a 7-yard loss and a 3-yard gain result in a 4-yard loss. This can be written symbolically as $-7 + 3 = -4$.

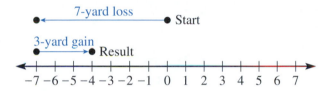

Use the number line to complete the following.

a. A 6-yard gain and a 3-yard gain result in a ___9-yard gain___.

b. A 5-yard gain and a 2-yard loss result in a ___3-yard gain___.

c. A 4-yard loss and a 3-yard loss result in a ___7-yard loss___.

d. A 1-yard gain and an 8-yard loss result in a ___7-yard loss___.

2. Write each of the situations in Problem 1 as an addition problem.

a. $6 + 3 = 9$ **b.** $5 + (-2) = 3$

c. $-4 + (-3) = -7$ **d.** $1 + (-8) = -7$

3. The San Francisco 49ers are famous for deciding their first 10 plays before the game begins. If the yardage gained or lost from each play is totaled, the sum represents the total yards gained (or lost) for those plays. One play sequence proceeded as follows.

Win Some, Lose Some

PLAY	1	2	3	4	5	6	7	8	9	10
YARDS	Lost 5	Gained 7	Gained 9	Lost 5	Gained 15	Gained 4	Lost 6	Gained 20	Lost 1	Gained 5

Find the total yards gained (or lost) for these 10 plays.

$-5 + 7 + 9 + (-5) + 15 + 4 + (-6) + 20 + (-1) + 5 = 43$

There was a 43-yard gain for these 10 plays.

Exercise numbers appearing in color are answered in the Selected Answers appendix.

4. Evaluate each of the following. Use a calculator to verify your result.

a. $-3 + (-7)$
$ = -10$

b. $-6 + 2$
$ = -4$

c. $4 + (-10)$
$ = -6$

d. $16 - 5$
$ = 11$

e. $7 - 12$
$ = -5$

f. $-2 - 8$
$ = -10$

g. $-5 + (-10)$
$ = -15$

h. $(-6) + 9$
$ = 3$

i. $-8 - 10$
$ = -18$

j. $-4 - (-5)$
$ = 1$

k. $0 - (-6)$
$ = 6$

l. $-3 + 0$
$ = -3$

m. $9 - 9$
$ = 0$

n. $9 - (-9)$
$ = 18$

o. $0 + (-7)$
$ = -7$

p. $-4 + (-5)$
$ = -9$

q. $-8 - (-3)$
$ = -5$

r. $(-9) - (-12)$
$ = 3$

s. $5 - 8$
$ = -3$

t. $(-7) - 4$
$ = -11$

u. $-\frac{5}{9} + \frac{4}{9}$
$ = -\frac{1}{9}$

v. $2.5 - 3.1$
$ = -0.6$

w. $-75 - 20$
$ = -95$

x. $33 - 66$
$ = -33$

5. The current temperature is $-12°C$. If it is forecast to be 5° warmer tomorrow, write a symbolic expression, and evaluate it to determine tomorrow's predicted temperature.

$-12 + 5 = -7$ Tomorrow's predicted temperature is $-7°C$.

6. The current temperature is $-4°C$. If it is forecast to be 7° colder tomorrow, write a symbolic expression, and evaluate it to determine tomorrow's predicted temperature.

$-4 - 7 = -11$ Tomorrow's predicted temperature is $-11°C$.

7. On a cold day in April, the temperature rose from $-5°C$ to $7°C$. Write a symbolic expression, and evaluate it to determine the change in temperature.

$7 - (-5) = 12$ The temperature rose 12°.

8. A heat wave hits New York City, increasing the temperature an average of 3°C per day for 5 days, starting with a high of 20°C on June 6. What is the approximate high temperature on June 11?

$3 \cdot 5 = 15; 20 + 15 = 35$

The approximate high temperature on June 11 is 35°C.

9. Insert the appropriate inequality symbol between the two numbers in each of the following number pairs.

a. $5 \; > \; 3$

b. $-5 \; < \; -3$

c. $-8 \; < \; -1$

d. $-4 \; > \; -7$

e. $-1 \; > \; -2$

Activity 1.13

Shedding the Extra Pounds

Objective

1. Multiply and divide signed numbers.

You have joined a health and fitness club to lose some weight. The club's registered dietitian and your personal trainer helped you develop a special diet and exercise program. Your weekly weight gain (positive value) or weekly weight loss (negative value) over the first 7 weeks of the program is recorded in the following table.

 **Weigh In**

WEEK NUMBER	1	2	3	4	5	6	7
CHANGE IN WEIGHT (lb)	−2	−2	3	−2	−2	3	−2

1. **a.** Identify the 2 weeks during which you gained weight. What was your total weight gain over these 2 weeks? Write your answer as a signed number and in words.

   I gained 6 pounds, which is expressed as a change of +6 pounds.

   b. Identify the 5 weeks during which you lost weight. What was your total weight loss over these 5 weeks? Write your answer as a signed number and in words.

   I lost 10 pounds, which is expressed as a change of − 10 pounds.

   c. Explain how you calculated the answers to parts a and b.

   In part a, I added $3 + 3 = 6$.
   In part b, I added $-2 + (-2) + (-2) + (-2) + (-2) = -10$.

   d. At the end of the first 7 weeks, what is the total change in your weight?

   The total change in weight is $6 + (-10) = -4$ pounds; that is, I lost 4 pounds.

Your answers to Problems 1a and 1b could be determined in two ways.

Solution to Problem 1a.

> **Repeated Addition:** The increase in weight in weeks 3 and 6 can be determined by repeated addition:
>
> $$3 + 3 = 6 \text{ pounds.}$$
>
> **Multiplication:** Recall that repeated addition can be expressed as multiplication. So, the increase in weight can be represented by the product
>
> $$2(3) = 6 \text{ pounds.}$$

Solution to Problem 1b.

> **Repeated Addition:** The decrease in weight in weeks 1, 2, 4, 5, and 7 can be determined by repeated addition:
>
> $$-2 + (-2) + (-2) + (-2) + (-2) = -10.$$
>
> **Multiplication:** Because multiplication of integers can be considered repeated addition, the loss in weight can be represented by the product:
>
> $$5(-2) = -10.$$

Note that 5 times the loss of 2 pounds per week means a loss of 10 pounds, represented by −10.

2. a. Fill in that part of the following table that requires multiplication of two positive numbers.

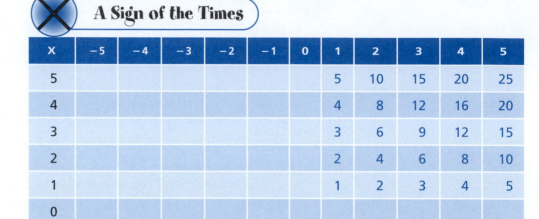

A Sign of the Times

X	−5	−4	−3	−2	−1	0	1	2	3	4	5
5							5	10	15	20	25
4							4	8	12	16	20
3							3	6	9	12	15
2							2	4	6	8	10
1							1	2	3	4	5
0											
−1											
−2											
−3											
−4											
−5											

b. Now, fill in that part of the table that requires the multiplication of numbers with opposite signs. Use what you observed in Problem 1b to help obtain the products.

X	−5	−4	−3	−2	−1	0	1	2	3	4	5
5	−25	−20	−15	−10	−5		5	10	15	20	25
4	−20	−16	−12	−8	−4		4	8	12	16	20
3	−15	−12	−9	−6	−3		3	6	9	12	15
2	−10	−8	−6	−4	−2		2	4	6	8	10
1	−5	−4	−3	−2	−1		1	2	3	4	5
0											
−1							−1	−2	−3	−4	−5
−2							−2	−4	−6	−8	−10
−3							−3	−6	−9	−12	−15
−4							−4	−8	−12	−16	−20
−5							−5	−10	−15	−20	−25

3. a. Use the patterns that you discovered so far in this activity to determine the sign of the product of two numbers that have opposite signs.

The sign of the product of two numbers with opposite signs is negative.

b. Explain how to determine the product of two numbers with opposite signs. Use a specific example in your explanation.

Multiply the absolute values of the numbers and attach a negative sign to the product. For example, to multiply 5 times -4, take the absolute value of each and multiply: 5 times 4 is 20. Then attach the negative sign determined in part a to obtain the product of -20.

- When you multiply two positive numbers, the result is always positive.

- When you multiply a positive number by a negative number, the answer is always negative.

Example 1   **a.** $5(-4) = -20$ **b.** $-3(4) = -12$ **c.** $-2(5) = -10$

4. Multiply each of the following. Use your calculator to check your answers.

a. $7(-3)$ **b.** $-5(12)$ **c.** $6(-8)$ **d.** $-9(7)$
   -21 -60 -48 -63

e. $-12(4)$ **f.** $10(-5)$ **g.** $-2.34(10)$ **h.** $7.5(-4.2)$
   -48 -50 -23.4 -31.5

i. $-\dfrac{1}{3} \cdot \left(\dfrac{7}{8}\right)$ **j.** $3\frac{2}{3}\left(-1\frac{1}{4}\right)$

   $-\dfrac{7}{24}$ $\dfrac{11}{3} \cdot \left(\dfrac{-5}{4}\right) = -\dfrac{55}{12} = -4\frac{7}{12}$

5. a. Fill in that part of the table on page 100 that involves multiplication by 0.

b. The remainder of the table requires multiplication of two negative numbers. Use the patterns in the table to complete the table.

c. Use the patterns that you discovered in Problem 5b to determine the sign of the product of two negative numbers that have the same sign.

The sign of the product of two numbers with the same sign is positive.

d. Explain how to determine the product of two numbers with the same sign. Use a specific example in your explanation.

Multiply the absolute values of the numbers to obtain the product. For example, to multiply -5 times -6, multiply 5 times 6 to obtain 30.

X	−5	−4	−3	−2	−1	0	1	2	3	4	5
5	−25	−20	−15	−10	−5	0	5	10	15	20	25
4	−20	−16	−12	−8	−4	0	4	8	12	16	20
3	−15	−12	−9	−6	−3	0	3	6	9	12	15
2	−10	−8	−6	−4	−2	0	2	4	6	8	10
1	−5	−4	−3	−2	−1	0	1	2	3	4	5
0	0	0	0	0	0	0	0	0	0	0	0
−1	5	4	3	2	1	0	−1	−2	−3	−4	−5
−2	10	8	6	4	2	0	−2	−4	−6	−8	−10
−3	15	12	9	6	3	0	−3	−6	−9	−12	−15
−4	20	16	12	8	4	0	−4	−8	−12	−16	−20
−5	25	20	15	10	5	0	−5	−10	−15	−20	−25

- When you multiply two negative numbers, the result is always positive.

Example 2   **a.** $(-5)(-4) = 20$ **b.** $(-2)(-3) = 6$ **c.** $(-4)(-1) = 4$

6. Multiply each of the following. Use your calculator to check.

a. $(-11)(-4)$

44

b. $(5)(13)$

65

c. $(-6)(-8)$

48

d. $(-12)(-3)$

36

e. $(-9)(-8)$

72

f. $(15)(4)$

60

g. $(-4.567)(-100)$

456.7

h. $(-7)(-4.50)$

31.5

i. $\left(-\frac{3}{8}\right)\left(-\frac{2}{21}\right)$

$\frac{1}{28}$

7. Multiply the following. Use your calculator to check.

a. $(-4)(-5)(-3)$

−60

b. $(-6)(-2)(-1)$

−12

c. $2(-3)(-5)$

30

d. $4(-2)(-3)(-1)$

−24

e. $3(-1)(-1)(-1)(-1)$

3

f. $(-1)(-1)(1)(-1)(-1)(-1)$

−1

g. $(-1)(-1)(-1)(-1)(-1)(-1)$

1

h. $(-2)(-3)(-4)(5)(-6)$

720

8. Refer to the results in Problem 7, and explain how you can determine the sign of the result of a sequence of multiplications.

The result of multiplying an even number of negative numbers is always positive. The result of multiplying an odd number of negative numbers is always negative.

Division of Signed Numbers

Suppose your friend is trying to put on weight and gains 8 pounds over a 4-week period. This is an average gain of $\dfrac{8 \text{ pounds}}{4 \text{ weeks}} = 8 \div 4 = 2$ pounds per week.

Suppose you lose 12 pounds during the same 4-week period. Your average loss per week can be represented by the quotient $\dfrac{-12 \text{ pounds}}{4 \text{ weeks}}$ or $-12 \div 4$. Using the multiplication check for division,

$$\frac{-12}{4} = -12 \div 4 = -3 \text{ because } 4 \cdot (-3) = -12.$$

A weight loss of 12 pounds over 4 weeks means that you lost an average of 3 pounds per week.

9. Use the multiplication check for division to obtain the only reasonable answer for each of the following. Then use your calculator to check your answers.

a. $33 \div 3$

 11

b. $-24 \div 6$

 −4

c. $15 \div (-5)$

 −3

d. $\dfrac{-36}{4}$

 −9

e. $\dfrac{-56}{-7}$

 8

f. $\dfrac{42}{-2}$

 −21

g. $(-64) \div (-8)$

 8

h. $\dfrac{-63}{-9}$

 7

i. $\dfrac{81}{-9}$

 −9

j. $\dfrac{-30.45}{2.1}$ −14.5

10. Use the patterns observed in Problem 9 to answer parts a and b.

a. The sign of the quotient of two numbers with the same sign is __positive__.

b. The sign of the quotient of two numbers with opposite signs is __negative__.

11. Determine the quotient of each of the following.

a. $\dfrac{-24}{8}$

 −3

b. $\dfrac{24}{-8}$

 −3

c. $-\dfrac{24}{8}$

 −3

d. Explain the effect of having one negative sign in fractions of the form in parts a–c. Does the result depend on the location of the negative sign?

When a single negative sign appears in a fraction, the numerical value of the fraction is negative. And it is always negative, whether the negative sign is in the numerator, in the denominator, or in front of the fraction.

A negative fraction can be written equivalently in three ways with the negative sign in three possible positions: in front of the fraction, in the numerator, or in the denominator. That is, $-\dfrac{6}{3}, \dfrac{-6}{3}$ and $\dfrac{6}{-3}$ are all equivalent fractions whose numerical value is -2.

Division Involving Zero

12. Evaluate each of the following, if possible.

a. $0 \div 9$

0

b. $0 \div (-3)$

0

c. $\dfrac{0}{-5}$

0

d. $9 \div 0$

undefined

e. $(-3) \div 0$

undefined

f. $\dfrac{-5}{0}$

undefined

g. Write a sentence explaining the result of a division that involves zero.

When 0 is divided by a nonzero number, such as $\frac{0}{9}$ or $0 \div 9$, the quotient is 0. When any number is divided by 0, such as $\frac{5}{0}$ or $5 \div 0$, the result has no numerical value—it is said to be undefined.

SUMMARY: ACTIVITY 1.13

1. When you multiply or divide two numbers with **opposite** signs, the result will always be negative.

$$(-3) \cdot 4 = -12; \quad 5 \cdot (-6) = -30; \quad 15 \div (-3) = -5; \quad (-21) \div 3 = -7$$

2. When you multiply or divide two numbers with the **same** sign, the result will always be positive.

$$3 \cdot 4 = 12; \quad (-5) \cdot (-6) = 30; \quad 15 \div 3 = 5; \quad (-21) \div (-3) = 7$$

3. When you multiply a string of signed numbers, the product will be positive if the number of negative factors is even and will be negative if the number of negative factors is odd.

$5 \cdot (-2) \cdot (-3) \cdot (-1) = -30$ since there are three (odd number) negative factors in the expression.

$5 \cdot (-2) \cdot (3) \cdot (-1) = 30$ since there are two (even number) negative factors in the expression.

4. When zero is divided by a nonzero number, the quotient is always zero. When any number is divided by zero, the result has no numerical value—it is undefined.

EXERCISES: ACTIVITY 1.13

1. Determine the product or quotient mentally. Use your calculator to check your answers.

a. $-5(-7)$

35

b. $(-3)(-6)$

18

c. $-4 \cdot 6$

-24

d. $(-8)(-2)$

16

e. $-1(2.718)$

-2.718

f. $-3.14(-1)$

3.14

g. $-6(0)$

0

h. $4(-7)$

-28

i. $(-3)(-8)$

24

j. $32 \div (-8)$

-4

k. $-25 \div (-5)$

5

l. $0 \div (-9)$

0

m. $-5 \div 0$

undefined

n. $(-12) \div 6$

-2

o. $16 \div (-2)$

-8

p. $-48 \div 6$

-8

q. $-63 \div (-1)$

63

r. $-6 \div 12$

-0.5

2. Evaluate each of the following. Use a calculator to verify your answers.

a. $\dfrac{1}{2} \div (-2)$

$-\frac{1}{4} = -0.25$

b. $-2 \div \left(\dfrac{1}{2}\right)$

-4

c. $(2.1)(-8)$

-16.8

d. $11(-3)(-4)$

132

e. $0.2(-0.3)$

-0.06

f. $-1.6 \div (-4)$

0.4

g. $\left(-\dfrac{3}{4}\right)\left(-\dfrac{2}{9}\right)$

$\frac{1}{6}$

h. $-4\frac{2}{3} \div \left(-\dfrac{7}{9}\right)$

$-\frac{14}{3} \cdot -\frac{9}{7} = 6$

i. $(-5)(-6)(-1)(3)(-2)$

180

3. Without actually performing the following calculations, determine the sign of the product.

a. $(-6)(-3)(2)(-4)$

negative

b. $(2)(-5)(2)(-5)(2)$

positive

c. $(-2)(-2)(-2)(-2)$

positive

4. Summarize the results of Exercise 3 in your own words.

The sign of a product will be positive if there is an even number of factors with negative signs; the sign of a product will be negative if there is an odd number of factors with negative signs.

5. A cold front moves through your hometown in January, dropping the temperature an average of 3° per day for 5 days. What is the change in temperature over a 5-day period? Write your answer as a signed number.

The change in temperature is $5(-3)$, which is $-15°$.

6. Yesterday, you wrote three checks, each for $21.86. How did these checks affect your balance? State your answer in words and as a signed number.

$(-21.86)(3) = -65.58$

My balance decreased by $65.58.

7. Your friend is very happy that he lost $21\frac{1}{2}$ pounds in 6 weeks. On the average, how many pounds did he lose per week?

He lost $3\frac{7}{12}$ (≈ 3.58) pounds per week.

8. During the last year, your friend lost $9876.34 in stocks. Express her average monthly loss as a signed number.

$-9876.34 \div 12 \approx -823.028$

On average, she lost $823.03 per month.

9. Stock values are reported daily in some newspapers. Reports usually state the values at both the opening and the closing of a business day. Suppose you own 220 shares of Corning stock.

a. At the highest point today, your stock is worth $26.03 per share, but by the close of the New York Stock Exchange, it had dropped to $24.25 per share. What is the total change in the value of your 220 shares from the highest point of the day to the close of business? State your answer in words and as a signed number.

$24.25 - $26.03 = -$1.78 per share

$220 \cdot (-$1.78) = -$391.60

Today my stock dropped $391.60 from the highest point to the end of the day.

b. Your stock opened today at $23.71 and gained $0.54 per share by the end of the day. What was the total value of your shares after the stock exchange closed?

$23.71 + $0.54 = $24.25

$220 \cdot $24.25 = $5335.00

At the end of the day, the total value of my shares was $5335.

10. You are one of three equal partners in a company.

a. In 2010, your company made a profit of $300,000. How much money did you and your partners each make? Write the answer in words and as a signed number.

$\frac{$300,000}{3} = $100,000$ Each made a profit of $100,000.

b. In 2009, your company experienced a loss of $150,000. What was your share of the loss? Write the answer in words and as a signed number.

$\frac{-150,000}{3} = -50,000$ My share of the loss was $50,000.

c. Over the 2-year period from January 1, 2009, to December 31, 2010, what was the net profit for each partner? Express your answer in words and as a signed number.

$100,000 + (-$50,000) = $50,000$ The net profit for each partner was $50,000 over the 2-year period.

d. Suppose that in 2011, your corporation suffered a further loss of $180,000. Over the 3-year period from January 1, 2009, to December 31, 2011, what was the net change for each partner? Express your answer in words and as a signed number.

$\frac{-180,000}{3} = -$60,000$ is the loss for each in 2011.

$100,000 + (-50,000) + (-60,000) = -$10,000$

Over the 3-year period, each partner suffered a $10,000 loss.

11. Each individual account in your small business has a current dollar balance, which can be either positive (a credit) or negative (a debit). To determine your current total account balance, you add all the individual balances.

Suppose your records show the following individual balances.

$$-220 \quad -220 \quad 350 \quad 350 \quad -220 \quad -220 \quad -220 \quad 350$$

a. What is the total of the individual positive balances?

$350 + 350 + 350 = 1050

The total of the individual positive balances is $1050.

   b. What is the total of the individual negative balances?

   $-220 + (-220) + (-220) + (-220) + (-220) = -\$1100$

   The total of the individual negative balances is $-\$1100$.

   c. In part b, did you sum the five negative balances? Or did you multiply -220 by 5? Which method is more efficient? What is the sign of the result of your calculation?

   (Answers will vary.)

   Multiplying is more efficient. The sign of the result is negative.

   d. What is the total balance of these accounts?

   $\$1050 + (-1100) = -\$50$ The balance for all the accounts is $-\$50$.

12. You own 180 shares of Verizon stock. The following table lists the changes (in dollars) during a given week for the price of a single share. Did you earn a profit or suffer a loss on the value of your shares during the given week? How much profit or loss?

DAY	Mon.	Tues.	Wed.	Thurs.	Fri.
DAILY CHANGE	+0.38	−0.75	−1.25	−1.13	2.55

$0.38 + (-0.75) + (-1.25) + (-1.13) + 2.55 = -0.2$

$180(-0.2) = -\$36.$

I suffered a $0.20 loss per share, for a total loss of $36.

Activity 1.14

Order of Operations Revisited

Objectives

1. Use the order of operations convention to evaluate expressions involving signed numbers.

2. Evaluate expressions that involve negative exponents.

3. Distinguish between such expressions as -5^4 and $(-5)^4$.

4. Write very small numbers in scientific notation.

You can combine the procedures for adding and subtracting signed numbers with those for multiplying and dividing signed numbers by using the order of operations convention you learned in Activity 1.3, Properties of Arithmetic. The order of operations convention is valid for all numbers, positive and negative, including decimals, fractions, and numerical expressions with exponents. Your ability to follow these procedures correctly with all numbers will help you use the formulas that you will encounter in applications.

1. Calculate the following expressions by hand. Then use your calculator to check your answers.

   a. $6 + 4(-2)$

   $6 + (-8) = -2$

   b. $-6 + 2 - 3$

   $-4 - 3 = -7$

   c. $(6 - 10) \div 2$

   $-4 \div 2 = -2$

   d. $-3 + (2 - 5)$

   $-3 + (-3) = -6$

   e. $(2 - 7)5$

   $(-5) \cdot 5 = -25$

   f. $-3(-3 + 4)$

   $-3(1) = -3$

   g. $-2 \cdot 3^2 - 15$

   $-2 \cdot 9 - 15 = -18 - 15 = -33$

   h. $(2 + 3)^2 - 10$

   $5^2 - 10 = 25 - 10 = 15$

   i. $-7 + 8 \div (5 - 7)$

   $-7 + 8 \div (-2)$

   $= -7 + (-4) = -11$

   j. $2.5 - (5.2 - 2.2)^2 + 8$

   $2.5 - (3)^2 + 8$

   $= 2.5 - 9 + 8 = -6.5 + 8 = 1.5$

   k. $4.2 \div 0.7 - (-5.6 + 8.7)$

   $6 - (3.1) = 2.9$

   l. $\dfrac{1}{4} - \left(\dfrac{2}{3} \cdot \dfrac{9}{8}\right)$

   $\frac{1}{4} - \left(\frac{3}{4}\right) = -\frac{2}{4} = -\frac{1}{2}$

   m. $\dfrac{5}{6} \div (-10) + \dfrac{7}{12}$

   $\frac{5}{6} \cdot -\frac{1}{10} + \frac{7}{12} = -\frac{1}{12} + \frac{7}{12} = \frac{6}{12} = \frac{1}{2}$

Negation and Exponentiation

2. Evaluate -3^2 by hand and then use your calculator to check your answer.

   $-3^2 = -(3)(3) = -9$

In Problem 2, two operations are to be performed on the number 3, exponentiation and negation. Negation can be considered as multiplication by -1. Therefore, by the order of operations convention, 3 is first squared to obtain 9 and then 9 is multiplied by -1 to determine the result, -9.

3. Evaluate $(-3)^2$ by hand and then check with your calculator.

   $(-3)^2 = (-3)(-3) = 9$

In Problem 3, parentheses are used to indicate that the number 3 is negated *first* and then -3 is squared to produce the result, 9.

Example 1   *Evaluate $(-2)^4$ and -2^4.*

SOLUTION

In $(-2)^4$, the parentheses indicate that 2 is first negated and then -2 is written as a factor 4 times.

$$(-2)^4 = (-2)(-2)(-2)(-2) = 16$$

In -2^4, the negative indicates multiplication by -1. Since exponentiation is performed before multiplication, the 2 is raised to the fourth power first and then the result is multiplied by -1 (negated).

$$-2^4 = -1 \cdot 2^4 = -1 \cdot 2 \cdot 2 \cdot 2 \cdot 2 = (-1) \cdot 16 = -16$$

4. Evaluate the following expressions by hand. Then use your calculator to check your answers.

   a. -5^2 -25 **b.** $(-5)^2$ 25 **c.** $(-3)^3$ -27

   d. -1^4 **e.** $2 - 4^2$ **f.** $(2-4)^2$
   $\quad -1$ $\quad 2 - 16 = -14$ $\quad (-2)^2 = 4$

   g. $-5^2 - (-5)^2$ **h.** $(-1)^8$ **i.** $-5^2 + (-5)^2$
   $\quad -25 - 25 = -50$ $\quad 1$ $\quad -25 + 25 = 0$

   j. $-\left(\dfrac{1}{2}\right)^2 + \left(\dfrac{1}{2}\right)^2$ $-\frac{1}{4} + \frac{1}{4} = 0$

Negative Exponents

So far in this book, you have encountered only zero or positive integers as exponents. Are numbers with negative exponents meaningful? How would you calculate an expression such as 10^{-5}? In the following problems, you will discover answers to these questions.

5. **a.** Complete the following table.

 Raised on Exponents

EXPONENTIAL FORM	EXPANDED FORM	VALUE
10^5	$10 \cdot 10 \cdot 10 \cdot 10 \cdot 10$	100,000
10^4	$10 \cdot 10 \cdot 10 \cdot 10$	10,000
10^3	$10 \cdot 10 \cdot 10$	1000
10^2	$10 \cdot 10$	100
10^1	10	10
10^0	1	1
10^{-1}	$\frac{1}{10}$	$\frac{1}{10}$ or 0.1
10^{-2}	$\frac{1}{10} \cdot \frac{1}{10}$	$\frac{1}{100}$ or 0.01
10^{-3}	$\frac{1}{10} \cdot \frac{1}{10} \cdot \frac{1}{10}$	$\frac{1}{1000}$ or 0.001
10^{-4}	$\frac{1}{10} \cdot \frac{1}{10} \cdot \frac{1}{10} \cdot \frac{1}{10}$	$\frac{1}{10,000}$ or 0.0001

b. Use the pattern developed in the preceding table to determine the relationship between each negative exponent in column 1 and the number of zeros in the denominator of the corresponding fraction and the decimal result in column 3. List and explain everything you observe.

If the exponent is negative, the number of zeros in the denominator is the same as the absolute value of the exponent. There is one less zero after the decimal point and before the 1.

The pattern you observed in the table in Problem 5 that involves raising base 10 to a negative exponent can be generalized to any nonzero number as base.

> A negative exponent always signifies a reciprocal. That is, $5^{-1} = \frac{1}{5}$ and $2^{-3} = \frac{1}{2^3} = \frac{1}{8}$. Symbolically,
>
> $$b^{-n} = \frac{1}{b^n}, \text{ where } b \neq 0.$$

Example 2 *Write each of the following without a negative exponent. Then write each answer as a simplified fraction and as a decimal.*

a. $10^{-5} = \dfrac{1}{10^5} = \dfrac{1}{100,000} = 0.00001$

b. $3^{-2} = \dfrac{1}{3^2} = \dfrac{1}{9} = 0.\overline{1}$

c. $(-4)^{-3} = \dfrac{1}{(-4)^3} = \dfrac{1}{-64} = -0.015625$

d. $6 \cdot 10^{-3} = \dfrac{6}{10^3} = \dfrac{6}{1000} = \dfrac{3}{500} = 0.006$

6. Write each of the following expressions without a negative exponent. Then write each result as a fraction and as a decimal.

a. $-2.72 \cdot 10^{-4}$

-0.000272

$-\dfrac{272}{1,000,000}$

b. 10^{-4}

$\dfrac{1}{10,000} = 0.0001$

c. $3.25 \cdot 10^{-3}$

$0.00325 = \dfrac{325}{100,000}$

d. 5^{-2}

$\dfrac{1}{5^2} = \dfrac{1}{25} = 0.04$

e. $(-5)^{-2}$

$\dfrac{1}{(-5)^2} = \dfrac{1}{25} = 0.04$

f. -5^{-2}

$-\dfrac{1}{5^2} = -\dfrac{1}{25} = -0.04$

g. 4^{-3}

$\dfrac{1}{4^3} = \dfrac{1}{64}$

0.015625

h. $-3^{-2} + 3^{-2}$

$-\dfrac{1}{3^2} + \dfrac{1}{3^2}$

$= -\dfrac{1}{9} + \dfrac{1}{9} = 0$

i. $-2^{-3} + 2^3$

$-\dfrac{1}{2^3} + 8$

$= -\dfrac{1}{8} + 8 = 7\dfrac{7}{8}$

$= 7.875$

Scientific Notation Revisited

In Activity 1.3, you learned that scientific notation provides a compact way of writing very large numbers. Recall that any number can be written in scientific notation by expressing the number as the product of a decimal number between 1 and 10 times an integer power of 10. For example, 53,000,000 can be written in scientific notation as 5.3×10^7.

$$\underbrace{N}_{\substack{\text{number from} \\ \text{1 to 10}}} \times \underbrace{10^n}_{\substack{\text{power of 10,} \\ n \text{ an integer}}}$$

7. The star nearest to Earth (excluding the Sun) is Proxima Centauri, which is 24,800,000,000,000 miles away. Write this number in scientific notation.

2.48×10^{13}

Very small numbers can also be written in scientific notation. For example, a molecule of DNA is about 0.000000002 meter wide. To write this number in scientific notation, you need to move the decimal point to the right in order to obtain a number between 1 and 10.

0.000000002.   **Move the decimal point nine places to the right.**

Therefore, 0.000000002 can be written as 2.0×10^{n}. You will determine the value of n in Problem 8.

8. a. If you multiply 2.0 by 10,000, or 10^{4}, you move the decimal point ___4___ places to the ___right___.

b. If you multiply 2.0 by 0.00001, you move the decimal point ___5___ places to the ___left___.

c. What is the relationship between each negative exponent in column 1 in the table in Problem 5 and the number of placeholding zeros in the decimal form of the number in column 3?

The number of zeros after the decimal point and before the 1 in the decimal form of the number is 1 less than the absolute value of the exponent.

d. Write 0.00001 as a power of 10. 10^{-5}

e. Use the results from parts b–d to determine the value of n in the following: $0.000000002 = 2.0 \times 10^{n}$. $n = -9$

f. How does your calculator display the number 0.000000002?

$2\text{E}-9$

9. Write each of the following in scientific notation.

a. 0.0000876 8.76×10^{-5} **b.** 0.00000018 1.8×10^{-7}

c. 0.0000000000000781 7.81×10^{-15}

10. The thickness of human hair ranges from fine to coarse, with coarse hair about 0.0071 inch in diameter. Describe how you would convert this measurement into scientific notation.

Move the decimal point in 0.0071 three places to the right to obtain 7.1. Three places to the right means the exponent is -3. Therefore, $0.0071 = 7.1 \times 10^{-3}$.

Calculations Involving Numbers Written in Scientific Notation

11. a. Recall from Activity 1.3 that you used the $\boxed{\text{EE}}$ key on your calculator to enter numbers written in scientific notation. For example, $10^{2} = 1 \times 10^{2}$ is entered as $\boxed{1}$ $\boxed{\text{EE}}$ $\boxed{2}$. Try it.

b. Now enter $10^{-2} = 1 \times 10^{-2}$ as $\boxed{1}$ $\boxed{\text{EE}}$ $\boxed{(-)}$ $\boxed{2}$. What is the result written as a decimal? Written as a fraction?

The result is 0.01 as a decimal and $\frac{1}{100}$ as a fraction.

c. Use the $\boxed{\text{EE}}$ key to enter your answers from Problem 9 into your calculator. Record your results, and compare them with the numbers given in Problem 9.

The results are the same.

12. The wavelength of X-rays is 1×10^{-8} centimeter, and that of ultraviolet light is 2×10^{-5} centimeter.

   a. Which wavelength is shorter?

   The wavelength of the X-ray is shorter.

   b. How many times shorter? Write your answer in standard notation and in scientific notation.

$$\frac{2 \times 10^{-5}}{1 \times 10^{-8}} = 2000 = 2.0 \times 10^3$$

   The wavelength of the X-ray is 2000 times shorter than that of ultraviolet light.

13. a. The mass of an electron is about $\frac{1}{2000}$ that of a proton. If the mass of a proton is 1.7×10^{-24} gram, determine the mass of an electron.

$$\frac{1}{2000} \times (1.7 \times 10^{-24}) = (5 \times 10^{-4})(1.7 \times 10^{-24}) = 8.5 \times 10^{-28}\,g$$

   The mass of an electron is 8.5×10^{-28} gram.

   b. The mass of one oxygen molecule is 5.3×10^{-23} gram. Determine the mass of 20,000 molecules of oxygen.

$$(2 \times 10^4)(5.3 \times 10^{-23}) = 1.06 \times 10^{-18}$$

   The mass of 20,000 molecules of oxygen is 1.06×10^{-18} gram.

14. The mass, m, of an object is determined by the formula

$$m = D \cdot v,$$

where D represents the density and v represents the volume of the object.

The density of water is 3.12×10^{-2} ton per cubic foot. The volume of Lake Superior is approximately 4.269×10^{14} cubic feet. Determine the mass (in tons) of the water in Lake Superior.

$$m = D \cdot v = \frac{3.12 \times 10^{-2}\,\text{ton}}{1\,\text{cu. ft.}} \cdot 4.269 \times 10^{14}\,\text{cu. ft.} \approx 1.33 \times 10^{13}\,\text{tons}$$

There are 1.33×10^{13} tons of water in Lake Superior.

15. Use your calculator to evaluate the following expressions.

   a. $\dfrac{16}{(4 \times 10^{-2}) - (2 \times 10^{-3})}$

   ≈ 421.05

   b. $\dfrac{3.2 \times 10^3}{(8.2 \times 10^{-2})(3.0 \times 10^2)}$

   ≈ 130.08

SUMMARY: ACTIVITY 1.14

1. To indicate that a negative number is raised to a power, the number must be enclosed in parentheses.

Example: "Square the number -4" is written as $(-4)^2$, which means $(-4)(-4)$ and equals 16.

2. An expression of the form $-a^2$ always signifies the negation (opposite) of a^2 and is *always* a negative quantity (for nonzero a).

Example: -4^2 represents the negation (opposite) of 4^2 and, therefore, has the value -16.

3. A negative exponent always signifies a reciprocal. Symbolically, this is written as

$b^{-n} = \dfrac{1}{b^n}$, where $b \neq 0$.

Examples: $6^{-1} = \dfrac{1}{6}$ and $5^{-3} = \dfrac{1}{5^3} = \dfrac{1}{125}$

EXERCISES: ACTIVITY 1.14

1. To discourage random guessing on a multiple-choice exam, the professor assigns 5 points for a correct response, deducts 2 points for an incorrect answer, and gives 0 points for leaving the question blank.

  a. What is your score on a 35-question exam if you have 22 correct and 7 incorrect answers?

    $22(5) + 7(-2) = 110 - 14 = 96$ I get 96 points on the exam.

  b. Express your score as a percentage of the total points possible.

    $35(5) = 175$ is the total points possible on the exam.

    $\frac{96}{175} \times 100\% = 54.86\% \approx 55\%$

    So my score is 55%.

2. You are remodeling your bathroom. To finish the bathroom floor, you decide to install molding around the base of each wall. You need the molding all the way around the room except for a 39-inch gap at the doorway. The dimensions of the room are $8\frac{2}{3}$ feet by $12\frac{3}{4}$ feet. You determine the number of feet of molding you need by evaluating the following expression: $2\left(8\frac{2}{3} + 12\frac{3}{4}\right) - 3\frac{1}{4}$. How much molding do you need?

$2\left(8\frac{8}{12} + 12\frac{9}{12}\right) - 3\frac{1}{4} = 2\left(20\frac{17}{12}\right) - 3\frac{1}{4} = 2\left(\frac{257}{12}\right) - 3\frac{1}{4} = \frac{257}{6} - 3\frac{1}{4} = 42\frac{5}{6} - 3\frac{1}{4} = 42\frac{10}{12}$
$-3\frac{3}{12} = 39\frac{7}{12}$

I need $39\frac{7}{12}$ feet of molding.

3.   **a.** Complete the following table.

EXPONENTIAL FORM	EXPANDED FORM	VALUE
2^5	$2 \cdot 2 \cdot 2 \cdot 2 \cdot 2$	32
2^4	$2 \cdot 2 \cdot 2 \cdot 2$	16
2^3	$2 \cdot 2 \cdot 2$	8
2^2	$2 \cdot 2$	4
2^1	2	2
2^0	1	1
2^{-1}	$\frac{1}{2}$	$\frac{1}{2}$
2^{-2}	$\frac{1}{2} \cdot \frac{1}{2}$	$\frac{1}{4}$
2^{-3}	$\frac{1}{2} \cdot \frac{1}{2} \cdot \frac{1}{2}$	$\frac{1}{8}$
2^{-4}	$\frac{1}{2} \cdot \frac{1}{2} \cdot \frac{1}{2} \cdot \frac{1}{2}$	$\frac{1}{16}$

Exercise numbers appearing in color are answered in the Selected Answers appendix.

b. Use the pattern in the table to determine the simplified fraction form of 2^{-7}.

$2^{-7} = \frac{1}{2} \cdot \frac{1}{2} \cdot \frac{1}{2} \cdot \frac{1}{2} \cdot \frac{1}{2} \cdot \frac{1}{2} \cdot \frac{1}{2} = \frac{1}{128}$

4. Evaluate the following expressions by hand. Verify your answer with a calculator.

a. $(8 - 17) \div 3 + 6$

$= -9 \div 3 + 6$
$= -3 + 6$
$= 3$

b. $-7 + 3(1 - 5)$

$= -7 + 3(-4)$
$= -7 + (-12)$
$= -19$

c. $-3^2 \cdot 2^2 + 25$

$= -9 \cdot 4 + 25$
$= -36 + 25$
$= -11$

d. $1.6 - (1.2 + 2.8)^2$

$= 1.6 - 4^2$
$= 1.6 - 16$
$= -14.4$

e. $\frac{5}{16} - 3\left(\frac{3}{16} - \frac{7}{16}\right)$

$= \frac{5}{16} - 3\left(-\frac{4}{16}\right)$
$= \frac{5}{16} - \left(-\frac{12}{16}\right)$
$= \frac{17}{16}$

f. $\frac{3}{4} \div \left(\frac{1}{2} - \frac{5}{12}\right)$

$= \frac{3}{4} \div \frac{1}{12}$
$= \frac{3}{4} \cdot \frac{12}{1}$
$= 9$

g. -1^{-2}

$= -1$

h. $4^3 - (-4)^3$

$= 64 - (-64)$
$= 128$

i. $(6)^2 - (6)^2$

$36 - 36 = 0$

5. Use your calculator to evaluate the following expressions.

a. $\dfrac{-3 \times 10^2}{(3 \times 10^{-2}) - (2 \times 10^{-2})}$

$-30,000$

b. $\dfrac{1.5 \times 10^3}{(-5.0 \times 10^{-1})(2.6 \times 10^2)}$

-11.538

6. Evaluate the following expressions:

a. 6^{-2}

$\dfrac{1}{6^2} = \dfrac{1}{36}$

b. $(-6)^2$

$(-6)(-6) = 36$

c. -6^2

$-1 \cdot 6 \cdot 6 = -36$

d. $(-6)^{-2}$

$\dfrac{1}{(-6)^2} = \dfrac{1}{36}$

7. In fiscal year 2008, the U.S. government paid $451.154 billion in interest on the national debt.

a. Write this number in standard notation and in scientific notation.

$451,154,000,000 = 4.51154 \times 10^{11}$

b. Assume that there were approximately 304 million people in the United States in 2008. How much of the interest could we say each person owed?

$\dfrac{4.51154 \times 10^{11}}{3.04 \times 10^8} = \$1484.06$ per person

8. You have found 0.4 gram of gold while panning for gold on your vacation.

   a. How many pounds of gold do you have ($1 \text{ g} = 2.2 \times 10^{-3}$ lb.)? Write your result in decimal form and in scientific notation.

   $$0.4 \text{ g} \left(\frac{2.2 \times 10^{-3} \text{ lb.}}{1 \text{ g}} \right)$$

   $$= 8.8 \times 10^{-4} \text{ lb.}$$

   $$= 0.00088 \text{ lb.}$$

   b. If you were to tell a friend how much gold you have, would you state the quantity in grams or in pounds? Explain.

   (Answers will vary.)

   I would use grams because the number is easier to say, and it may sound larger and impress my friend.

Cluster 4   What Have I Learned?

1. Explain how you would add two signed numbers and how you would subtract two signed numbers. Use examples to illustrate each.

   If the signs of two numbers being added are alike, add the numerical parts, and the result also has that same sign. If the signs of the two numbers being added are opposite, subtract the numerical parts and take the sign of the larger numerical part. To subtract a signed number, add its opposite. For example:

   $-7 + (-2) = -9$
   $-7 + 2 = -5$
   $-7 - 2 = -7 + (-2) = -9$
   $-7 - (-2) = -7 + 2 = -5$

2. Explain how you would multiply two signed numbers and how you would divide two signed numbers. Use examples to illustrate each.

   When you multiply or divide two signed numbers, the result is positive if the numbers have the same sign and negative if they have opposite signs. For example:

   $3(-2) = -6$
   $-3(-2) = 6$
   $\dfrac{-36}{-4} = 9$
   $-36 \div 4 = -9$

3. What would you suggest to your classmates to avoid confusing addition and subtraction procedures with multiplication and division procedures?

   (Answers will vary.)

4. a. Without actually performing the calculation, determine the *sign* of the product $(-0.1)(+3.4)(6.87)(-0.5)(+4.01)(3.9)$. Explain.

      The result is positive because there are two negatives being multiplied, resulting in a positive value.

   b. Determine the *sign* of the product $(-0.2)(-6.5)(+9.42)(-0.8)(1.73)(-6.72)$. Explain.

      The sign of the product is positive because there is an even number of negative factors.

   c. Determine the *sign* of the product $(-1)(-5.37)(-3.45)$. Explain.

      The sign of the product is negative because there is an odd number of negative factors.

   d. Determine the *sign* of the product $(-4.3)(+7.89)(-69.8)(-12.5)(+4.01)(-3.9)(-78.03)$. Explain.

      The sign of the product is negative because there is an odd number of negative factors.

   e. What rule is suggested by the results you obtained in parts a–d?

      The product of any number of positive values is positive. The product of an even number of negative values is positive. The product of an odd number of negative values is negative.

5. a. If -2 is raised to the power 4, what is the sign of the result? What is the sign if -2 is raised to the sixth power?

$(-2)^4$ is positive; $(-2)^6$ is positive.

b. Raise -2 to the eighth power, tenth power, and twelfth power. What are the signs of the numbers that result?

$(-2)^8 = 256; (-2)^{10} = 1024; (-2)^{12} = 4096$

All are positive.

c. What general rule involving signs is suggested by the preceding results?

A negative number raised to an even power produces a positive result.

6. a. Raise -2 to the third power, fifth power, seventh power, and ninth power.

$(-2)^3 = -8; (-2)^5 = -32; (-2)^7 = -128; (-2)^9 = -512$

b. What general rule involving signs is suggested by the preceding results?

A negative number raised to an odd power produces a negative result.

7. The following table contains the daily midnight temperatures (in °F) in Batavia, New York, for a week in January.

 It's Cold Outside

DAY	Sun.	Mon.	Tues.	Wed.	Thurs.	Fri.	Sat.
TEMPERATURE (°F)	-7	11	11	11	-7	-7	11

a. Determine the average daily temperature for the week.

$3(-7) + 4(11) = -21 + 44 = 23$

$\frac{23}{7} \approx 3.29°\text{F}$

The average daily temperature was about 3.3°F.

b. Did you use the most efficient way to do the calculation? Explain.

(Answers will vary.)

Yes, I grouped like values and multiplied. This is more efficient than adding.

8. a. What are the values of $3^2, 3^0,$ and 3^{-2}?

$3^2 = 3 \cdot 3 = 9; 3^0 = 1; 3^{-2} = \frac{1}{3} \cdot \frac{1}{3} = \frac{1}{9}$

b. What is the value of any nonzero number raised to the zero power?

Any number other than zero raised to the zero power equals 1.

c. What is the meaning of any nonzero number raised to a negative power? Give two examples using -2 as the exponent.

A number raised to a negative power is the reciprocal of the number raised to the absolute value of the power. For example, $3^{-2} = \frac{1}{3^2} = \frac{1}{9}$ and $7^{-2} = \frac{1}{7^2} = \frac{1}{49}$.

9. a. The diameter of raindrops in drizzle near sea level is reported to be approximately 30×10^{-2} millimeter. Write this number in standard decimal notation and in scientific notation.

$30 \times 10^{-2} = 0.3 = 3.0 \times 10^{-1}$ mm

b. What is the measurement of these raindrops in centimeters (10 mm $= 1$ cm)? Write your answer in standard notation and in scientific notation.

0.03 cm $= 3.0 \times 10^{-2}$ cm

c. From your experience with scientific notation, explain how to convert numbers from standard notation to scientific notation and vice versa.

(Answers will vary.)

To convert a number from standard notation to scientific notation, insert a decimal point to the right of the first nonzero digit to obtain a number between 1 and 10. Then determine how many places and in which direction you need to move the decimal point to obtain the original number. Scientific notation consists of the decimal form of the number between 1 and 10 multiplied by a power of 10. The exponent of the 10 is equal to the number of places you have to move the decimal point and is positive if the decimal point is moved to the right to obtain the original number but negative if it is moved to the left.

To convert a number from scientific notation to standard notation, move the decimal point to the right (if the exponent of 10 is positive) or left (if the exponent of 10 is negative) as many places as the value of the exponent.

Cluster 4   How Can I Practice?

Calculate by hand. Use your calculator to check your answers.

1. $15 + (-39)$
   -24

2. $-43 + (-28)$
   -71

3. $-0.52 + 0.84$
   0.32

4. $-7.8 + 2.9$
   -4.9

5. $-32 + (-45) + 68$
   $-77 + 68 = -9$

6. $-46 - 63$
   -109

7. $53 - (-64)$
   117

8. $8.9 - (-12.3)$
   21.2

9. $-75 - 47$
   -122

10. $-34 - (-19)$
   -15

11. $-4.9 - (-2.4) + (-5.6) + 3.2$
   -4.9

12. $16 - (-28) - 82 + (-57)$
   -95

13. $-1.7 + (-0.56) + 0.92 - (-2.8)$
   1.46

14. $\frac{2}{3} + \left(-\frac{3}{5}\right)$
   $\frac{1}{15}$

15. $-\frac{3}{7} + \left(\frac{4}{5}\right) + \left(-\frac{2}{7}\right)$
   $\frac{3}{35}$

16. $-\frac{5}{9} - \left(-\frac{8}{9}\right)$
   $\frac{1}{3}$

17. $2 - \left(\frac{3}{2}\right)$
   $\frac{1}{2}$

18. $1 - \left(\frac{3}{4}\right) + 1 - \frac{3}{4}$
   $\frac{1}{2}$

19. $0 - \left(-\frac{7}{10}\right) + \left(-\frac{1}{2}\right) - \frac{1}{5}$
   0

20. $\frac{-48}{12}$
   -4

21. $\frac{63}{-9}$
   -7

22. $\frac{121}{-11}$
   -11

23. $\frac{-84}{-21}$
   4

24. $-125 \div -25$
   5

25. $-2.4 \div 6$
   -0.4

26. $\frac{24}{-6}$
   -4

27. $-\frac{24}{6}$
   -4

28. $-0.8(12)$
   -9.6

29. $4(-0.06)$
   -0.24

30. $-9(-11)$
   99

31. $-4(-0.6)(-5)(-0.01)$
   0.12

32. $0.5(-7)(-2)(-3)$
   -21

33. $-4(-4)$
   16

34. $(-4)(-4)$
   16

35. $4(-4)$
   -16

Exercise numbers appearing in color are answered in the Selected Answers appendix.

36. You have $85.30 in your bank account. You write checks for $23.70 and $35.63. You then deposit $325.33. Later, you withdraw $130.00. What is your final balance?

$85.30 - 23.70 - 35.63 + 325.33 - 130 = 221.30$

My final balance is $221.30.

37. You lose 4 pounds, then gain 3 pounds back. Later you lose another 5 pounds. If your aim is to lose 12 pounds, how many more pounds do you need to lose? State your answer in words and as a signed number.

$-4 + 3 - 5 = -6$ I still need to lose 6 pounds.

38. You and your friend Patrick go on vacation. Patrick decides to go scuba diving, but you prefer to do some mountain climbing. You decide to separate for part of the day and do your individual activities. While you climb to an altitude of 2567 feet above sea level, Patrick is 49 feet below sea level. What is the vertical distance between you and Patrick?

$2567 - (-49) = 2616$ Patrick and I are 2616 feet apart.

39. In parts a–e, state your answer as a signed number and in words.

a. The temperature is 2°C in the morning but drops to -5°C by night. What is the change in temperature?

$-5°C - 2°C = -7°C$

The temperature drops 7°C.

b. The temperature is -12°C in the morning but is expected to drop 7°C during the day. What will be the evening temperature?

$-12°C - 7°C = -19°C$

The evening temperature is expected to be -19°C.

c. The temperature is -17°C this morning but is expected to rise 9°C by noon. What will be the noon temperature?

$-17°C + 9°C = -8°C$

The noon temperature will be -8°C.

d. The temperature is -8°C in the morning but drops to -17°C by night. What is the change in temperature?

$-17°C - (-8°C) = -9°C$

There was a 9°C drop in temperature.

e. The temperature is -14°C in the morning and -6°C at noon. What is the change in temperature?

$-6°C - (-14°C) = 8°C$

The temperature rises 8°C.

40. Evaluate the following expressions. Use your calculator to check your result.

a. $-6 \div 2 \cdot 3$

$-3 \cdot 3$
$= -9$

b. $(-3)^2 + (-7) \div 2$

$9 + (-7) \div 2$
$= 9 + (-3.5)$
$= 5.5$

c. $(-2)^3 + (-9) \div (-3)$

$\quad -8 + (-9) \div (-3)$
$\quad = -8 + 3$
$\quad = -5$

d. $(-14 - 4) \div 3 \cdot 2$

$\quad -18 \div 3 \cdot 2$
$\quad = -6 \cdot 2$
$\quad = -12$

e. $(-4)^2 - [-8 \div (2 + 6)]$

$\quad 16 - [-8 \div 8]$
$\quad = 16 - (-1)$
$\quad = 17$

f. $-75 \div (-5)^2 + (-8)$

$\quad -75 \div 25 + (-8)$
$\quad = -3 + (-8)$
$\quad = -11$

g. $(-3)^3 \div 9 - 6$

$\quad -27 \div 9 - 6$
$\quad = -3 - 6$
$\quad = -9$

41. a. On the average, fog droplets measure 20×10^{-3} millimeter in diameter at sea level. How many inches is this $(1 \text{ mm} = 0.03937 \text{ in.})$?

$$20 \times 10^{-3} \text{mm} \left(\frac{0.03937 \text{ in.}}{1 \text{ mm}} \right) = 7.874 \times 10^{-4} = 0.0007874 \text{ in.}$$

b. At sea level, the average diameter of raindrops is approximately 1 millimeter. How many inches is this?

$0.03937 = 3.937 \times 10^{-2} \text{ in.}$

c. How many times larger are raindrops than fog droplets?

$$\frac{3.937 \times 10^{-2}}{7.874 \times 10^{-4}} = 50$$

Raindrops are 50 times the size of fog droplets.

Skills Check 2

1. Which number is greater, 1.0001 or 1.001?   **1.001**

2. Evaluate: $32.09 \cdot 0.0006$ **0.019254**

3. Evaluate 27^0.   **1**

4. Evaluate $|-9|$.   **9**

5. Determine the length of the hypotenuse of a right triangle if each of the legs has a length of 6 meters. Use the formula $c = \sqrt{a^2 + a^2}$, where c represents the length of the hypotenuse and a represents the length of the leg.
   $c = \sqrt{6^2 + 6^2} = \sqrt{72} \approx 8.49$ m   The hypotenuse is approximately 8.5 meters.

6. Evaluate $|9|$.   **9**

7. Write 203,000,000 in scientific notation.   **2.03×10^8**

8. How many quarts are there in 8.3 liters?
   There are approximately 8.8 quarts in 8.3 liters.

9. Write 2.76×10^{-6} in standard form.   **0.00000276**

10. How many ounces are there in a box of crackers that weighs 283 grams?
    The box of crackers weighs approximately 9.96 ounces.

11. Determine the area of the front of a U.S. $1 bill.
    The area is $A = 2\frac{5}{8}$ in. $\times 6\frac{1}{8}$ in. $= \frac{21}{8} \times \frac{49}{8} = \frac{1029}{64} \approx 16.1$ sq. in.

12. Determine the circumference of a quarter. Use the formula $C = \pi d$, where C represents the circumference and d represents the diameter of the circle.
    I measured the diameter to be $d = \frac{15}{16}$ inches. Then $C = \pi\left(\frac{15}{16}\right) \approx 2.95$ inches. The circumference is approximately 2.95 inches.

13. Determine the area of a triangle with a base of 14 inches and a height of 18 inches. Use the formula $A = \frac{1}{2}bh$, where A represents area, b represents the base, and h represents the height.
    The area is $A = \frac{1}{2}(14)(18) = 126$ square inches.

14. Twenty-five percent of the students in your history class scored between 70 and 80 on the last exam. In a class of 32 students, how many students does this include?
    $0.25(32) = 8$ students scored between 70 and 80 on the last exam.

15. You have decided to accept word processing jobs to earn some extra money. Your first job is to type a 420-page thesis. If you can type 12 pages per hour, how long will it take you to finish the job?
    $420 \text{ pages} \left(\frac{1 \text{ hr.}}{12 \text{ pages}}\right) = 35$ hr.
    It will take me 35 hours to complete the job.

16. How many pieces of string 1.6 yards long can be cut from a length of string 9 yards long?

$\frac{9}{1.6} = 5.625$ You can cut 5 pieces that are 1.6 yards long.

17. Reduce $\frac{16}{36}$ to lowest terms.

$\frac{4 \cdot 4}{4 \cdot 9} = \frac{4}{9}$

18. Change $\frac{16}{7}$ to a mixed number.

$$7\overline{)16} \quad 2\frac{2}{7}$$
$$\underline{14}$$
$$2$$

19. Change $4\frac{6}{7}$ to an improper fraction.

$\frac{4 \cdot 7 + 6}{7} = \frac{34}{7}$

20. Add: $\frac{2}{9} + \frac{3}{5}$

$\frac{10}{45} + \frac{27}{45} = \frac{37}{45}$

21. Subtract: $4\frac{2}{5} - 2\frac{7}{10}$

$\frac{44}{10} - \frac{27}{10} = \frac{17}{10} = 1\frac{7}{10}$

22. Evaluate: $\frac{4}{21} \cdot \frac{14}{9}$

$\frac{8}{27}$

23. Calculate: $1\frac{3}{4} \div \frac{7}{12}$

$\frac{7}{4} \div \frac{7}{12} = \frac{7}{4} \cdot \frac{12}{7} = \frac{12}{4} = 3$

24. You mix $2\frac{1}{3}$ cups of flour, $\frac{3}{4}$ cup of sugar, $\frac{3}{4}$ cup of mashed bananas, $\frac{1}{2}$ cup of walnuts, and $\frac{1}{2}$ cup of milk to make banana bread. How many cups of mixture do you have?

$\frac{7}{3} + \frac{3}{4} + \frac{3}{4} + \frac{1}{2} + \frac{1}{2} = \frac{28}{12} + \frac{9}{12} + \frac{9}{12} + \frac{6}{12} + \frac{6}{12} = \frac{58}{12} = 4\frac{5}{6}$ cups

I have almost 5 cups of mixture.

25. Your stock starts the day at $30\frac{1}{2}$ points and goes down $\frac{7}{8}$ of a point during the day. What is its value at the end of the day?

$30\frac{1}{2} - \frac{7}{8} = \frac{244}{8} - \frac{7}{8} = \frac{237}{8} = 29\frac{5}{8}$

My stock is worth $29\frac{5}{8}$ points at the end of the day.

26. You just opened a container of orange juice that has 64 fluid ounces. You have three people in your household who drink one 8-ounce serving of orange juice per day. In how many days will you need a new container?

$64\ \text{oz.} \cdot \dfrac{1\ \text{day}}{24\ \text{oz.}} = 2.\overline{6}$ days

Buy more on the second day to have enough for the next day.

27. You want to serve quarter-pound hamburgers at your barbecue. There will be seven adults and three children at the party, and you estimate that each adult will eat two hamburgers and each child will eat one. How much hamburger meat should you buy?

$$(7 \cdot 2 + 3 \cdot 1) \cdot \frac{1}{4} = \frac{17}{4} = 4\frac{1}{4} \text{ lb. of hamburger}$$

I should buy at least $4\frac{1}{4}$ pounds of hamburger.

28. Your friend tells you that her height is 1.67 meters and her weight is 58 kilograms. Convert her height and weight to feet and pounds, respectively.

$$\frac{1.67 \text{ m}}{1} \cdot \frac{1 \text{ ft.}}{0.3048 \text{ m}} \approx 5.48 \text{ ft.}; \frac{58 \text{ kg}}{1} \cdot \frac{2.21 \text{ lb.}}{1 \text{ kg}} \approx 128.2 \text{ lb.}$$

My friend is about 5.5 feet (5 feet 6 inches) tall and weighs about 128 pounds.

29. Convert the following measurements to the indicated units:

   a. 10 miles = _____52,800_____ feet

   b. 3 quarts = _____96_____ ounces

   c. 5 pints = _____80_____ ounces

   d. 6 gallons = _____768_____ ounces

   e. 3 pounds = _____48_____ ounces

30. The following table lists the minimum distance of each listed planet from Earth, in millions of miles. Convert each distance into scientific notation.

Way Out There

PLANET	DISTANCE (IN MILLIONS OF MILES)	DISTANCE IN MILES (IN SCIENTIFIC NOTATION)
Mercury	50	5×10^7
Venus	25	2.5×10^7
Mars	35	3.5×10^7
Jupiter	368	3.68×10^8
Saturn	745	7.45×10^8
Uranus	1606	1.606×10^9
Neptune	2674	2.674×10^9

31. A recipe you obtained on the Internet indicates that you need to melt 650 grams of chocolate in $\frac{1}{5}$ liter of milk to prepare icing for a cake. How many pounds of chocolate and how many ounces of milk do you need?

$$\frac{650 \text{ g}}{1} \cdot \frac{0.0022 \text{ lb.}}{1 \text{ g}} = 1.43 \text{ lb. chocolate}$$

$$\frac{1 \text{ ℓ}}{5} \cdot \frac{33.92 \text{ oz.}}{1 \text{ ℓ}} \approx 6.78 \text{ oz. milk}$$

I will need about 1.5 pounds of chocolate and 7 ounces of milk.

32. You ask your friend if you may borrow her eraser. She replies that she has only a tiny piece of eraser, 2.5×10^{-5} kilometer long. Does she really have such a tiny piece that she cannot share it with you?

$$2.5 \times 10^{-5} \text{ km} \cdot \frac{1000 \text{ m}}{1 \text{ km}} \cdot \frac{100 \text{ cm}}{1 \text{ m}} \cdot \frac{1 \text{ in.}}{2.54 \text{ cm}} \approx 0.98 \text{ in.}$$

A 1-inch eraser is big enough to share.

33. $\left(3\frac{2}{3} + 4\frac{1}{2}\right) \div 2$

$= \left(\frac{11}{3} + \frac{9}{2}\right) \cdot \frac{1}{2} = \left(\frac{22}{6} + \frac{27}{6}\right) \cdot \frac{1}{2} = \frac{49}{6} \cdot \frac{1}{2} = \frac{49}{12} = 4\frac{1}{12}$

34. $18 \div 6 \cdot 3$

$= 3 \cdot 3 = 9$

35. $18 \div (6 \cdot 3)$

$= 18 \div 18 = 1$

36. $\left(3\frac{3}{4} - 2\frac{1}{3}\right)^2 + 7\frac{1}{2}$

$= \left(\frac{15}{4} - \frac{7}{3}\right)^2 + \frac{15}{2}$

$= \left(\frac{45}{12} - \frac{28}{12}\right)^2 + \frac{15}{2} = \left(\frac{17}{12}\right)^2 + \frac{15}{2}$

$= \frac{289}{144} + \frac{1080}{144} = \frac{1369}{144} = 9\frac{73}{144}$

37. $\left(5\frac{1}{2}\right)^2 - 8\frac{1}{3} \cdot 2$

$= \left(\frac{11}{2}\right)^2 - \frac{25}{3} \cdot 2 = \frac{121}{4} - \frac{50}{3}$

$= \frac{363}{12} - \frac{200}{12} = \frac{163}{12} = 13\frac{7}{12}$

38. $6^2 \div 3 \cdot 2 + 6 \div (-3 \cdot 2)^2$

$= 36 \div 3 \cdot 2 + 6 \div 36$

$= 12 \cdot 2 + 6 \div 36$

$= 24\frac{1}{6}$

39. $6^2 \div 3 \cdot (-2) + 6 \div 3 \cdot 2^2$

$= 36 \div 3 \cdot -2 + 6 \div 3 \cdot 4$

$= 12 \cdot -2 + 6 \div 3 \cdot 4$

$= -24 + 2 \cdot 4 = -24 + 8 = -16$

40. $-4 \cdot 9 + (-9)(-8)$

$= -36 + 72 = 36$

41. $-6 \div 3 - 3 + 5^0 - 14 \cdot (-2)$

$= -2 - 3 + 1 - (-28) = 24$

The bracketed numbers following each concept indicate the activity in which the concept is discussed.

CONCEPT/SKILL	DESCRIPTION	EXAMPLE
Problem solving [1.2]	Problem solving strategies include: • discussing the problem • organizing information • drawing a picture • recognizing patterns • doing a simpler problem	You drive for 2 hours at an average speed of 47 mph. How far do you travel? $d = rt$ $d = \dfrac{47 \text{ mi.}}{\text{hr.}} \cdot 2 \text{ hr.} = 94 \text{ mi.}$
The commutative property of addition [1.3]	Changing the order of the addends yields the same sum.	$9 + 8 = 17; \quad 8 + 9 = 17$
The commutative property of multiplication [1.3]	Changing the order of the factors yields the same product.	$13 \cdot 4 = 52; \quad 4 \cdot 13 = 52$
Subtraction is not commutative [1.3]	The result is different if the order is reversed, so subtraction is not commutative.	$13 - 7 = 6; \quad 7 - 13 = -6$
Division is not commutative [1.3]	The result is different if the order is reversed, so division is not commutative.	$8 \div 10 = 0.8;$ $10 \div 8 = 1.25$
Distributive property [1.3]	$c \cdot (a + b) = c \cdot a + c \cdot b$ for addition, or $c \cdot (a - b) = c \cdot a - c \cdot b$ for subtraction	$7 \cdot (10 + 2) = 7 \cdot 10 + 7 \cdot 2$
Order of operations [1.3] and [1.14]	Use the order of operations convention for evaluating arithmetic expressions containing parentheses, addition, subtraction, multiplication, division, and exponentiation. First, perform all operations contained within parentheses before performing any operations outside the parentheses. Perform all operations in the following order. 1. Evaluate all exponents as you read the expression from left to right. 2. Do all multiplication and division as you read the expression from left to right. 3. Do all addition and subtraction as you read the expression from left to right.	$2 \cdot (3 + 4^2 \cdot 5) - 9$ $= 2 \cdot (3 + 16 \cdot 5) - 9$ $= 2 \cdot (3 + 80) - 9$ $= 2 \cdot (83) - 9$ $= 166 - 9$ $= 157$ $-4^2 - 18 \div 3 \cdot (-2)$ $= -16 - 18 \div 3 \cdot (-2)$ $= -16 - 6 \cdot (-2)$ $= -16 - (-12)$ $= -16 + 12$ $= -4$
Exponential form [1.3]	A number written as 10^4 is in exponential form; 10 is the base and 4 is the exponent. The exponent indicates how many times the base appears as a factor.	$10^4 = \underbrace{10 \cdot 10 \cdot 10 \cdot 10}$ 10 is used as a factor 4 times

CONCEPT/SKILL	DESCRIPTION	EXAMPLE
A number raised to the zero power [1.3]	Any nonzero number raised to the zero power equals 1.	$2^0 = 1$; $50^0 = 1$; $200^0 = 1$; $(-2)^0 = 1$; $(5 \cdot 6)^0 = 1$
Scientific notation [1.3] and [1.14]	When a number is written as the product of a decimal number between 1 and 10 and a power of 10, it is expressed in scientific notation.	$0.0072 = 7.2 \times 10^{-3}$ $43,100 = 4.31 \times 10^4$
Formulas [1.3]	A formula shows the arithmetic relationship between two or more quantities; each quantity is represented by a letter or symbol.	$F = ma$, for $m = 120$ and $a = 25$ $F = 120 \cdot 25 = 3000$
Simple average [1.5]	Add all your scores, and divide the sum by the number of exams.	$\dfrac{71 + 66 + 82 + 86}{4} = 76.25$
Weighted average [1.5]	To compute a weighted average of several components: 1. Multiply each component by its respective weight. 2. Sum these weighted components to obtain the weighted average.	Weights of tests 1, 2, and 3: $\dfrac{1}{5}$ Weight of final exam: $\dfrac{2}{5}$ Weighted average: $\dfrac{1}{5} \cdot 71 + \dfrac{1}{5} \cdot 66 +$ $\dfrac{1}{5} \cdot 82 + \dfrac{2}{5} \cdot 86 = 78.2$
Sum of the weights [1.5]	The sum of the weights used to compute a weighted average will always be equal to 1.	$\dfrac{1}{5} + \dfrac{1}{5} + \dfrac{1}{5} + \dfrac{2}{5} = 1$
Ratio [1.6]	A ratio is a quotient that compares two similar numerical quantities, such as "part" to "total."	4 out of 5 is a ratio. It can be expressed as a fraction, $\frac{4}{5}$, a decimal, 0.8, or a percent, 80%.
Converting from decimal to percent format [1.6]	Move the decimal point two places to the right, and then attach the % symbol.	0.125 becomes 12.5%. 0.04 becomes 4%. 2.50 becomes 250%.
Converting from percent to decimal format [1.6]	Drop the percent symbol, and move the decimal point two places to the left.	35% becomes 0.35. 6% becomes 0.06. 200% becomes 2.00.
Applying a known ratio to a given piece of information [1.7]	Total · known ratio = unknown part Part ÷ known ratio = unknown total	Forty percent of the 350 children play an instrument; 40% is the known ratio, 350 is the total. $350 \cdot 0.40 = 140$ A total of 140 children play an instrument. Twenty-four children, constituting 30% of the marching band, play the saxophone; 30% is the known ratio, 24 is the part. $24 \div 0.30 = 80$ A total of 80 children are in the marching band.

CONCEPT/SKILL	DESCRIPTION	EXAMPLE
Relative change [1.8]	$$\text{relative change} = \frac{\text{actual change}}{\text{original value}}$$	A quantity changes from 25 to 35. The actual change is 10; the relative change is $\frac{10}{25}$, or 40%.
Growth factor [1.9]	When a quantity increases by a specified percent, the ratio of its new value to the original value is called the *growth factor* associated with the specified percent increase. The growth factor is formed by adding the specified percent increase to 100% and then changing this percent into its decimal form.	To determine the growth factor associated with a 25% increase, add 25% to 100% to obtain 125%. Change 125% into a decimal to obtain the growth factor 1.25.
Applying a growth factor to an original value [1.9]	original value · growth factor = new value	A $120 item increases by 25%. Its new value is $120 · 1.25 = $150.
Applying a growth factor to a new value [1.9]	new value ÷ growth factor = original value	An item has already increased by 25% and is now worth $200. Its original value was $200 ÷ 1.25 = $160.
Decay factor [1.9]	When a quantity decreases by a specified percent, the ratio of its new value to the original value is called the *decay factor* associated with the specified percent decrease. The decay factor is formed by subtracting the specified percent decrease from 100%, and then changing this percent into its decimal form.	To determine the decay factor associated with a 25% decrease, subtract 25% from 100%, to obtain 75%. Change 75% into a decimal to obtain the decay factor 0.75.
Applying a decay factor to an original value [1.9]	original value · decay factor = new value	A $120 item decreases by 25%. Its new value is $120 · 0.75 = $90.
Applying a decay factor to a new value [1.9]	new value ÷ decay factor = original value	An item has already decreased by 25% and is now worth $180. Its original value was $180 ÷ 0.75 = $240.
Applying a sequence of percent changes [1.10]	The cumulative effect of a sequence of percent changes is the product of the associated growth or decay factors.	The effect of a 25% increase followed by a 25% decrease is 1.25 · 0.75 = 0.9375. That is, the item is now worth only 93.75% of its original value. The item's value has decreased by 6.25%.

CONCEPT/SKILL	DESCRIPTION	EXAMPLE
Using unit (or dimensional) analysis to solve conversion problems [1.11]	1. Identify the unit of the result. 2. Set up the sequence of multiplications so the appropriate units divide out, leaving the unit of the result. 3. Multiply and divide the numbers as indicated to obtain the numerical part of the result. 4. Check that the appropriate units divide out, leaving the expected unit of the result.	To convert your height of 70 inches to centimeters: $70 \text{ in.} \cdot \frac{2.54 \text{ cm}}{1 \text{ in.}} = 177.8 \text{ cm}$
Integers [1.12]	Integers consist of the positive and negative counting numbers and zero.	$\ldots -3, -2, -1, 0, 1, 2, 3, \ldots$
Number line [1.12]	Use a number line scaled with integers to visualize relationships between positive and negative numbers.	
Order on the number line [1.12]	If $a < b$, then a is left of b on a number line. If $a > b$, then a is right of b on a number line.	$-3 < -2; \qquad 4 < 7; \qquad 8 > -2$
Change in value [1.12]	Calculate a change in value by subtracting the original value from the final value.	$10 - 8 = 2$
Absolute value [1.12]	The absolute value of a number, a, written $|a|$, is the distance of a from zero on a number line. It is always positive or zero.	$|-2| = 2$ $|2| = 2$
Opposites [1.12]	Different numbers that have the same absolute value are called *opposites*. The sum of two opposite numbers is zero.	$-9 + 9 = 0$
Adding signed numbers [1.12]	When adding two numbers with the same sign, add the absolute values of the numbers. The sign of the sum is the same as the sign of the original numbers. When adding two signed numbers with different signs, find their absolute values and then subtract the smaller from the larger. The sign of the sum is the sign of the number with the larger absolute value.	$11 + 6 = 17$ $-7 + (-8) = -15$ $-9 + 12 = 3$ $14 + (-18) = -4$
Subtracting signed numbers [1.12]	To subtract two signed numbers, change the operation of subtraction to addition, and change the number being subtracted to its opposite; then follow the rules for adding signed numbers.	$4 - (-9) = 4 + 9 = 13$ $-6 - 9 = -6 + (-9) = -15$ $-8 - (-2) = -8 + 2 = -6$
Multiplying or dividing two numbers with opposite signs [1.13]	When you multiply or divide two numbers with *opposite* signs, the result will always be negative.	$-9 \cdot 5 = -45$ $56 \div (-7) = -8$

CONCEPT/SKILL	DESCRIPTION	EXAMPLE
Multiplying or dividing two numbers with the same signs [1.13]	When you multiply or divide two numbers with the *same* signs, the result will always be positive.	$(-3) \cdot (-8) = 24$ $(-72) \div (-6) = 12$
Multiplying or dividing a string of numbers [1.13]	When you multiply a string of signed numbers, the product will be positive if the number of negative factors is even and will be negative if the number of negative factors is odd.	$(-5)(-4)(-3) = -60$ $(-5)(-4)(-3)(-2) = 120$
Powers of signed numbers [1.14]	To indicate that a negative number is raised to a power, the number *must* be enclosed in parentheses. An expression of the form $-a^2$ *always* signifies the negation (opposite) of a^2 and is *always* a negative quantity (for nonzero a).	$(-4)^2 = (-4)(-4) = 16$ $-4^2 = -(4)(4) = -16$
Whole numbers raised to negative exponents [1.14]	A negative exponent always signifies a reciprocal. Symbolically, this is written as $b^{-n} = \dfrac{1}{b^n}$, where $b \neq 0$.	$7^{-2} = \dfrac{1}{7^2} = \dfrac{1}{49}$

1. Write the number 2.0202 in words.

   two and two hundred two ten-thousandths

2. Write the number fourteen and three thousandths in standard notation.

   14.003

3. What is the place value of the 4 in the number 3.06704?

   hundred thousandths

4. Evaluate: $3.02 + 0.5 + 7 + 0.004$

   10.524

5. Subtract 9.04 from 21.2.

   12.16

6. Evaluate: $6.003 \cdot 0.05$

   0.30015

7. Divide 0.0063 by 0.9.

   0.007

8. Round 2.045 to the nearest hundredth.

   2.05

9. Change 4.5 to a percent.

   450%

10. Change 0.3% to a decimal.

    0.003

11. Change 7.3 to a fraction.

    $7\frac{3}{10}$ or $\frac{73}{10}$

12. Change $\frac{3}{5}$ to a percent.

    60%

13. Write the following numbers in order from largest to smallest: $1.001, 1.1, 1.01, 1\frac{1}{8}$.

    $1\frac{1}{8}, 1.1, 1.01, 1.001$

14. Evaluate: 3^3

    27

15. Evaluate: 6^0

    1

16. Evaluate: 4^{-2}

    $\frac{1}{4^2} = \frac{1}{16}$

17. Evaluate: $(-4)^3$

    -64

18. Evaluate: $|16|$

    16

19. What number has the same absolute value as 25?

    -25

20. Evaluate: $|-12|$

    12

21. Write 0.0000543 in scientific notation.

    5.43×10^{-5}

22. Write 3.7×10^4 in standard notation.

    37,000

23. Determine the average of your exam scores: 25 out of 30, 85 out of 100, and 60 out of 70. Assume that each exam counts equally.

    $\frac{1}{3}\left(\frac{25}{30} + \frac{85}{100} + \frac{60}{70}\right) = 0.8468 \approx 85\%$

24. Change $\frac{9}{2}$ to a mixed number.

    $4\frac{1}{2}$

25. Change $5\frac{3}{4}$ to an improper fraction.

    $\frac{23}{4}$

26. Reduce $\frac{15}{25}$ to lowest terms.

    $\frac{3}{5}$

27. Evaluate: $\frac{1}{6} + \frac{5}{8}$

    $\frac{4}{24} + \frac{15}{24} = \frac{19}{24}$

Answers to all Gateway exercises are included in the Selected Answers appendix.

28. Calculate: $5\frac{1}{4} - 3\frac{3}{4}$

$\frac{21}{4} - \frac{15}{4} = \frac{6}{4} = \frac{3}{2} = 1\frac{1}{2}$

29. Evaluate: $\frac{2}{9} \cdot 3\frac{3}{8}$

$\frac{\cancel{2}}{\cancel{9}} \cdot \frac{\overset{3}{\cancel{27}}}{\underset{4}{\cancel{8}}} = \frac{3}{4}$

30. Divide: $\dfrac{6}{11}$ by $\dfrac{8}{22}$

$\frac{\overset{3}{\cancel{6}}}{\cancel{11}} \cdot \frac{\overset{2}{\cancel{22}}}{\underset{4}{\cancel{8}}} = \frac{6}{4} = \frac{3}{2} = 1\frac{1}{2}$

31. Evaluate: $2\frac{5}{8} \cdot 2\frac{2}{7}$

$\frac{\overset{3}{\cancel{21}}}{\cancel{8}} \cdot \frac{\overset{2}{\cancel{16}}}{\cancel{7}} = 6$

32. What is 20% of 80?

$0.2(80) = 16$

20% of 80 is 16.

33. On one exam you answered 12 out of 20 questions correctly. On a second, you answered 16 out of 25 correctly. On which exam did you do better?

12 out of 20 represents 60% correct. 16 out of 25 represents 64% correct.

I did better on the second exam.

34. This year, the number of customer service complaints to the cable company increased by 40% over last year's number. If there were 3500 complaints this year, how many were made last year?

$3500 \div 1.40 = 2500$

There were 2500 complaints last year.

35. Twenty percent of the students in your math class scored between 80 and 90 on the last test. In a class of 30, how many students does this include?

$(0.20)30 = 6$ students

36. Determine the circumference, C, of a circle whose radius, r, is 3 inches. Use the formula $C = 2\pi r$.

$C = 2\pi r \approx 2(3.14)(3) \approx 18.84$ in.

The circumference is about 19 inches.

37. Determine the perimeter, P, of the following figure.

5 in.

2 in.

2 in.

5 in.

$P = 2 + 5 + 2 + 5 = 14$ in.

The perimeter is 14 inches.

38. Determine the area, A, of a triangle with a base, b, of 5 feet and a height, h, of 7 feet. Use the formula $A = \frac{1}{2}bh$.

The area of the triangle is $A = \frac{1}{2}bh = \frac{1}{2}(5)(7) = \frac{35}{2} = 17.5$ square feet.

39. Determine the area, A, of a circle whose radius, r, is 1 inch. Round your answer to the nearest tenth. Use $A = \pi r^2$.

The area of the circle is $A = \pi r^2 \approx (3.14)(1^2) \approx 3.1$ square inches.

40. Determine the volume, V, of a road cone with a radius, r, of 10 centimeters and a height, h, of 20 centimeters. Round your answer to the nearest whole number. Use $V = \dfrac{\pi r^2 h}{3}$.

$V = \dfrac{\pi r^2 h}{3} = \dfrac{\pi 10^2 (20)}{3} \approx \dfrac{3.14(100)(20)}{3} = 2093.\overline{3} \approx 2093$ cm^3

The volume of the cone is approximately 2093 cubic centimeters.

41. In one summer, the number of ice-cream stands in your town decreased from 10 to 7. What is the percent decrease in the number of stands?

$\dfrac{3}{10} = 0.30$ decrease

The number of stands decreased by 30%.

42. A campus survey reveals that 30% of the students are in favor of the proposal for no smoking in the cafeteria. If 540 students responded to the survey, how many students are in favor of the proposal?

$0.30(540) = 162$ students are in favor of the proposal to ban smoking in the cafeteria.

43. You read 15 pages of your psychology text in 90 minutes. At this rate, how many pages can you read in 4 hours?

$\dfrac{15 \text{ pages}}{1.5 \text{ hr.}} \cdot 4 \text{ hr.} = 40 \text{ pages}$

I should be able to read 40 pages in 4 hours.

44. The scale at a grocery store registered 0.62 kilogram. Convert this weight to pounds (1 kg = 2.2 lb.).

$\dfrac{0.62 \text{ kg}}{1} \cdot \dfrac{2.2 \text{ lb.}}{1 \text{ kg}} = 1.364 \text{ lb.}$

45. Your turtle walks at the rate of 2 centimeters per minute. What is the turtle's rate in inches per second (1 in. = 2.54 cm)?

$\dfrac{2 \text{ cm}}{1 \text{ min.}} \cdot \dfrac{1 \text{ in.}}{2.54 \text{ cm}} \cdot \dfrac{1 \text{ min}}{60 \text{ sec}} \approx 0.013 \dfrac{\text{in.}}{\text{sec.}}$

46. Use the formula $C = \frac{5}{9}(F - 32)$ to determine the Celsius temperature, C, when the Fahrenheit temperature, F, is 50°.

$C = \dfrac{5}{9}(50 - 32) = \dfrac{5}{\cancel{9}} \cdot \cancel{18}^{2} = 10°C$

47. $-5 + 4 =$

-1

48. $2(-7) =$

-14

49. $-3 - 4 - 6 =$

-13

50. $-12 \div (-4) =$

3

51. $-5^2 =$

-25

52. $(-1)(-1)(-1) =$

-1

53. $-5 - (-7) =$

2

54. $-3 + 2 - (-3) - 4 - 9 =$

-11

55. $3 - 10 + 7 =$

0

56. $\left(-\dfrac{1}{6}\right)\left(-\dfrac{3}{5}\right) =$

$\dfrac{1}{10}$

57. $7 \cdot 3 - \dfrac{4}{2} =$

$21 - 2 = 19$

58. $4(6 + 7 \cdot 2) =$

$4(6 + 14)$

$= 4 \cdot 20$

$= 80$

59. $3 \cdot 4^3 - \dfrac{6}{2} \cdot 3 =$

$3 \cdot 64 - 3 \cdot 3$

$= 192 - 9 = 183$

60. $5(7 - 3) =$

$5 \cdot 4 = 20$

61. A deep-sea diver dives from the surface to 133 feet below the surface. If the diver swims down another 27 feet, determine his depth.

$-133 - 27 = -160$ ft.

The diver reaches a depth 160 feet below surface.

62. You lose $200 on each of 3 consecutive days in the stock market. Represent your total loss as a product of signed numbers, and write the result.

$(-200)(3) = -\$600$, which represents a loss of $600 in stocks.

63. You have $85 in your checking account. You write a check for $53, make a deposit of $25, and then write another check for $120. How much is left in your account?

$85 - 53 + 25 - 120 = -63$

I am overdrawn and am $63 in debt.

64. The mass of the hydrogen atom is about 0.00000000000000000000002 gram. If 1 gram is equal to 0.0022 pound, determine the mass of a hydrogen atom in pounds. Express the result in scientific notation.

$$\frac{0.00000000000000000000002 \text{ g}}{1} \left(\frac{0.0022 \text{ lb}}{1 \text{ g}} \right) \approx 4.4 \times 10^{-27} \text{ lb.}$$

The mass of a hydrogen atom is 4.4×10^{-27} pound.

Variable Sense

Arithmetic is the branch of mathematics that deals with counting, measuring, and calculating. Algebra is the branch that deals with variables and the relationships between and among variables. Variables and their relationships, expressed in table, graph, verbal, and symbolic forms, will be the central focus of this chapter.

Cluster 1 — Interpreting and Constructing Tables and Graphs

Activity 2.1

Blood-Alcohol Levels

Objectives

1. Identify input and output in situations involving two variable quantities.

2. Determine the replacement values for a variable within a given situation.

3. Use a table to numerically represent a relationship between two variables.

4. Represent a relationship between two variables graphically.

5. Identify trends in data pairs that are represented numerically and graphically.

Suppose you are asked to give a physical description of yourself. What categories might you include? You probably would include gender, ethnicity, and age. Can you think of any other categories? Each of these categories represents a distinct variable.

> **Definition**
>
> A **variable**, usually represented by a letter, is a quantity or quality that may change, or vary, in value from one particular instance to another.

The particular responses are the **values** of each variable. The first category, or variable, gender, has just two possible values: male or female. The second variable, ethnicity, has several possible values: Caucasian, African American, Hispanic, Asian, Native American, and so on. The third variable, age, has a large range of possible values: from 17 years up to possibly 110 years. Note that the values of the variable age are numerical but the values of gender and ethnicity are not. The variables you will deal with in this text are numerical (quantitative rather than qualitative).

Blood-Alcohol Concentration—Numerical Representation

In 2000, the U.S. Congress moved to reduce highway funding to states that did not implement a national standard of 0.08% blood-alcohol concentration as the minimum legal limit for drunk driving. By 2004, every state plus the District of Columbia had adopted the 0.08% legal limit. To understand the relationship between the amount of alcohol consumed and the concentration of alcohol in the blood, variables are quite useful.

The number of 12-ounce beers someone consumes in an hour is a variable (that is, the number consumed varies from person to person). Let this variable be represented by the letter n.

1. Can n be reasonably replaced by a negative value, such as −2? Explain.

   No, since −2 beers makes no sense.

2. Can n be reasonably replaced by zero? Explain.

   Yes, since having no beers is perfectly possible.

3. What is the possible collection of replacement values for *n* in this situation?

Replacement values can be any rational number greater than or equal to zero, but the largest value would depend on the individual.

The concentration of alcohol in the blood (expressed as a percent) also varies. Use the letter *B* to represent this variable.

4. Can *B* be reasonably replaced by zero? Explain.

Yes, since having no alcohol in the blood is the normal state.

5. Can *B* be reasonably replaced by 100? Explain.

No, since 100% would mean the blood is all alcohol, quite impossible unless someone is dead.

6. What is the possible collection of reasonable replacement values for *B* in this situation?

Depending on the maximum possible amount of alcohol in the blood before death, any decimal number between zero and that maximum.

One way the relationship between these two variables can be represented by using a **table** of paired data values. Typically, one variable is designated as the input and the other is called the output. The **input** is the value that is considered first. The **output** is the number that is associated with or results from the input value. A table can be arranged either horizontally (as in the following table, with inputs in the first row) or vertically (with inputs in the first column).

The following table presents a **numerical** description of the relationship between the two variables introduced in Problems 1–6.

Number of Beers in an Hour, *n*	1	2	3	4	5	6	7	8	9	10
Blood-Alcohol Concentration (%),* *B*	0.018	0.035	0.053	0.070	0.087	0.104	0.121	0.138	0.155	0.171

*Based on body weight of 200 pounds

7. Why is it reasonable in this situation to designate the number of beers consumed in an hour as the input and the blood-alcohol concentration as the output?

It is reasonable because the blood-alcohol concentration depends on the number of beers consumed in an hour.

8. What is the blood-alcohol concentration for a 200-pound person who has consumed four beers in 1 hour? Nine beers in 1 hour?

The blood-alcohol concentration is 0.070% for 4 beers and 0.155% for 9 beers.

9. What is the blood-alcohol concentration when eight beers are consumed in an hour?

The blood-alcohol level is 0.138% when 8 beers are consumed in an hour.

Notice that a table can reveal *numerical* patterns and relationships between the input and output variables. You can also describe these patterns and relationships *verbally* using such terms as *increases*, *decreases*, or *remains the same*.

10. According to the preceding table, as the number of beers consumed in 1 hour increases, what happens to the blood-alcohol concentration?

The blood-alcohol level also increases.

11. From the preceding table, determine the number of beers a 200-pound person can consume in 1 hour without exceeding the recommended legal measure of drunk driving.

A 200-pound person can consume at most 4 beers in 1 hour for a blood-alcohol level of 0.070%, less than the recommended 0.08% cutoff.

Graphical Representation

A visual display (**graph**) of data on a **rectangular coordinate system** is often helpful in detecting trends or other information not apparent in a table.

> On a graph, the input is referenced on the **horizontal axis**, and the output is referenced on the **vertical axis**.

The following graph provides a visual representation of the beer consumption and blood-alcohol level data from the preceding table, based on a body weight of 200 pounds. Note that each of the 10 input/output data pairs in that table corresponds to a plotted point on the graph. For example, drinking seven beers in an hour is associated with a blood-alcohol concentration of 0.121%. If you read across to 7 along the input (horizontal) axis and move up to 0.121 on the output (vertical) axis, you locate the point that represents the **ordered pair** of numbers (7, 0.121). Similarly, you would label the other points on the graph by ordered pairs of the form (n, B), where n is the input value and B is the output value. Such ordered pairs are called the **coordinates** of the point.

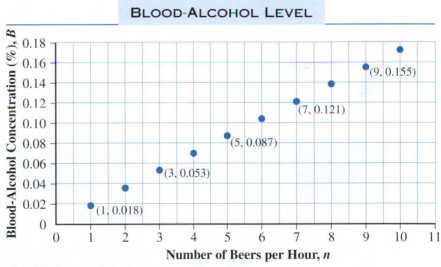

BLOOD-ALCOHOL LEVEL

Based on body weight of 200 pounds

12. a. From the graph, approximate the blood-alcohol concentration of a 200-pound person who consumes two beers in an hour.

A 200-pound person who consumes 2 beers in an hour would have a blood-alcohol concentration of approximately 0.035%.

Write this input/output correspondence using ordered-pair notation. (2, 0.035)

b. Estimate the number of beers that a 200-pound person must consume in 1 hour to have a blood-alcohol concentration of 0.104.

A 200-pound person must consume approximately 6 beers in 1 hour to have a blood-alcohol concentration of 0.104%.

Write this input/output correspondence using ordered-pair notation. (6, 0.104)

    c. When $n = 4$, what is the approximate value of B from the graph?

      $B \approx 0.07$

    Write this input/output correspondence using ordered-pair notation.

    $(4, 0.07)$

    d. As the number of beers consumed in an hour increases, what happens to the blood-alcohol concentration?

    Blood-alcohol concentration also increases.

13. What are some advantages and disadvantages of using a graph when you are trying to describe the relationship between the number of beers consumed in an hour and blood-alcohol concentration?

(Answers will vary.)

Advantage: You can immediately see that the concentration rises as n increases.

Disadvantage: Individual data values are more difficult to determine from the graph.

14. What advantages and disadvantages do you see in using a table?

(Answers will vary.)

Advantage: Data values can be reported and read more precisely.

Disadvantage: General trends are usually not immediately obvious.

Effect of Weight on Blood-Alcohol Concentration

You know that weight is a determining factor in a person's blood-alcohol concentration. The following table presents a numerical representation of the relationship between the number of beers consumed in an hour by a 130-pound person (input) and her corresponding blood-alcohol concentration (output). Other factors may influence blood-alcohol concentration, such as the rate at which the individual's body processes alcohol, the amount of food eaten prior to drinking, and the concentration of alcohol in the drink, but these effects are not considered here.

Number of Beers, n	1	2	3	4	5	6	7	8	9	10
Blood-Alcohol Concentration (%), B	0.027	0.054	0.081	0.108	0.134	0.160	0.187	0.212	0.238	0.264

*Based on body weight of 130 pounds

15. According to the preceding table, how many beers can a 130-pound person consume in an hour without exceeding the recommended legal limit for driving while intoxicated?

A 130-pound person can consume at most 2 beers in an hour and stay within DWI guidelines.

16. Use the data in the previous two tables to explain how body weight affects the blood-alcohol concentration for a given number of beers consumed in an hour.

The blood-alcohol concentration is higher in a person with lower body weight who consumes the same number of beers as a heavier person.

SUMMARY: ACTIVITY 2.1

1. A **variable**, usually represented by a letter, is a quantity or quality that may change, or vary, in value from one particular instance to another.

2. The set of **replacement values** for a variable is the collection of numbers that make sense within the given context.

3. The **input** is the value that is given first in an input/output relationship.

4. The **output** is the second number in an input/output situation. It is the number that is associated with or results from the input value.

5. An input/output relationship can be represented **numerically** by a table of paired data values.

6. An input/output relationship can be represented **graphically** as plotted points on a rectangular coordinate system. The **ordered pair** (input, output) are called the **coordinates** of the point.

7. The input variable is referenced on the **horizontal axis**.

8. The output variable is referenced on the **vertical axis**.

EXERCISES: ACTIVITY 2.1

1. Medicare is a government program that helps senior citizens pay for medical expenses. As the U.S. population ages and greater numbers of senior citizens join the Medicare rolls each year, the expense and quality of health service becomes an increasing concern. The following graph presents Medicare expenditures from 1967 through 2007. Use the graph to answer the following questions.

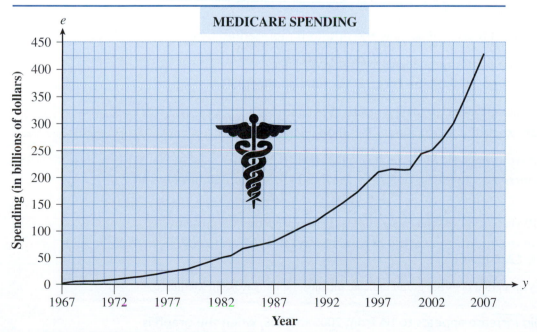

Source: Centers for Medicare and Medicaid Services

Exercise numbers appearing in color are answered in the Selected Answers appendix.

a. Identify the input variable and the output variable in this situation.

The input variable is *year*.

The output variable is *Medicare expenditures*, in billions of dollars.

b. Use the graph to estimate the Medicare expenditures for the years in the following table. (Answers may vary by ±5.)

The Cost of Care

YEAR, y	MEDICARE EXPENDITURES, e (IN BILLIONS OF DOLLARS)
1967	3
1972	9
1977	23
1984	70
1989	100
1994	160
1999	210
2001	240
2003	273
2005	340
2007	430

c. What letters in the table in part b are used to represent the input variable and the output variable?

The input variable is represented by *y*; the output is represented by *e*.

d. Estimate the year in which the expenditures reached $100 billion.

Expenditures reached $100 billion for the first time during 1989.

e. Estimate the year in which the expenditures reached $25 billion.

It appears that expenditures reached $25 billion in 1977.

f. During which 10-year period did Medicare expenditures change the least?

Expenditures increased the least between 1967 and 1977, approximately $20 billion.

g. During which 10-year period was the change in Medicare expenditures the greatest?

The increase in expenditures was the greatest between 1997 and 2007, approximately $220 billion.

h. For which one-year period does the graph indicate the most rapid increase in Medicare expenditures?

The most rapid increase appears to be from 2005 to 2006, when the graph is the steepest.

i. In what period was there almost no change in Medicare expenditures?

There was almost no change in expenditures from 1997 to 2000.

j. From the graph, predict the Medicare expenditures in 2010.

Continuing at the same rate as from 2006 to 2007, the expenditures will rise to approximately $520 billion in 2010.

k. What assumptions are you making about the change in Medicare expenditures from 2007 to 2010?

The assumption in part j is that Medicare expenditures will continue to grow at the same rate as they grew between 2006 to 2007, approximately $30 billion per year.

2. When the input variable is measured in units of time, the relationship between input (time) and output indicates how the output *changes* over time. For example, the U.S. oil-refining industry has undergone many changes since the Middle East oil embargo of the 1970s. Among the factors that have had a major impact on the industry are changing crude oil prices, changing demand, increased imports, economics, and quality control. Since 1981, 175 U.S. oil refineries have been dismantled. Most were small and inefficient. The following graph shows the fluctuations in the number of U.S. refineries from 1974 to 2007.

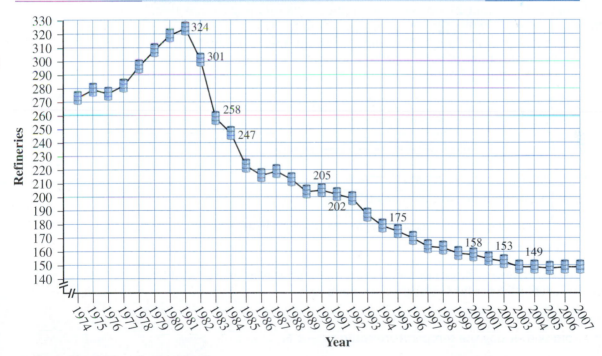

OPERATING U.S. REFINERIES: 1974–2007

Source: Energy Information Administration

a. Identify the input variable.   The input variable is year.

b. Which axis represents the input variable?

Values of the input variable are marked on the horizontal axis.

c. Identify the output variable.

The output variable is the number of operational U.S. oil refineries.

d. Which axis represents the output variable?

Values of the output variable are marked on the vertical axis.

e. Complete the following table, listing the number of refineries (output) in operation each given year (input).

(Answers may vary slightly.)

YEAR	NUMBER OF OPERATING REFINERIES	YEAR	NUMBER OF OPERATING REFINERIES
1974	273	1985	223
1976	276	1990	205
1978	296	1995	175
1980	319	2002	153

f. Summarize the changes that the U.S. oil-refining industry has undergone since the 1970s. Include specific information. For example, indicate the intervals during which the number of U.S. oil refineries increased, decreased, and remained about the same. Also, indicate the years in which the number of refineries increased and decreased the most.

There was a steady and moderately steep increase of oil refineries from 1974 to 1981: 273 refineries to 324 refineries, an increase of about 51 refineries. A very steep decline occurred from 1981 to 1986, when the number of refineries dropped from 324 to about 216, a decrease of about 108 refineries. From 1986 to 1988, the number of refineries remained approximately the same, around 216, but still about 60 below the 1974 number. Through the 1990s, and until 2002, the number of refineries continued to decline. Between 2002 and 2007, the number of operating refineries was about the same.

3. Suppose the variable n represents the number of DVDs a student owns.

a. Can n be reasonably replaced by a negative value, such as -2? Explain.

No. Owning a negative number of DVDs doesn't make sense.

b. Can n be reasonably replaced by the number 0? Explain.

Yes. It means the student doesn't own any DVDs.

c. What is a possible collection of replacement values for the variable n in this situation?

(Answers will vary.) 0, 1, 2, ... , 1000.

4. Suppose the variable t represents the average daily Fahrenheit temperature in Oswego, a city in upstate New York, during the month of February in any given year.

a. Can t be reasonably replaced with a temperature of $-3°F$? Explain.

Yes, on a cold winter day the temperature could be $-3°F$.

b. Can t be reasonably replaced with a temperature of $-70°F$? Explain.

No. Records show that daily temperatures are not this low, even in Oswego in winter.

c. Can t be reasonably replaced with a temperature of $98°F$? Explain.

$98°F$ would not be a reasonable temperature in winter.

d. What is a possible collection of replacement values for the variable t that would make this situation realistic?

(Answers will vary.) Temperatures from $-5°$ to $45°F$ would be reasonable.

Activity 2.2

Earth's Temperature

Objectives

1. Construct a graph of data pairs using an appropriately scaled and labeled rectangular coordinate system.

2. Determine the coordinates of a point on a graph.

3. Identify points that lie in a given quadrant or on a given axis.

Living on Earth's surface, you experience a relatively narrow range of temperatures. You may know what $-20°F$ (or $-28.9°C$) feels like on a bitterly cold winter day. Or you may have sweated through $100°F$ (or $37.8°C$) during summer heat waves. If you were to travel below Earth's surface and above Earth's atmosphere, you would discover a wider range of temperatures. The following graph displays a relationship between the altitude (input) and temperature (output). Note that the altitude is measured from Earth's surface. That is, Earth's surface is at altitude 0.

TEMPERATURE VERSUS ALTITUDE

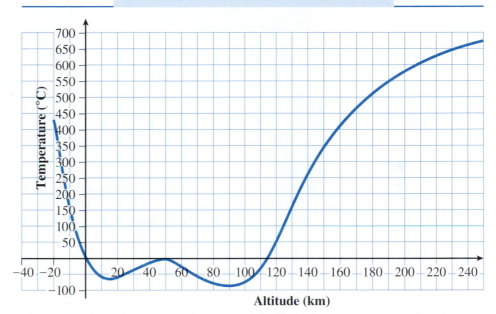

This graph uses a **rectangular coordinate system**, also known as a Cartesian coordinate system. It contains both a horizontal axis and a vertical axis. Recall that the input variable is referenced on the horizontal axis. The output variable is referenced on the vertical axis.

1. a. With what variable and units of measure is the horizontal axis labeled?

The horizontal axis is labeled with values of the altitude, in kilometers.

b. What is the practical significance of the positive values of this quantity?

A positive altitude represents a height above Earth's surface.

c. What is the practical significance of the negative values of this quantity?

A negative altitude represents a depth below Earth's surface.

d. How many kilometers are represented between the tick marks on the horizontal axis?

Adjacent tick marks are 10 kilometers apart.

2. a. With what variable and units of measure is the vertical axis labeled?

The vertical axis is labeled with values of the temperature, in degrees Celsius.

b. What is the practical significance of the positive values of this quantity?

A positive value represents a temperature above the freezing point of water (melting point of ice).

c. What is the practical significance of the negative values of this quantity?

A negative value represents a temperature below the freezing point of water (melting point of ice).

d. How many degrees Celsius are represented between the tick marks on the vertical axis?

Adjacent tick marks on the vertical axis are 50°C apart.

3. Consult the graph to determine at which elevations or depths the temperature is high enough to cause water to boil. (Water boils at 100°C.)

The graph indicates that at depths below −5 kilometers and at elevations above 125 kilometers, the temperature is high enough to cause water to boil.

Observations

- The distance represented by space between adjacent tick marks on an axis is determined by the range of particular replacement values for the variable represented on that axis. The process of determining and labeling an appropriate distance between tick marks is called **scaling**.

- The axes are often labeled and scaled differently. The way axes are labeled and scaled depends on the context of the problem. Note that the tick marks to the left of zero on the horizontal axis are negative; similarly, the tick marks below zero on the vertical axis are negative.

- On each axis, equal distance between adjacent pairs of tick marks must be maintained. For example, if the distance between the first two tick marks is 10 units, then all adjacent tick marks on that axis must also differ by 10 units.

Rectangular Coordinate System

In the rectangular (Cartesian) coordinate system, the horizontal axis (commonly called the *x*-axis) and the vertical axis (commonly called the *y*-axis) are number lines that intersect at their respective zero values at a point called the **origin**. The two perpendicular coordinate axes divide the plane into four **quadrants**. The quadrants are labeled counterclockwise, using Roman numerals, with quadrant I being the upper-right quadrant.

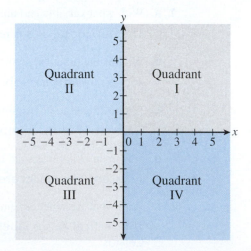

Each point in the plane is identified by an **ordered pair** of numbers (x, y) that can be thought of as the point's "address" relative to the origin. The ordered pair representing the origin is $(0, 0)$. The first number, x, of an ordered pair (x, y) is called the **horizontal coordinate** because it represents the point's horizontal distance (to the right if x is positive, to the left if x is negative) from the *y*-axis. Similarly, the second number, y, is called the **vertical coordinate** because it represents the point's vertical distance (up if y is positive, down if y is negative) from the *x*-axis.

4. The following graph displays eight points selected from the temperature graph on page 141.

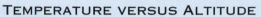

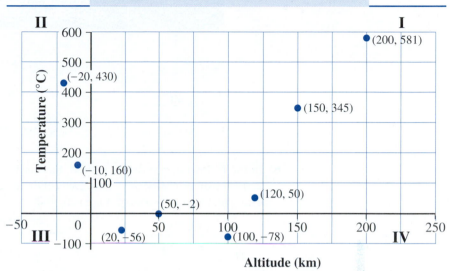

TEMPERATURE VERSUS ALTITUDE

a. What is the practical meaning of the point with coordinates (150, 345)?

At an elevation 150 kilometers above Earth's surface, the average temperature is 345°C.

b. What is the practical meaning of the point with coordinates (100, −78)?

At an elevation 100 kilometers above Earth's surface, the average temperature is −78°C.

c. What is the practical meaning of the point with coordinates (−20, 430)?

At a depth 20 kilometers below Earth's surface, the average temperature is 430°C.

d. In which quadrant are the points (120, 50), (150, 345), and (200, 581) located?

The points are located in quadrant I.

e. In which quadrant are the points (−10, 160) and (−20, 430) located?

The points are located in quadrant II.

f. In which quadrant are the points (20, −56) and (100, −78) located?

The points are located in quadrant IV.

g. Are there any points located in quadrant III? What is the significance of your answer?

There are no points located in quadrant III because temperatures are positive below Earth's surface.

5. Consider ordered pairs of the form (x, y). Determine the sign (positive or negative) of the x- and y-coordinates of a point in each quadrant. For example, any point located in quadrant I has a positive x-coordinate and a positive y-coordinate.

QUADRANT	SIGN (+ OR −) OF x-COORDINATE	SIGN (+ OR −) OF y-COORDINATE
I	+	+
II	−	+
III	−	−
IV	+	−

6. In the following coordinate system, the horizontal axis is labeled the x-axis, and the vertical axis is labeled the y-axis. Determine the coordinates (x, y) of points A to N. For example, the coordinates of A are $(30, 300)$.

A $(30, 300)$ B $(90, 200)$ C $(70, 100)$

D $(40, 150)$ E $(50, 0)$ F $(80, -100)$

G $(20, -75)$ H $(0, -50)$ I $(-60, -200)$

J $(-90, -175)$ K $(-70, 0)$ L $(-90, 100)$

M $(-50, 175)$ N $(-80, 225)$

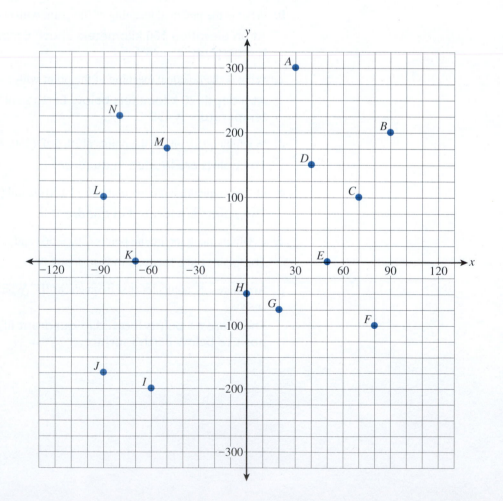

7. Which points (*A* to *N*) from Problem 6 are located

    a. in the first quadrant?   *A, B, C, D* are in the first quadrant.

    b. in the second quadrant?   *L, M, N* are in the second quadrant.

    c. in the third quadrant?   *I, J* are in the third quadrant.

    d. in the fourth quadrant?   *F, G* are in the fourth quadrant.

Points that lie on either the horizontal or vertical axis are not in any of the four quadrants. They lie on a boundary between quadrants.

    8. a. Which points (*A* to *N*) from Problem 6 are on the horizontal axis? What are their coordinates?

        E (50, 0) and *K*(− 70, 0) are on the horizontal axis.

    b. What is the *y*-value of any point located on the *x*-axis?

        Any point located on the *x*-axis has a *y*-value of zero.

    c. Which points are on the vertical axis? What are their coordinates?

        H (0, − 50) is the only point on the vertical axis.

    d. What is the *x*-value of any point located on the *y*-axis?

        The *x*-value of any point located on the *y*-axis is zero.

Additional Relationships Represented Graphically

    9. You work for the National Weather Service and are asked to study the average daily temperatures in Fairbanks, Alaska. You calculate the mean of the average daily temperatures for each month. You decide to place the information on a graph in which the date is the input and the temperature is the output. You also decide that January 2000 will correspond to the month zero. Determine the quadrant in which you would plot the points that correspond to the following data.

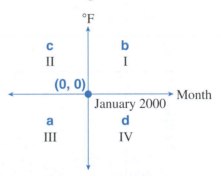

    a. The average daily temperature for January 1990 was − 13°F.

        The point representing the ordered pair (January 1990, − 13°F) would be located in quadrant III.

    b. The average daily temperature for June 2005 was 62°F.

        The point representing (June 2005, 62°F) would be located in quadrant I.

    c. The average daily temperature for July 1995 was 63°F.

        The point representing (July 1995, 63°F) would be located in quadrant II.

    d. The average daily temperature for January 2009 was − 13°F.

        The point representing (January 2009, − 13°F) would be located in quadrant IV.

10. Measurements in wells and mines have shown that the temperatures within Earth generally increase with depth. The following table shows average temperatures for several depths below sea level.

A Hot Topic

D, Depth (km) Below Sea Level	0	25	50	75	100	150	200
T, Temperature (°C)	20	600	1000	1250	1400	1700	1800

a. Choose an appropriate scale to represent the data from the table graphically on the grid following part d. The data points should be spread out across most of the grid, both horizontally and vertically. Place depth (input) along the horizontal axis and temperature (output) along the vertical axis.   (Answers will vary.)

b. How many units does each tick mark on the horizontal axis represent?
(Answers will vary, based on graph.)

c. How many units does each tick mark on the vertical axis represent?
(Answers will vary, based on graph.)

d. Explain your reasons for selecting the particular scales that you used.
(Answers will vary.)

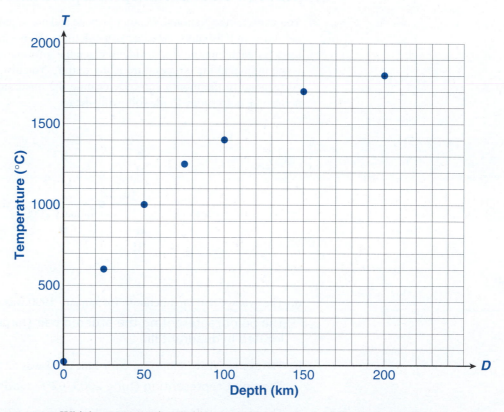

e. Which representation (table or graph) presents the information and trends in this data more clearly? Explain your choice.

(Answers will vary.) Trends are more easily seen on a graph.

SUMMARY: ACTIVITY 2.2

1. Input/output data pairs can be represented graphically as plotted points on a grid called a **rectangular coordinate system**.

2. The rectangular coordinate system consists of two perpendicular number lines (called **coordinate axes**) that intersect at their respective zero values. The point of intersection is called the **origin** and has coordinates $(0, 0)$.

3. The **input variable** is referenced on the horizontal axis. The **output variable** is referenced on the vertical axis.

4. Each point in the plane is identified by its horizontal and vertical directed distance from the axes. The distances are listed as an **ordered pair** of numbers (x, y) in which the horizontal coordinate, x, is written first and the vertical coordinate, y, second.

5. The two perpendicular coordinate axes divide the plane into four **quadrants**. The quadrants are labeled counterclockwise, using roman numerals, with quadrant I being the upper-right quadrant.

6. The distance represented by the space between adjacent tick marks on an axis is determined by the particular replacement values for the variable represented on the axis. This is called **scaling**. On each axis, equal distance between adjacent pairs of tick marks must be maintained.

EXERCISES: ACTIVITY 2.2

Points in a coordinate plane lie in one of the four quadrants or on one of the axes. In Exercises 1–18, place the letter corresponding to the phrase that best describes the location of the given point.

a. The point lies in the first quadrant.   **b.** The point lies in the second quadrant.

c. The point lies in the third quadrant.   **d.** The point lies in the fourth quadrant.

e. The point lies at the origin.   **f.** The point lies on the positive *x*-axis.

g. The point lies on the negative *x*-axis.   **h.** The point lies on the positive *y*-axis.

i.   The point lies on the negative *y*-axis.

1. $(2, -5)$ d   **2.** $(-3, -1)$ c   **3.** $(-4, 0)$ g

4. $(0, 0)$ e   **5.** $(0, 3)$ h   **6.** $(-2, 4)$ b

7. $(8, 6)$ a   **8.** $(0, -6)$ i   **9.** $(5, 0)$ f

10. $(4, -6)$ d   **11.** $(2, 0)$ f   **12.** $(0, 6)$ h

Exercise numbers appearing in color are answered in the Selected Answers appendix.

13. $(-7, -7)$ c   **14.** $(12, 5)$ a   **15.** $(0, -2)$ i

16. $(12, 0)$ f   **17.** $(-10, 2)$ b   **18.** $(-13, 0)$ g

19. The following table presents the average recommended weights for given heights for 25- to 29-year-old medium-framed women (wearing 1-inch heels and 3 pounds of clothing). Consider height to be the input variable and weight to be the output variable. As ordered pairs, height and weight take on the form (h, w). Designate the horizontal (input) axis as the h-axis, and the vertical (output) axis as the w-axis. Since all values of the data are positive, the points will lie in quadrant I only.

h, Height (in.)	58	60	62	64	66	68	70	72
w, Weight (lb.)	115	119	125	131	137	143	149	155

Plot the ordered pairs in the height-weight table on the following grid. Note the consistent spacing between tick marks on each axis. The distance between tick marks on the horizontal axis represents 2 inches. On the vertical axis the distance between tick marks represents 5 pounds.

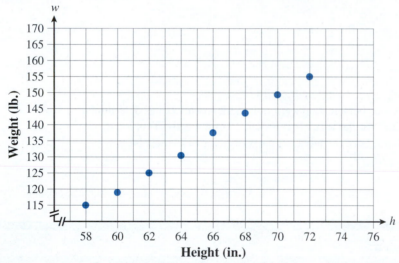

Note: The slash marks (//) near the origin indicate that the interval from 0 to 56 on the horizontal axis and the interval from 0 to 115 on the vertical axis are not shown. That is, only the part of the graph containing the plotted points is shown.

Activity 2.3

College Expenses

Objectives

1. Identify input variables and output variables.

2. Determine possible replacement values for the input.

3. Write verbal rules that represent relationships between input and output variables.

4. Construct tables of input/output values.

5. Construct graphs from input/output tables.

You are considering taking some courses at your local community college on a part-time basis for the upcoming semester. For a student who carries fewer than 12 credits (the full-time minimum), the tuition is $143 for each credit hour taken. You have a limited budget, so you need to calculate the total cost based on the number of credit hours.

Whenever you use a letter or symbol to represent a variable in the process of solving a problem, you must state what quantity (with units of measure) the variable represents and what collection of numbers are meaningful replacement values for the variable.

1. **a.** Describe in words the input variable and the output variable in the college tuition situation. Remember, the inputs are the values that are given or considered first. The outputs are the values that are determined by the given input values.

   The *input* variable is the number of credit hours you are carrying. The *output* variable is your tuition cost in dollars.

   b. Choose a letter or symbol to represent the input variable and a letter or symbol to represent the output variable.

   (Answers will vary.) Let *n* represent the number of credit hours (input). Let *T* represent the tuition (output).

   c. Would a replacement value of zero be reasonable for the input? Explain.

   An input value of zero would represent a decision not to enroll. There would not be a tuition bill.

   d. Would a replacement value of 15 be reasonable for the input? Explain.

   No. An input value of 15 would represent a full-time student's credit-hour load.

   e. What are reasonable replacement values for the input?

   Reasonable replacement values would be whole numbers from 1 to 11 (at most colleges).

   f. What is the tuition bill if you only register for a 3-credit-hour accounting course?

   $143 · 3 = $429   The tuition bill for 3 credits is $429.

   g. Use the replacement values you determined in part e to complete the following table.

Part-Time Tuition

Number of Credit Hours	1	2	3	4	5	6	7	8	9	10	11
Tuition ($)	143	286	429	572	715	858	1001	1144	1287	1430	1573

   h. Write a rule in words (called a **verbal rule**) that describes how to calculate the total tuition bill for a student carrying fewer than 12 credit hours.

   To determine the total tuition bill for part-time students, multiply $143 times the number of credit hours taken.

   i. What is the change in the tuition bill as the number of credit hours increases from 2 to 3? Is the change in the tuition bill the same amount for each unit increase in credit hours?

   The tuition increases by $143 for each additional credit hour taken.

Additional College Expenses

Other possible college expenses include parking fines, library fines, and food costs. Students whose first class is later in the morning often do not find a parking space and park illegally. Campus police have no sympathy and readily write tickets. Last semester, your first class began at 11:00 A.M.

2. a. The following table shows your cumulative number of parking tickets as the semester progressed and the total in fines you owed the college at that time. Use the given data pairs to determine how you can calculate the total fine, given your cumulative number of tickets. Describe this relationship in words (write a verbal rule).

The total fine equals the number of tickets times $12.50.

Parking Expense

Cumulative Number of Tickets Last Semester, T	2	3	5	8	10	12
Total Fines, F	$25	$37.50	$62.50	$100	$125	$150

b. Use the verbal rule to complete the table in part a.

c. Graph the information from the preceding table. Choose a scale for each axis that will enable you to plot all six points. Label the horizontal axis to represent the cumulative number of tickets received and the vertical axis to represent the total in fines owed to the college at that time.

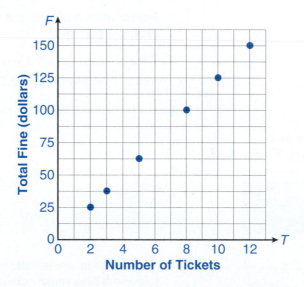

3. The college library fine for an overdue book is $0.25 per day with a maximum charge of $10.00 per item. You leave a library book at a friend's house and totally forget about it.

a. The number of days the book is overdue represents the input. What are the possible replacement values?

Possible replacement values are any counting number.

b. Determine the total fine (output) for selected values of the input given in the following table.

NUMBER OF DAYS OVERDUE	TOTAL FINE
4	$1.00
8	$2.00
16	$4.00
20	$5.00
28	$7.00
40	$10.00
56	$10.00

c. Write a verbal rule describing the arithmetic relationship between the input variable and the output variable.

The total fine is calculated by multiplying the number of days overdue by $0.25. If the number of days exceeds 40, however, the maximum fine of $10.00 is assessed.

d. Graph the information from the table in part b.

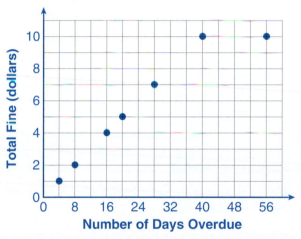

4. At the beginning of the semester, you buy a cafeteria meal ticket worth $150. The daily lunch special costs $4 in the college cafeteria.

a. Fill in the following table. Use the number of lunch specials you purchase as the input variable.

NUMBER OF LUNCH SPECIALS PURCHASED	REMAINING BALANCE ON YOUR MEAL TICKET
0	$150
10	$110
20	$70
30	$30
35	$10

b. Write a verbal rule describing the relationship between the input variable, number of lunch specials, and the output variable, remaining balance.

Multiply the number of lunch specials purchased by $4 and then subtract the result from $150 to obtain the remaining balance on your meal ticket.

c. What are possible replacement values for the input variable? Explain.

The input variable may be replaced by any whole number between 0 and 37, inclusive. There is not enough money to pay for more than 37 lunches since 4 · 38 = $152.

d. Graph the information from the table in part a.

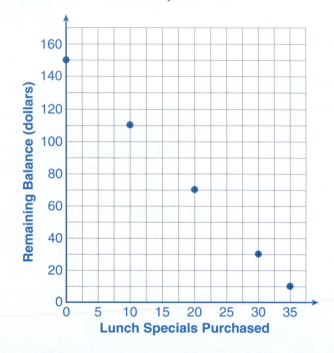

SUMMARY: ACTIVITY 2.3

1. A **verbal rule** is a statement that describes (in words) the arithmetic steps used to calculate the output corresponding to any input value.

2. The set of **replacement values** for the input is the collection of all numbers for which a meaningful output value can be determined.

3. Input/output data can be visually presented as points on an appropriately scaled rectangular coordinate system. Input is referenced along the horizontal axis and output along the vertical axis.

EXERCISES: ACTIVITY 2.3

1. Suppose the variable n represents the number of notebooks purchased by a student in the college bookstore for the semester.

a. Can n be reasonably replaced by a negative value, such as -2? Explain.

$n = -2$ would not be reasonable. You cannot purchase a negative number of notebooks.

Exercise numbers appearing in color are answered in the Selected Answers appendix.

b. Can *n* be reasonably replaced by the number zero? Explain.

Yes, *n* = 0 is reasonable. You might not purchase any notebooks at the college bookstore.

c. Can *n* be reasonably replaced by 1.5? Explain.

No, because the number of notebooks must be a whole number.

d. What is a possible collection of replacement values for the variable *n* in this situation?

(Answers will vary.) Possible replacement values are 0, 1, 2, 3, ... (Basically, any whole number that is reasonable.)

2. Determine the arithmetic relationship common to all input and output pairs in each of the following tables. For each table, write a verbal rule that describes how to calculate the output (*y*) from its corresponding input (*x*). Then complete the tables.

a.

INPUT, x	OUTPUT, y
2	4
4	8
6	12
8	16
10	20
20	40
25	50

The output is 2 times the input.

b.

INPUT, x	OUTPUT, y
2	4
3	9
4	16
5	25
6	36
7	49
8	64

The output is the square of the input.

c.

INPUT, x	OUTPUT, y
−2	−4
0	−2
2	0
4	2
6	4
8	6
10	8

Subtract 2 from the input to obtain the output.

d. Graph the input/output data from part a on an appropriately scaled and labeled set of coordinate axes.

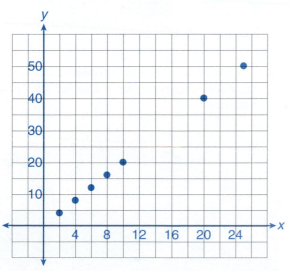

e. Graph the input/output data from part b on an appropriately scaled and labeled set of coordinate axes.

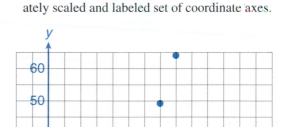
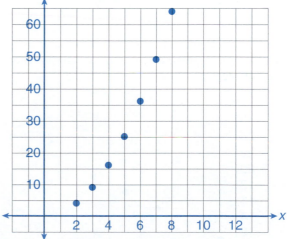

f. Graph the input/output data from part c on an appropriately scaled and labeled set of coordinate axes.

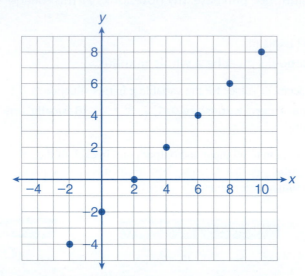

3. Complete the following tables using the verbal rule in column 2.

a.

INPUT	OUTPUT IS 10 MORE THAN THE INPUT
1	11
3	13
5	15
7	17
9	19
12	22
14	24

b.

INPUT	OUTPUT IS 3 TIMES THE INPUT, PLUS 2
−5	−13
−4	−10
−3	−7
−2	−4
−1	−1
0	2
1	5
2	8

c. Graph the information in the completed table in part a on an appropriately scaled and labeled set of coordinate axes.

d. Graph the information in the completed table in part b on an appropriately scaled and labeled set of coordinate axes.

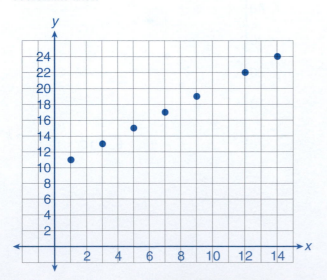

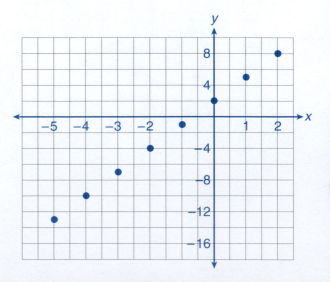

4. You are considering taking a part-time job that pays $9.50 per hour.

  a. What would be your gross pay if you work 22 hours one week?

  22 hrs · ($9.50 per hour) = $209 gross pay

  b. Write a verbal rule that describes how to determine your gross weekly pay based on the number of hours worked.

  Gross pay is calculated by multiplying the number of hours worked by $9.50, the pay per hour.

  c. Complete the following table using the rule in part b.

NUMBER OF HOURS WORKED (INPUT), h	WEEKLY PAY (OUTPUT), P
6	$57.00
12	$114.00
18	$171.00
20	$190.00
22	$209.00
28	$266.00
30	$285.00

  d. What are realistic replacement values for the input variable, the number of hours worked?

  (Answers will vary.) Values from 0 to 30, rounded to the nearest quarter hour, would be reasonable.

  e. Graph the information given in the table in part c. Make sure you use properly scaled and labeled axes.

  (Graphs will vary.)

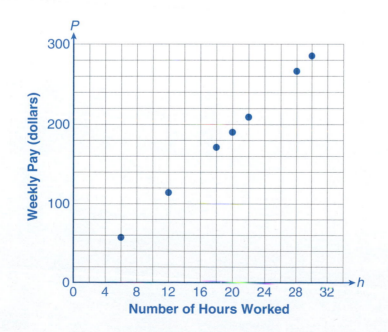

f. Explain your reasons for selecting the particular scales that you used.

(Answers will vary.) The scales I chose allow me to graph all the given information on the available grid.

g. Which representation (table or graph) presents the information and trends more clearly? Explain your choice.

(Answers will vary.) Trends are more easily seen on a graph.

Activity 2.4

Symbolizing Arithmetic

Objectives

1. Generalize from an arithmetic calculation to a symbolic representation by utilizing variables.

2. Evaluate algebraic expressions.

To communicate in a foreign language, you must learn the language's alphabet, grammar, and vocabulary. You must also learn to translate between your native language and the foreign language you are learning. The same is true for algebra, which, with its symbols and grammar, is the language of mathematics. To become confident and comfortable with algebra, you must practice speaking and writing it and translating between it and the natural language you are using (English, in this book).

In this activity you will note the close connections between algebra and arithmetic. Arithmetic involves determining a numerical value using a given sequence of operations. Algebra focuses on this sequence of operations, not on the numerical result. The following problems will highlight this important connection.

Arithmetic Calculations

1. Suppose you use a wire cutter to snip a 24-inch piece of wire into two parts. One of the parts measures 8 inches.

   a. What is the length of the second part?

   It measures 16 inches.

   b. Write down and describe in words the arithmetic operation you used to determine your answer.

   $24 - 8 = 16$; I subtracted 8 from 24.

2. Now, you use a wire cutter to snip another 24-inch piece of wire into two parts. This time one of the parts measures 10 inches.

   a. What is the length of the second part?

   It measures 14 inches.

   b. Write down and describe in words the arithmetic operation you used to determine your answer.

   $24 - 10 = 14$; I subtracted 10 from 24.

Symbolic Representations

Using the arithmetic calculations in Problems 1 and 2, you can generalize with a symbolic representation, utilizing a variable.

3. Consider the situation in which a 24-inch piece of wire is cut into two parts. Suppose you have no measuring device available, so you will denote the length of the first part by the symbol x, representing the variable length.

   a. Using Problems 1 and 2 as a guide, represent the length of the second part symbolically in terms of x.

   It measures $24 - x$.

   b. What are the reasonable replacement values for the variable x?

   Any number between 0 and 24.

Notice that your symbolic representation in Problem 3 provides a general method for calculating the length of the second part given the length x of the first part. The method is all about the operation, subtracting the length of the first part from 24. This symbolic representation, $24 - x$, is often called an **algebraic expression** in the variable x.

Now suppose you take the two parts of the snipped wire and bend each to form a square as shown in the following figure.

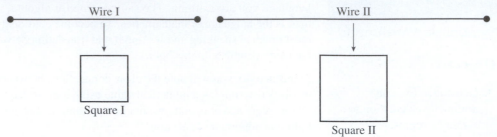

Remember, the area of a square equals the length of a side squared. The perimeter of a square is four times the length of one side.

4. Refer back to Problem 1, in which one piece of wire measuring 8 inches is cut from a 24-inch piece of wire.

a. Explain in words how you can use the known length of the wire to determine the area of the first square. What is this area?

The perimeter of the first square is 8 inches, so the side of the square is 2 inches.

I calculated the area by squaring the side length; 4 square inches.

b. Write down the sequence of arithmetic operation(s) you used to determine your answer in part a.

I divided 8 by 4 to obtain 2; then I squared the result. $(8 \div 4)^2 = 4$

c. Explain in words how to determine the area of the second square.

The perimeter of the square is $24 - 8 = 16$ inches, so the side of the square is 4 inches.

I calculated the area by squaring the side length; 16 square inches.

d. Write down the sequence of arithmetic operation(s) you used to determine your answer in part c.

I divided 16 by 4 to obtain 4; then I squared the result. $(16 \div 4)^2 = 16$

5. Refer back to Problem 3, in which one piece of wire cut from a 24-inch piece measured x inches.

a. Using Problem 4b as a guide, represent the area of the first square symbolically using x.

Divide x by 4 to obtain $\dfrac{x}{4}$; then square to obtain the area, $\left(\dfrac{x}{4}\right)^2$.

b. Similarly, represent the area of the second square symbolically using x.

Divide $(24 - x)$ by 4 to obtain $\dfrac{24 - x}{4}$; then square to obtain the area, $\left(\dfrac{24 - x}{4}\right)^2$.

c. Replace x by 8 in your algebraic expressions in parts a and b to calculate the areas of the two squares formed when a length of 8 inches is cut from a 24-inch piece of wire. Compare your results with the answers in Problem 4a and c.

$x = 8$, so the first area is $\left(\dfrac{8}{4}\right)^2 = 2^2 = 4$ square inches, and the area of

the second square is $\left(\dfrac{24-8}{4}\right)^2 = 4^2 = 16$ square inches. The answers are

the same.

When a specific number (an input value) replaces a variable in an algebraic expression, as in Problem 5c, the resulting calculation produces a single output value. This process is often called **evaluating** an algebraic expression.

Here are some additional problems to explore arithmetically and algebraically.

6. Suppose you have 8 dimes and 6 quarters in a desk drawer.

    a. Explain how you can use this information to determine the total value of these coins. What is the value?

       Each dime is worth $0.10 and each quarter is worth $0.25. The value of the dimes is $0.80 and the value of the quarters is $1.50, for a total of $2.30.

    b. Write down and describe in words the sequence of arithmetic operations you used to determine your answer.

       I multiplied the 8 dimes by $0.10, the 6 quarters by $0.25, and then added.

       $8 \cdot \$0.10 + 6 \cdot \$0.25 = \$2.30$

    c. Suppose you have d dimes and q quarters in a desk drawer. Represent the total value of these coins symbolically in terms of d and q.

       The total value is represented symbolically by $0.10d + 0.25q$.

    d. Evaluate your algebraic expression in part c when there are 14 dimes and 21 quarters.

       Replacing d with 14 and q with 21, the value is $0.10(14) + 0.25(21) = \$6.65$.

The algebraic expression in Problem 6c contained a variable multiplied by a constant number. In such a situation, it is not necessary to include a multiplication symbol. The arithmetic operation between a constant and a variable is always understood to be multiplication when no symbol is present.

Some special terminology is often used to describe important features of an algebraic expression. The **terms** of an algebraic expression are the parts of the expression that are added or subtracted. For example, the expression in Problem 6c has two terms, namely $0.10d$ and $0.25q$. A numerical constant that is multiplied by a variable is called the **coefficient** of the variable. For example, in the expression in Problem 6c, the coefficient of the variable d is 0.10 and the coefficient of the variable q is 0.25.

7. a. Suppose a rectangle has length 8 inches and width 5 inches. Determine its perimeter and its area.

       The perimeter is the sum of the lengths of its sides = 2 times the length + 2 times the width = $2(8) + 2(5) = 26$ in.

       Its area is the length times the width = 8 times 5 = 40 square inches.

    b. If the length of the rectangle is represented by l and its width is represented by w, write an expression that represents its perimeter and its area.

       perimeter = $2l + 2w$;   area = lw

    c. Suppose the length and width are each increased by 3 inches. Determine the new length, width, perimeter, and area of the expanded rectangle.

       The new length is $8 + 3 = 11$ in.; the new width is $5 + 3 = 8$ in.

       The perimeter is now $2(11) + 2(8) = 38$ in.; the area is now $11(8) = 88$ square inches.

    d. Represent the new length and width of the expanded rectangle in part c symbolically using the original dimensions l and w.

       The new length is $l + 3$; the new width is $w + 3$.

e. Represent the perimeter and area of the expanded rectangle symbolically in terms of the original length l and width w.

The new perimeter is $2(l + 3) + 2(w + 3)$; the new area is $(l + 3)(w + 3)$.

f. Evaluate the expressions in part e for the dimensions of the original rectangle. How do they compare to your results in part c?

Replacing l and w with the original values, 8 and 5, the perimeter is $2(8 + 3) + 2(5 + 3) = 38$ inches and the area is $(8 + 3)(5 + 3) = 88$ square inches. The results are the same.

g. Suppose the length and width of the original rectangle are doubled and tripled, respectively. Determine the new length, width, perimeter and area of the expanded rectangle.

The new length is 16 in.; the new width is 15 in.
The perimeter is now 62 in.; the area is now 240 square inches.

h. Represent the new length and width symbolically using the original dimensions l and w.

The new length is $2l$; the new width is $3w$.

i. Represent the new perimeter and new area symbolically using the original dimensions w and l. Is your representation consistent with your answer to part g?

The new perimeter is $2(2l) + 2(3w)$; the new area is $(2l)(3w)$.
Yes, it is consistent; replacing l and w by their original values, 8 and 5, the perimeter is $32 + 30 = 62$ inches and the area is $(16)(15) = 240$ square inches.

8. In Activity 2.3 (College Expenses, Problem 4), you purchased a meal ticket at the beginning of the semester for $150. The daily lunch special costs $4 in the college cafeteria. You want to be able to determine the balance on your meal ticket after you have purchased a given number of lunch specials.

a. Identify the input and output variables in the meal ticket situation.

The input variable is the number of lunch specials; the output variable is the balance on your meal ticket.

b. Write a verbal description of the sequence of operations that you must perform on the input value to obtain the corresponding output value.

Multiply the number of lunches by $4 and subtract that result from $150.

c. Translate the verbal description in part b to a symbolic expression using x to represent the input.

$150 - 4x$

d. Suppose you have purchased 16 lunch specials so far in the semester. Use the algebraic expression from part c to determine the remaining balance.

The remaining balance is $150 - 4(16) = \$86$.

An algebraic expression gives us a very powerful symbolic tool to represent relationships between an input variable and output variable. This symbolic rule generalizes the many possible numerical pairs, as might be represented in a table or graph. The remainder of Chapter 2 will further your fluency in this language of algebra.

SUMMARY: ACTIVITY 2.4

1. An **algebraic expression** represents symbolically the arithmetic calculations to be performed on given input values that will determine corresponding output values.

2. An algebraic expression is **evaluated** when a replacement value is substituted for the variable (or variables), and the sequence of arithmetic operations is performed to produce a single output value.

3. The **terms** of an algebraic expression are the parts of the expression that are added or subtracted.

4. A numerical constant that is multiplied by a variable is called the **coefficient** of the variable.

EXERCISES: ACTIVITY 2.4

1. **a.** You have 4 nickels and 8 quarters in a jar. What is the total value of these coins?

   The value of the nickels is \$0.20 and the value of the quarters is \$2.00, for a total of \$2.20.

   b. Suppose you have n nickels and q quarters in a jar. Represent the total value of these coins symbolically in terms of n and q.

   The total value is represented symbolically by $0.05n + 0.25q$.

   c. Now, suppose you have p pennies, n nickels, and d dimes in your pocket. Represent the total value of these coins symbolically in terms of p, n, and d.

   The total value is represented symbolically by $0.01p + 0.05n + 0.10d$.

2. You and your friend are standing facing each other 100 yards apart.

   a. Suppose he walks y yards toward you and stops. Represent his distance from you symbolically in terms of y.

   His distance from me is $100 - y$.

   b. Suppose, instead, he walks y yards away from you and stops. Represent his distance from you symbolically in terms of y.

   His distance from me is $100 + y$.

3. An equilateral triangle has sides of length s.

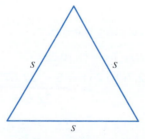

   a. If each side is increased by 5, represent the new perimeter in terms of s.

   The new perimeter is $3s + 15$.

   b. If each original side is doubled, represent the new perimeter in terms of s.

   The new perimeter is $6s$.

4. a. The markdown of every piece of clothing in the college bookstore is 20% of the original price of the item. If p represents the original price, write an algebraic expression that can be used to determine the markdown.   **0.20p**

b. Complete the following table by evaluating the expression in part a with the given input values.

Input Price, p ($)	50	100	150	200
Output Markdown ($)	10	20	30	40

5. a. The perimeter of a rectangle is twice the length plus twice the width. If l represents the length and w represents the width, write an algebraic expression that represents the perimeter.

The perimeter is represented by the expression 2l + 2w.

b. Calculate the perimeter of a rectangle with length 10 inches and width 17 inches.

2 · 10 in. + 2 · 17 in. = 54 in.   The perimeter is 54 inches.

c. The area of a rectangle is the product of its length and width. Write an algebraic expression in l and w that represents its area.

The area is represented by the product l · w.

d. Calculate the area of a rectangle with length 10 inches and width 17 inches.

10 in. · 17 in. = 170 sq. in.   The area is 170 square inches.

6. Let x represent the input variable. Translate each of the phrases into an algebraic expression.

a. The input decreased by 7 $x - 7$

b. Five times the difference between the input and 6 $5(x - 6)$

c. Seven increased by the quotient of the input and 5 $7 + \frac{x}{5}$

d. Twelve more than one-half of the square of the input $\frac{1}{2}x^2 + 12$

e. Twenty less than the product of the input and -2

$-2x - 20$

f. The sum of three-eighths of the input and five

$\frac{3}{8}x + 5$

7. Evaluate each of the following algebraic expressions for the given value(s).

a. $3x - 10$, for $x = -2.5$

$3(-2.5) - 10 = -7.5 - 10$
$= -17.5$

b. $5 - 2.7x$, for $x = 10.25$

$5 - 2.7(10.25) = 5 - 27.675$
$= -22.675$

c. $\frac{1}{2}bh$, for $b = 10$ and $h = 6.5$

$\frac{1}{2}(10)(6.5) = 32.5$

d. $5x - 7$, for $x = -2.5$

$5(-2.5) - 7 = -12.5 - 7$
$= -19.5$

e. $55 - 6.7x$, for $x = 10$

$55 - 6.7(10) = 55 - 67$
$= -12$

f. $\frac{1}{2}bh$, for $b = 5$ and $h = 5$

$\frac{1}{2}(5)(5) = (2.5)(5)$
$= 12.5$

g. $\frac{1}{2}h(b + B)$, for $h = 10, b = 5, B = 7$

$$\frac{1}{2}(10)(5 + 7) = \frac{1}{2}(10)(12)$$
$$= (5)(12)$$
$$= 60$$

8. A cardboard box used for flat rate shipping by a postal service has length 18 inches, width 13 inches, and height 5 inches.

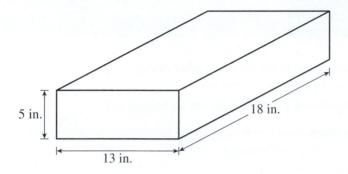

a. The volume of a rectangular box is the product of the length times the width times the height. Determine the volume of the box.

Volume = length · width · height = 18 · 13 · 5 = 1170 cubic inches

b. Suppose each dimension is increased by x inches. Represent the length, width, and height of the box symbolically in terms of x.

The length is 18 + x, the width is 13 + x, and the height is 5 + x.

c. Write a single algebraic expression that symbolically expresses the volume of the box in terms of x.

Using the expressions from part b, the volume is $(18 + x)(13 + x)(5 + x)$.

d. Now evaluate your algebraic expression for the volume when $x = 3$ to determine the volume when each side of the box is increased by three inches.

Replacing x with 3 results in $(18 + 3)(13 + 3)(5 + 3) = 21 · 16 · 8 = 2688$ cubic inches.

Lab Activity 2.5

How Many Cups Are in That Stack?

Objectives

1. Collect input/output data.

2. Represent input/output data numerically in tables.

3. Construct tables of data pairs for graphing.

4. Graph input/output data pairs.

In this lab, you will stack and measure the heights of disposable Styrofoam or plastic cups placed inside one another. With a centimeter ruler, measure and record the height (to the nearest tenth of a centimeter) of stacks containing increasing numbers of cups.

(Answers will vary.) We present a sample here.

Heightened Importance

Number of Cups	1	2	3	4	5	6	7	8
Height of Stack (cm)	8.9	10.1	11.4	12.7	14.0	15.3	16.7	18.0

1. Graph the data with the inputs (number of cups) on the horizontal axis and the outputs (height of stack) on the vertical axis.

   (Graphs will vary, depending on scaling of axes.)

2. Use the information you have gathered to *estimate* the height (to the nearest tenth of a centimeter) of 10 stacked cups.

   approximately 20.6 cm

3. Now construct a stack of 10 cups. Measure the stack and compare it with your estimate. Was your estimate reasonable? If not, why not?

   (Answers will vary, depending on student's measurement accuracy and estimate in Problem 2.)

4. Estimate the height of 16 cups. Is a stack of 16 cups twice as tall as a stack of 8 cups? Justify your answer.

   Watch out for a quick 'yes'. The 16-cup stack will *not* be twice as tall. The pattern indicates it will be approximately 28.4 centimeters tall.

5. Estimate the height of 20 cups. Is a stack of 20 cups twice as tall as a stack of 10 cups? Justify your answer.

   The argument is similar to that in Problem 4; the 20-cup stack is approximately 33.6 centimeters tall.

6. Describe the formula or method you used to estimate the height of a stack of cups.

   (Answers will vary.) Students may use a graph or a formula or pattern involving "lip" height.

7. If your shelf has clearance of 40 centimeters, how many cups can you stack to fit on the shelf? Explain how you obtained your answer.

   (Reasoning will vary. Encourage creative thinking.) Approximately 24 cups will fit.

Cluster 1   **What Have I Learned?**

1. Describe several ways that relationships between variables may be represented. What are some advantages and disadvantages of each?

   Relationships between variables may be represented (1) numerically in a table, (2) graphically on a grid, (3) symbolically in an equation, and (4) verbally in a natural language. Each representation has advantages and disadvantages. Sometimes, two or more of the representations may be used together to improve the description of the relationship.

   1. A table can precisely present information about data. However, trends are often difficult to see, especially if a data set is large.

   2. Graphs can readily picture long-term and short-term trends in data. Yet sometimes a useful graph is difficult to achieve because of limitations imposed by data values and available grid sizes.

   3. A symbolic rule can conveniently generate and determine precise information in some relationships. However, some data sets do not lend themselves easily to representation by a symbolic relationship. For example, see the data sets for Medicare costs and for oil refineries in Activity 2.1.

   4. In some cases, it is possible to describe the relationship verbally in natural language that is more familiar than the other representations. However, if the relationships are complex and/or data sets have more than a few values, a verbal representation may not be appropriate.

2. Obtain a graph from a local newspaper, magazine, or textbook in your major field. Identify the input and output variables. The input variable is referenced on which axis? Describe any trends in the graph.

   (Answers will vary.)

3. You will be graphing some data from a table in which the input values range from 0 to 150 and the output values range from 0 to 2000. Assume that your grid is a square with 16 tick marks across and up.

   a. Will you use all four quadrants? Explain.

      Since input and output values are nonnegative, only the first quadrant and positive axes are necessary.

   b. How many units does the distance between tick marks on the horizontal axis represent?

      (Answers will vary.) If the distance between tick marks represents 10 units, the input values would fit.

   c. How many units does the distance between tick marks on the vertical axis represent?

      (Answers will vary.) If the distance between tick marks represents 150 units, the output values would fit.

Exercise numbers appearing in color are answered in the Selected Answers appendix.

Cluster 1   How Can I Practice?

1. You are a scuba diver and plan a dive in the St. Lawrence River. The water depth in the diving area does not exceed 150 feet. Let x represent your depth in feet *below* the surface of the water.

   a. What are the possible replacement values to represent your depth from the surface?

   Replacement values include any number between 0 and 150.

   b. Would a replacement value of zero feet be reasonable? Explain.

   Yes, you could be floating on the surface.

   c. Would a replacement value of -200 be reasonable? Explain.

   No, negative numbers would represent distance above the surface.

   d. Would a replacement value of 12 feet be reasonable? Explain.

   Yes. This would place you 12 feet below the surface.

2. You bought stock in a company in 2007 and have tracked the company's profits and losses from its beginning in 2001 to the present. You decide to graph the information where the number of years since 2007 is the input variable and profit or loss for the year is the output variable. Note that the year 2007 corresponds to zero on the horizontal axis. Determine the quadrant in which, or axis on which, you would plot the points that correspond to the following data. If your answer is on an axis, indicate between which quadrants the point is located.

   a. The loss in 2004 was $1500.

   The point having coordinates $(-3, -1500)$ is in quadrant III.

   b. The profit in 2008 was $6000.

   The point having coordinates $(1, 6000)$ is in quadrant I.

   c. The loss in 2010 was $1000.

   The point representing a $1000 loss in 2010 is located in quadrant IV.

   d. In 2003, there was no profit or loss.

   The point representing the fact that there was no profit or loss in 2003 is located on the horizontal axis, between quadrant II and quadrant III.

   e. The profit in 2001 was $500.

   The point representing a $500 profit in 2001 is in quadrant II.

   f. The loss in 2007 was $800.

   The point representing an $800 loss in 2007 is located on the vertical axis, between quadrant III and quadrant IV.

3. Fish need oxygen to live, just as you do. The amount of dissolved oxygen (D.O.) in water is measured in parts per million (ppm). Trout need a minimum of 6 ppm to live.

   The data in the table shows the relationship between the temperature of the water and the amount of dissolved oxygen present.

Temp (°C)	11	16	21	26	31
D.O. (in ppm)	10.2	8.6	7.7	7.0	6.4

Exercise numbers appearing in color are answered in the Selected Answers appendix.

a. Represent the data in the table graphically. Place temperature (input) along the horizontal axis and dissolved oxygen (output) along the vertical axis.

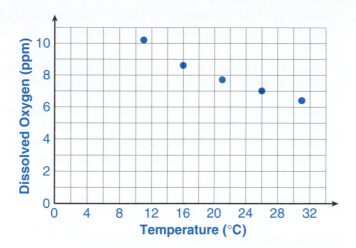

b. What general trend do you notice in the data?

The D.O. decreases as temperature increases.

c. In which of the 5° temperature intervals given in the table does the dissolved oxygen content change the most?

The D.O. content changes the most between 11°C and 16°C, when it drops 1.6 ppm.

d. Which representation (table or graph) presents the information and trends more clearly?

(Answers will vary.) The graph shows general trends more clearly.

4. When you were born, your uncle invested $1000 for you in a local bank. The following graph shows how your investment grows.

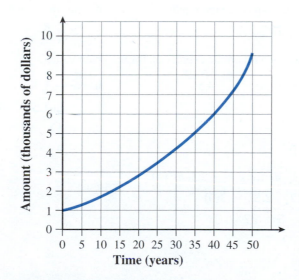

a. Which variable is the input variable?

The input variable is time, in number of years.

b. How much money did you have when you were 10 years old?

From the graph, it appears to be approximately $1700.

c. Estimate in what year your original investment will have doubled.

The investment would double by approximately age 13.

d. If your college bill is estimated to be $3000 in the first year of college, will you have enough to pay the bill with these funds? (Assume that you attend when you are 18.) Explain.

It appears that when I am 18, the investment should be worth between $2500 and $2800. So, there *will not* be enough.

e. Assume that you expect to be married when you are 30 years old. You figure that you will need about $5000 for your share of the wedding and honeymoon expenses. Assume also that you left the money in the bank and did not use it for your education. Will you have enough money to pay your share of the wedding and honeymoon expenses? Explain.

The amount at age 30 is less than $5000, so it is not likely that there will be enough.

5. A square has perimeter 80 inches.

a. What is the length of each side?

$80 \div 4 = 20$ in.

b. If the length of each side is increased by 5 inches, what will the new perimeter equal?

$4(20 + 5) = 100$ in.

c. If each side of the original square is increased by x inches, represent the new perimeter in terms of x.

The new perimeter is $4(20 + x)$ or $80 + 4x$.

6. Let x represent the input variable. Translate each of the following phrases into an algebraic expression.

a. 20 less than twice the input

$2x - 20$

b. the sum of half the input and 6

$\frac{1}{2}x + 6$

7. Evaluate each algebraic expression for the given value(s).

a. $2x + 7$, for $x = -3.5$

$2(-3.5) + 7$

$= -7 + 7$

$= 0$

b. $8x - 3$, for $x = 1.5$

$8(1.5) - 3$

$= 12 - 3$

$= 9$

c. $2k + 3h$, for $k = -2.5, h = 9$

$2(-2.5) + 3(9)$

$= -5 + 27$

$= 22$

Cluster 2   Solving Equations

Activity 2.6

Let's Go Shopping

Objectives

1. Translate verbal rules into symbolic rules.

2. Solve an equation of the form $ax = b$, $a \neq 0$, for x using an algebraic approach.

3. Solve an equation of the form $x + a = b$ for x using an algebraic approach.

The sales tax collected on taxable items in Allegany County in western New York is 8.5%. The tax you must pay depends on the price of the item you are purchasing.

1. What is the sales tax on a dress shirt that costs $20?

The sales tax is $20(0.085) = \$1.70$

2. a. Because you are interested in determining the sales tax given the price of an item, which variable is the input?

The price of the item is the input.

b. What are its units of measurement?

The units of measurement are dollars.

3. a. Which variable is the output?

The amount of sales tax paid is the output.

b. What are its units of measurement?

The units of measurement are dollars.

4. Determine the sales tax you must pay on the following items.

 Power Shopping

ITEM	PRICE ($)	SALES TAX ($)
Calculator	12.00	1.02
Shirt	25.00	2.13
Microwave Oven	200.00	17.00
Car	15,000.00	1275.00

The rule you followed to calculate sales tax (the output) for a given price (the input) can be described either **verbally** (in words) or **symbolically** (in an algebraic equation).

Definition

A **symbolic rule** is a mathematical statement that defines an **output** variable as an algebraic expression in terms of the **input** variable. The symbolic rule is essentially a recipe that describes how to determine the output value corresponding to a given input value.

---

5. a. Explain in your own words how to determine the sales tax on an item of a given price.

Obtain the sales tax by multiplying the price by 0.085.

b. Translate the verbal rule in part a into a symbolic rule, with x representing the input (price) and y representing the output (sales tax).

$y = 0.085x$

Solving Equations of the Form $ax = b$, $a \neq 0$, Using an Algebraic Approach

You can diagram the symbolic rule $y = 0.085x$ in the following way.

Start with x (price) $\longrightarrow$ multiply by 0.085 $\longrightarrow$ to obtain y (sales tax).

How might you determine the price of a fax machine for which you paid a sales tax of $21.25? In this situation you know the sales tax (output), and want to determine the price (input). To accomplish this, "reverse the direction" in the preceding diagram, and replace the operation (multiplication) by its inverse (division).

Start with y (sales tax) $\longrightarrow$ divide by 0.085 $\longrightarrow$ to obtain x (price).
Start with 21.25 $\longrightarrow$ $21.25 \div 0.085$ $\longrightarrow$ to obtain 250.

Therefore, the price of a fax machine for which you paid $21.25 sales tax is $250.

6. Use this reverse process to determine the price of a scanner for which you paid a sales tax of $61.20.

$61.20 \div 0.085 = \$720$ The scanner's price was $720.

The reverse process illustrated above can also be done in a more common and formal way as demonstrated in Example 1.

Example 1   *The sales tax for the fax machine was $21.25. In the symbolic rule $y = 0.085x$, replace y by 21.25 and proceed as follows:*

$21.25 = 0.085x$ This is the equation to be solved.

$\dfrac{21.25}{0.085} = \dfrac{0.085x}{0.085}$ Divide both sides of the equation by 0.085 or, equivalently, multiply by $\dfrac{1}{0.085}$.

$250 = x$

The approach demonstrated in Example 1 is the **algebraic method** of solving the equation $21.25 = 0.085x$. In this situation, you needed to undo the multiplication of x and 0.085 by dividing each side of the equation by 0.085. The value obtained for x (here, 250) is called the **solution** of the equation. Note the solution of the equation is complete when the variable x is isolated on one side of the equation, with coefficient 1.

7. Use the symbolic rule $y = 0.085x$ to determine the price of a Blu-Ray DVD player for which you paid a sales tax of $34.85.

$34.85 = 0.085x$

$\dfrac{34.85}{0.085} = \dfrac{0.085x}{0.085}$

$410 = x$

The price of the Blu-Ray player was $410.

Solving Equations of the Form $x + a = b$ Using an Algebraic Approach

8. Suppose you have a $15 coupon that can be used on any purchase over $100 at your favorite clothing store.

a. Determine the discount price of a sports jacket having a retail price of $116.

The discount price is $116 - 15 = \$101$.

b. Identify the input variable and the output variable.

The input is the retail price, in dollars, and the output is the discount price, in dollars.

c. Let *x* represent the input and *y* represent the output. Write a symbolic rule that describes the relationship between the input and output. Assume that your total purchase will be greater than $100.

$y = x - 15$, for $x > \$100$

d. Complete the following table using the symbolic rule determined in part c.

x, RETAIL PRICE ($)	y, DISCOUNT PRICE ($)
105	90
135	120
184	169
205	190

You can also diagram this symbolic rule in the following way:

Start with *x* (retail price) $\longrightarrow$ subtract 15 $\longrightarrow$ to obtain *y* (discount price).

9. Suppose you are asked to determine the retail price if the discount price of a suit is $187.

a. Is 187 an input value or an output value?

$187 is an output value.

b. Reverse the direction in the diagram above, replacing the operation (subtraction) with its inverse (addition).

Start with $y = 187$ (output) $\longrightarrow$ ___add 15___ $\longrightarrow$ to obtain *x* (input).

c. Use this reverse process to determine the retail price of the suit.

$187 + 15 = \$202$ The retail price of the suit is $202.

The equation in Problem 9 can be solved using an algebraic approach.

Example 2   *Solve the equation $187 = x - 15$ for x.*

$187 = x - 15$	Equation to be solved.
$\underline{+15 \qquad + 15}$	Add 15 to each side, to isolate the variable.
$202 = x$	

---

10. Use the symbolic rule $y = x - 15$ to determine the retail price, *x*, of a trench coat whose discounted price is $203.

$$203 = x - 15$$
$$203 + 15 = x - 15 + 15$$
$$218 = x$$

The retail price of the coat is $218.

In this activity you have practiced solving equations involving a single operation—addition, subtraction, multiplication, or division—by performing the appropriate inverse operation. Recall that addition and subtraction are inverse operations, as are multiplication and division. Use this algebraic method to solve the equations arising in the following problems.

11. Set up and solve the appropriate equation in each case.

a. $y = 7.5x$

Determine x when $y = 90$.

$$90 = 7.5x$$

$$\frac{90}{7.5} = \frac{7.5x}{7.5}$$

$$12 = x$$

b. $z = -5x$

Determine x when $z = -115$.

$$-115 = -5x$$

$$\frac{-115}{-5} = \frac{-5x}{-5}$$

$$23 = x$$

c. $y = \dfrac{x}{4}$

Determine x when $y = 2$.

$$2 = \frac{x}{4}$$

$$4(2) = 4\left(\frac{x}{4}\right)$$

$$8 = x$$

d. $p = \dfrac{2}{3}x$

Determine x when $p = -18$.

$$-18 = \tfrac{2}{3}x$$

$$\tfrac{3}{2}(-18) = \tfrac{3}{2}\left(\tfrac{2}{3}x\right)$$

$$-27 = x$$

12. Use an algebraic approach to determine the input x for the given output value.

a. $y = x - 10$

Determine x when $y = -13$.

$$\begin{array}{rcl} -13 &=& x - 10 \\ +10 && +10 \\ \hline -3 &=& x \end{array}$$

b. $y = 13 + x$

Determine x when $y = 7$.

$$\begin{array}{rcl} 7 &=& 13 + x \\ -13 && -13 \\ \hline -6 &=& x \end{array}$$

c. $p = x + 4.5$

Determine x when $p = -10$.

$$\begin{array}{rcl} -10 &=& x + 4.5 \\ -4.5 && -4.5 \\ \hline -14.5 &=& x \end{array}$$

d. $x - \dfrac{1}{3} = s$

Determine x when $s = 8$.

$$\begin{array}{rcl} x - \tfrac{1}{3} &=& 8 \\ +\tfrac{1}{3} && +\tfrac{1}{3} \\ \hline x &=& 8\tfrac{1}{3} \end{array}$$

SUMMARY: ACTIVITY 2.6

1. A **solution** of an equation containing one variable is a replacement value for the variable that produces equal values on both sides of the equation.

2. An **algebraic approach** to solving an equation for a given variable is complete when the variable (such as x) is isolated on one side of the equation with coefficient 1. To isolate the variable, you apply the appropriate inverse operation, as follows.

a. Multiplication and division are inverse operations. Therefore,

i. to undo multiplication of the input by a nonzero number, divide each side of the equation by that number.

ii. to undo division of the input by a number, multiply each side of the equation by that number

b. Addition and subtraction are inverse operations. Therefore,

i. to undo addition of a number to the input, subtract that number from each side of the equation

ii. to undo subtraction of a number from the input, add that number to each side of the equation

EXERCISES: ACTIVITY 2.6

1. You've decided to enroll in a local college as a part-time student (taking fewer than 12 credit hours). Full-time college work does not fit into your present financial or personal situation. The cost per credit hour at your college is $175.

   a. What is the cost of a 3-credit-hour literature course?

   A 3-credit-hour literature course costs $175(3) = $525.

   b. Write a symbolic rule to determine the total tuition for a given number of credit hours. Let n represent the number of credit hours taken (input) and y represent the total tuition paid (output).

   $y = 175n$

   c. What are the possible replacement values for n?

   n can be any counting number up to 11.

   d. Complete the following table.

CREDIT HOURS	TUITION PAID ($)
1	175
2	350
3	525
4	700

   e. Suppose you have enrolled in a psychology course and a computer course, each of which is 3 credit hours. Use the symbolic rule in part b to determine the total tuition paid.

   $y = 175(6) = 1050 is the total tuition for 6 credit hours.

   f. Use the symbolic rule from part b to determine algebraically the number of credit hours carried by a student with the following tuition bill.

   i. $875

   $175x = 875$

   $x = \dfrac{875}{175}$

   $x = 5$

   A student with an $875 tuition bill carries 5 credit hours.

   ii. $1400

   $175x = 1400$

   $x = \dfrac{1400}{175}$

   $x = 8$

   A student with a $1400 tuition bill carries 8 credit hours.

2. a. The average amount, A, of precipitation in Boston during March is four times the average amount, P, of precipitation in Phoenix. Translate the verbal rule into a symbolic rule.

   $A = 4P$

   b. What is the amount of precipitation in Boston if there are 2 inches in Phoenix?

   $A = 4(2)$

   $= 8$

   There are 8 inches of precipitation in Boston when there are 2 inches in Phoenix.

Exercise numbers appearing in color are answered in the Selected Answers appendix.

c. What is the amount of precipitation in Phoenix if there are 24 inches in Boston?

$24 = 4P$

$\frac{24}{4} = P$

$6 = P$

There are 6 inches of precipitation in Phoenix when there are 24 inches in Boston.

3. a. The depth (inches) of water that accumulates in the spring soil from melted snow can be determined by dividing the cumulative winter snowfall (inches) by 12. Translate the verbal statement into a symbolic rule, using I for the accumulated inches of water and n for the inches of fallen snow.

$I = \frac{n}{12}$

b. Determine the amount of water that accumulates in the soil if 25 inches of snow falls.

$I = \frac{25}{12} = 2\frac{1}{12} = 2.08\overline{3}$ in.

Approximately 2.08 inches of water accumulates in the soil when 25 inches of snow falls.

c. Determine the total amount of winter snow that accumulates 6 inches of water in the soil.

$6 = \frac{n}{12}$

$6(12) = n$

$72 = n$

Approximately 72 inches of snow produces 6 inches of water.

4. Have you ever played a game based on the popular TV game show *The Price Is Right*? The idea is to guess the price of an item. You win the item by coming closest to the correct price without going over. If your opponent goes first, a good strategy is to overbid her regularly by a small amount, say, $15. Then your opponent can win only if the item's price falls in that $15 region between her bid and yours.

You can model this strategy by defining two variables: The input, x, will represent your opponent's bid, and the output, y, will represent your bid.

a. Write a symbolic rule to represent the input/output relationship in this strategy.

$y = x + 15$

b. Determine your bid if your opponent's bid is $475.

$y = 475 + 15 = \$490$ Your bid is $490.

c. Complete the following table.

OPPONENT'S BID	YOUR BID
$390	$405
$585	$600
$1095	$1110

d. Use the symbolic rule determined in part a to calculate your opponent's bid if you have just bid $605. Stated another way, determine x if $y = 605$.

$605 = x + 15$

$\underline{-15 \qquad -15}$

$590 = x$

Your opponent's bid was $590.

5. a. Profit is calculated as revenue minus expenses. If a company's expenses were $10 million, write a symbolic rule that expresses profit, P (in millions of dollars), in terms of revenue, R (in millions of dollars).

$P = R - 10$

b. Determine the profit when revenue is $25 million.

$P = 25 - 10$

$= 15$

Revenue of $25 million produces a profit of $15 million.

c. Determine the revenue that will produce a profit of $5 million.

$5 = R - 10$

$5 + 10 = R$

$15 = R$

Revenue of $15 million is needed to produce a profit of $5 million.

For part a of Exercises 6–11, determine the output when you are given the input. For part b, use an algebraic approach to solve the equation for the input when you are given the output.

6. $3.5x = y$

a. Determine y when $x = 15$.

$y = 3.5(15) = 52.5$

b. Determine x when $y = 144$.

$3.5x = 144$

$\dfrac{3.5x}{3.5} = \dfrac{144}{3.5}$

$x \approx 41.14$

7. $z = -12x$

a. Determine z when $x = -7$.

$z = -12(-7) = 84$

b. Determine x when $z = 108$.

$-12x = 108$

$\dfrac{-12x}{-12} = \dfrac{108}{-12}$

$x = -9$

8. $y = 15.3x$

a. Determine y when $x = -13$.

$y = 15.3(-13) = -198.9$

b. Determine x when $y = 351.9$.

$15.3x = 351.9$

$\dfrac{15.3x}{15.3} = \dfrac{351.9}{15.3}$

$x = 23$

9. $y = x + 5$

a. Determine y when $x = -11$.

$y = -11 + 5 = -6$

b. Determine x when $y = 17$.

$17 = x + 5$

$\underline{-5 \qquad -5}$

$12 = x$

10. $y = x + 5.5$

a. Determine y when $x = -3.7$.

$y = -3.7 + 5.5 = 1.8$

b. Determine x when $y = 13.7$.

$13.7 = x + 5.5$

$\underline{-5.5 \qquad -5.5}$

$8.2 = x$

11. $z = x - 11$

a. Determine z when $x = -5$.

$z = -5 - 11 = -16$

b. Determine x when $z = -4$.

$$-4 = x - 11$$
$$\underline{+11 \qquad\quad +11}$$
$$7 = x$$

In some situations the input/output relationship is given in a table or a graph. You can use the table or graph to estimate the input required to produce a given output. It is not necessary to write the corresponding symbolic rule (if one even exists!). The following problems illustrate this idea.

12. A ball is dropped from the top of the Willis Tower (formerly called the Sears Tower) in Chicago. The table below gives the ball's distance from the ground (the output) at a given time after it is dropped (the input).

Time (sec.)	1	2	3	4	5	6	7
Distance (ft.)	1435	1387	1307	1195	1051	875	667

a. Estimate the number of seconds it takes for the ball to be 1000 feet above the ground.

The ball will be 1051 feet above the ground in 5 seconds, so it will take slightly more than 5 seconds to reach 1000 feet above the ground.

b. The Willis Tower is 1451 feet tall. Approximately how many seconds will it take the ball to be halfway to the ground?

Halfway would be about 725 feet above the ground. From the table, it will take between 6 and 7 seconds, so a likely estimate is approximately 6.5 seconds.

c. If you knew the symbolic rule for this input/output relation, how might your answers be improved?

If I solved the equations algebraically, I could get more exact answers instead of approximations.

13. The following graph displays a Booster Club's share (the output) from fund-raiser sales (total gross receipts, the input).

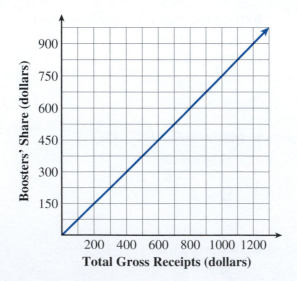

a. Use the graph to estimate the total receipts when the Booster's share is $750.

The total gross receipts were $1000.

b. Was it necessary to know the equation to solve the problem in part a?

The symbolic rule is not necessary, but it might lead to more exact answers.

c. Can the graph as shown be used to determine the gross receipts if the goal was to raise $1050 for the Booster Club?

The answer cannot be read directly from the graph as shown. However, if the graph were extended vertically and horizontally so $1050 is included on the graph and the line graph is assumed to continue in the same direction, the answer could be read to be $1400.

(Note to instructors: Some students may be savvy enough to see that every $100 dollars in gross receipts will produce $75 for the Boosters and determine the answer from that.)

Exercises 12 and 13 illustrate the important roles that numerical and graphical methods play in solving input/output problems. These methods will be more fully explored and developed in Chapters 3 and 4, including their connections to symbolic rules.

Activity 2.7

Leasing a Copier

Objectives

1. Model contextual situations with symbolic rules of the form $y = ax + b, a \neq 0$.

2. Solve equations of the form $ax + b = c, a \neq 0$.

As part of your college program, you are a summer intern in a law office. You are asked by the office manager to gather some information about leasing a copy machine for the office. The sales representative at Eastern Supply Company recommends a 50-copy/minute copier to satisfy the office's copying needs. The copier leases for $455 per month, plus 1.5 cents a copy. Maintenance fees are covered in the monthly charge. The lawyers would own the copier after 39 months.

1. **a.** The total monthly cost depends upon the number of copies made. Identify the input and output variables.

   The input is the number of copies made; the output is the total monthly cost.

   b. Write a verbal rule to determine the total monthly cost in terms of the number of copies made during the month.

   Multiply the number of copies by 0.015 and then add 455 to the product to obtain the total monthly cost.

   c. Complete the following table.

The Cost of Leasing

Number of Copies	5000	10,000	15,000	20,000
Monthly Cost ($)	530	605	680	755

   d. Translate the verbal rule in part b into a symbolic rule. Let n represent the number of copies (input) made during the month and c represent the total monthly cost (output).

   $c = 0.015n + 455$

2. **a.** Use the symbolic rule obtained in Problem 1d to determine the monthly cost if 12,000 copies are made.

   $635 is the monthly cost, since $c = (0.015)(12,000) + 455 = 635$.

   b. Is 12,000 a replacement value for the input or for the output variable?

   12,000 is a replacement value for the input variable.

In Problem 2, you determined the cost by evaluating the symbolic rule $c = 0.015n + 455$ for $n = 12,000$. This was accomplished by performing the sequence of operations shown in the diagram.

   Start with n $\longrightarrow$ multiply by 0.015 $\longrightarrow$ add 455 $\longrightarrow$ to obtain c.

3. Suppose the monthly budget for leasing the copier is $800.

   a. Is 800 an input value for n or an output value for c?

   It is the output value for c.

   b. To determine the input, n, for a given output value of c, reverse the sequence of operations in the preceding diagram and replace each operation with its inverse. Recall that addition and subtraction are inverse operations and that multiplication and division are inverse operations. Complete the following:

   Start with c $\longrightarrow$ subtract 455 $\longrightarrow$ divide by 0.015 $\longrightarrow$ to obtain n.

   c. Use the sequence of operations in part b to determine the number of copies that can be made for a monthly budget of $c = 800$.

   $800 - 455 = 345$

   $345 \div 0.015 = 23,000$

   You can make 23,000 copies for $800.

Solving Equations Algebraically

Once you understand the concept of using the reverse process to solve equations (see Problem 3), you will want a more systematic algebraic procedure to follow.

4. Use the symbolic rule in Problem 1 to write an equation that can be used to determine the number, n, of copies that can be made with a monthly budget of $c = 800$.

$$800 = 0.015n + 455$$

In Problem 1 you constructed a symbolic rule involving *two* operations.

In Problem 2 you evaluated this rule using the indicated sequence of operations.

In Problem 3 you solved an equation by performing the sequence in reverse order, replacing each original operation with its inverse. This algebraic solution is commonly written in a vertical format in which each operation is performed on a separate line. Example 1 illustrates this format.

Example 1 *The following example illustrates a systematic algebraic procedure by undoing two operations to solve an equation.*

Solve for n:

$$800 = 0.015n + 455$$

$$\underline{-455 \qquad\qquad -455}$$

$$345 = 0.015n$$

Step 1: To undo the addition of 455, subtract 455 from each side of the equation.

$$\frac{345}{0.015} = \frac{0.015n}{0.015}$$

Step 2: To undo the multiplication by 0.015, divide each side of the equation by 0.015.

$$23000 = n$$

Check:

$$800 \stackrel{?}{=} 0.015(23000) + 455$$

$$800 \stackrel{?}{=} 345 + 455$$

$$800 = 800$$

Note that the algebraic procedure emphasizes that an equation can be thought of as a scale whose arms are in balance. The equal sign can be thought of as the balancing point.

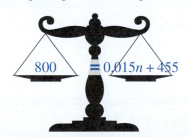

As you perform the appropriate inverse operation to solve the equation $800 = 0.015n + 455$ for n, you must maintain the balance as you perform each step in the process. If you subtract 455 from one side of the equal sign, then you must subtract 455 from the other side. Similarly, if you divide one side of the equation by 0.015, then you must divide the other side by 0.015.

5. Use the algebraic procedure outlined in Example 1 to solve the equation $120 = 3x + 90$.

$$120 = 3x + 90$$

$$\underline{-90 \qquad\qquad -90}$$

$$30 = 3x$$

$$\frac{30}{3} = \frac{3x}{3}$$

$$10 = x$$

6. In each of the following, replace y by the given value and solve the resulting equation for x.

a. If $y = 3x - 5$ and $y = 10$, determine x.

$$10 = 3x - 5$$
$$15 = 3x$$
$$x = 5$$

b. If $y = 30 - 2x$ and $y = 24$, determine x.

$$24 = 30 - 2x$$
$$-6 = -2x$$
$$x = 3$$

c. If $y = \frac{3}{4}x - 21$ and $y = -9$, determine x.

$$-9 = \frac{3}{4}x - 21$$
$$12 = \frac{3}{4}x$$
$$\frac{12 \cdot 4}{3} = x$$
$$x = 16$$

d. If $-2x + 15 = y$ and $y = -3$, determine x.

$$-2x + 15 = -3$$
$$-2x = -18$$
$$x = 9$$

7. Complete the following tables.

a. $y = 4x - 11$

$$y = 4 \cdot 6 - 11 \qquad 53 = 4x - 11$$
$$y = 13 \qquad\qquad 64 = 4x$$
$$\qquad\qquad\qquad x = 16$$

x	y
6	13
16	53

b. $y = -5x - 80$

$$y = -5 \cdot 12 - 80 \qquad 35 = -5x - 80$$
$$y = -140 \qquad\qquad 115 = -5x$$
$$\qquad\qquad\qquad x = -23$$

x	y
12	−140
−23	35

SUMMARY: ACTIVITY 2.7

1. To solve an equation containing two or more operations, perform the sequence of operations in *reverse* order, replacing each original operation with its *inverse*.

EXERCISES: ACTIVITY 2.7

1. In 2010, the number of full-time students at your local community college was 1850. The admissions office anticipated that the enrollment will increase at a constant rate of 90 students per year for the next several years.

a. Write a verbal rule to determine the enrollment of full-time students in a given number of years after 2010.

Multiply the number of years after 2010 by 90, then add 1850 to determine the total enrollment of full-time students in the given year.

b. Translate the verbal rule in part a into a symbolic rule. Use N to represent the total enrollment of full-time students and t to represent the number of years after 2010.

$N = 90t + 1850$

Exercise numbers appearing in color are answered in the Selected Answers appendix.

c. Determine the total enrollment of full-time students in the year 2013.

Let $t = 3$; $N = 90(3) + 1850 = 270 + 1850 = 2120$ full-time students

d. In what year is the total enrollment of full-time students expected to be 2300?

$2300 = 90t + 1850$

$450 = 90t$

$5 = t$

The enrollment of full-time students is expected to be 2300 in the year 2015.

2. You are considering taking some courses at your local community college on a part-time basis for the upcoming semester. For a student who carries fewer than 12 credits (the full-time minimum), the tuition is $175 for each credit hour taken. All students, part-time or full-time, must pay a $50 parking fee for the semester. This fixed fee is added directly to your tuition and is included in your total bill.

a. Complete the following table, where t represents the total bill and n represents the number of credit hours taken.

Number of Credit Hours, n	1	2	3	4	5	6
Total Tuition, t ($)	225	400	575	750	925	1100

b. Write a symbolic rule to determine the total bill, t, for a student carrying fewer than 12 credit hours. Use n to represent the number of hours taken for the semester.

$t = 175n + 50$

c. Determine the total bill if you take 9 credit hours.

$t = 175 \cdot 9 + 50 = \$1625$

The total bill for 9 credit hours is $1625.

d. Suppose you have $1650 to spend on the total bill. How many credit hours can you carry for the semester?

$1650 = 175n + 50$

$1600 = 175n$

$n \approx 9.1$

I can take 9 credits.

3. The social psychology class is organizing a campus-entertainment night to benefit charities in the community. You are a member of the budget committee for the class project. The committee suggests a $10 per person admission donation for food, nonalcoholic beverages, and entertainment. The committee determines that the fixed costs for the event (food, drinks, posters, tickets, etc.) will total $2100. The college is donating the use of the gymnasium for the evening.

a. The total revenue (gross income before expenses are deducted) depends on the number, n, of students who attend. Write an expression in terms of n that represents the total revenue if n students attend.

$10n$

b. Profit is the net income after expenses are deducted. Write a symbolic rule expressing the profit, p, in terms of the number, n, of students who attend.

$p = 10n - 2100$

c. If the gymnasium holds a maximum of 700 people, what is the maximum amount of money that can be donated to charity?

$p = 10 \cdot 700 - 2100 = \$4900$

The maximum amount of money that can be donated to charity is $4900.

d. Suppose that the members of the committee want to be able to donate at least $1500 to community charities. How many students must attend in order to have a profit of $1500?

$1500 = 10n - 2100$

$3600 = 10n$

$n = 360$

In order to have a profit of $1500, at least 360 students must attend.

4. The value of most assets, such as a car, computer, or house, depreciates, or drops, over time. When an asset depreciates by a fixed amount per year, the depreciation is called straight-line depreciation. Suppose a car has an initial value of $12,400 and depreciates $820 per year.

a. Let v represent the value of the car after t years. Write a symbolic rule that expresses v in terms of t.

$v = 12,400 - 820t$

b. What is the value of the car after 4 years?

$v = 12,400 - 820 \cdot 4 = \$9120$

After 4 years, the car is worth $9120.

c. How long will it take for the value of the car to decrease to $2000?

$2000 = 12,400 - 820t$

$-10,400 = -820t$

$t \approx 12.68$

It will take about 13 years for the value of the car to decrease to $2000.

5. The cost, c, in dollars, of mailing a priority overnight package weighing 1 pound or more is given by the formula $c = 2.085x + 15.08$, where x represents the weight of the package in pounds.

a. Determine the cost of mailing a package that weighs 10 pounds.

$c = 2.085 \cdot 10 + 15.08 = \$35.93$

The mailing cost of a 10-pound package is $35.93.

b. Determine the weight of a package that costs $56.78 to mail.

$56.78 = 2.085x + 15.08$

$41.70 = 2.085x$

$x = 20 \text{ lb.}$

A package that costs $56.78 to mail weighs 20 pounds.

6. Archaeologists and forensic scientists use the length of human bones to estimate the height of individuals. A person's height, h, in centimeters, can be determined from the length of the femur, f (the bone from the knee to the hip socket), in centimeters, using the following formulas:

Man: $h = 69.089 + 2.238f$

Woman: $h = 61.412 + 2.317f$

a. A partial skeleton of a man is found. The femur measures 50 centimeters. How tall was the man?

$h = 69.089 + 2.238(50) = 180.989$ cm ≈ 180.99 cm

The man was approximately 181 centimeters tall.

b. What is the length of the femur for a woman who is 150 centimeters tall?

$150 = 61.412 + 2.317f$

$88.588 = 2.317f$

$f \approx 38.23$ cm

The woman's femur is approximately 38.23 centimeters long.

7. Let p represent the perimeter of an isosceles triangle that has two equal sides of length a and a third side of length b. The formula for the perimeter is $p = 2a + b$. Determine the length of the equal side of an isosceles triangle having perimeter of $\frac{3}{4}$ yard and a third side measuring $\frac{1}{3}$ yard.

$\frac{3}{4} = 2a + \frac{1}{3}$

$3 = 8a + \frac{4}{3}$

$9 = 24a + 4$

$\frac{5}{24} = a$

The equal sides are $\frac{5}{24}$ yard long.

8. The recommended weight of an adult male is given by the formula $w = \frac{11}{2}h - 220$, where w represents his recommended weight in pounds and h represents his height in inches. Determine the height of a man whose recommended weight is 165 pounds.

$165 = \frac{11}{2}h - 220$

$385 = \frac{11}{2}h$

$h = \frac{385 \cdot 2}{11}$

$h = 70$ in.

A man whose recommended weight is 165 pounds is 70 inches tall.

9. Even though housing prices have been declining in many parts of the country, in your neighborhood they have been increasing steadily since you bought your home in 2002. You purchased your home in 2002 for $130,000 and it steadily increased its value by $3500 a year. Let V represent the market value (in dollars) of your home and x, the length of time you have lived there (in years).

a. Complete the following table.

YEAR	x	V, MARKET VALUE
2002	0	$130,000
2007	5	$147,500
2010	8	$158,000

b. Write a symbolic rule that expresses V in terms of x.

$V = 130{,}000 + 3500x$

c. Determine the expected value of your home in 2012.

$2012 \Rightarrow x = 10$

$V = 130{,}000 + 3500(10)$

$V = \$165{,}000$

The value of my home will be $165,000 in 2012.

d. In which year is the value of your home expected to reach $186,000?

$$186,000 = 130,000 + 3500x$$
$$56,000 = 3500x$$
$$16 = x$$

Year: $2002 + 16 = 2018$

The value of my home is expected to reach $186,000 in 2018.

10. Solve each of the following equations for x.

a. $10 = 2x + 12$
$$-2 = 2x$$
$$x = -1$$

b. $-27 = -5x - 7$
$$-20 = -5x$$
$$x = 4$$

c. $3x - 26 = -14$
$$3x = 12$$
$$x = 4$$

d. $24 - 2x = 38$
$$-2x = 14$$
$$x = -7$$

e. $5x - 12 = 15$
$$5x = 27$$
$$x = \frac{27}{5} = 5.4$$

f. $-4x + 8 = 8$
$$-4x = 0$$
$$x = 0$$

g. $12 + \frac{1}{5}x = 9$
$$\frac{1}{5}x = -3$$
$$x = -15$$

h. $\frac{2}{3}x - 12 = 0$
$$\frac{2}{3}x = 12$$
$$x = \frac{12 \cdot 3}{2} = 18$$

i. $0.25x - 14.5 = 10$
$$0.25x = 24.5$$
$$x = 98$$

j. $5 = 2.5x - 20$
$$25 = 2.5x$$
$$x = 10$$

11. Complete the following tables using algebraic methods. Indicate the equation that results from replacing x or y by its assigned value.

a. $y = 2x - 10$

x	y
4	−2
12	14

$$y = 2 \cdot 4 - 10$$
$$y = 8 - 10$$
$$y = -2$$

$$14 = 2x - 10$$
$$24 = 2x$$
$$x = 12$$

b. $y = 20 + 0.5x$

x	y
3.5	21.75
−60	−10

$$y = 20 + 0.5 \cdot 3.5 = 21.75$$

$$-10 = 20 + 0.5x$$
$$-30 = 0.5x$$
$$-60 = x$$

c. $y = -3x + 15$

x	y
$\frac{2}{3}$	13
6	-3

$y = -3 \cdot \frac{2}{3} + 15$

$y = 13$

$-3 = -3x + 15$

$-18 = -3x$

$x = 6$

d. $y = 12 - \frac{3}{4}x$

x	y
-8	18
24	-6

$y = 12 - \frac{3}{4} \cdot (-8) = 12 + 6 = 18$

$-6 = 12 - \frac{3}{4}x$

$-18 = -\frac{3}{4}x$

$18 \cdot \frac{4}{3} = 24 = x$

e. $y = \frac{x}{5} - 2$

x	y
24	2.8
100	18

$y = \frac{24}{5} - 2$

$y = 4.8 - 2$

$y = 2.8$

$18 = \frac{x}{5} - 2$

$20 = \frac{x}{5}$

$100 = x$

Activity 2.8

The Algebra of Weather

Objectives

1. Evaluate formulas for specified input values.

2. Solve a formula for a specified variable.

Windchill

On Monday morning you listen to the news and weather before going to class. The meteorologist reports that the temperature is 25°F, a balmy February day on the campus of SUNY Oswego in New York State, but he adds that a 30-mile-per-hour wind makes it feel like 8°F.

Curious, you do some research and learn that windchill is the term commonly used to describe how cold your skin feels if it is exposed to the wind. You also discover a formula that allows you to calculate windchill as a temperature relative to air temperature and a 30-mile-per-hour wind.

1. Windchill temperature, w (°F), produced by a 30-mile-per-hour wind at various air temperatures, t (°F), can be modeled by the formula

$$w = 1.36t - 26.$$

Complete the following table using the given formula.

Air temperature, t (°F)	− 15	5	30
Windchill temperature, w (°F) (30-mph wind)	− 46	− 19	15

2. **a.** Use the formula $w = 1.36t - 26$ to determine the windchill temperature if the air temperature is 7°F.

$$w = 1.36(7) - 26 = -16.48$$

The windchill temperature is − 16°F if the air temperature is 7°F.

b. On a cold day in New York City, the wind is blowing at 30 miles per hour. If the windchill temperature is reported to be − 18°F, then what is the air temperature on that day? Use the formula $w = 1.36t - 26$.

$$-18 = 1.36t - 26$$
$$8 = 1.36t$$
$$t = \frac{8}{1.36} \approx 6$$

If the windchill is − 18°F, then the air temperature would be approximately 6°F.

The formula $w = 1.36t - 26$ is said to be solved for w in terms of t because the variable w is isolated on one side of the equation. To determine the windchill temperature, w, for air temperature $t = 7°F$ (Problem 2a), you substitute 7 for t in $1.36t - 26$ and do the arithmetic:

$$w = 1.36(7) - 26 = -16.48°F$$

In Problem 2b, you were asked to determine the air temperature, t, for a windchill temperature of − 18°F. In this case, you replaced w by − 18 and solved the resulting equation

$$-18 = 1.36t - 16$$

for t.

Each time you are given a windchill temperature and asked to determine the air temperature you need to set up and solve a similar equation. Often it is more convenient and efficient to solve the original formula $w = 1.36t - 26$ for t symbolically and then evaluate the new rule to determine values of t.

Example 1   *Solving the formula* $w = 1.36t - 26$ *for t is similar to solving the equation* $-18 = 1.36t - 26$ *for t.*

$$-18 = 1.36t - 26$$
$$\underline{+26 = +26}$$
$$8 = 1.36t$$
$$\frac{8}{1.36} = \frac{1.36t}{1.36}$$
$$6 \approx t$$

$$w = 1.36t - 26$$
$$\underline{+26 = +26}$$
$$w + 26 = 1.36t$$
$$\frac{w + 26}{1.36} = \frac{1.36t}{1.36}$$
$$\frac{w + 26}{1.36} = t$$

The new formula is $t = \dfrac{w + 26}{1.36}$ or $t = \dfrac{1}{1.36}w + \dfrac{26}{1.36}$, which is approximately equivalent to $t = 0.735w + 19.12$, after rounding.

> To solve the equation $w = 1.36t - 26$ for t means to rewrite the equation so that t becomes the output and w the input.

3. Redo Problem 2b using the new formula derived in Example 1 that expresses t in terms of w.

$$t = \frac{-18 + 26}{1.36} = \frac{8}{1.36} = 5.88 \approx 6$$

4. **a.** If the wind speed is 15 miles per hour, the windchill temperature, w, can be approximated by the formula $w = 1.28t - 19$, where t is the air temperature in degrees Fahrenheit. Solve the formula for t.

$$w = 1.28t - 19$$
$$w + 19 = 1.28t$$
$$\frac{w + 19}{1.28} = t$$

   b. Use the new formula from part a to determine the air temperature, t, if the windchill temperature is $-10°F$.

$$\frac{-10 + 19}{1.28} = t$$

$$t \approx 7$$

The air temperature is approximately 7°F if the windchill temperature is $-10°F$.

Weather Balloon

A weather balloon is launched at sea level. The balloon is carrying instruments that measure temperature during the balloon's trip. After the balloon is released, the data collected shows that the temperature dropped $0.0117°F$ for each meter that the balloon rose.

5. **a.** If the temperature at sea level is 50°F, determine the temperature at a height of 600 meters above sea level.

$50 - 0.0117(600) = 42.98$ The temperature is 43°F at a height of 600 meters above sea level.

b. Write a verbal rule to determine the temperature at a given height above sea level on a 50°F day.

Multiply the height by 0.0117 and then subtract this product from 50 to determine the temperature.

c. If t represents the temperature (°F) a height of m meters above sea level, translate the verbal rule in part b into a symbolic rule.

$t = 50 - 0.0117m$

d. Complete the following table.

Meters Above Sea Level, m	500	750	1000
Temperature, t (50°F Day)	44.2	41.2	38.3

Recall that subtracting a number (or symbol) can be understood as adding its opposite. It is helpful to think of the symbolic rule $t = 50 - 0.0117m$ equivalently as

$$t = 50 + (-0.0117)m$$

6. a. Solve the formula $t = 50 - 0.0117m$ for m.

$\frac{t - 50}{-0.0117} = m$

b. Water freezes at 32°F. Determine the height above sea level at which water will freeze on a 50°F day. Use the formula from part a.

$\frac{32 - 50}{-0.0117} \approx 1538.5$

On a 50°F day, water will freeze at 1538.5 meters above sea level.

Temperature to What Degree?

Since the 1970s, for almost all countries in the world, the Celsius scale is the choice for measuring temperature. The Fahrenheit scale is still the preference for nonscientific work in just a few countries, among them Belize, Jamaica, and the United States.

The scales are each based on the freezing point and boiling point properties of water and a straightforward relationship exists between the two scales. If F represents Fahrenheit temperature and C, Celsius temperature, the conversion formula for C in terms of F is

$$C = \frac{5}{9}(F - 32)$$

7. a. Average normal body temperature for humans is given to be 98.6°F. A body temperature 100°F or above usually indicates a fever. A patient in a hospital had a temperature of 39°C. Did she have a fever?

The following calculation shows that the temperature of 100°F is 37.8°C. $39 > 37.8$, so yes, the patient had a fever.

$C = \frac{5}{9}(100 - 32)$

$C = \frac{5}{9}(68)$

$C = 37.8°$

b. Solve the equation $C = \dfrac{5}{9}(F - 32)$ for F.

$$\frac{9}{5}C = \frac{9}{5} \cdot \frac{5}{9}(F - 32)$$

$$\frac{9}{5}C = F - 32$$

$$F = \frac{9}{5}C + 32$$

c. What is the patient's temperature of 39°C in terms of Fahrenheit degrees?

$$F = \frac{9}{5}(39) + 32$$

$$F = 102.2°$$

The patient's temperature is 102.2°F.

8. Solve each of the following formulas for the indicated variable.

a. $A = lw$, for w

$\dfrac{A}{l} = w$

b. $p = c + m$, for m

$p - c = m$

c. $P = 2l + 2w$, for l

$\dfrac{P - 2w}{2} = l$

d. $R = 165 - 0.75a$, for a

$\dfrac{R - 165}{-0.75} = a$

e. $V = \pi r^2 h$, for h

$\dfrac{V}{\pi r^2} = h$

SUMMARY: ACTIVITY 2.8

To solve a **formula** for a variable, isolate that variable on one side of the equation with all other terms on the "opposite" side.

EXERCISES: ACTIVITY 2.8

1. The profit that a business makes is the difference between its revenue (the money it takes in) and its costs.

a. Write a formula that describes the relationship between the profit, p, revenue, r, and costs, c.

$p = r - c$

b. It costs a publishing company \$85,400 to produce a textbook. The revenue from the sale of the textbook is \$315,000. Determine the profit.

$p = 315,000 - 85,400$

$= 229,600$

The profit is \$229,600.

c. The sales from another textbook amount to $877,000, and the company earns a profit of $465,000. Use your formula from part a to determine the cost of producing the book.

$r = \$877,000$

$p = \$465,000$

$p = r - c$

$\quad 465,000 = 877,000 - c$

$-412,000 = -c$

$\quad 412,000 = c$

The cost of producing the book is $412,000.

2. The distance traveled is the product of the rate (speed) at which you travel and the amount of time you travel at that rate.

a. Write a symbolic rule that describes the relationship between the distance, d, rate, r, and time, t.

$d = r \cdot t$

b. A gray whale can swim 20 hours a day at an average speed of approximately 3.5 miles per hour. How far can the whale swim in a day?

$d = 3.5(20) = 70$

A gray whale can swim 70 miles in a day.

c. A Boeing 747 flies 1950 miles at an average speed of 575 miles per hour. Use the formula from part a to determine the flying time. (Round to nearest tenth of an hour.)

$1950 = 575t$

$\dfrac{1950}{575} = \dfrac{575t}{575}$

$3.39 \approx t$

Flying time is approximately 3.4 hours.

d. Solve the formula in part a for the variable t. Then use this new formula to rework part c.

$d = rt$

$\dfrac{d}{r} = \dfrac{rt}{r}$

$t = \dfrac{d}{r} = \dfrac{1950}{575} \approx 3.39$

Flying time is approximately 3.4 hours.

3. The speed, s, of an ant (in centimeters per second) is related to the temperature, t (in degrees Celsius), by the formula

$$s = 0.167t - 0.67.$$

a. If an ant is moving at 4 centimeters per second, what is the temperature? (Round to nearest degree.)

$\quad 4 = 0.167t - 0.67$

$4.67 = 0.167t$

$\quad t = 27.96 \approx 28$

If an ant is moving 4 centimeters per second, then the temperature is approximately 28°C.

b. Solve the equation $s = 0.167t - 0.67$ for t.

$$\frac{s + 0.67}{0.167} = t$$

c. Use the new formula from part b to answer part a.

$$\frac{4 + 0.67}{0.167} = t = \frac{4.67}{0.167} \approx 28$$

If an ant is moving 4 centimeters per second, then the temperature is approximately 28°C.

4. The number of women enrolled in college has steadily increased. The following table gives the enrollment, in millions, of women in a given year.

Women in College

Year	1970	1975	1980	1985	1990	1995	2000	2005	2007
Number Enrolled in Millions	3.54	5.04	6.22	6.43	7.54	7.92	8.59	10.0	10.4

Source: U.S. Department of Education, National Center for Education Statistics.

Let t represent the number of years since 1970. The number N (in millions) of women enrolled in college can be modeled by the formula

$$N = 0.17t + 3.99.$$

a. Estimate in what year women's college enrollment will reach 12 million.

$12 = 0.17t + 3.99$

$8.01 = 0.17t$

$47 \approx t$

Year: $1970 + 47 = 2017$

According to the model, women's enrollment should reach 12 million in the year 2017.

b. Solve the equation $N = 0.17t + 3.99$ for t.

$$t = \frac{N - 3.99}{0.17}$$

c. Use the new formula in part b to estimate the year in which women's enrollment will reach 13 million.

$$t = \frac{13 - 3.99}{0.17} = 53$$

Year: $1970 + 53 = 2023$

According to the model, women's enrollment should reach 13 million in the year 2023.

5. The pressure, p, of water (in pounds per square foot) at a depth of d feet below the surface is given by the formula

$$p = 15 + \frac{15}{33}d.$$

a. On November 14, 1993, Francisco Ferreras reached a record depth for breath-held diving. During the dive, he experienced a pressure of 201 pounds per square foot. What was his record depth?

$$201 = 15 + \tfrac{15}{33}d$$
$$186 = \tfrac{15}{33}d$$
$$186 \cdot \tfrac{33}{15} = d$$
$$d = 409.2$$

Ferreras's record breath-held diving depth was approximately 409 feet.

b. Solve the equation $p = 15 + \dfrac{15}{33}d$ for d.

$$\frac{33(p - 15)}{15} = d$$

c. Use the formula from part b to determine the record depth and compare your answer to the one you obtained in part a.

$\frac{33(201 - 15)}{15} = 409.2$ feet, the same depth I obtained in part a.

Solve each of the following formulas for the specified variable.

6. $E = IR$, for I

$\frac{E}{R} = I$

7. $C = 2\pi r$, for r

$\frac{C}{2\pi} = r$

8. $P = 2a + b$, for b

$P - 2a = b$

9. $P = 2l + 2w$, for w

$\frac{P - 2l}{2} = w$

10. $R = 143 - 0.65a$, for a

$\frac{R - 143}{-0.65} = a$

11. $A = P + Prt$, for r

$\frac{A - P}{Pt} = r$

12. $y = mx + b$, for m

$\frac{y - b}{x} = m$

13. $m = g - vt^2$, for g

$m + vt^2 = g$

Activity 2.9

Four out of Five Dentists Prefer Crest

Objectives

1. Recognize that equivalent fractions lead to proportions.

2. Use proportions to solve problems involving ratios and rates.

3. Solve proportions.

Manufacturers of retail products often conduct surveys to see how well their products are selling compared to competing products. In some cases, they use the results in advertising campaigns.

One company, Proctor and Gamble, ran a TV ad in the 1970s claiming that "four out of five dentists surveyed preferred Crest toothpaste over other leading brands." The ad became a classic, and the phrase, *four out of five prefer*, has become a popular cliché in advertising, in appeals, and in one-line quips.

1. Suppose Proctor and Gamble asked 250 dentists what brand toothpaste he or she preferred and 200 dentists responded that they preferred Crest.

    a. What is the ratio of the number of dentists who preferred Crest to the number who were asked the question? Write your answer in words, and as a fraction.

    200 out of 250: $\dfrac{200}{250}$

    b. Reduce the ratio in part a to lowest terms and write the result in words.

    $\dfrac{200 \div 50}{250 \div 50} = \dfrac{4}{5}$; 4 out of 5

Note that part b of Problem 1 shows that the ratio $\dfrac{200}{250}$ is equivalent to the ratio $\dfrac{4}{5}$, or $\dfrac{200}{250} = \dfrac{4}{5}$. This is an example of a **proportion**.

Definition

The mathematical statement that two ratios are equivalent is called a **proportion**.

In fraction form, a proportion is written $\dfrac{a}{b} = \dfrac{c}{d}$.

If one of the component numbers a, b, c, or d in a proportion is unknown and the other three are known, the equation can be solved algebraically for the unknown component.

2. Suppose that Proctor and Gamble had interviewed 600 dentists and reported that $\dfrac{4}{5}$ of the 600 dentists preferred Crest. However, the report did not say how many dentists in this survey preferred Crest.

    a. Let x represent the number of dentists who preferred Crest. Write a ratio of the number of dentists who preferred Crest to the number of dentists who were in the survey.

    $\dfrac{x}{600}$

    b. According to the report, the ratio in part a must be equivalent to $\dfrac{4}{5}$. Write a corresponding proportion.

    $\dfrac{x}{600} = \dfrac{4}{5}$

One method of solving a proportion uses the fact that

$$\frac{a}{b} = \frac{c}{d} \quad \text{and} \quad a \cdot d = b \cdot c \quad \text{are equivalent.}$$

Transforming the equation on the left (containing two ratios) into the equation on the right is commonly called **cross multiplication**. The numerator of the first fraction is multiplied by the denominator of the second, and the numerator of the second fraction is multiplied by the denominator of the first.

The following example demonstrates how to use cross multiplication to solve a proportion.

Example 1   *Use cross multiplication to solve the proportion* $\dfrac{2}{5} = \dfrac{x}{6000}$.

SOLUTION

$\dfrac{2}{5} = \dfrac{x}{6000}$ **Cross multiply**

$2 \cdot 6000 = 5x$ **Solve for x**

$12{,}000 = 5x$

$2400 = x$

3. Use cross multiplication to solve the proportion in Problem 2b.

$\dfrac{x}{600} = \dfrac{4}{5}$

$5x = 2400$

$x = 480$

Therefore, 480 dentists in the survey preferred Crest.

4. The best season for the New York Mets baseball team was 1986 when they won $\dfrac{2}{3}$ of the games they played and won the World Series, beating the Boston Red Sox. If the Mets played 162 games, how many did they win?

Let x represent the number of games the Mets won.

$\dfrac{2}{3} = \dfrac{x}{162}$

$3x = 324$

$x = 108$

The Mets won 108 games in 1986.

Sometimes, the unknown in a proportion is in the denominator of one of the fractions. The next example shows a method of solving a proportion that is especially useful in this situation.

Example 2   *The Los Angeles Dodgers were once the Brooklyn Dodgers team that made it to the World Series seven times from 1941 to 1956, each time opposing the New York Yankees. The Brooklyn Dodgers won only one World Series Championship against the Yankees and that was in 1955.*

During the 1955 regular season, the Dodgers won about 16 out of every 25 games that they played. They won a total of 98 games. How many games did they play during the regular season?

SOLUTION

Let x represent the total number of games that the Dodgers played in the 1955 regular season. To write a proportion, the units of the numerators in the proportion must be the same. Similarly, the units of the denominators must be the same. In this case, the proportion is

$$\frac{16 \text{ games won}}{25 \text{ total games played}} = \frac{98 \text{ games won}}{x \text{ total games played}} \quad \text{or} \quad \frac{16}{25} = \frac{98}{x}$$

$$\frac{16}{25} = \frac{98}{x}$$

$$16x = 98 \cdot 25$$

$$x = \frac{98 \cdot 25}{16} \approx 153$$

The Dodgers played 153 games in the 1955 season.

5. During the economic recession a company was forced to lay off 12 workers, representing $\frac{3}{8}$ of its employees. How many workers had been employed before these layoffs?

$$\frac{12}{x} = \frac{3}{8}$$

$$3 \cdot x = 12 \cdot 8$$

$$\frac{3 \cdot x}{3} = \frac{12 \cdot 8}{3}$$

$$x = 32$$

Therefore, 32 workers had been employed before the layoff.

Additional Applications

6. Solve each proportion for x.

a. $\frac{7}{10} = \frac{x}{24}$

$$10x = 7 \cdot 24$$

$$x = \frac{7 \cdot 24}{10}$$

$$x = 16.8$$

b. $\frac{9}{x} = \frac{6}{50}$

$$\frac{9}{x} = \frac{6}{50}$$

$$6x = 50 \cdot 9$$

$$x = \frac{50 \cdot 9}{6} = 75$$

c. $\frac{2}{3} = \frac{x}{48}$

$$3x = 2 \cdot 48$$

$$x = \frac{2 \cdot 48}{3} = 32$$

d. $\frac{5}{8} = \frac{120}{x}$

$$5x = 8 \cdot 120$$

$$x = \frac{8 \cdot 120}{5} = 192$$

e. $\frac{3}{20} = \frac{x}{3500}$

$$20x = 3 \cdot 3500$$

$$x = \frac{3 \cdot 3500}{20} = 525$$

7. As a volunteer for a charity, you were given a job stuffing envelopes for an appeal for donations. After stuffing 240 envelopes, you were informed that you are two-thirds done. How many envelopes in total are you expected to stuff?

Let x represent the number of envelopes to be stuffed.

$$\frac{2}{3} = \frac{240}{x}$$

$$2x = 3 \cdot 240$$

$$x = \frac{3 \cdot 240}{2} = 360$$

I am expected to stuff 360 envelopes.

8. You are entertaining a group of six friends who are joining you to watch a figure skating competition and everyone wants soup. The information on a can of soup states that one can of soup contains about 2.5 servings. Use a proportion to determine the number of cans of soup to open so that all seven of you have one serving of soup.

Let x represent the number of cans to be opened.

$$\frac{1 \text{ can}}{2.5 \text{ servings}} = \frac{x \text{ cans}}{7 \text{ servings}}$$

$$2.5x = 7$$

$$x = \frac{7}{2.5} = 2.8$$

I will have to open 3 cans of soup so that we will each have one serving.

SUMMARY: ACTIVITY 2.9

1. The mathematical statement that two ratios are equivalent is called a **proportion**. In fraction form, a proportion is written as $\frac{a}{b} = \frac{c}{d}$.

2. Shortcut (cross-multiplication) procedure for solving proportions: Rewrite the proportion $\frac{a}{b} = \frac{c}{d}$ as $a \cdot d = b \cdot c$ and solve for the unknown quantity.

EXERCISES: ACTIVITY 2.9

1. A company that manufactures optical products estimated that three out of five people in the United States and Canada wore eyeglasses in 2006. The population estimate for the United States and Canada in 2006 was about 334 million. Use a proportion to determine how many people in this part of the world were estimated to wear eyeglasses in 2006.

$$\frac{3}{5} = \frac{x}{334}$$

$$5x = 3 \cdot 334$$

$$x = 3 \cdot 334 \div 5$$

$$x = 200.4$$

The company estimates that in 2006 about 200 million Americans and Canadians wore glasses.

2. Solve each proportion for x.

a. $\frac{2}{9} = \frac{x}{108}$

$9x = 2 \cdot 108$

$x = \frac{2 \cdot 108}{9} = 24$

b. $\frac{8}{7} = \frac{120}{x}$

$8x = 7 \cdot 120$

$x = \frac{7 \cdot 120}{8} = 105$

c. $\frac{x}{20} = \frac{70}{100}$

$100x = 20 \cdot 70$

$x = \frac{20 \cdot 70}{100} = 14$

Exercise numbers appearing in color are answered in the Selected Answers appendix.

3. The Center for Education Reform is an organization founded in 1993 to help foster better education opportunities in American communities. It spends 4 cents out of every dollar in its spending budget for administrative expenses. In 2004, the center's spending budget was approximately 2.8 million dollars. Use a proportion to estimate the administrative expenses for 2004.

$$\frac{4}{100} = \frac{x}{2,800,000}$$

$$x = \frac{4 \cdot 2,800,000}{100} = 112,000$$

The center spent approximately $112,000 on administrative costs.

4. Powdered skim milk sells for $6.99 a box in the supermarket. The amount in the box is enough to make 8 quarts of skim milk. What is the price for 12 quarts of skim milk prepared in this way?

$$\frac{\$6.99}{8 \text{ qt.}} = \frac{\$x}{12 \text{ qt.}}$$

$$x = \frac{12 \cdot 6.99}{8}$$

$$x \approx 10.49$$

Twelve quarts of skim milk prepared this way cost $10.49.

5. A person who weighs 120 pounds on Earth would weigh 42.5 pounds on Mars. What would be a person's weight on Mars if she weighs 150 pounds on Earth?

$$\frac{42.5 \text{ Mars pounds}}{120 \text{ Earth pounds}} = \frac{x \text{ Mars pounds}}{150 \text{ Earth pounds}}$$

$$x = \frac{42.5 \cdot 150}{120}$$

$$x \approx 53.1$$

A 150-pound woman on Earth would weigh about 53.1 pounds on Mars.

6. You want to make up a saline (salt) solution in chemistry lab that has 12 grams of salt per 100 milliliters of water. How many grams of salt would you use if you needed 15 milliliters of solution for your experiment?

$$\frac{x}{15} = \frac{12}{100}$$

$$x = \frac{12 \cdot 15}{100} = 1.8$$

I would use 1.8 grams of salt.

7. Tealeaf, a company that offers management solutions to companies that sell online, announced in a 2005 consumer survey that 9 out of 10 customers reported problems with transactions online. The survey sampled 1859 adults in the United States, 18 years and older, who had conducted an online transaction in the past year. How many of those adults reported problems with transactions online?

$$\frac{x}{1859} = \frac{9}{10}$$

$$x = \frac{9 \cdot 1859}{10}$$

$$x \approx 1673$$

1673 consumers in the sample reported problems with transactions online.

8. In 2005, the birth rate in the United States was estimated to be 14.14 births per 1000 persons. If the population estimate for the United States was 296.7 million, how many births were expected that year?

$$\frac{x \text{ births}}{296.7 \text{ million people}} = \frac{14.14 \text{ births}}{1000 \text{ people}}$$

$$x = \frac{14.14 \cdot 296.7}{1000}$$

$$x \approx 4.2$$

4.2 million births were expected in 2005, based on this data.

9. A conservationist can estimate the total number of a certain type of fish in a lake by using a technique called "capture-mark-recapture." Suppose a conservationist marks and releases 100 rainbow trout into a lake. A week later she nets 50 trout in the lake and finds 3 marked fish. An estimate of the number of rainbow trout in the lake can be determined by assuming that the proportion of marked fish in the sample taken is the same as the proportion of marked fish in the total population of rainbow trout in the lake. Use this information to estimate the population of rainbow trout in the lake.

Let x represent the total population of rainbow trout in the lake.

$$\frac{3}{50} = \frac{100}{x}$$

$$3x = 5000$$

$$x = 1666.6$$

Therefore, there are approximately 1667 rainbow trout in the lake.

10. The school taxes on a house assessed at $210,000 are $2340. At the same tax rate, what are the taxes (to the nearest dollar) on a house assessed at $275,000?

Let x represent the amount of school taxes.

$$\frac{2340}{210,000} = \frac{x}{275,000}$$

$$210,000x = 2340(275,000)$$

$$x = 3064.285$$

Therefore, the school tax is $3064.

11. The designers of sport shoes assume that the force exerted on the soles of shoes in a jump shot is proportional to the weight of the person jumping. A 140-pound athlete exerts a force of 1960 pounds on his shoe soles when he returns to the court floor after a jump. Determine the force that a 270-pound professional basketball player exerts on the soles of his shoes when he returns to the court floor after shooting a jump shot.

Let x represent the force on the shoe soles:

$$\frac{140}{1960} = \frac{270}{x}$$

$$140x = 529,200$$

$$x = 3780$$

The player exerts 3780 pounds of force.

Cluster 2 | What Have I Learned?

1. Describe how solving the equation $4x - 5 = 11$ for x is similar to solving the equation $4x - 5 = y$ for x.

In both cases, you need to isolate x by performing the inverse of the operations indicated by $4x - 5$ in reverse order; that is, add 5, and then divide by 4.

2. In the formula $d = rt$, assume that the rate, r, is 60 miles per hour. The formula then becomes the equation $d = 60t$.

a. Which variable is the input variable?

t is the input variable.

b. Which is the output variable?

d is the output variable.

c. Which variable from the original formula is now a constant?

r is a constant.

d. Discuss the similarities and differences in the equations $d = 60t$ and $y = 60x$.

The relationship between the input and output variables is the same. The variables just have different names.

3. Describe in words the sequence of operations that must be performed to solve for x in each of the following.

a. $10 = x - 16$

add 16

b. $-8 = \dfrac{1}{2}x$

divide by $\frac{1}{2}$ (or multiply by 2)

c. $-2x + 4 = -6$

subtract 4

divide by -2

d. $ax - b = c$

add b

divide by a

4. The area, A, of a triangle is given by the formula $A = \frac{1}{2}bh$, where b represents the base and h represents the height. The formula can be rewritten as $b = \frac{2A}{h}$ or $h = \frac{2A}{b}$. Which formula would you use to determine the base, b, of a triangle, given its area, A, and height, h? Explain.

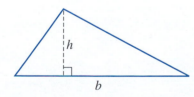

You would use the formula, $b = \frac{2A}{h}$, which contains the given values on the right side. It is easier to evaluate than to solve.

5. Explain how you would solve the proportion equation $\dfrac{5}{x} = \dfrac{35}{168}$.

By cross multiplication,

$$\frac{x}{5} = \frac{168}{35}$$

$$35 \cdot x = 168 \cdot 5$$

$$x = \frac{168 \cdot 5}{35} = 24$$

Exercise numbers appearing in color are answered in the Selected Answers appendix.

Cluster 2   How Can I Practice?

1. Let x represent the input variable. Translate each of the following phrases into an algebraic expression.

  a. input increased by 10

  $x + 10$

  b. the input subtracted from 10

  $10 - x$

  c. twelve divided by the input

  $\frac{12}{x}$

  d. eight less than the product of the input and -4

  $-4x - 8$

  e. the quotient of the input and 4, increased by 3

  $\frac{x}{4} + 3$

  f. one-half of the square of the input, decreased by 2

  $\frac{1}{2}x^2 - 2$

2. a. Write a symbolic rule that represents the relationship in which the output variable y is 35 less than the input variable x.

  $y = x - 35$

  b. What is the output corresponding to an input value of 52?

  $y = 52 - 35 = 17$ (output)

  c. What is the input corresponding to an output value of 123?

  $x - 35 = 123$

  $x = 158$ (input)

3. a. Write a symbolic rule that represents the relationship in which the output variable t is 10 more than 2.5 times the input variable r.

  $t = 2.5r + 10$

  b. What is the output corresponding to an input of 8?

  $t = 2.5(8) + 10 = 30$ (output)

  c. What is the input corresponding to an output of -65?

  $2.5r + 10 = -65$

  $2.5r = -75$

  $r = -30$ (input)

4. Use an algebraic approach to solve each of the following equations for x. In each case, check your result in the original equation.

  a. $x + 5 = 2$

  $\underline{-5 \quad -5}$

  $x = -3$

  b. $2x = -20$

  $\frac{2x}{2} = \frac{-20}{2}$

  $x = -10$

Exercise numbers appearing in color are answered in the Selected Answers appendix.

c. $x - 3.5 = 12$

$\underline{+3.5 +3.5}$

$x = 15.5$

d. $-x = 9$

$-1(-x) = -1(9)$

$x = -9$

e. $13 = x + 15$

$\underline{-15 -15}$

$-2 = x$

f. $4x - 7 = 9$

$\underline{+7 +7}$

$4x = 16$

$x = 4$

g. $10 = -2x + 3$

$7 = -2x$

$-\frac{7}{2} = x \text{or} x = -3.5$

h. $\frac{3}{5}x - 6 = 1$

$\frac{3}{5}x = 7$

$x = 7\left(\frac{5}{3}\right) = \frac{35}{3} = 11.\overline{6}$

5. Your car needs a few new parts to pass inspection. The labor cost is $68 an hour, charged by the half hour, and the parts cost a total of $148. Whether you can afford these repairs depends on how long it will take the mechanic to install the parts.

a. Write a verbal rule that will enable you to determine a total cost for repairs.

Multiply the number of hours of labor by $68, and add $148 for parts to obtain the total cost of the repair.

b. Write the symbolic form of this verbal rule, letting x be the input variable and y be the output variable. What does x represent? (Include its units.) What does y represent? (Include its units.)

$y = 148 + 68x$

Let x represent the number of hours worked, rounded up to the next half hour.

Let y represent the cost of the job, in dollars.

c. Use the symbolic rule in part b to create a table of values.

HOURS, x	TOTAL COST ($), y
$\frac{1}{2}$	182
1	216
$1\frac{1}{2}$	250
2	284
$2\frac{1}{2}$	318

d. How much will it cost if the mechanic works 4 hours?

It will cost $420 if the mechanic works 4 hours.

e. You have $350 available in your budget for car repair. Determine if you have enough money if the mechanic says that it will take him $3\frac{1}{2}$ hours to install the parts.

$3.5(68) + 148 = \$386$

No, 3.5 hours work plus parts will cost $386. You do not have enough.

f. You decide that you can spend an additional $100. How long can you afford to have the mechanic work?

$68x + 148 = 450$

$68x = 302$

$x \approx 4.44$

You can afford to have the mechanic work 4 hours.

g. Solve the symbolic rule from part b for the input variable x. Why would it ever be to your advantage to do this?

$$148 + 68x = y$$
$$68x = y - 148$$
$$x = \frac{y - 148}{68}$$

Solving for x allows you to answer questions such as part f more quickly. It would be especially useful if you wanted to determine the number of mechanic hours for several different amounts of money.

6. In 1966, the U.S. Surgeon General's health warnings began appearing on cigarette packages. The following data seems to demonstrate that public awareness of the health hazards of smoking has had some effect on consumption of cigarettes.

The percentage, p, of the total population (18 and older) who smoke t years after 1965 can be approximated by the model $p = -0.555t + 41.67$, where t is the number of years after 1965. Note that $t = 0$ corresponds to the year 1965, $t = 9$ corresponds to 1974, etc.

a. Determine the percentage of the population who smoked in 1981 ($t = 16$).

$$p = -0.555(16) + 41.67 = 32.79$$

Approximately 32.8% of the population smoked in 1981.

b. Using the formula, in what year would the percentage of smokers be 15% ($p = 15$)?

$$15 = -0.555t + 41.67$$
$$-26.67 = -0.555t$$
$$t \approx 48$$

Year: $1965 + 48 = 2013$

The formula predicts that the percentage of smokers will be 15% in 2013.

7. Medical researchers have determined that, for exercise to be beneficial, a person's desirable heart rate, R, in beats per minute, can be approximated by the formulas

$$R = 143 - 0.65a \text{ for women}$$
$$R = 165 - 0.75a \text{ for men,}$$

where a (years) represents the person's age.

a. If the desirable heart rate for a woman is 130 beats per minute, how old is she?

$$130 = 143 - 0.65a$$
$$-13 = -0.65a$$
$$a = 20$$

A woman whose desirable heart rate is 130 beats per minute is age 20.

b. If the desirable heart rate for a man is 135 beats per minute, how old is he?

$$135 = 165 - 0.75a$$
$$-30 = -0.75a$$
$$a = 40$$

A man whose desirable heart rate is 135 beats per minute is 40 years old.

8. The basal energy rate is the daily amount of energy (measured in calories) needed by the body at rest to maintain body temperature and the basic life processes of respiration, cell metabolism, circulation, and glandular activity. As you may suspect, the basal energy rate differs for individuals, depending on their gender, age, height, and weight. The formula for the basal energy rate for men is

$$B = 655.096 + 9.563W + 1.85H - 4.676A,$$

where B is the basal energy rate (in calories), W is the weight (in kilograms), H is the height (in centimeters), and A is the age (in years).

a. A male patient is 70 years old, weighs 55 kilograms, and is 172 centimeters tall. A total daily calorie intake of 1000 calories is prescribed for him. Determine if he is being properly fed.

$B = 655.096 + 9.563(55) + 1.85(172) - 4.676(70) \approx 1171.9$

He should be fed about 1172 calories a day. He is not properly fed; his prescribed diet is short 172 calories.

b. A man is 178 centimeters tall and weighs 84 kilograms. If his basal energy rate is 1500 calories, how old is the man?

$1500 = 655.096 + 9.563(84) + 1.85(178) - 4.676A$

$1500 = 1787.688 - 4.676A$

$-287.688 = -4.676A$

$61.52 \approx A$

He is about 62 years old.

9. Solve each of the following equations for the given variable.

a. $d = rt$, for r

$\frac{d}{t} = r$

b. $P = a + b + c$, for b

$P - a - c = b$

c. $A = P + Prt$, for r

$A - P = Prt$

$\frac{A - P}{Pt} = r$

d. $y = 4x - 5$, for x

$y + 5 = 4x$

$\frac{y + 5}{4} = x$

e. $w = \frac{4}{7}h + 3$, for h

$w - 3 = \frac{4}{7}h$

$\frac{7}{4}(w - 3) = h$

10. You are designing a cylindrical container as new packaging for a popular brand of coffee. The current package is a cylinder with a diameter of 4 inches and a height of 5.5 inches. The volume of a cylinder is given by the formula

$$V = \pi r^2 h,$$

where V is the volume (in cubic inches), r is the radius (in inches), and h is the height (in inches).

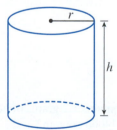

How much coffee does the current container hold? Round your answer to the nearest tenth of a cubic inch.

$r = \frac{d}{2} = \frac{4}{2} = 2$ in.

$h = 5.5$ in.

$V = \pi r^2 h = \pi(2)^2(5.5) \approx 69.1$ cu in.

The current container holds approximately 69.1 cubic inches of coffee.

11. You have been asked to alter the dimensions of the container in Problem 10 so that the new package will contain less coffee. To save money, the company plans to sell the new package for the same price as before.

You will do this in one of two ways:

i. By increasing the diameter and decreasing the height by $\frac{1}{2}$ inch each (resulting in a slightly wider and shorter can), or

ii. By decreasing the diameter and increasing the height by $\frac{1}{2}$ inch each (resulting in a slightly narrower and taller can).

a. Determine which new design, if either, will result in a package that holds less coffee than the current one.

   i. $d = 4.5$ in. $r = 2.25$ in. $h = 5$ in. $V \approx 79.5$ cu in.

   ii. $d = 3.5$ in. $r = 1.75$ in. $h = 6$ in. $V \approx 57.7$ cu in.

The second package holds 11.4 cubic inches less than the original one.

b. By what percent will you have decreased the volume?

Volume decreases by 11.4 cubic inches, which is approximately a 16.5% decrease.

$\frac{11.4}{69.1} \approx 16.5\%$

12. Solve each proportion for the unknown quantity.

a. $\frac{2}{3} = \frac{x}{48}$

$3x = 96$

$x = 32$

b. $\frac{5}{8} = \frac{120}{x}$

$5x = 960$

$x = 192$

13. According to the 2000 U.S. Census, 7 out of every 25 homes are heated by electricity. At this rate, predict how many homes in a community of 12,000 would be heated by electricity.

Let x represent the number of homes heated by electricity.

$\frac{7}{25} = \frac{x}{12,000}$

$25x = 84,000$

$x = 3,360$ homes

14. A hybrid car can travel 70 miles on one gallon of gas. Determine the amount of gas needed for a 500 mile trip.

Let x represent the amount of gas needed.

$\frac{1}{70} = \frac{x}{500}$

$x \approx 7.14$ gal. of gas

15. A normal 10-cubic centimeter specimen of human blood contains 1.2 grams of hemoglobin. How much hemoglobin would 16 cubic centimeters of the same blood contain?

Let h represent the amount of hemoglobin in the blood.

$\frac{10}{1.2} = \frac{16}{h}$

$10h = 19.2$

$h = 1.92$ g of hemoglobin

16. The ratio of the weight of an object on Mars to the weight of an object on Earth is 0.4 to 1. How much would a 170 pound astronaut weigh on Mars?

Let w represent the weight of astronaut on Mars.

$\frac{0.4}{1} = \frac{w}{170}$

$w = 68$ lb.

Cluster 3 | Problem Solving Using Algebra

Activity 2.10

Are They the Same?

Objectives

1. Translate verbal rules into symbolic (algebraic) rules.

2. Write algebraic expressions that involve grouping symbols.

3. Evaluate algebraic expressions containing two or more operations.

4. Identify equivalent algebraic expressions by examining their outputs.

Throughout this book, you will be solving problems from everyday life—in science, business, sports, social issues, finances, and so on—that require mathematics. This approach illustrates the fact that searching for solutions to everyday problems resulted in the development of the mathematics we use today.

When used to solve problems or to describe a situation, algebra is a powerful tool. In this cluster, you will be acquiring new skills working with algebraic expressions. You will also learn additional techniques to solve equations. These skills and techniques will provide you with valuable tools in developing your problem-solving abilities.

Equivalent Expressions

A major road-construction project in your neighborhood is forcing you to take a 3-mile detour each way when you leave and return home. To compute the round-trip mileage for a routine trip, you will have to double the usual one-way mileage, adding in the 3-mile detour.

Does it matter in which order you perform these operations? That is, to determine the round-trip mileage (output), do you

a. double the usual one-way mileage (input) and then add 3, or

b. add 3 to the usual one-way mileage (input) and then double the result?

Complete the following table to determine whether there is a difference in the results from using these two methods.

0 4 2 9 5 1 6 Mile After Mile

USUAL MILEAGE (INPUT)	RULE 1: TO OBTAIN THE OUTPUT, DOUBLE THE INPUT, THEN ADD 3.	RULE 2: TO OBTAIN THE OUTPUT, ADD 3 TO THE INPUT, THEN DOUBLE.
8	$16 + 3 = 19$	$11 \cdot 2 = 22$
10	$20 + 3 = 23$	$13 \cdot 2 = 26$
15	$30 + 3 = 33$	$18 \cdot 2 = 36$

1. Do rules 1 and 2 generate the same output values (round-trip mileage)?

   no

2. From which sequence of operations do you obtain the correct round-trip mileage? Explain why.

   The correct results are from rule 2. The 3-mile detour occurs in both one-way trips and has to be added in twice.

3. For each of the rules given in the preceding table, let x represent the input and y represent the output. Translate each verbal rule into a symbolic rule.

   $y_1 = 2x + 3$

   $y_2 = (x + 3) \cdot 2$

Algebraic expressions are said to be **equivalent** if identical inputs always produce the same outputs. The expressions $2x + 3$ (from rule 1) and $(x + 3) \cdot 2$ (from rule 2) are not equivalent. The table preceding Problem 1 shows that an input such as $x = 8$ produces different outputs.

4. The graphs of rules 1 and 2 are given on the same axes. Explain how the graphs demonstrate that the expressions $2x + 3$ and $(x + 3) \cdot 2$ are not equivalent.

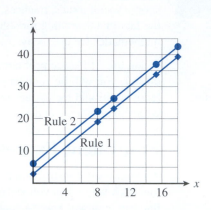

For a given input, such as $x = 8$, the outputs are different, leading to different points when graphed. Two different graphs mean the two expressions (rules) are not equivalent.

5. **a.** For each of the following, translate the verbal rule into a symbolic rule. Let x represent the input and y the output. Then, complete the given table.

Rule 3: To obtain y, multiply x by 2, and then add 10 to the product.

$y_3 = 2x + 10$

x	y
−1	8
0	10
2	14
5	20
10	30

Rule 4: To obtain y, add 5 to x, and then multiply the sum by 2.

$y_4 = 2(x + 5)$

x	y
−1	8
0	10
2	14
5	20
10	30

b. What do you notice about the outputs generated by rules 3 and 4?

The outputs are identical.

c. Is the expression $2x + 10$ equivalent to the expression $2(x + 5)$? Explain.

These expressions are equivalent because they generate the same output.

d. If you were to graph rule 3 and rule 4, how would you expect the graphs to compare?

I would expect the graphs to be identical.

6. a. For each of the following rules, translate the given verbal rule into a symbolic rule. Let x represent the input and y the output. Then, complete the tables.

Rule 5:   To obtain y, square the sum of x and 3.

$y = (x + 3)^2$

x	y
−1	4
0	9
2	25
5	64
10	169

Rule 6:   To obtain y, add 9 to the square of x.

$y = x^2 + 9$

b. What do you notice about the outputs generated by rules 5 and 6?

The outputs are different.

c. Is the expression $(x + 3)^2$ equivalent to the expression $x^2 + 9$? Explain.

No. They are *not* equivalent. They generate different outputs.

x	y
−1	10
0	9
2	13
5	34
10	109

d. If you were to graph rule 5 and rule 6, how would you expect the graphs to compare?

I would expect the graphs to be different.

SUMMARY: ACTIVITY 2.10

1. Algebraic expressions are said to be **equivalent** if identical inputs always produce the same output.

2. The order in which arithmetic operations are performed on the input affects the output.

3. The graphs of rules that contain equivalent expressions are identical.

EXERCISES: ACTIVITY 2.10

Let x represent the input variable and y represent the output variable. Translate each of the verbal rules in Exercises 1–6 into a symbolic rule.

1. The output is the input decreased by ten.

$y = x - 10$

2. The output is three times the difference between the input and four.

$y = 3(x - 4)$

3. The output is nine increased by the quotient of the input and six.

$y = 9 + \frac{x}{6}$

Exercise numbers appearing in color are answered in the Selected Answers appendix.

4. The output is seven more than one-half of the square of the input.

$y = \frac{1}{2}x^2 + 7$

5. The output is fifteen less than the product of the input and -4.

$y = -4x - 15$

6. The output is the sum of one-third of the input and ten.

$y = \frac{1}{3}x + 10$

7. **a.** Complete the following table.

x	x·3	3x
−4	−12	−12
−1	−3	−3
0	0	0
3	9	9

b. Do the expressions $x \cdot 3$ and $3x$ produce the same output value when given the same input value?

yes

c. In part b, the input value times 3 gives the same result as 3 times the input value. What property of multiplication does this demonstrate?

This demonstrates the *commutative* property of multiplication.

 d. Use a graphing calculator to sketch a graph of $y_1 = x \cdot 3$ and $y_2 = 3x$ on the same coordinate axes. How do the graphs compare?

 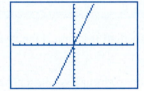

The graphs are the same. Since $x \cdot 3$ and $3x$ are the same, then their graphs coincide (identical inputs produce identical outputs).

8. **a.** Is the expression $x - 3$ equivalent to $3 - x$? Complete the following table to help justify your answer.

x	x − 3	3 − x
−5	−8	8
−3	−6	6
0	−3	3
1	−2	2
3	0	0

No, the expressions are not equivalent because they generate different output values.

b. What correspondence do you observe between the output values in columns 2 and 3? How is the expression $3 - x$ related to the expression $x - 3$?

The output values in columns 2 and 3 are opposites. $3 - x$ is the opposite (or negative) of $x - 3$.

c. Is the operation of subtraction commutative?

No, subtraction is *not* commutative. Reversing the order in subtraction changes the sign of the result.

 d. Use your graphing calculator to sketch a graph of $y_1 = x - 3$ and $y_2 = 3 - x$. How do the graphs compare? Why does this show that $x - 3$ and $3 - x$ are not equivalent expressions?

 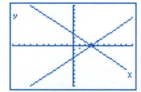

The graphs are different. If $x - 3$ and $3 - x$ were the same, then their graphs would coincide (identical inputs produce identical outputs).

9. a. Write a verbal rule that describes the sequence of operations indicated by the given symbolic rule. Then complete the tables.

i. $y_1 = x^2$

To obtain the output, square the input.

x	y_1
−2	4
−1	1
0	0
2	4
3	9

ii. $y_2 = -x^2$

To obtain the output, square the input and then change the sign.

x	y_2
−2	−4
−1	−1
0	0
2	−4
3	−9

iii. $y_3 = (-x)^2$

To obtain the output, change the sign of the input, and then square that result.

x	y_3
−2	4
−1	1
0	0
2	4
3	9

b. Which, if any, of the expressions x^2, $-x^2$, and $(-x)^2$ are equivalent? Explain.

The expressions x^2 and $(-x)^2$ are equivalent because they generate identical outputs.

The symbolic rules in the following exercises each contain two or more arithmetic operations. In each exercise, diagram the rule, complete the table, and then answer the question about your results.

10.

x	$y_1 = 2x + 1$	$y_2 = 2(x + 1)$
-1	-1	0
0	1	2
2	5	6
5	11	12

y_1: Start with $x \rightarrow$ multiply by 2 $\rightarrow$ add 1 $\rightarrow$ to obtain y.

y_2: Start with $x \rightarrow$ add 1 $\rightarrow$ multiply by 2 $\rightarrow$ to obtain y.

Are the expressions $2x + 1$ and $2(x + 1)$ equivalent? Why or why not?

No. They generate different outputs.

11.

x	$y_5 = 1 + x^2$	$y_6 = (1 + x)^2$
-1	2	0
0	1	1
2	5	9
5	26	36

y_5: Start with $x \rightarrow$ square $\rightarrow$ add 1 $\rightarrow$ to obtain y.

y_6: Start with $x \rightarrow$ add 1 $\rightarrow$ square $\rightarrow$ to obtain y.

Are the expressions $1 + x^2$ and $(1 + x)^2$ equivalent? Why or why not?

No. They generate different outputs.

12.

x	$y_7 = 2x^2 - 4$	$y_8 = 2(x^2 - 2)$
-3	14	14
-1	-2	-2
0	-4	-4
1	-2	-2
3	14	14

y_7: Start with $x \rightarrow$ square $\rightarrow$ multiply by 2 $\rightarrow$ subtract 4 $\rightarrow$ to obtain y.

y_8: Start with $x \rightarrow$ square $\rightarrow$ subtract 2 $\rightarrow$ multiply by 2 $\rightarrow$ to obtain y.

Are the expressions $2x^2 - 4$ and $2(x^2 - 2)$ equivalent? Why or why not?

Yes. They generate identical outputs (and graphs).

13.

x	$y_9 = 3x^2 + 1$	$y_{10} = (3x)^2 + 1$
-1	4	10
0	1	1
2	13	37
5	76	226

y_9: Start with $x \rightarrow$ square $\rightarrow$ multiply by 3 $\rightarrow$ add 1 $\rightarrow$ to obtain y.

y_{10}: Start with $x \rightarrow$ multiply by 3 $\rightarrow$ square $\rightarrow$ add 1 $\rightarrow$ to obtain y.

Are the expressions $3x^2 + 1$ and $(3x)^2 + 1$ equivalent? Why or why not?

No. They generate different outputs.

Activity 2.11

Do It Two Ways

Objectives

1. Apply the distributive property.

2. Use areas of rectangles to interpret the distributive property geometrically.

3. Identify equivalent expressions.

4. Identify the greatest common factor in an expression.

5. Factor out the greatest common factor in an expression.

6. Recognize like terms.

7. Simplify an expression by combining like terms.

1. You earn $8 per hour at your job and are paid every other week. You work 25 hours the first week and 15 hours the second week. Use two different approaches to compute your gross salary for the pay period. Explain in a sentence each of the approaches you used.

   i. $8(25) + 8(15) = 200 + 120 = \$320$

      I multiplied the hours from each week by 8 and then added the products to obtain $320 as the gross salary for the pay period.

   ii. $8(25 + 15) = 8(40) = \$320$

      I added the hours for the two weeks and then multiplied the sum by 8 to obtain $320 as the gross salary for the pay period.

Problem 1 demonstrates the **distributive property** of multiplication over addition. In the problem, the distributive property asserts that adding the hours first and then multiplying the sum by $8 produces the same gross salary as does multiplying separately each week's hours by $8 and then adding the weekly salaries.

> The **distributive property** is expressed algebraically as
>
> $$a \cdot (b + c) = a \cdot b + a \cdot c$$
>
> factored form   expanded form
>
> Note that in factored form, you add first, and then multiply. In the expanded form, you calculate the individual products first and then add the results.

Geometric Interpretation of the Distributive Property

The distributive property can also be interpreted geometrically. Consider the following diagram:

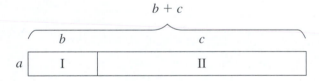

2. **a.** Write an expression for the area of rectangle I in the diagram.

   $a \cdot b$

   b. Write an expression for the area of rectangle II in the diagram.

   $a \cdot c$

   c. Write an expression for the area of the rectangle having width a and total top length $b + c$.

   $a(b + c)$

   d. Explain in terms of the areas in the geometric diagram why $a(b + c)$ equals $ab + ac$.

   The area of the large rectangle represented by $a(b + c)$ is equal to the sum of the areas of the two smaller rectangles represented by ab and ac.

The distributive property is frequently used to transform one algebraic expression into an equivalent expression.

Example 1   *$2(x + 5)$ can be transformed into $2 \cdot x + 2 \cdot 5$, which is usually written as $2x + 10$.*

The *factored form*, $2(x + 5)$, indicates that you start with x, add 5, and then multiply by 2. The *expanded form*, $2x + 10$, indicates that you start with x, multiply by 2, and then add 10.

A few definitions (some you will recall from Activity 2.4) will be helpful for understanding Example 1 and all that follows.

- A **term** is either a number or a product of a number and one or more variables. Terms can be easily identified as the parts of an algebraic expression that are added or subtracted.

  For example, the expression $13x + y - 7$ contains the three terms $13x$, y, and -7. Notice that the subtraction of the last term should be understood as addition of its opposite.

- When two or more mathematical expressions are multiplied to form a product, each of the original expressions is called a **factor** of the product.

  For example, the product $13x$ contains the two factors 13 and x.
  The product $5(x + 2)(x + 4)$ contains the three factors 5, $x + 2$, and $x + 4$.

- A numerical factor that multiplies a variable term is called the **coefficient** of the variable.

  For example, in the expression $13x + y - 7$, the coefficient of the first term is 13. The coefficient of the second term is understood to be 1. Because the third term has no variable factor, we simply call it a **constant** term.

The distributive property transforms a product such as $2(x + 5)$ into a sum of two terms $2x + 10$.

3. a. Complete the following table to demonstrate numerically that the expression $2(x + 5)$ is equivalent to the expression $2x + 10$:

INPUT	OUTPUT 1	OUTPUT 2
x	$y_1 = 2(x + 5)$	$y_2 = 2x + 10$
1	$y_1 = 2(1 + 5) = 12$	$y_2 = 2(1) + 10 = 12$
2	$y_1 = 2(2 + 5) = 14$	$y_2 = 2(2) + 10 = 14$
4	$y_1 = 2(4 + 5) = 18$	$y_2 = 2(4) + 10 = 18$
10	$y_1 = 2(10 + 5) = 30$	$y_2 = 2(10) + 10 = 30$

b. Explain how the table illustrates the equivalence of the two expressions.

The output of each rule agrees for every input in the table.

c. Use your graphing calculator to sketch the graphs of both symbolic rules. How are the graphs related? What does this indicate about the expressions in the two symbolic rules?

The graphs are identical, indicating that the two expressions will always produce equal outputs for a specified input, and thus are equivalent.

Using the Distributive Property to Expand a Factored Expression

There are two ways in which you can visualize the process of writing an expression such as $4(3x - 5)$ in expanded form using the distributive property.

First, you can make a diagram that looks very similar to a rectangular area problem. (See below.) Place the factor 4 on the left of the diagram. Place the terms of the expression $3x - 5$ along the top. Multiply each term along the top by 4, and then add the resulting products. Note that the terms do not actually represent lengths of the sides of the rectangle. The diagram is an organizational tool to help you apply the distributive property properly.

	$3x$	-5
4	$12x$	-20

Therefore, $4(3x - 5) = 12x - 20$.

Second, you can draw arrows to emphasize that each term within the parentheses is multiplied by the factor 4:

$$4(3x - 5) = 4(3x) - 4(5) = 12x - 20.$$

4. Use the distributive property to multiply and simplify each of the following expressions, writing in expanded form.

   a. $5(x + 6)$

   $5 \cdot x + 5 \cdot 6 = 5x + 30$

   b. $-10x(y + 11)$

   $-10x(y) + (-10x)(11) = -10xy - 110x$

   c. *Note:* A negative sign preceding parentheses indicates multiplication by -1. Use this fact to expand $-(2x - 7)$.

   $-1(2x) - (-1)(7) = -2x + 7$

   d. $3x(4 - 2x)$

   $12x - 6x^2$

Extension of the Distributive Property

The distributive property can be extended to sums of more than two terms within the parentheses. For example, it can be used to multiply $5(3x - 2y + 6)$.

To help you apply the distributive property, you can use the diagram approach,

	$3x$	$-2y$	6
5	$15x$	$-10y$	30

or you can use the arrows approach,

$$5(3x - 2y + 6) = 5 \cdot 3x + 5 \cdot (-2y) + 5 \cdot 6 = 15x - 10y + 30.$$

5. Verbally describe the procedure used to multiply $5(3x - 2y + 6)$.

   Multiply each term within the parentheses by 5. Then, sum the products.

6. Use the distributive property to write each of the expressions in expanded form.

   a. $-3(2x^2 - 4x - 5)$

   $-3(2x^2) - (-3)(4x) - (-3)(5) = -6x^2 + 12x + 15$

   b. $2x(3a + 4b - x)$

   $6xa + 8xb - 2x^2$

Using the Distributive Property to Factor an Expanded Expression

Consider the expression $5 \cdot x + 5 \cdot 3$, whose two terms both contain the common factor 5. The distributive property allows you to rewrite this expression in factored form by

i. dividing each term by the common factor to remove it from the term, and

ii. placing the common factor outside parentheses that contain the sum of the remaining factors.

Therefore, $5 \cdot x + 5 \cdot 3$ can be written equivalently as $5 \cdot (x + 3)$.

The challenge in transforming an expanded expression into factored form is identifying the common factor.

7. Identify the common factor in each of the following expressions and then rewrite the expression in factored form.

a. $2a + 6$

$2(a + 3)$

b. $5x + 3x - 7x$

$x(5 + 3 - 7)$

c. $3x + 12$

$3(x + 4)$

d. $2xy - 5x$

$x(2y - 5)$

Greatest Common Factor

A common factor is called a **greatest common factor** if there are no additional factors common to the terms in the expression. In the expression $12x + 30$, the numbers 2, 3, and 6 are common factors, but 6 is the greatest common factor. When an expression is written in factored form and the remaining terms in parentheses have no factors in common, the expression is said to be in **completely factored form**. Therefore, $12x + 30$ is written in completely factored form as $6(2x + 5)$.

8. a. What is the greatest common factor of $8x + 20$? Rewrite the expression in completely factored form.

The greatest common factor is 4.

$8x + 20 = 4(2x + 5)$

b. What are common factors in the sum $10x + 6x$? What is the greatest common factor? Rewrite the expression as an equivalent product.

The common factors are 2, x, and 2x. The greatest common factor is 2x.

$10x + 6x = 2x(5 + 3)$

Procedure

Factoring a Sum of Terms Containing a Common Factor

1. Identify the common factor.

2. Divide each term of the sum by the common factor. (Factor out the common factor.)

3. Place the sum of the remaining factors inside the parentheses, and place the common factor outside the parentheses.

9. Factor each of the following completely by factoring out the greatest common factor.

a. $6x + 18y - 24$

$6(x + 3y - 4)$

b. $5x^2 - 3x^2$

$x^2(5 - 3)$

c. $9x - 12x + 15x$

$3x(3 - 4 + 5)$

Like Terms

Consider the expression $2x + 5x + 3x$, in which each term contains the identical variable factor x. Because of this common variable factor, the three terms are called like terms. They differ only by their numerical coefficients.

Definition

Like terms are terms that contain identical variable factors, including exponents.

Example 2

a. $4x$ and $6x$ are like terms.

b. $4xy$ and $-10xy$ are like terms.

c. x^2 and $-10x^2$ are like terms.

d. $4x$ and $9y$ are not like terms.

e. $-3x^2$ and $-3x$ are not like terms.

Combining Like Terms

The distributive property provides a way to simplify expressions containing like terms by combining the like terms into a single term. You can do this by factoring out the common variable factor, as follows:

$$2x + 5x + 3x = (2 + 5 + 3)x = 10x,$$

where the coefficient 10 in the simplified term is precisely the sum of the coefficients of the original like terms.

Like terms can be combined by adding their coefficients. This is a direct result of the distributive property.

Example 3

15xy and $-8xy$ *are like terms with coefficients* **15** *and* -8, *respectively. Thus,* $15xy - 8xy = (15 - 8)xy = 7xy.$

10. Identify the like terms, if any, in each of the following expressions, and combine them.

a. $3x - 5y + 2z - 2x$

$3x, -2x$ are like terms.

$x - 5y + 2z$

b. $13s^2 + 6s - 4s^2$

$13s^2, -4s^2$ are like terms.

$9s^2 + 6s$

c. $2x + 5y - 4x + 3y - x$

$2x, -4x, -x$ are like terms.

$5y, 3y$ are like terms.

$-3x + 8y$

d. $3x^2 + 2x - 4x^2 - (-4x)$

$2x, 4x$ are like terms.

$3x^2, -4x^2$ are like terms.

$-1x^2 + 6x$

When there is no coefficient written immediately to the left of a set of parentheses, the number 1 is understood to be the coefficient of the expression in parentheses. For example, $45 - (x - 7)$ can be understood as $45 - 1(x - 7)$. Therefore, you can use the distributive property to multiply each term inside the parentheses by -1 and then combine like terms:

$$45 - 1(x - 7) = 45 - x + 7 = 52 - x.$$

11. Professor Sims brings calculators to class each day. There are 30 calculators in the bag she brings. She never knows how many students will be late for class on a given day. Her routine is to first remove 15 calculators and then to remove one additional calculator for each late arrival. If x represents the number of late arrivals on any given day, then the expression $30 - (15 + x)$ represents the number of calculators left in the bag after the late arrivals remove theirs.

a. Simplify the expression for Professor Sims so that she can more easily keep track of her calculators.

$30 - 15 - x = 15 - x$

b. Suppose there are four late arrivals. Evaluate both the original expression and the simplified expression. Compare your two results.

$x = 4$

$30 - (15 + 4) = 30 - 19 = 11$

$15 - 4 = 11$

The results are the same.

12. Use the distributive property to simplify and combine like terms.

a. $20 - (10 - x)$

$20 - 10 + x$

$= 10 + x$

b. $4x - (-2x + 3)$

$4x + 2x - 3$

$= 6x - 3$

c. $2x - 5y - 3(5x - 6y)$

$2x - 5y - 15x + 18y$

$= -13x + 13y$

d. $2(x - 3) - 4(x + 7)$

$2x - 6 - 4x - 28$

$= -2x - 34$

SUMMARY: ACTIVITY 2.11

1. The **distributive property** is expressed algebraically as $a \cdot (b + c) = a \cdot b + a \cdot c$, where $a, b,$ and c are any real numbers.

2. A **term** is either a number or a product of a number and one or more variables. Terms can be easily identified as the parts of an algebraic expression that are added or subtracted.

3. When two or more mathematical expressions are multiplied to form a product, each of the original expressions is called a **factor** of the product.

4. A numerical factor that multiplies a variable term is called the **coefficient** of the variable.

5. The **distributive property** can be extended to sums of more than two terms, as follows: $a \cdot (b + c + d) = a \cdot b + a \cdot c + a \cdot d.$

6. The process of writing a sum equivalently as a product is called **factoring**.

7. A factor common to each term of an algebraic expression is called a **common factor**.

8. The process of dividing each term by a common factor and placing it outside the parentheses containing the sum of the remaining terms is called **factoring out a common factor**.

9. A common factor of an expression involving a sum of terms is called a **greatest common factor** if the terms remaining inside the parentheses have no factor in common other than 1.

10. An algebraic expression is said to be in **completely factored form** when it is written in factored form and none of its factors can themselves be factored any further.

11. Procedure for factoring a sum of terms containing a common factor:

    a. Identify the common factor.

    b. Divide each term of the sum by the common factor. (Factor out the common factor.)

    c. Place the sum of the remaining factors inside the parentheses, and place the common factor outside the parentheses.

12. **Like terms** are terms that contain identical variable factors, including exponents. They differ only in their numerical coefficients.

13. **To combine like terms** of an algebraic expression, add or subtract their coefficients.

EXERCISES: ACTIVITY 2.11

1. Use the distributive property to expand the algebraic expression $10(x - 8)$. Then evaluate the factored form and the expanded form for these values of x: 5, -3, and $\frac{1}{2}$. What do you discover about these two algebraic expressions? Explain.

$10(x - 8) = 10x - 80$

$x = 5$: $10(5 - 8) = 10(-3) = -30$

$\qquad 10(5) - 10(8) = 50 - 80 = -30$

$x = -3$: $10(-3 - 8) = 10(-11) = -110$

$\qquad 10(-3) - 10(8) = -30 - 80 = -110$

$x = \frac{1}{2}$: $10(\frac{1}{2} - 8) = 10(-7.5) = -75$

$\qquad 10(\frac{1}{2}) - 10(8) = 5 - 80 = -75$

The values of the factored expression and the expanded expression are equal.

Use the distributive property to expand each of the algebraic expressions in Exercises 2–13.

2. $6(4x - 5)$

   $24x - 30$

3. $-7(t + 5.4)$

   $-7t - 37.8$

4. $2.5(4 - 2x)$

   $10 - 5x$

5. $3(2x^2 + 5x - 1)$

   $6x^2 + 15x - 3$

6. $-(3p - 17)$

   $-3p + 17$

7. $-(-2x - 3y)$

   $2x + 3y$

8. $-3(4x^2 - 3x + 7)$

   $-12x^2 + 9x - 21$

9. $-(4x + 10y - z)$

   $-4x - 10y + z$

10. $\frac{5}{6}(\frac{3}{4}x + \frac{2}{3})$

    $\frac{5}{8}x + \frac{5}{9}$

11. $-\frac{1}{2}(\frac{6}{7}x - \frac{2}{5})$

    $-\frac{3}{7}x + \frac{1}{5}$

Exercise numbers appearing in color are answered in the Selected Answers appendix.

12. $3x(5x - 4)$

$15x^2 - 12x$

13. $4a(2a + 3b - 6)$

$8a^2 + 12ab - 24a$

14. Consider the algebraic expression $8x^2 - 10x + 9y - z + 7$.

a. How many terms are in this expression?

There are five terms.

b. What is the coefficient of the first term?

The coefficient of the first term is 8.

c. What is the coefficient of the fourth term?

The coefficient of the fourth term is -1.

d. What is the coefficient of the second term?

The coefficient of the second term is -10.

e. What is the constant term?

The constant term is 7.

f. What are the factors in the first term?

The factors are 8 and x^2 (or 8 and x and x).

15. In chemistry, the ideal gas law is given by the equation $PV = n(T + 273)$, where P is the pressure, V the volume, T the temperature, and n the number of moles of gas. Write the right-hand side of the equation in expanded form (without parentheses).

$PV = nT + 273n$

16. In business, an initial deposit of P dollars, invested at a simple interest rate, r (in decimal form), will grow after t years, to amount A given by the formula

$$A = P(1 + rt).$$

Write the right side of the equation in expanded form (without parentheses).

$A = P + Prt$

17. a. The width, w, of a rectangle is increased by 5 units. Write an expression that represents the new width.

$w + 5$

b. If l represents the length of the rectangle, write a product that represents the area of the new rectangle.

$l(w + 5)$

c. Use the distributive property to write the expression in part b in expanded form (without parentheses).

$lw + 5l$

18. A rectangle has width w and length l. If the width is increased by 4 units and the length is decreased by 2 units, write a formula in expanded form that represents the perimeter of the new rectangle.

Let p represent the perimeter.

$p = 2(w + 4) + 2(l - 2)$

$p = 2w + 8 + 2l - 4$

$p = 2w + 2l + 4$

19. The manager of a local discount store reduces the regular retail price of a certain cell phone brand by $5.

  a. Use y to represent the regular retail price and write an expression that represents the discounted price of the cell phone.

  $y - 5$

  b. If 12 of these phones are sold at the reduced price, write an expression in expanded form (without parentheses) that represents the store's total receipts for the 12 phones.

  $12(y - 5) = 12y - 60$

20. Identify the greatest common factor and rewrite the expression in completely factored form.

  a. $3x + 15$

  $3(x + 5)$

  b. $5w - 10$

  $5(w - 2)$

  c. $3xy - 7xy + xy$

  $xy(3 - 7 + 1)$

  d. $6x + 20xy - 10x$

  $2x(3 + 10y - 5)$

  e. $4 - 12x$

  $4(1 - 3x)$

  f. $2x^2 + 3x^2y$

  $x^2(2 + 3y)$

  g. $4srt^2 - 3srt^2 + 10st$

  $st(4rt - 3rt + 10)$

  h. $10abc + 15abd + 35ab$

  $5ab(2c + 3d + 7)$

21. a. How many terms are in the expression $2x^2 + 3x - x - 3$ as written?

  There are four terms in the expression.

  b. How many terms are in the simplified expression $2x^2 + 2x - 3$?

  There are three terms in the simplified expression.

22. Are $3x^2$ and $3x$ like terms? Explain.

  No, the exponents of x are different.

23. Simplify the following expressions by combining like terms.

  a. $5a + 2ab - 3b + 6ab$

  $5a + 8ab - 3b$

  b. $3x^2 - 6x + 7$

  The expression cannot be simplified further.

  c. $100r - 13s^2 + 4r - 18s^3$

  $104r - 13s^2 - 18s^3$

  d. $3a + 7b - 5a - 10b$

  $-2a - 3b$

  e. $2x^3 - 2y^2 + 4x^2 + 9y^2$

  $2x^3 + 7y^2 + 4x^2$

  f. $7ab - 3ab + ab - 10ab$

  $-5ab$

  g. $xy^2 + 3x^2y - 2xy^2$

  $3x^2y - xy^2$

  h. $9x - 7x + 3x^2 + 5x$

  $7x + 3x^2$

  i. $2x - 2x^2 + 7 - 12$

  $2x - 2x^2 - 5$

  j. $3mn^3 - 2m^2n + m^2n - 7mn^3 + 3$

  $-4mn^3 - m^2n + 3$

  k. $5r^2s - 6rs^2 + 2rs + 4rs^2 - 3r^2s + 6rs - 7$

  $2r^2s - 2rs^2 + 8rs - 7$

24. Simplify the expression $2x - (4x - 8)$.

  $2x - 4x + 8 = -2x + 8$

25. Use the distributive property, and then combine like terms to simplify the expressions.

a. $30 - (x + 6)$

$30 - x - 6$

$= 24 - x$

b. $18 - (x - 8)$

$18 - x + 8$

$= 26 - x$

c. $2x - 3(25 - x)$

$2x - 75 + 3x$

$= 5x - 75$

d. $27 - 6(4x + 3y)$

$27 - 24x - 18y$

e. $12.5 - (3.5 - x)$

$12.5 - 3.5 + x$

$= 9 + x$

f. $3x + 2(x + 4)$

$3x + 2x + 8$

$= 5x + 8$

g. $x + 3(2x - 5)$

$x + 6x - 15$

$= 7x - 15$

h. $4(x + 2) + 5(x - 1)$

$4x + 8 + 5x - 5$

$= 9x + 3$

i. $7(x - 3) - 2(x - 8)$

$7x - 21 - 2x + 16$

$= 5x - 5$

j. $11(0.5x + 1) - (0.5x + 6)$

$5.5x + 11 - 0.5x - 6$

$= 5x + 5$

k. $2x^2 - 3x(x + 3)$

$2x^2 - 3x^2 - 9x$

$= -x^2 - 9x$

l. $y(2x - 2) - 3y(x + 4)$

$2xy - 2y - 3xy - 12y$

$= -xy - 14y$

26. You will be entering a craft fair with your latest metal wire lawn ornament. It is metal around the exterior and hollow in the middle, in the shape of a bird, as illustrated. However, you can produce different sizes, and all will be in proportion, depending on the length of the legs, x.

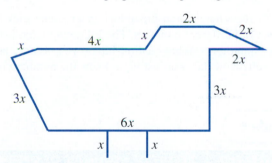

a. Write an algebraic expression that would represent the amount of wire you would need to create the bird. Be sure to include its legs.

$x + 4x + x + 2x + 2x + 2x + 3x + 6x + x + x + 3x$

b. Simplify the expression by combining like terms.

$26x$

c. The amount of wire you need for each lawn ornament is determined from the expression in part b plus an extra 2 inches for the eye. If you have a spool containing 400 inches of metal wire, write an expression showing how much wire you will have left after completing one ornament.

$400 - (26x + 2) = 400 - 26x - 2$

$= 398 - 26x$

d. Exactly how much wire is needed for an ornament whose legs measure 3 inches? How much wire will be left on the spool?

$x = 3$

$26(3) + 2 = 80$ inches of wire needed

$398 - 26(3) = 398 - 78 = 320$ inches remain on the spool.

Lab Activity 2.12

Math Magic

Objectives

1. Recognize an algebraic expression as a code of instruction.

2. Simplify algebraic expressions.

Algebraic expressions arise every time a sequence of arithmetic operations (instructions) is applied to a variable. For example, if your instructions are to double a variable quantity and then add 5, you would express this algebraically as $2x + 5$. If, however, you start with an algebraic expression, say $3x - 2$, you can "decode" the expression to determine the sequence of operations that is applied to x:

Multiply by 3, and then subtract 2.

Perhaps you have seen magicians on TV who astound their audiences by guessing numbers that a volunteer has secretly picked. Consider the following examples of "math magic" and see if you can decode the tricks.

Select a number, preferably an integer for ease of calculation. Don't tell anyone your number. Perform the following sequence of operations using the number you selected:

> Add 1 to the number.
>
> Triple the result.
>
> Subtract 6.
>
> Divide by 3.
>
> Tell your instructor your result.

You might be very surprised that your instructor, a "math magician" of sorts, can tell you the number that you originally selected. There is a hidden pattern in the sequence of operations that causes the number selected and the result of the sequence of operations to always have the same relationship.

1. One way to guess the relationship hidden in a number trick is to use the code of instructions on several different numbers. The following table lists results from this trick for four selected numbers. What relationship do you observe between the number selected and the result? How did your instructor know the number you selected?

 Pick a Number

NUMBER SELECTED	RESULT OF SEQUENCE OF OPERATIONS
0	−1
2	1
5	4
10	9

The result always seems to be 1 less than the original number. Therefore, my instructor added 1 to my result to tell me the number I selected.

2. **a.** To show how and why this trick works, you first generalize the situation by choosing a variable to represent the number selected. Use x as your variable and translate the first instruction into an algebraic expression. Then simplify it by removing parentheses (if necessary) and combining like terms.

   Step 1: $x + 1$

b. Now, use the result from part a and translate the second instruction into an algebraic expression, and simplify where possible. Continue until you complete all the steps.

Step 2: $3(x + 1) = 3x + 3$ (triple the result)

Step 3: $3x + 3 - 6 = 3x - 3$ (subtract 6)

Step 4: $\frac{3x - 3}{3} = x - 1$ (divide by 3)

c. Use your result from part b to interpret verbally (in words) the relationship between the number selected and the result. Also explain how to obtain the original number.

The result is always 1 less than the number originally selected, so the original number is 1 more than the result.

3. Rather than simplifying after each step, you can wait until after you have written a single algebraic expression using all the steps. As in Problem 2, the resulting expression will represent the algebraic code for the number trick.

a. Write all the steps for this number trick as a single algebraic expression without simplifying at each step.

$$x + 1 \longrightarrow 3(x + 1) \longrightarrow 3(x + 1) - 6 \longrightarrow \frac{3(x + 1) - 6}{3}$$

Step 1 Step 2 Step 3 Step 4

b. Simplify the expression you obtained in part a. How does this simplified expression compare with the result you obtained in Problem 2b?

$$\frac{3x + 3 - 6}{3} = \frac{3x - 3}{3} = x - 1$$

They are the same.

4. Here is another trick you can try on a friend. The instructions for your friend are:

Think of a number.

Double it.

Subtract 5.

Multiply by 3.

Add 9.

Divide by 6.

Add 1.

Tell me your result.

a. Try this trick with a friend several times. Record the result of your friend's calculations and the numbers he or she originally selected.

i. Start with 5: $10 \longrightarrow 5 \longrightarrow 15 \longrightarrow 24 \longrightarrow 4 \longrightarrow 5$

ii. Start with 8: $16 \longrightarrow 11 \longrightarrow 33 \longrightarrow 42 \longrightarrow 7 \longrightarrow 8$

iii. Start with 10: $20 \longrightarrow 15 \longrightarrow 45 \longrightarrow 54 \longrightarrow 9 \longrightarrow 10$

b. Explain how you know your friend's numbers.

The result is the same as the number originally chosen.

c. Show how your trick works no matter what number your friend chooses. Use an algebraic approach as you did in either Problem 2 or Problem 3.

$$\frac{3(2x - 5) + 9}{6} + 1 = \frac{6x - 15 + 9}{6} + 1$$

$$= \frac{6x - 6}{6} + 1 = x - 1 + 1 = x, \text{ the original number}$$

5. You are a magician, and you need a new number trick for your next show at the Magic Hat Club. Make up a trick like the ones in Problems 1 and 4. Have a friend select a number and do the trick. Explain how you figured out the number. Show how your trick will work using any number.

(Answers will vary.)

6. Your magician friend had a show to do tonight but was suddenly called out of town due to an emergency. She asked you and some other magician friends to do her show for her. She wanted you to do the mind trick, and she left you the instructions in algebraic code. Translate the code into a sequence of arithmetic operations to be performed on a selected number. Try the trick on a friend before the show. Write down how you will figure out any chosen number.

Code: $[2(x - 1) + 8] \div 2 - 5$

Start with any number, x.

Subtract 1: $x - 1$

Multiply by 2: $2(x - 1) = 2x - 2$

Add 8: $2x - 2 + 8 = 2x + 6$

Divide by 2: $\frac{2x + 6}{2} = x + 3$

Subtract 5: $x + 3 - 5 = x - 2$

The result is 2 less than the original number, so I will add 2 to the reported result to obtain the original number.

7. Here are more codes you can use in the show. Try them out as you did in Problem 6. State what the trick is.

   a. Code: $[(3n + 8) - n] \div 2 - 4$

$\frac{3n + 8 - n}{2} - 4$	**Trick: Pick a number.**
$\frac{2n + 8}{2} - 4$	**Multiply it by 3.**
$n + 4 - 4$	**Add 8.**
n	**Subtract the original number.**
	Divide by 2.
	Subtract 4.

   The result is the same as the original number.

   b. Code: $\dfrac{(4x - 5) + x}{5} + 2$

$\frac{4x - 5 + x}{5} + 2$	**Trick: Pick a number.**
$\frac{5x - 5}{5} + 2$	**Multiply it by 4.**
$x - 1 + 2$	**Subtract 5.**
$x + 1$	**Add the original number.**
	Divide by 5.
	Add 2.

   The result is 1 more than the original number, so subtract 1 from the result to obtain the original.

8. Simplify the algebraic expressions.

a. $[3 - 2(x - 3)] \div 2 + 9$

$(3 - 2x + 6) \div 2 + 9$

$\frac{9 - 2x}{2} + 9$

$13.5 - x$

b. $[3 - 2(x - 1)] - [-4(2x - 3) + 5] + 4$

$(3 - 2x + 2) - (-8x + 12 + 5) + 4$

$3 - 2x + 2 + 8x - 12 - 5 + 4$

$6x - 8$

c. $\dfrac{2(x - 6) + 8}{2} + 9$

$\frac{2x - 12 + 8}{2} + 9$

$x - 6 + 4 + 9$

$x + 7$

SUMMARY: ACTIVITY 2.12

Procedure for simplifying algebraic expressions:

1. Simplify the expression from the innermost parentheses outward using the order of operations convention.

2. Use the distributive property when it applies.

3. Combine like terms.

EXERCISES: ACTIVITY 2.12

1. You want to boast to a friend about the stock that you own without telling him how much money you originally invested in the stock. Let your original stock value be represented by x dollars. You watch the market once a month for 4 months and record the following.

Month	1	2	3	4
Stock Value	Increased $50	Doubled	Decreased $100	Tripled

a. Use x to represent the value of your original investment, and write an algebraic expression to represent the value of your stock after the first month.

$x + 50$

b. Use the result from part a to determine the value at the end of the second month, simplifying when possible. Continue until you determine an expression that represents the value of your stock at the end of the fourth month.

2nd month: $2(x + 50) = 2x + 100$

3rd month: $2x + 100 - 100 = 2x$

4th month: $3(2x) = 6x$

c. Do you have good news to tell your friend? Explain to him what has happened to the value of your stock over 4 months.

Yes, the news is good. The value of my stock increased sixfold.

d. Instead of simplifying expressions after each step, write a single algebraic expression that represents the 4-month period.

$$x + 50 \longrightarrow 2(x + 50) \longrightarrow 2(x + 50) - 100 \longrightarrow 3[2(x + 50) - 100]$$

    Step 1   Step 2   Step 3   Step 4

e. Simplify the expression in part d. How does the simplified algebraic expression compare with the result in part b?

$3[2x + 100 - 100]$

$3(2x)$

$6x$

The result is the same as in part b.

2. a. Write a single algebraic expression for the following sequence of operations. Begin with a number represented by n.

Multiply by -2.

Add 4.

Divide by 2.

Subtract 5.

Multiply by 3.

Add 6.

$$3\left[\frac{-2n + 4}{2} - 5\right] + 6$$

b. Simplify the algebraic expression $3(-n + 2 - 5) + 6$.

$3(-n - 3) + 6$

$-3n - 9 + 6$

$-3n - 3$

c. Your friend in Alaska tells you that if you replace n with the value 10 in your answer to Problem 2b, you will discover the average Fahrenheit temperature in Alaska for the month of December. Does that seem reasonable? Explain.

(Explanations may vary.)

$-3(10) - 3 = -33°$

It seems too cold to be reasonable.

3. Show how you would simplify the expression $2\{3 - [4(x - 7) - 3] + 2x\}$.

$2\{3 - [4x - 28 - 3] + 2x\}$

$2\{3 - 4x + 31 + 2x\}$

$2\{34 - 2x\}$

$68 - 4x$

4. Simplify the following algebraic expressions.

a. $5x + 2(4x + 9)$

$5x + 8x + 18$

$13x + 18$

b. $2(x - y) + 3(2x + 3) - 3y + 4$

$2x - 2y + 6x + 9 - 3y + 4$

$8x - 5y + 13$

c. $-(x - 4y) + 3(-3x + 2y) - 7x$

$-x + 4y - 9x + 6y - 7x$

$10y - 17x$

d. $2 + 3[3x + 2(x - 3) - 2(x + 1) - 4x]$

$2 + 3[3x + 2x - 6 - 2x - 2 - 4x]$

$2 + 3(-x - 8)$

$2 - 3x - 24$

$-3x - 22$

e. $6[3 + 2(x - 5)] - [2 - (x + 1)]$

$6[3 + 2x - 10] - [2 - x - 1]$

$6[2x - 7] - [1 - x]$

$12x - 42 - 1 + x$

$13x - 43$

f. $\dfrac{5(x - 2) - 2x + 1}{3}$

$\dfrac{5x - 10 - 2x + 1}{3}$

$\dfrac{3x - 9}{3}$

$x - 3$

5. a. Complete the following table by performing the following sequence of operations on the input value to produce a corresponding output value: Multiply the input by 4, add 12 to the product, divide the sum by 2, and subtract 6 from the quotient.

INPUT VALUE	OUTPUT VALUE
2	4
5	10
10	20
15	30

b. There is a direct connection between each input number and corresponding output obtained by performing the sequence of operations in part a. Describe it.

Each input value is multiplied by 2 to produce the corresponding output value.

c. Confirm your observation in part b algebraically by performing the sequence of operations given in part a. Use the variable x to represent the original number. Simplify the resulting algebraic expression.

$\dfrac{4x + 12}{2} - 6 = 2x + 6 - 6$

$= 2x$

d. Does your result in part b confirm the pattern you observed in part a?

Yes, the expression $2x$ means that the sequence of operations given in part a produces a result that is always twice the original number.

Activity 2.13

Comparing
Energy Costs

Objectives

1. Translate verbal rules into symbolic rules.

2. Write and solve equations of the form $ax + b = cx + d$.

3. Use the distributive property to solve equations involving grouping symbols.

4. Develop mathematical models to solve problems.

5. Solve formulas for a specified variable.

You have hired an architect to design a home. She gives you the following information regarding the installation and operating costs of two types of heating systems: solar and electric.

 Some Like It Hot

TYPE OF HEATING SYSTEM	INSTALLATION COST	OPERATING COST PER YEAR
Solar	$25,600	$200
Electric	$5500	$1600

1. **a.** Determine the total cost of the solar heating system after 5 years of use.

$25,600 + 200(5) = 25,600 + 1000 = 26,600$

The total cost of solar heating for 5 years is $26,600.

b. Write a verbal rule for the total cost of the solar heating system (output) in terms of the number of years of use (input).

Multiply the number of years of use by 200, and then add 25,600 to the product to get the total cost of solar heating.

c. Let x represent the number of years of use and S represent the total cost of the solar heating system. Translate the verbal rule in part b into a symbolic rule.

$S = 200x + 25,600$

d. Use the formula from part c to complete the following table:

Number of Years in Use, x	5	10	15	20
Total Cost, $S(\$)$	26,600	27,600	28,600	29,600

2. **a.** Determine the total cost of the electric heating system after 5 years of use.

$5500 + (1600)5 = 13,500$

The total cost of the electric heating system after 5 years is $13,500.

b. Write a verbal rule for the total cost of the electric heating system (output) in terms of the number of years of use (input).

Multiply the number of years by 1600, and then add 5500 to the product to obtain the total cost of electric heating.

c. Let x represent the number of years of use and E represent the total cost of the electric heating system. Translate the verbal rule in part b into a symbolic rule.

$E = 1600x + 5500$

d. Use the equation from part c to complete the following table:

Number of Years in Use, x	5	10	15	20
Total Cost, $E(\$)$	13,500	21,500	29,500	37,500

The installation cost of solar heating is much more than that of the electric system, but the operating cost per year of the solar system is much lower. Therefore, it is reasonable to think that the total cost for the electric system will eventually "catch up" and surpass the total cost of the solar system.

3. Use the tables in Problems 1d and 2d to compare costs.

  a. When was the total cost for electric heating less than the cost for solar heating?

  Electric heating cost less up to at least 10 years.

  b. In which years did the cost for electric heating surpass that of solar heating?

  Electric heating cost more after 15 years.

  c. For which year were the costs the closest?

  The costs were the closest at about 15 years when the electric costs first surpassed the solar costs.

  d. Use your observations from parts a–c to estimate when the cost of electric heating will catch up to and surpass the cost of solar heating.

  From the observations in parts a–c, a reasonable estimate is about 14 to 15 years.

The year in which the total costs of the two heating systems would be equal can be determined exactly by an algebraic process. The costs will be equal when the algebraic expressions representing their respective costs are equal. That is, when

$$1600x + 5500 = 200x + 25,600$$

The challenge in solving such an equation is to combine the like terms containing x. Because these two terms appear on different sides of the equation, the first step must be to subtract one of these terms from *both* sides of the equation. This will eliminate the term from its side of the equation and will allow you to combine like terms on the other side. You can then proceed as usual in completing the solution.

Example 1 illustrates this procedure in solving a similar equation.

Example 1   *Solve for x:* $2x + 14 = 8x + 2.$

SOLUTION

$$\begin{array}{ll} 2x + 14 = 8x + 2 & \text{Subtract } 8x \text{ from both sides and combine like terms.} \\ \underline{-8x \qquad\quad -8x} \\ -6x + 14 = 2 \\ \underline{\qquad\quad -14 = -14} & \text{Subtract 14 from each side and combine like terms.} \\ -6x \qquad = -12 \\ \dfrac{-6x}{-6} = \dfrac{-12}{-6} & \text{Divide each side by } -6 \text{, the coefficient of } x. \\ x = 2 \end{array}$$

Check: $2(2) + 14 = 8(2) + 2$

$\qquad\qquad 4 + 14 = 16 + 2$

$\qquad\qquad\qquad 18 = 18$

4. In Example 1, the variable terms are combined on the left side of the equation. To solve the equation $2x + 14 = 8x + 2$, combine the variable terms on the right side. Does it matter on which side you choose to place the variable terms?

$$2x + 14 = 8x + 2$$
$$\underline{-2x \qquad\qquad -2x}$$
$$14 = 6x + 2$$
$$\underline{-2 \qquad -2}$$
$$12 = 6x$$
$$\frac{12}{6} = \frac{6x}{6}$$
$$2 = x$$

No, it does not matter.

5. **a.** Solve the equal-cost equation, $1600x + 5500 = 200x + 25{,}600$.

$$1600x + 5500 = 200x + 25{,}600$$
$$\underline{-200x \qquad\qquad -200x}$$
$$1400x + 5500 = 25{,}600$$
$$\underline{-5500 \quad -5{,}500}$$
$$\frac{1400x}{1400} = \frac{20{,}100}{1400}$$
$$x \approx 14.36 \approx 14$$

b. Interpret what your answer in part a represents in the context of the heating system situation.

After 14 years, the money you spent over that time (installation and monthly charges) would be equal for solar and electric heating.

Purchasing a Car

You are interested in purchasing a new car and have narrowed the choice to a Honda Accord LX (4 cylinder) and a Passat GLS (4 cylinder). Being concerned about the value of the car depreciating over time, you search the Internet and obtain the following information:

 Driven Down . . .

MODEL OF CAR	MARKET SUGGESTED RETAIL PRICE (MSRP) ($)	ANNUAL DEPRECIATION ($)
Accord LX	20,925	1730
Passat GLS	24,995	2420

6. **a.** Complete the following table:

YEARS THE CAR IS OWNED	VALUE OF ACCORD LX ($)	VALUE OF PASSAT GLS ($)
1	19,195	22,575
2	17,465	20,155
3	15,735	17,735

b. Will the value of the Passat GLS ever be lower than the value of the Accord LX? Explain.

Yes, because its value is depreciating more rapidly.

c. Let v represent the value of the car after x years of ownership. Write a symbolic rule to determine v in terms of x for the Accord LX.

$v = 20{,}925 - 1730x$

d. Write a symbolic rule to determine v in terms of x for the Passat GLS.

$v = 24{,}995 - 2420x$

e. Write an equation to determine when the value of the Accord LX will equal the value of the Passat GLS.

$20{,}925 - 1730x = 24{,}995 - 2420x$

f. Solve the equation in part e.

$$\begin{array}{l} 20{,}925 - 1730x = 24{,}995 - 2420x \\ \underline{\qquad + 1730x \qquad\qquad + 1730x} \\ 20{,}925 \qquad\quad = 24{,}995 - 690x \\ \underline{-24{,}995 \qquad\qquad = -24{,}995} \\ \qquad -4070 = -690x \\ \qquad\quad 5.90 \approx x \end{array}$$

After approximately 5.9 years, the cars will have the same value.

NBA Basketball Court

You and your friend are avid professional basketball fans and discover that your mathematics instructor shares your enthusiasm for basketball. During a mathematics class, your instructor tells you that the perimeter of an NBA basketball court is 288 feet and the length is 44 feet more than its width. He challenges you to use your algebra skills to determine the dimensions of the court. To solve the problem, you and your friend use the following plan.

7. a. Let w represent the width of the court. Write an expression for the length in terms of the width, w.

$w + 44$

b. Use the formula for the perimeter of a rectangle, $P = 2l + 2w$, and the information given to obtain an equation containing just the variable w.

$P = 2(w + 44) + 2w \quad 288 = 2(w + 44) + 2w$

c. To solve the equation you obtained in part b, first apply the distributive property and then combine like terms in the expression involving w.

$288 = 2w + 88 + 2w$

$288 = 4w + 88$

$$\begin{array}{l} \underline{-88 \qquad\quad -88} \\ 200 = 4w \\ \quad w = 50 \end{array}$$

d. What are the dimensions of an NBA basketball court?

The width is 50 feet and the length is $50 + 44 = 94$ feet.

8. Solve each of the following formulas for the specified variable.

(The form of the answers may vary.)

a. $P = 2l + 2w$ for w

$P - 2l = 2w$

$\dfrac{P - 2l}{2} = w$

b. $V(P + a) = k$ for P

$VP + Va = k$

$VP = k - Va$

$P = \dfrac{k - Va}{V}$

c. $w = 110 + \frac{11}{2}(h - 60)$ for h

$$w - 110 = \frac{11}{2}(h - 60)$$

$$\tfrac{2}{11}(w - 110) = h - 60$$

$$\tfrac{2}{11}(w - 110) + 60 = h$$

$$h = \tfrac{2}{11}w + 40$$

9. Solve each of the following equations for x. Remember to check your result in the original equation.

a. $2x + 9 = 5x - 12$

$$\underline{-2x \qquad\quad -2x}$$
$$9 = 3x - 12$$
$$\underline{+12 \qquad\quad +12}$$
$$21 = 3x$$
$$x = 7$$

b. $21 - x = -3 - 5x$

$$\underline{+5x \qquad\quad + 5x}$$
$$21 + 4x = -3$$
$$\underline{-21 \qquad\qquad -21}$$
$$4x = -24$$
$$x = -6$$

c. $2(x - 3) = -8$

$$2x - 6 = -8$$
$$\underline{+6 \quad +6}$$
$$2x = -2$$
$$x = -1$$

d. $2(x - 4) + 6 = 4x - 7$

$$2x - 8 + 6 = 4x - 7$$
$$2x - 2 \quad\;\; = 4x - 7$$
$$\underline{-2x \qquad\qquad -2x}$$
$$-2 \quad\;\; = 2x - 7$$
$$\underline{+7 \qquad\qquad\quad + 7}$$
$$5 = 2x$$
$$x = 2.5$$

10. Three friends worked together on a homework assignment that included two equations to solve. Although they are certain that they solved the equations correctly, they are puzzled by the results they obtained. Here they are.

a. $3(x + 4) + 2x = 5(x - 2)$

$$3x + 12 + 2x = 5x - 10$$

$$5x + 12 = 5x - 10$$

$$5x - 5x = -12 - 10$$

$$0 = -22?$$

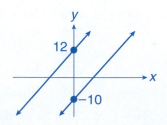

Graph the equations $y_1 = 3(x + 4) + 2x$ and $y_2 = 5(x - 2)$. Do the graphs intersect?

The graphs are nonintersecting parallel lines. They have no points in common.

What does this indicate about the solution to the original equation?

There is no solution to the original equation.

b. $2(x + 3) + 4x = 6(x + 1)$

$2x + 6 + 4x = 6x + 6$

$6x + 6 = 6x + 6$

$6x - 6x = 6 - 6$

$0 = 0?$

Graph the equations $y_1 = 2(x + 3) + 4x$ and $y_2 = 6(x + 1)$. Do the graphs intersect?

Only one graph appears. y_1 and y_2 have identical graphs.

What does this indicate about the solution to the original equation?

y_1 and y_2 are equivalent, so for every input, their outputs match. All real numbers are solutions.

11. Solve the following equations, if possible.

a. $12x - 9 = 4(6 + 3x)$

$12x - 9 = 24 + 12x$

$0 = 33$

no solution

b. $2(8x - 4) + 10 = 22 - 4(5 - 4x)$

$16x - 8 + 10 = 22 - 20 + 16x$

$16x + 2 = 2 + 16x$

$0 = 0$

Solution is all real numbers.

In the process of solving these equations in Problems 10 and 11, the identical variable term appeared on each side of the simplified equation. Subtracting this term from both sides eliminated the variable completely. Two outcomes are possible:

1. The final line of the solution is an equation that is true. For example, $0 = 0$. When this occurs, the equation is called an **identity**. **All real numbers** are solutions to the original equation.

2. The final line of the solution is an equation that is false. For example, $0 = 30$. When this occurs, the equation is called a **contradiction**. There is **no solution** to the original equation.

SUMMARY: ACTIVITY 2.13

General strategy for solving equations for an unknown quantity, such as x:

1. If necessary, apply the distributive property to remove parentheses.

2. Combine like terms that appear on the same side of the equation.

3. Isolate x so that only a single term in x remains on one side of the equation and all other terms, not containing x, are moved to the other side. This is generally accomplished by adding (or subtracting) appropriate terms to both sides of the equation.

4. Solve for x by dividing each side of the equation by its coefficient.

5. Check your result by replacing x by this value in the original equation. A correct solution will produce a true statement.

EXERCISES: ACTIVITY 2.13

1. Finals are over, and you are moving back home for the summer. You need to rent a truck to move your possessions from the college residence hall back to your home. You contact two local rental companies and acquire the following information for the 1-day cost of renting a truck:

Company 1: $19.99 per day, plus $0.79 per mile

Company 2: $29.99 per day, plus $0.59 per mile

Let x represent the number of miles driven in 1 day and C represent that total daily rental cost ($).

a. Write a symbolic rule that represents the total daily rental cost from company 1 in terms of the number of miles driven.

$C = 0.79x + 19.99$

b. Write a symbolic rule that represents the total daily rental cost from company 2 in terms of the number of miles driven.

$C = 0.59x + 29.99$

c. Write an equation to determine for what mileage the 1-day rental cost would be the same.

$0.79x + 19.99 = 0.59x + 29.99$

d. Solve the equation you obtained in part c.

$$0.79x + 19.99 = 0.59x + 29.99$$
$$\underline{-0.59x \qquad\qquad -0.59x}$$
$$0.20x + 19.99 = 29.99$$
$$\underline{\;\; -19.99 \quad -19.99}$$
$$0.20x = 10$$
$$x = 50$$

e. For which mileages would company 2 have the lower price?

Company 2 is lower if I drive more than 50 miles.

2. You are considering installing a security system in your new house. You gather the following information about similar security systems from two local home security dealers:

Dealer 1: $3560 to install and $15 per month monitoring fee

Dealer 2: $2850 to install and $28 per month for monitoring

You see that the initial cost of the security system from dealer 1 is much higher than that of the system from dealer 2, but that the monitoring fee is lower. You want to determine which is the better system for your needs.

Let x represent the number of months that you have the security system and C represent the cumulative cost ($).

a. Write a symbolic rule that represents the total cost of the system from dealer 1 in terms of the number of months you have the system.

$C = 3560 + 15x$

b. Write a symbolic rule that represents the total cost of the system from dealer 2 in terms of the number of months you have the system.

$C = 2850 + 28x$

c. Write an equation to determine the number of months for which the total cost of the systems will be equal.

$3560 + 15x = 2850 + 28x$

d. Solve the equation in part c.

$$3560 + 15x = 2850 + 28x$$
$$\underline{\quad -15x \qquad\qquad -15x}$$
$$3560 \qquad = 2850 + 13x$$
$$\underline{-2850 \qquad\qquad -2850}$$
$$710 = 13x$$
$$\frac{710}{13} = \frac{13x}{13}$$
$$x \approx 55$$

At about 55 months or approximately $4\frac{1}{2}$ years, both systems would cost me the same.

e. If you plan to live in the house and use the system for 10 years, which system would be less expensive?

Ten years is equivalent to 120 months. Dealer 1 is better for more than 55 months' use.

3. You are able to get three summer jobs to help pay for college expenses. In your job as a cashier, you work 20 hours per week and earn $8.00 per hour. Your second and third jobs are both at a local hospital. There you earn $9.50 per hour as a payroll clerk and $7.00 per hour as an aide. You always work 10 hours less per week as an aide than you do as a payroll clerk. Your total weekly salary depends on the number of hours that you work at each job.

a. Determine the input and output variables for this situation.

The input variable represents the number of hours I work in payroll.

The output variable represents my total weekly salary.

b. Explain how you calculate the total amount earned each week.

(hours worked at payroll) $\cdot$ $9.50 $+$

(hours worked as aide) $\cdot$ $7.00 $+$ 20(8.00)

c. If x represents the number of hours that you work as a payroll clerk, represent the number of hours that you work as an aide in terms of x.

$x - 10$

d. Write a symbolic rule that describes the total amount you earn each week. Use x to represent the input variable and y to represent the output variable. Simplify the expression as much as possible.

$y = 9.50x + 7(x - 10) + 160 = 16.50x + 90$

e. If you work 12 hours as a payroll clerk, how much will you make in 1 week?

$x = 12$

$y = 16.5(12) + 90 = 288$

If I work 12 hours as a payroll clerk, I will earn $288 for the week.

f. What are realistic replacement values for x? Would 8 hours at your payroll job be a realistic replacement value? What about 50 hours?

x must be at least 10, or else the aide hours will be negative. 50 is too large because that gives $50 + 40 + 20 = 110$ total hours out of a possible 168 hours. Realistic values fall between 10 and 25.

g. When you don't work as an aide, what is your total weekly salary?

0 hr. as aide	$x = 10$
10 hr. in payroll	$y = 16.5(10) + 90$
20 hr. as cashier	$y = 255$

If I don't work as an aide, my total weekly salary will be $255.

h. If you plan to earn a total of $500 in 1 week from all jobs, how many hours would you have to work at each job? Is the total number of hours worked realistic? Explain.

(Use rule from part d to obtain the equation.)

$$16.50x + 90 = 500$$
$$16.50x = 410$$

$x \approx 24.85 \approx 25$ hours at payroll, $x - 10 = 15$ hours as an aide, and 20 hours as a cashier.

Total hours worked would be 60 hours, a heavy schedule but not impossible.

i. Solve the equation in part d for x in terms of y. When would it be useful to have the equation in this form?

$$x = \frac{y - 90}{16.5}$$

This form of the equation would be useful when I am setting a salary goal and need to determine the number of hours I must work.

4. A florist sells roses for $1.50 each and carnations for $0.85 each. Suppose you purchase a bouquet of 1 dozen flowers consisting of roses and carnations.

a. Let x represent the number of roses purchased. Write an expression in terms of x that represents the number of carnations purchased.

The number of carnations is $12 - x$.

b. Write an expression that represents the cost of purchasing x roses.

The cost of purchasing x roses is $1.50x$.

c. Write an expression that represents the cost of purchasing the carnations.

The cost of purchasing the carnations is $0.85(12 - x)$.

d. What does the sum of the expressions in parts b and c represent?

The sum represents the total cost of the dozen flowers.

e. Suppose you are willing to spend $14.75. Write an equation that can be used to determine the number of roses that can be included in a bouquet of 1 dozen flowers consisting of roses and carnations.

$$1.50x + 0.85(12 - x) = 14.75$$

f. Solve the equation in part e to determine the number of roses and the number of carnations in the bouquet.

$$1.5x + 0.85(12 - x) = 14.75$$
$$1.5x + 10.20 - 0.85x = 14.75$$
$$0.65x = 4.55$$
$$x = 7$$

Carnations: $12 - 7 = 5$

Therefore, the bouquet would contain 7 roses and 5 carnations.

5. The viewing window of a certain calculator is in the shape of a rectangle.

a. Let w represent the width of the viewing window in centimeters. If the window is 5 centimeters longer than it is wide, write an expression in terms of w for the length of the viewing window.

The length of the viewing window is given by the expression $w + 5$.

b. Write a symbolic rule that represents the perimeter, P, of the viewing window in terms of w.

$P = 2(w + 5) + 2w$, or $P = 4w + 10$

c. If the perimeter of the viewing window is 26 centimeters, determine the dimensions of the window.

$$26 = 4w + 10$$
$$\underline{-10 \qquad\quad -10}$$
$$16 = 4w$$
$$w = 4 \qquad\qquad l = w + 5 = 9$$

So, the dimensions are 4 centimeters by 9 centimeters.

Solve the equations in Exercises 6–19.

6. $5x - 4 = 3x - 6$

$$\underline{-3x \qquad\quad -3x}$$
$$2x - 4 = -6$$
$$\underline{+4 \quad +4}$$
$$2x = -2$$
$$x = -1$$

7. $3x - 14 = 6x + 4$

$$\underline{-3x \qquad\quad -3x}$$
$$-14 = 3x + 4$$
$$\underline{-4 \qquad\quad -4}$$
$$-18 = 3x$$
$$x = -6$$

8. $0.5x + 9 = 4.5x + 17$

$$\underline{-0.5x \qquad -0.5x}$$
$$9 = 4x + 17$$
$$\underline{-17 \qquad -17}$$
$$-8 = 4x$$
$$x = -2$$

9. $4x - 10 = -2x + 8$

$$\underline{+2x \qquad\quad +2x}$$
$$6x - 10 = 8$$
$$\underline{+10 \; +10}$$
$$6x = 18$$
$$x = 3$$

10. $0.3x - 5.5 = 0.2x + 2.6$

$$\underline{-0.2x \qquad\quad -0.2x}$$
$$0.1x - 5.5 = 2.6$$
$$\underline{+5.5 + 5.5}$$
$$0.1x \qquad = 8.1$$
$$x = 81$$

11. $4 - 0.025x = 0.1 - 0.05x$

$$\underline{+0.05x \qquad +0.05x}$$
$$4 + 0.025x = 0.1$$
$$\underline{-4 \qquad\qquad -4}$$
$$0.025x = -3.9$$
$$\frac{0.025x}{0.025} = \frac{-3.9}{0.025}$$
$$x = -156$$

12. $5t + 3 = 2(t + 6)$

$$5t + 3 = 2t + 12$$
$$\underline{-2t \qquad -2t}$$
$$3t + 3 = 12$$
$$\underline{-3 \quad -3}$$
$$3t = 9$$
$$t = 3$$

13. $3(w + 2) = w - 14$

$$3w + 6 = w - 14$$
$$\underline{-w \qquad -w}$$
$$2w + 6 = -14$$
$$\underline{-6 \quad -6}$$
$$2w = -20$$
$$w = -10$$

14. $21 + 3(x - 4) = 4(x + 5)$

$\quad 21 + 3x - 12 = 4x + 20$

$\quad\quad 9 + 3x \quad\quad = 4x + 20$

$\quad\quad\quad -3x \quad\quad\quad -3x$

$\quad\quad\quad\quad\quad 9 = x + 20$

$\quad\quad\quad\quad -20 \quad\quad -20$

$\quad\quad\quad\quad -11 = x$

15. $2(x + 3) = 5(2x + 1) + 4x$

$\quad 2x + 6 = 10x + 5 + 4x$

$\quad 2x + 6 = 14x + 5$

$\quad -2x \quad\quad\quad -2x$

$\quad\quad\quad 6 = 12x + 5$

$\quad\quad -5 \quad\quad\quad -5$

$\quad\quad\quad 1 = 12x$

$\quad\quad\quad x = \frac{1}{12}$

16. $500 = 0.75x - (750 + 0.25x)$

$\quad 500 = 0.75x - 750 - 0.25x$

$\quad 500 = 0.5x - 750$

$\quad +750 \quad\quad\quad +750$

$\quad 1250 = 0.5x$

$\quad \frac{1250}{0.5} = \frac{0.5x}{0.5}$

$\quad 2500 = x$

17. $1.5x + 3(22 - x) = 70$

$\quad 1.5x + 66 - 3x = 70$

$\quad -1.5x + 66 \quad\quad = 70$

$\quad\quad\quad -66 \quad\quad\quad -66$

$\quad\quad\quad -1.5x = 4$

$\quad\quad \frac{-1.5x}{-1.5} = \frac{4}{-1.5}$

$\quad\quad\quad x = -2.\overline{6}$

18. $18 + 2(4x - 3) = 8x + 12$

$\quad 18 + 8x - 6 = 8x + 12$

$\quad 12 + 8x = 8x + 12$

$\quad\quad\quad 0 = 0$

The solution is all real numbers.

19. $3 - 2(x - 4) = 5 - 2x$

$\quad 3 - 2x + 8 = 5 - 2x$

$\quad 11 - 2x = 5 - 2x$

$\quad\quad\quad 6 = 0$

no solution

20. You are asked to grade some of the questions on a skills test. Here are five results you are asked to check. If an example is incorrect, find the error and show the correct solution.

a. $34 = 17 - (x - 5)$

$\quad 34 = 17 - x \boxed{-} 5$ error: $-$

$\quad 34 = 12 - x$ should be $+$

$\quad 22 = -x$

$\quad\quad x = -22$

Correct solution:

$34 = 17 - x + 5$

$34 = 22 - x$

$\quad x = -12$

b. $-47 = -6(x - 2) + 25$

$\quad -47 = -6x + 12 + 25$

$\quad -47 = -6x + 37$

$\quad \boxed{-}6x = 84$ error: $-$

$\quad\quad x = 14$ should be $+$

With the negative 6, the answer would have been $x = -14$ (another error). The two errors cancel each other out. The correct answer is $x = 14$.

c. $-93 = -(x - 5) - 13x$

$\quad -93 = -x + 5 - 13x$

$\quad -93 = -14x + 5$

$\quad -14x = -98$

$\quad\quad x = 7$

Correct as is.

d. $3(x + 1) + 9 = 22$

$\quad 3x + \boxed{4} + 9 = 22$ error: 4

$\quad 3x + 13 = 22$ should

$\quad\quad 3x = 9$ be 3

$\quad\quad x = 3$

Correct solution:

$3x + 3 + 9 = 22$

$3x + 12 = 22$

$3x = 10$

$x = \frac{10}{3}$

e. $83 = -(x + 19) - 41$

$83 = -x - 19 - 41$

$83 = -x - \widecircle{50}$ error: should be 60

$133 = -x$

$x = -133$

Correct solution:

$83 = -x - 60$

$x = -143$

21. Explain how to solve for y in terms of x in the equation $2x + 4y = 7$.

Isolate y by subtracting $2x$ from both sides, $4y = 7 - 2x$ and then dividing both sides by 4. $y = \frac{7 - 2x}{4}$

22. Solve each formula for the specified variable.

a. $y = mx + b$, for x

$x = \frac{y - b}{m}$

b. $A = \dfrac{B + C}{2}$, for B

$B = 2A - C$

c. $A = 2\pi r^2 + 2\pi rh$, for h

$h = \frac{A - 2\pi r^2}{2\pi r}$

d. $F = \dfrac{9}{5}C + 32$, for C

$C = \frac{5}{9}(F - 32)$

e. $3x - 2y = 5$, for y

$y = \frac{3x - 5}{2}$

f. $12 = -x + \frac{y}{3}$, for y

$y = 3(12 + x)$

g. $A = P + Prt$, for P
(*Hint:* Factor first.)

$A = P(1 + rt)$

$P = \frac{A}{1 + rt}$

h. $z = \dfrac{x - m}{s}$, for x

$x = sz + m$

*Project
Activity 2.14*

Summer Job
Opportunities

Objective

1. Use critical-thinking skills
to make decisions based
on solutions of systems of
two linear equations.

It can be very difficult keeping up with college expenses, so it is important for you to find a summer job that pays well. Luckily, the classified section of your newspaper lists numerous summer job opportunities in sales, road construction, and food service. The advertisements for all these positions welcome applications from college students. All positions involve the same 10-week period from early June to mid-August.

Sales

A new electronics store opened recently. There are several sales associate positions that pay an hourly rate of $7.25 plus a 5% commission based on your total weekly sales. You would be guaranteed at least 30 hours of work per week, but not more than 40 hours.

Construction

Your state's highway department hires college students every summer to help with road construction projects. The hourly rate is $13.50 with the possibility of up to 10 hours per week in overtime, for which you would be paid time and a half. Of course, the work is totally dependent on good weather, and so the number of hours that you would work per week could vary.

Restaurants

Local restaurants experience an increase in business during the summer. There are several positions for wait staff. The hourly rate is $3.60, and the weekly tip total ranges from $300 to $850. You are told that you can expect a weekly average of approximately $520 in tips. You would be scheduled to work five dinner shifts of 6.5 hours each for a total of 32.5 hours per week. However, on slow nights you might be sent home early, perhaps after working only 5 hours. Thus, your total weekly hours might be fewer than 32.5.

All of the jobs would provide an interesting summer experience. Your personal preferences might favor one position over another. Keep in mind that you have a lot of college expenses.

1. At the electronics store, sales associates average $8000 in sales each week.

   a. Based on the expected weekly average of $8000 in sales, calculate your gross weekly paycheck (before any taxes or other deductions) if you worked a full 40-hour week in sales.

   $7.25(40) + 0.05(8000) = 690$

   My gross weekly paycheck would be $690.

   b. Use the average weekly sales figure of $8000 and write a symbolic rule for your weekly earnings, s, where x represents the total number of hours you would work.

   5% of $8000 = 0.05(8000) = $400 commission

   $s = 7.25x + 400$

   c. What would be your gross paycheck for the week if you worked 30 hours and still managed to sell $8000 in merchandise?

   $x = 30; s = 7.25(30) + 400 = 617.50$

   The amount of my gross paycheck would be $617.50 if I worked 30 hours and had $8000 in merchandise sales.

   d. You are told that you would typically work 35 hours per week if your total electronic sales do average $8000. Calculate your typical gross paycheck for a week.

   $x = 35; s = 7.25(35) + 400 = 653.75$

   My typical gross paycheck would be $653.75.

e. You calculate that to pay college expenses for the upcoming academic year, you need to gross at least $675 a week. How many hours would you have to work in sales each week? Assume that you would sell $8000 in merchandise.

$675 = 7.25x + 400$

$275 = 7.25x$

$x \approx 37.93$

To earn at least $675 per week, I would have to work at least 38 hours in the sales position.

2. In the construction job, you would average a 40-hour workweek.

a. Calculate your gross paycheck for a typical 40-hour workweek.

$13.50(40) = 540$

I would earn $540 for a typical 40-hour workweek on the construction job.

b. Write a symbolic rule for your weekly salary, s, for a week with no overtime. Let x represent the total number of hours worked.

$s = 13.50x$

c. If the weather is ideal for a week, you can expect to work 10 hours in overtime (over and above the regular 40-hour workweek). Determine your total gross pay for a week with 10 hours of overtime.

$540 + (1.5)(13.50)(10) = 540 + 202.50 = 742.50$

If I work 10 hours overtime, my total gross paycheck would be $742.50.

d. The symbolic rule in part b can be used to determine the weekly salary, s, when x, the total number of hours worked, is less than or equal to 40 (no overtime). Write a symbolic rule to determine your weekly salary, s, if x is greater than 40 hours.

$s = 540 + (1.5)(13.50)(x - 40)$

$s = 540 + 20.25(x - 40)$

$s = 20.25x - 270$

e. Suppose it turns out to be a gorgeous summer and your supervisor says that you can work as many hours as you want. If you are able to gross $800 a week, you will be able to afford to buy a computer. How many hours would you have to work each week to achieve your goal?

$800 = 20.25x - 270$

$1070 = 20.25x$

$x = 52.84$

I would need to work 53 hours each week to gross $800.

3. The restaurant job involves working a maximum of five dinner shifts of 6.5 hours each.

a. Calculate what your gross paycheck would be for an exceptionally busy week of five 6.5-hour dinner shifts and $850 in tips.

$3.60(5)(6.5) + 850 = 967$

In a busy week, my pay would be $967.

b. Calculate what your gross paycheck would be for an exceptionally slow week of five 5-hour dinner shifts and only $300 in tips.

$3.60(5)(5) + 300 = 390$

For a slow week, my gross pay would be $390.

c. Calculate what your gross paycheck would be for a typical week of five 6.5-hour dinner shifts and $520 in tips.

$3.60(5)(6.5) + 520 = 637$

In a typical week, my gross pay would be $637.

d. Use $520 as your typical weekly total for tips, and write a symbolic rule for your gross weekly salary, s, where x represents the number of hours.

$s = 3.60x + 520$

e. Calculate what your gross paycheck would be for a 27-hour week and $520 in tips.

$s = 3.60(27) + 520 = 617.20$

For a 27-hour workweek with $520 in tips, I would earn $617.20.

f. During the holiday week of July 4, you would be asked to work an extra dinner shift. You are told to expect $280 in tips for that night alone. Assuming a typical workweek for the rest of the week, would working that extra dinner shift enable you to gross at least $950?

$\$637 + 280 + 6.5(3.60) = 940.40$

My salary that week is just short of $950.

4. You would like to make an informed decision in choosing one of the three positions. Based on all the information you have about the three jobs, fill in the following table.

Comparing Gross Pay

	LOWEST WEEKLY GROSS PAYCHECK	TYPICAL WEEKLY GROSS PAYCHECK	HIGHEST WEEKLY GROSS PAYCHECK
Sales Associate	617.50	653.75	690
Construction Worker	0	540	742.50
Wait Staff	390	637	967

5. Money may be the biggest factor in making your decision. But it is summer, and it would be nice to enjoy what you are doing. Discuss the advantages and disadvantages of each position. What would your personal choice be? Why?

(Answers will vary.)

6. You decide that you would prefer an indoor job. Use the algebraic rules you developed for the sales job in Problem 1b and for the restaurant position in Problem 3d to calculate how many hours you would have to work in each job to receive the same weekly salary.

$7.25x + 400 = 3.60x + 520$

$3.65x = 120$

$x = \dfrac{120}{3.65} \qquad \approx 32.88$

I would have to work about 33 hours in each job to receive the same salary.

Cluster 3   What Have I Learned?

1. a. Explain the difference between the two expressions $-x^2$ and $(-x)^2$. Use an example to illustrate your explanation.

By the order of operations convention, $-x^2$ indicates to square x first, then negate. For example, $-3^2 = -(3)(3) = -9$.

The parentheses in $(-x)^2$ indicate to negate x and then square the result. For example, $(-3)^2 = (-3)(-3) = 9$.

b. What role does the negative sign to the left of the parentheses play in simplifying the expression $-(x - y)$? Simplify this expression.

The negative sign can be interpreted as -1 and by the distributive property reverses the signs of the terms in parentheses. The simplified expression is $-x + y$.

2. a. Are $2x$ and $2x^2$ like terms? Why or why not?

No. The x factors have different exponents. By definition, a variable factor that appears in like terms must possess the same exponent.

b. A student simplified the expression $6x^2 - 2x + 5x$ and obtained $9x^2$. Is he correct? Explain.

No, the first term is not like the second two terms, which are like terms. The correct combination should be $6x^2 + 3x$.

3. Can the distributive property be used to simplify the expression $3(2xy)$? Explain.

No, $3(2xy) = 3 \cdot 2 \cdot x \cdot y = 6xy$, that is, 3 multiplies only one term, $2xy$. The distributive property is used when a factor multiplies another factor that is a sum or difference. For example, $3(2x + y) = 6x + 3y$ by the distributive property.

4. a. Is the expression $2x(5y + 15x)$ completely factored? Explain.

No, the expression in parentheses contains a common factor of 5. A complete factoring is $10x(y + 3x)$.

b. One classmate factored the expression $12x^2 - 18x + 12$ as $2(6x^2 - 9x + 6)$ and another factored it as $6(2x^2 - 3x + 2)$. Which is correct?

They are both correct. However, $6(2x^2 - 3x + 2)$ is a complete factoring because the terms in the parentheses have no other factors in common. The terms in the parentheses of the expression $2(6x^2 - 9x + 6)$ have the factor 3 in common.

5. Another algebraic code of instructions you can use in your number trick show is

$$\frac{4x + 4(x - 1)}{4} + 1.$$

a. Write all the steps you will use to figure out any chosen number.
(*Hint*: See Activity 2.12, Math Magic.)

(Answers will vary.)

The code implies the following steps:

1. Specify a value for x.

2. Multiply the chosen value by 4.

3. Subtract 1 from the value of x and multiply by 4.

4. Add the result of step 2 to the result of step 3.

5. Divide the result of step 4 by 4.

6. Add 1 to the result of step 5.

b. Simplify the original expression and state how you will determine the chosen number when given a resulting value.

$$\frac{4x + 4(x - 1)}{4} + 1$$

$$\frac{4x + 4x - 4}{4} + 1$$

$$\frac{8x - 4}{4} + 1$$

$$2x - 1 + 1$$

$$2x$$

The given code of instruction leads to doubling the starting value. Therefore, I would divide the resulting value by 2 to obtain the starting value.

6. For extra credit on exams, your mathematics instructor permits you to locate and circle your errors and then correct them. To get additional points, you are to show all correctly worked steps alongside the incorrect ones. On your last math test, the following problem was completed incorrectly. Show the work necessary to obtain the extra points.

$$2(x - 3) = 5x + 3x - 7(x + 1)$$

$$2x - ⑤ = 8x - 7x ⊕ 7$$

$$2x - 5 = x + 7$$

$$③x = 12$$

$$x = 4$$

$$2x - 6 = 8x - 7x - 7$$

$$2x - 6 = x - 7$$

$$x = -1$$

Cluster 3 How Can I Practice?

1. Complete the following table. Then determine numerically and algebraically which of the following expressions are equivalent.

   a. $13 + 2(5x - 3)$ **b.** $10x + 10$ **c.** $10x + 7$

x	13 + 2(5x − 3)	10x + 10	10x + 7
1	17	20	17
5	57	60	57
10	107	110	107

   Column 2 and column 4 produce the same results. Therefore, expressions a and c seem to be equivalent.

   Algebraically,

   $13 + 2(5x - 3)$

   $13 + 10x - 6$

   $10x + 7$

   Therefore, expression a and expression c are equivalent.

2. Use the distributive property to expand each of the following algebraic expressions.

   a. $6(x - 7)$ **b.** $3x(x + 5)$ **c.** $-(x - 1)$

       $6x - 42$ $3x^2 + 15x$ $-x + 1$

   d. $-2.4(x + 1.1)$ **e.** $4x(a - 6b - 1)$ **f.** $-2x(3x + 2y - 4)$

       $-2.4x - 2.64$ $4xa - 24xb - 4x$ $-6x^2 - 4xy + 8x$

3. Write each expression in completely factored form.

   a. $5x - 30$ **b.** $6xy - 8xz$

       $5(x - 6)$ $2x(3y - 4z)$

   c. $-6y - 36$ **d.** $2xa - 4xy + 10xz$

       $-6(y + 6)$ $2x(a - 2y + 5z)$

   e. $2x^2 - 6x$

       $2x(x - 3)$

4. Consider the expression $5x^3 + 4x^2 - x - 3$.

   a. How many terms are there?

       There are four terms.

   b. What is the coefficient of the first term?

       The coefficient of the first term is 5.

   c. What is the coefficient of the third term?

       The coefficient of the third term is -1.

d. If there is a constant term, what is its value?

The constant term is -3.

e. What are the factors of the second term?

The factors of the second term are 4 and x^2 (or 4, x, x).

5. a. Let n represent the input. The output is described by the following verbal phrase.

Six times the square of the input, decreased by twice the input and then increased by eleven

Translate the above phrase into an algebraic expression.

$6n^2 - 2n + 11$

b. How many terms are in this expression?

There are three terms in the expression.

c. List the terms in the expression from part a.

$6n^2, -2n, 11$

6. Combine like terms to simplify the following expressions.

a. $5x^3 + 5x^2 - x^3 - 3$

$4x^3 + 5x^2 - 3$

b. $xy^2 - x^2y + x^2y^2 + xy^2 + x^2y$

$x^2y^2 + 2xy^2$

c. $3ab - 7ab + 2ab - ab$

$-3ab$

7. For each of the following algebraic expressions, list the specific operations indicated, in the order in which they are to be performed.

a. $10 + 3(x - 5)$

i. Subtract 5 from x.

ii. Multiply the result by 3.

iii Add 10.

b. $(x + 5)^2 - 15$

i. Add 5 to x.

ii. Square the result.

iii. Subtract 15.

c. $(2x - 4)^3 + 12$

i. Multiply x by 2.

ii. Subtract 4.

iii. Raise to the third power.

iv. Add 12.

8. Simplify the following algebraic expressions.

a. $4 - (x - 2)$

$4 - x + 2$

$= 6 - x$

b. $4x - 3(4x - 7) + 4$

$4x - 12x + 21 + 4$

$= -8x + 25$

c. $x(x - 3) + 2x(x + 3)$

$\quad x^2 - 3x + 2x^2 + 6x$

$= 3x^2 + 3x$

d. $2[3 - 2(a - b) + 3a] - 2b$

$\quad 2(3 - 2a + 2b + 3a) - 2b$

$= 6 - 4a + 4b + 6a - 2b$

$= 6 + 2a + 2b$

e. $3 - [2x + 5(x + 3) - 2] + 3x$

$\quad 3 - [2x + 5x + 15 - 2] + 3x$

$= 3 - [7x + 13] + 3x$

$= 3 - 7x - 13 + 3x$

$= -4x - 10$

f. $\dfrac{7(x - 2) - (2x + 1)}{5}$

$\quad \dfrac{7x - 14 - 2x - 1}{5}$

$= \dfrac{5x - 15}{5}$

$= x - 3$

9. You own 25 shares of a certain stock. The share price at the beginning of the week is x dollars. By the end of the week, the share's price doubles and then drops \$3. Write a symbolic rule that represents the total value, V, of your stock at the end of the week.

$V = 25(2x - 3)$

10. A volatile stock began the last week of the year worth x dollars per share. The following table shows the changes during that week. If you own 30 shares, write a symbolic rule that represents the total value, V, of your stock at the end of the week.

Day	1	2	3	4	5
Change in Value/Share	Doubled	Lost 10	Tripled	Gained 12	Lost half its value

$x \longrightarrow 2x \longrightarrow 2x - 10 \longrightarrow 6x - 30 \longrightarrow 6x - 18 \longrightarrow 3x - 9$

$V = 30(3x - 9)$

11. Physical exercise is most beneficial for fat burning when it increases a person's heart rate to a target level. The symbolic rule

$$T = 0.6(220 - a)$$

describes how to calculate an individual's target heart rate, T, measured in beats per minute, in terms of age, a, in years.

a. Use the distributive property to rewrite the right-hand side of this symbolic rule in expanded form.

$T = 132 - 0.6a$

b. Use both the factored and expanded rules to determine the target heart rate for an 18-year-old person during aerobics.

$T = 132 - 0.6(18) = 121.2$

$T = 0.6(220 - 18) = 121.2$

12. The manager of a clothing store decides to reduce the price of a leather jacket by \$25.

a. Use x to represent the regular cost of the jacket, and write an expression that represents the discounted price of the leather jacket.

$x - 25$

b. If eight jackets are sold at the reduced price, write an expression in factored form that represents the total receipts for the jackets.

$8(x - 25)$

c. Write the expression in part b as an equivalent expression without parentheses (expanded form).

$8x - 200$

13. You planned a trip with your best friend from college. You had only 4 days for your trip and planned to travel x hours each day.

The first day, you stopped for sightseeing and lost 2 hours of travel time. The second day, you gained 1 hour because you did not stop for lunch. On the third day, you traveled well into the night and doubled your planned travel time. On the fourth day, you traveled only a fourth of the time you planned because your friend was sick. You averaged 45 miles per hour for the first 2 days and 48 miles per hour for the last 2 days.

a. How many hours, in terms of x, did you travel the first 2 days?

$x - 2 + x + 1 = (2x - 1)$ hr.

b. How many hours, in terms of x, did you travel the last 2 days?

$2x + 0.25x = 2.25x$ hr.

c. Express the total distance, D, traveled over the 4 days as a symbolic rule in terms of x. Simplify the rule.
Recall that distance = average rate $\cdot$ time.

$D = 45(2x - 1) + 48(2.25x)$
$\quad = 90x - 45 + 108x$
$\quad = 198x - 45$ mi.

d. Write a symbolic rule that expresses the total distance, y, you would have traveled had you traveled exactly x hours each day at the average speeds indicated above. Simplify the rule.

Traveling an equal number of hours each day, you would have covered
$y = 45(2x) + 48(2x) = 186x$ miles.

e. If you had originally planned to travel 7 hours each day, how many miles did you actually travel?

$198(7) - 45 = 1341$

I have traveled 1341 miles.

f. How many miles would you have gone had you traveled exactly 7 hours each day?

$186(7) = 1302$ mi.

I would have traveled 1302 miles.

14. You read about a full-time summer position in sales at the Furniture Barn. The job pays $280 per week plus 20% commission on sales over $1000.

a. Explain how you would calculate the total amount earned each week.

Multiply the amount of sales over $1000 by 0.20, and then add $280 to the product to obtain the total amount earned.

b. Let x represent the dollar amount of sales for the week. Write a symbolic rule that expresses your earnings, E, for the week in terms of x.

$E = 280 + 0.20(x - 1000), x > 1000$

c. Write an equation to determine how much furniture you must sell to have a gross salary of $600 for the week.

$600 = 280 + 0.20(x - 1000)$

d. Solve the equation in part c.

$600 = 280 + 0.20x - 200$

$600 = 80 + 0.20x$

$520 = 0.20x$

$x = 2600$

I must sell $2600 worth of furniture in order to have a gross salary of $600.

e. Is the total amount of sales reasonable?

(Answers will vary.) I will have to be a pretty good salesperson to sell $2600 worth of furniture every week.

15. As a prospective employee in a furniture store, you are offered a choice of salary. The following table shows your options.

Option 1	$200 per week	Plus 30% of all sales
Option 2	$350 per week	Plus 15% of all sales

a. Write a symbolic rule to represent the total salary, S, for option 1 if the total sales are x dollars per week.

$S = 200 + 0.30x$

b. Write a symbolic rule to represent the total salary, S, for option 2 if the total sales are x dollars per week.

$S = 350 + 0.15x$

c. Write an equation that you could use to determine how much you would have to sell in a week to earn the same salary under both plans.

$200 + 0.30x = 350 + 0.15x$

d. Solve the equation in part c. Interpret your result.

$200 + 0.30x = 350 + 0.15x$

$0.15x = 150$

$x = \frac{150}{0.15} = 1000$

I would have to sell about $1000 worth of furniture per week to earn the same weekly salary from either option.

e. What is the common salary for the amount of sales found in part d?

$200 + 0.30(1000) = 500$

Therefore, the common salary is approximately $500.

16. A triathlon includes swimming, long-distance running, and cycling.

a. Let x represent the number of miles the competitors swim. If the long-distance run is 10 miles longer than the distance swum, write an expression that represents the distance the competitors run in the event.

$x + 10$

b. The distance the athletes cycle is 55 miles longer than they run. Use the result in part a to write an expression in terms of x that represents the cycling distance of the race.

$x + 10 + 55 = x + 65$

c. Write an expression that represents the total distance of all three phases of the triathlon. Simplify the expression.

$x + (x + 10) + (x + 65) = 3x + 75$

d. If the total distance of the triathlon is 120 miles, write and solve an equation to determine x. Interpret the result.

$120 = 3x + 75$

$45 = 3x$

$x = 15$

The competitors will swim 15 miles.

e. What are the lengths of the running and cycling portions of the race?

Run: $15 + 10 = 25$

Cycle: $25 + 55 = 80$

The competitors will run 25 miles and cycle 80 miles.

The bracketed numbers following each concept indicate the activity in which the concept is discussed.

CONCEPT/SKILL	DESCRIPTION	EXAMPLE
Variable [2.1]	A quantity or quality that may change, or vary, in value from one particular instance to another, usually represented by a letter. When a variable describes an actual quantity, its values must include the unit of measurement of that quantity.	The number of miles you drive in a week is a variable. Its value may (and usually does) change from one week to the next. x and y are commonly used to represent variables.
Input [2.1]	The value that is given first in an input/output relationship.	Your weekly earnings depend on the number of hours you work. The two variables in this relationship are hours worked (input) and total earnings (output).
Output [2.1]	The second number in an input/output relationship. It is the number that corresponds to or is matched with the input.	
Set of replacement values for the input [2.1]	The set of replacement values for the input is the collection of all numbers for which a corresponding output value can be determined.	The amount of lawn you can mow depends on whether you can mow for 2, 3, or 4 hours. The set of input values is {2, 3, 4}.
Input/output relationship [2.1]	Relationship between two variables that can be represented numerically by a table of values (ordered pairs) or graphically as plotted points in a rectangular coordinate system.	<table><tr><th>NUMBER OF HOURS WORKED</th><th>EARNINGS</th></tr><tr><td>15</td><td>$90</td></tr><tr><td>25</td><td>$150</td></tr></table>
Horizontal axis [2.1]	In graphing an input/output relationship, the input is referenced on the horizontal axis.	Output ↑ → Input
Vertical axis [2.1]	In graphing an input/output relationship, the output is referenced on the vertical axis.	
Rectangular coordinate system [2.1] and [2.2]	Allows every point in the plane to be identified by an ordered pair of numbers, the coordinates of the point, determined by the distance of the point from two perpendicular number lines (called coordinate axes) that intersect at their respective 0 values, the origin.	y -----•(x, y) 0 x
Scaling [2.2]	Setting the same distance between each pair of adjacent tick marks on an axis.	−4 −3 −2 −1 0 1 2 3 4 Scale is 1

CONCEPT/SKILL	DESCRIPTION	EXAMPLE
Quadrants [2.2]	Two perpendicular coordinate axes divide the plane into four quadrants, labeled counterclockwise with Quadrant I being the upper-right quadrant.	
Point in the plane [2.2]	Points are identified by an ordered pair of numbers (x, y) in which x represents the horizontal distance from the origin and y represents the vertical distance from the origin.	$(2, 30)$ are the coordinates of a point in the first quadrant located 2 units to the right and 30 units above the origin.
Verbal rule [2.3]	A statement that describes in words the arithmetic relationship between the input and output variables.	Your payment (in dollars) for mowing a lawn is 8 times the number of hours worked.
Algebraic expression [2.4]	An algebraic expression is a shorthand code for a sequence of arithmetic operations to be performed on a variable.	The algebraic expression $2x + 3$ indicates that you start with a value of x, multiply by 2, and then add 3.
Evaluate an algebraic expression [2.4]	To evaluate an algebraic expression, replace the variable(s) by its (their) assigned value(s) and perform the indicated arithmetic operation(s).	Evaluate $3x^2 - 2x + 4$ when $x = 2$. $3(2)^2 - 2(2) + 4$ $= 3 \cdot 4 - 4 + 4$ $= 12 - 4 + 4$ $= 12$
Equation [2.6]	An equation is a statement that two algebraic expressions are equal.	$2x + 3 = 5x - 9$
Symbolic rule [2.6]	A symbolic rule is a mathematical statement that defines an output variable as an algebraic expression in terms of the input variable. The symbolic rule is essentially a recipe that describes how to determine the output value corresponding to a given input value.	The symbolic rule $y = 3x - 5$ indicates that the output, y, is obtained by multiplying x by 3 and then subtracting 5.

CONCEPT/SKILL	DESCRIPTION	EXAMPLE
Solution of an equation [2.6]	The solution of an equation is a replacement value for the variable that makes both sides of the equation equal in value.	3 is a solution of the equation $4x - 5 = 7$.
Solve equations of the form $ax + b = c$ algebraically [2.7]	Solve equations containing more than one arithmetic operation by performing the inverse operations in reverse order of the operations shown in the equation.	To solve $3x - 6 = 15$ for x, add 6, then divide by 3 as follows: $$3x - 6 = 15$$ $$3x - 6 + 6 = 15 + 6$$ $$3x = 21$$ $$\frac{3x}{3} = \frac{21}{3}$$ $$x = 7$$
Solve a formula for a given variable [2.8]	To solve a formula, isolate the term containing the variable of interest on one side of the equation, with all other expressions on the other side. Then, divide both sides of the equation by the coefficient of the variable.	$2(a + b) = 8 - 5a$, for a $$2a + 2b = 8 - 5a$$ $$2a + 5a = 8 - 2b$$ $$7a = 8 - 2b$$ $$a = \frac{8 - 2b}{7}$$
Proportion [2.9]	An equation stating that two ratios are equivalent.	$\frac{a}{b} = \frac{c}{d}$
Cross multiply [2.9]	A procedure for solving a proportion. Multiply both sides of the proportion by the denominators and then solve. $\frac{a}{b} = \frac{c}{d} \Rightarrow bd \cdot \frac{a}{b} = \frac{c}{d} \cdot bd \Rightarrow da = cb$	$\frac{15}{x} = \frac{3}{20}$ becomes $15 \cdot 20 = 3x$, so $x = 100$.
Equivalent expressions [2.10]	Equivalent expressions are two algebraic expressions that always produce the same output for identical inputs.	$4(x + 6)$ and $4x + 24$ are equivalent expressions. For any value of x, adding 6 and then multiplying by 4 *always* gives the same result as multiplying by 4 and then adding 24.
Constant [2.11]	A constant is a quantity that does not change in value within the context of a problem.	A number, such as 2, or a symbol, such as k, that is understood to have a constant value within a problem.
Numerical coefficient [2.11]	A numerical coefficient is a number that multiplies a variable or expression.	The number 8 in $8x$.
Factors [2.11]	Factors are numbers, variables, and/or expressions that are multiplied together to form a product.	$2 \cdot x \cdot (x - 1)$ Here, 2, x, and $x - 1$ are all factors.
Terms [2.11]	Terms are parts of an expression that are separated by plus or minus signs.	The expression $3x + 6y - 8z$ contains three terms.

CONCEPT/SKILL	DESCRIPTION	EXAMPLE
Distributive property [2.11]	$\underbrace{a \cdot (b + c)}_{\text{factored form}} = \underbrace{a \cdot b + a \cdot c}_{\text{expanded form}}$	$4(x + 6) = 4 \cdot x + 4 \cdot 6$ $= 4x + 24$
Geometric interpretation of the distributive property [2.11]	The area of the large rectangle is equal to the sum of the areas of the two smaller rectangles.	$a \cdot (b + c)$ equals $a \cdot b + a \cdot c$
Factored form of an algebraic expression [2.11]	An algebraic expression is in factored form when it is written as a product of factors.	The expression $4(x + 6)$ is in factored form. The two factors are 4 and $x + 6$.
Expanded form of an algebraic expression [2.11]	An algebraic expression is in expanded form when it is written as a sum of distinct terms.	The expression $4x + 24$ is in expanded form. The two distinct terms are $4x$ and 24.
Extension of the distributive property [2.11]	The distributive property extended to sums or differences of more than two terms is $a(b + c + d) = a \cdot b + a \cdot c + a \cdot d.$	$4(3x + 5y - 6)$ $= 4 \cdot 3x + 4 \cdot 5y - 4 \cdot 6$ $= 12x + 20y - 24$
Factoring [2.11]	Factoring is the process of writing a sum of distinct terms equivalently as a product of factors.	$10xy + 15xz = 5x(2y + 3z)$
Common factor [2.11]	The common factor is one that is a factor contained in every term of an algebraic expression.	In the expression $12abc + 3abd - 21ab$, 3, a, and b are common factors.
Factoring out a common factor [2.11]	This is the process of dividing each term by a common factor and placing this factor outside parentheses containing the sum of remaining terms.	$12abc + 3abd - 21ab$ $= 3b(4ac + ad - 7a)$
Greatest common factor [2.11]	A common factor such that there are no additional factors (other than 1) common to the terms in the expression is the greatest common factor.	The greatest common factor in the expression $12abc + 6abd - 21ab$ is $3ab$.

CONCEPT/SKILL	DESCRIPTION	EXAMPLE
Completely factored form [2.11]	An algebraic expression is in completely factored form when none of its factors can themselves be factored any further.	$4x(2y + 10z)$ is not in completely factored form because the expression in the parentheses has a common factor of 2. $8x(y + 5z)$ is the completely factored form of this expression.
Procedure for factoring an algebraic expression whose terms contain a common factor [2.11]	The procedure for factoring an algebraic expression whose terms contain a common factor: a. Identify the common factor. b. Divide each term by the common factor. c. Place the sum of the remaining factors inside the parentheses, and place the common factor outside the parentheses.	Given $20x + 35y - 45$: The common factor of the terms in this expression is 5. Dividing each term by 5 yields remaining terms $4x$, $7y$, -9. The factored form is therefore $5(4x + 7y - 9)$.
Like terms [2.11]	Like terms are terms that contain identical variable factors, including exponents. They differ only in their numerical coefficients.	$5y^2$ and $5y$ are not like terms. $7xy$ and $-3xy$ are like terms.
Combining like terms [2.11]	To combine like terms into a single term, add or subtract the coefficients of like terms.	$12cd^2 - 5c^2d + 10 + 3cd^2 - 6c^2d$ $= 15cd^2 - 11c^2d + 10$
Procedure for simplifying algebraic expressions [2.12]	Steps for simplifying algebraic expressions: 1. Simplify the expression from the innermost grouping outward using the order of operations convention. 2. Use the distributive property when applicable. 3. Combine like terms.	$[5 - 2(x - 7)] + 5(3x - 10)$ $= [5 - 2x + 14] + 5(3x - 10)$ $= -2x + 19 + 5(3x - 10)$ $= -2x + 19 + 15x - 50$ $= 13x - 31$
Algebraic goal of solving an equation [2.13]	The goal of solving an equation is to isolate the variable on one side of the equation.	$3x - 5 = x - 2$ $2x = 3$ $x = \dfrac{3}{2}$

CONCEPT/SKILL	DESCRIPTION	EXAMPLE
General strategy for solving equations algebraically [2.13]	General strategy for solving an equation algebraically: 1. Simplify the expressions on each side of the equation (as discussed above). 2. Isolate the variable term on one side of the equation. 3. Divide each side of the equation by the coefficient of the variable. 4. Check your result in the original equation.	$3(2x - 5) + 2x = 9$ $6x - 15 + 2x = 9$ $8x - 15 = 9$ $8x = 24$ $x = 3$ Check: $3(2 \cdot 3 - 5) + 2 \cdot 3$ $= 3(6 - 5) + 6$ $= 3 \cdot 1 + 6$ $= 9$
Identity [2.13]	An equation that is true for all real number replacement values is called an identity. This occurs algebraically when both sides of the equation become identical in the process of solving.	Solve $2x - (6x - 5) = 5 - 4x$. $2x - 6x + 5 = 5 - 4x$ $-4x + 5 = 5 - 4x$ $0 = 0$ *all real numbers*
Contradiction [2.13]	An equation that is never true for any real number replacement value is called a contradiction. This occurs algebraically when the variable terms are eliminated from both sides of the equation in the process of solving it and the resulting equation is a false statement.	Solve $3(2x + 4) = 6x - 5$. $6x + 12 = 6x - 5$ $12 = -5$ *no solution*

1. Scale each axis on the Cartesian coordinate plane below, and plot the points whose coordinates are given in parts a–g.

a. $(1, 2)$ **b.** $(3, -4)$ **c.** $(-2, 5)$ **d.** $(-4, -6)$

e. $(0, 4)$ **f.** $(-5, 0)$ **g.** $(0, 0)$

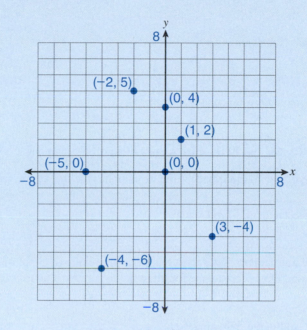

2. Let x represent any number. Translate each of the following phrases into an algebraic expression.

a. 5 more than x

$x + 5$

b. x less than 18

$18 - x$

c. Double x

$2x$

d. Divide 4 by x

$\frac{4}{x}$

e. 17 more than the product of 3 and x

$3x + 17$

f. 12 times the sum of 8 and x

$12(8 + x)$

g. 11 times the difference of 14 and x

$11(14 - x)$

h. 49 less than the quotient of x and 7

$\frac{x}{7} - 49$

3. Determine the values requested in the following.

a. If $y = 2x$ and $y = 18$, what is the value of x?

$x = 9$

b. If $y = -8x$ and $x = 3$, what is the value of y?

$y = -24$

c. If $y = 15$ and $y = 6x$, what is the value of x?

$x = 2.5$

d. If $y = -8$ and $y = 4 + x$, what is the value of x?

$x = -12$

Answers to all Gateway exercises are included in the Selected Answers appendix.

257

e. If $x = 19$ and $y = x - 21$, what is the value of y?

$y = -2$

f. If $y = -71$ and $y = x - 87$, what is the value of x?

$x = 16$

g. If $y = 36$ and $y = \dfrac{x}{4}$, what is the value of x?

$x = 144$

h. If $x = 5$ and $y = \dfrac{x}{6}$, what is the value of y?

$y = \frac{5}{6}$

i. If $y = \dfrac{x}{3}$ and $y = 24$, what is the value of x?

$x = 72$

4. Tiger Woods had scores of 63, 68, and 72 for three rounds of a golf tournament.

a. Let x represent his score on the fourth round. Write a symbolic rule that expresses his average score after four rounds of golf.

$$A = \frac{63 + 68 + 72 + x}{4} = \frac{203 + x}{4}$$

b. To be competitive in the tournament, Tiger must maintain an average of about 66. Use the symbolic rule from part a to determine what he must score on the fourth round to achieve a 66 average for the tournament.

He must score 61.

5. Determine the values requested in the following.

a. If $y = 2x + 8$ and $y = 18$, what is the value of x?

$x = 5$

b. If $x = 14$ and $y = 6x - 42$, determine the value of y.

$y = 42$

c. If $y = -33$ and $y = -5x - 3$, determine the value of x.

$x = 6$

d. If $y = -38$ and $y = 24 + 8x$, what is the value of x?

$x = -7.75$

e. If $x = 72$ and $y = -54 - \dfrac{x}{6}$, what is the value of y?

$y = -66$

f. If $y = 66$ and $y = \dfrac{2}{3}x - 27$, what is the value of x?

$x = 139.5$

g. If $y = 39$ and $y = -\dfrac{x}{4} + 15$, what is the value of x?

$x = -96$

h. If $y = 32.56x + 27$ and $x = 0$, what is the value of y?

$y = 27$

6. You must drive to Syracuse to take care of some legal matters. The cost of renting a car for a day is $25, plus 15 cents per mile.

   a. Identify the input variable.

   The input variable is the number of miles driven.

   b. Identify the output variable.

   The output variable is the cost of rental for a day.

   c. Use x to represent the input and y to represent the output. Write a symbolic rule that describes the daily rental cost in terms of the number of miles driven.

   $y = 25 + 0.15x$

   d. Complete the following table.

Input, x (mi)	100	200	300	400	500
Output, y ($)	40	55	70	85	100

   e. Plot the points from the table in part d. Then draw the line through all five points. (Make sure you label your axes and use appropriate scaling.)

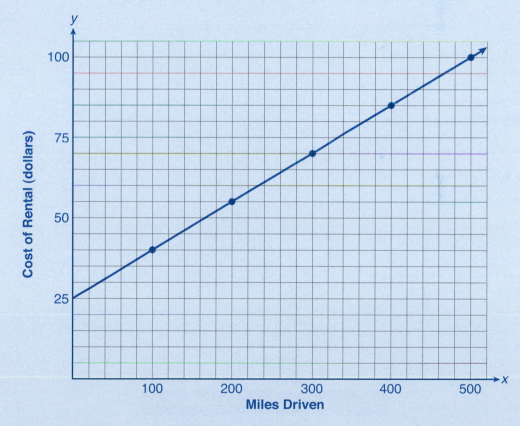

   f. The distance from Buffalo to Syracuse is 153 miles. Estimate from the graph in part e how much it will cost to travel from Buffalo to Syracuse and back.

   The cost to travel round-trip from Buffalo to Syracuse is just under $75.

   g. Use the symbolic rule in part c to determine the exact cost of this trip.

   $25 + 0.15(306) = 70.90$ The cost is $70.90.

h. You have budgeted $90 for car rental for your day trip. Use your graph to estimate the greatest number of miles you can travel in the day and not exceed your allotted budget. Estimate from your graph.

I can drive approximately 425 miles

i. Use the symbolic rule in part c to determine the exact number of miles you can travel in the day and not exceed $90.

$90 = 25 + 0.15x$

$x = \frac{65}{0.15} = 433.33$

I can drive approximately 433 miles.

j. What are realistic replacement values for the input if you rent the car for only 1 day?

(Answers will vary.)

7. As a real estate salesperson, you earn a small salary, plus a percentage of the selling price of each house you sell. If your salary is $100 a week, plus 3.5% of the selling price of each house sold, what must your total annual home sales be for you to gross $30,000 in 1 year? Assume that you work 50 weeks per year.

$5000 + 0.035s = 30,000$

$s = \frac{25,000}{0.035} \approx 714,286$

My total annual home sales must be $714,286 in order to gross $30,000.

8. Use the formula $I = Prt$ to evaluate I, given the following information.

a. $P = \$2000, r = 5\%, t = 1$

$I = 2000(0.05)(1) = \$100$

b. $P = \$3000, r = 6\%, t = 2$

$I = 3000(0.06)(2) = \$360$

9. Use the formula $P = 2(w + l)$, to evaluate P for the following information.

a. $w = 2.8$ and $l = 3.4$

$P = 2(2.8 + 3.4) = 12.4$

b. $w = 7\frac{1}{3}$ and $l = 8\frac{1}{4}$

$P = 2(7\frac{1}{3} + 8\frac{1}{4}) = 31\frac{1}{6}$

10. Complete the following table and use your results to determine numerically which, if any, of the following expressions are equivalent.

a. $(4x - 3)^2$

b. $4x^2 - 3$

c. $(4x)^2 - 3$

x	$(4x - 3)^2$	$4x^2 - 3$	$(4x)^2 - 3$
-1	49	1	13
0	9	-3	-3
3	81	33	141

None are equivalent.

11. Simplify the following expressions.

    a. $3(x + 1)$

       $3x + 3$

    c. $x(4x - 7)$

       $4x^2 - 7x$

    e. $4 + 2(6x - 5) - 19$

       $12x - 25$

    b. $-6(2x^2 - 2x + 3)$

       $-12x^2 + 12x - 18$

    d. $6 - (2x + 14)$

       $-8 - 2x$

    f. $5x - 4x(2x - 3)$

       $17x - 8x^2$

12. Factor the following expressions completely.

    a. $4x - 12$

       $4(x - 3)$

    c. $-12x - 20$

       $-4(3x + 5)$

    b. $18xz + 60x - xy$

       $x(18z + 60 - y)$

13. Simplify the following expressions.

    a. $4x^2 - 3x - 2 + 2x^2 - 3x + 5$

       $6x^2 - 6x + 3$

    b. $(3x^2 - 7x + 8) - (2x^2 - 4x + 1)$

       $x^2 - 3x + 7$

14. Use the distributive property to expand each of the algebraic expressions.

    a. $5(2x - 7y)$

       $10x - 35y$

    c. $-10(x - 2w + z)$

       $-10x + 20w - 10z$

    b. $\dfrac{1}{3}(6a + 3b - 9)$

       $2a + b - 3$

    d. $4\left(\dfrac{1}{2}c - \dfrac{1}{4}\right)$

       $2c - 1$

15. Solve the given equations, and check your results.

    a. $\dfrac{5}{28} = \dfrac{x}{49}$

       $x = 8.75$

    b. $\dfrac{18}{x} = \dfrac{90}{400}$

       $x = 80$

    c. $4(x + 5) - x = 80$

       $x = 20$

    d. $-5(x - 3) + 2x = 6$

       $x = 3$

    e. $38 = 57 - (x + 32)$

       $x = -13$

    f. $-13 + 4(3x + 5) = 7$

       $x = 0$

    g. $5x + 3(2x - 8) = 2(x + 6)$

       $x = 4$

    h. $-4x - 2(5x - 7) + 2 = -3(3x + 5) - 4$

       $x = 7$

    i. $-32 + 6(3x + 4) = -(-5x + 38) + 3x$

       $18x - 8 = 8x - 38$

       $x = -3$

    j. $4(3x - 5) + 7 = 2(6x + 7)$

       $12x - 13 = 12x + 14$

       $0 = 27$

       Since 0 does not equal 27, there is no solution.

    k. $2(9x + 8) = 4(3x + 4) + 6x$

       $18x + 16 = 18x + 16$

       All real numbers are solutions.

16. Your sister has just found the perfect dress for a wedding. The price of the dress is reduced by 30%. She is told that there will be a huge sale next week, so she waits to purchase it. When she goes to buy the dress, she finds that it has been reduced again by 30% of the already reduced price.

a. If the original price of the dress is $400, what is the price of the dress after the first reduction?

$400(0.7) = 280

The price of the dress after the first reduction is $280.

b. What is the price of the dress after the second reduction?

After the second reduction, the price of the dress is $196.

c. Let x represent the original price (input). Determine an expression that represents the price of the dress after the first reduction. Simplify this expression.

$0.70x$

d. Use the result from part c to write an expression that represents the price of the dress after the second reduction. Simplify this expression.

$0.7(0.7x) = 0.49x$

e. Use the result from part d to determine the price of the dress after the two reductions if the original price is $400. How does this compare with your answer to part b?

$0.49(400) = 196

The price is the same.

f. She sees another dress that is marked down to $147 after the same two reductions. She wants to know the original price of the dress. Write and solve the equation to determine the original price.

$0.49x = 147$

$x = 300

The original price is $300.

17. The proceeds from your college talent show to benefit a local charity totaled $1550. Since the seats were all taken, you know that 500 people attended. The cost per ticket was $2.50 for students and $4.00 for adults. Unfortunately, you misplaced the ticket stubs that would indicate how many students and how many adults attended. You need this information for accounting purposes and future planning.

a. Let n represent the number of students who attended. Write an expression in terms of n to represent the number of adults who attended. $500 - n$

b. Write an expression in terms of n that will represent the proceeds from the student tickets. $2.50n$

c. Write an expression in terms of n that will represent the proceeds from the adult tickets.

$4.00(500 - n)$

d. Write an equation that indicates that the total proceeds from the student and adult tickets totaled $1550.

$2.50n + 4.00(500 - n) = 1550$

e. How many student tickets and how many adult tickets were sold?

$n = 300$ student tickets and 200 adult tickets were sold.

18. You have an opportunity to be the manager of a day camp for the summer. You know that your fixed costs for operating the camp are $1200 per week, even if there are no campers. Each camper who attends costs the management $25 per week. The camp charges each camper $60 per week.

Let x represent the number of campers.

a. Write a symbolic rule in terms of x that represents the total cost, C, of running the camp per week.

$C = 1200 + 25x$

b. Write a symbolic rule in terms of x that represents the total income (revenue), R, from the campers per week.

$R = 60x$

c. Write a symbolic rule in terms of x that represents the total profit, P, from the campers per week.

$P = 60x - (1200 + 25x)$

$= 35x - 1200$

d. How many campers must attend for the camp to break even with revenue and costs?

$35x - 1200 = 0$

$x \approx 34.29 = 35$

Thirty-five campers must attend to break even.

e. The camp would like to make a profit of \$620. How many campers must enroll to make that profit?

$35x - 1200 = 620$

$35x = 1820$

$x = 52$

Fifty-two campers must enroll for the camp to make \$620 profit.

f. How much money would the camp lose if only 20 campers attend?

\$500 If only 20 campers attend, the camp would lose \$500.

19. Solve each of the following equations for the specified variable.

a. $I = Prt$, for P

$P = \frac{I}{rt}$

b. $f = v + at$, for t

$t = \frac{f - v}{a}$

c. $2x - 3y = 7$, for y

$y = \frac{2x - 7}{3}$

20. You contact the local print shop to produce a commemorative booklet for your college theater group. It is the group's twenty-fifth anniversary, and in the booklet you want a short history plus a description of all the theater productions for the past 25 years. It costs \$750 to typeset the booklet and 25 cents for each copy produced.

a. Write a symbolic rule that gives the total cost, C, in terms of the number, x, of booklets produced.

$C = 750 + 0.25x$

b. Use the symbolic rule from part a to determine the total cost of producing 500 booklets.

$C = 750 + 0.25(500) = 750 + 125 = 875$

The total cost of producing 500 booklets is \$875.

c. How many booklets can be produced for \$1000?

$1000 = 750 + 0.25x$

$250 = 0.25x$

$x = \frac{250}{0.25} = 1000$

A thousand booklets can be produced for \$1000.

d. Suppose the booklets are sold for 75 cents each. Write a symbolic rule for the total revenue, R, from the sale of x booklets.

$R = 0.75x$

e. How many booklets must be sold to break even? That is, for what value of x is the total cost of production equal to the total amount of revenue?

$0.75x = 750 + 0.25x$

$0.50x = 750$

$x = \frac{750}{0.50} = 1500$

To break even, 1500 booklets must be sold.

f. How many booklets must be sold to make a $500 profit?

$0.75x - (750 + 0.25x) = 500$

$0.75x - 750 - 0.25x = 500$

$0.50x - 750 = 500$

$0.50x = 1250$

$x = \frac{1250}{0.50} = 2500$

To make a $500 profit, 2500 booklets must be sold.

21. You live 7.5 miles from work, where you have free parking. Some days, you must drive to work. On other days, you can take the bus. It costs you 30 cents per mile to drive the car and $2.50 round-trip to take the bus. Assume that there are 22 working days in a month.

a. Let x represent the number of days that you take the bus. Express the number of days that you drive in terms of x.

$22 - x$

b. Express the cost of taking the bus in terms of x.

$2.50x$

c. Express the cost of driving in terms of x.

$(22 - x)(0.30)(15) = 4.50(22 - x)$

d. Express the total cost of transportation in terms of x.

$2.50x + 4.5(22 - x) = 2.50x + 99 - 4.5x = 99 - 2x$

e. How many days can you drive if you budget $70 a month for transportation?

$70 = 99 - 2x$

$-29 = -2x$

$x = \frac{-29}{-2} = 14.5$

$22 - 14.5 = 7.5$

The number of driving days is 7.5.
I can drive 7 days.

f. How much should you budget for the month if you would like to take the bus only half of the time?

I would take the bus for $x = \frac{1}{2}(22) = 11$ days.

$2.5(11) + 4.5(11) = 27.50 + 49.50 = 77$

I must budget $77 if I take the bus half the time.

Function Sense and Linear Functions

Chapter 3 continues the study of relationships between input and output variables. The focus here is on functions, which are special relationships between input and output variables. You will learn how functions are represented verbally, numerically, graphically, and symbolically. You will also study a special type of function—the linear function.

Cluster 1   Function Sense

Activity 3.1

Graphs Tell Stories

Objectives

1. Describe in words what a graph tells you about a given situation.

2. Sketch a graph that best represents a situation that is described in words.

3. Identify increasing, decreasing, and constant parts of a graph.

4. Identify minimum and maximum points on a graph.

5. Define a function.

6. Use the vertical line test to determine whether a graph represents a function.

"A picture is worth a thousand words" may be a cliché, but nonetheless, it is frequently true. Numerical relationships are often easier to understand when presented in visual form. Understanding graphical pictures requires practice going in both directions—from graphs to words and from words to graphs.

Graphs are always constructed so that as you read the graph from left to right, the input variable increases in value. The graph shows the change (increasing, decreasing, or constant) in the output values as the input values increase.

a.

As the output value increases, the graph rises to the right.

b.

As the output value decreases, the graph falls to the right.

c.

If the output values are constant, the graph remains horizontal.

The point where a graph changes from rising to falling is called a **local maximum point**. The *y*-value of this point is called a **local maximum value**. The point where a graph changes from falling to rising is called a **local minimum point.** The *y*-value of this point is called a **local minimum value.** (See graphs on page 266.)

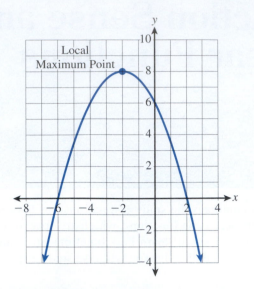

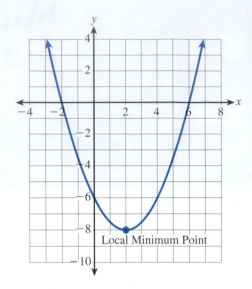

Graphs to Stories

The graphs in Problems 1–4 present visual images of several situations. Each graph shows how the outputs change in relation to the inputs. In each situation, identify the input variable and the output variable. Then, interpret the situation; that is, describe, in words, what the graph is telling you about the situation. Indicate whether the graph rises, falls, or is constant and whether the graph reaches either a minimum (smallest) or maximum (largest) output value.

1. A person's core body temperature (°F) in relation to time of day

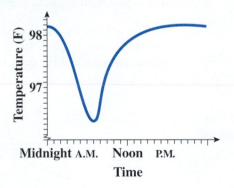

    a. Input: _time of day_   Output: _temperature_

    b. Interpretation:

       This person's body temperature was normal at midnight and then dropped below normal as she slept, reaching a minimum of approximately 96.3°F at 7 A.M. During the day her temperature returned to normal and remained at that level.

2. Performance of a simple task in relation to interest level

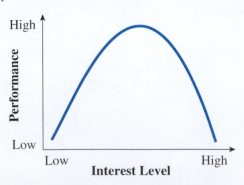

a. Input: <u>interest level</u>   Output: <u>performance</u>

b. Interpretation: (Answers may vary.)

Performance increases as interest level increases. As interest level reaches a certain point, the performance reaches a maximum. Very high interest possibly causes anxiety, and performance declines.

3. Net profit of a particular business in relation to time

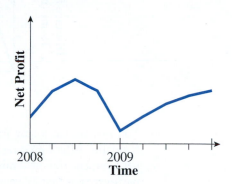

a. Input: <u>time</u>   Output: <u>net profit</u>

b. Interpretation:

Profit increases during the first quarter of 2008 and then increases more slowly to midyear, when the profit reaches a maximum. During the third quarter, profit decreases and then drops sharply. Profit begins to increase during 2009 (but at a slower rate each quarter) until the profit reaches the same level as in 2008.

Possibly this business sells or services summer merchandise.

4. Annual gross income in relation to number of years

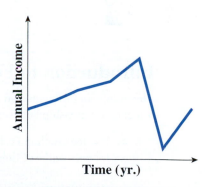

a. Input: <u>time in years</u>   Output: <u>annual income</u>

b. Interpretation:

For a few years, income rises steadily, then more sharply, perhaps due to a raise, and reaches a maximum income. Income drops rather quickly and reaches a minimum. When work begins again, income begins to rise.

Stories to Graphs

In Problems 5 and 6, sketch a graph that best represents the situation described. Note that in many cases, the actual values are unknown, so you will need to estimate what seems reasonable to you.

5. You drive to visit your parents, who live 100 miles away. Your average speed is 50 miles per hour. On arrival, you stay for 5 hours and then return home, again at an average speed of 50 miles per hour. Graph your distance in miles from home, from the time you leave until you return home.

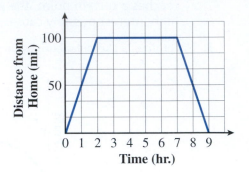

6. You just started a new job that pays $8 per hour, with a raise of $2 per hour every 6 months. After 1½ years, you receive a promotion that gives you a wage increase of $5 per hour, but your next raise won't come for another year. Sketch a graph of your hourly pay over your first 2½ years.

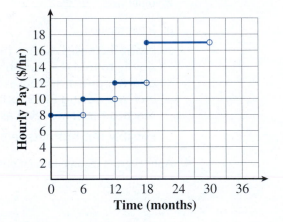

Introduction to Functions

The graphs in Problems 1–6 have different shapes and different properties, but they all share a special relationship between the input and output variables, called a **function**.

7. a. Use the graph in Problem 1 to complete the following table. Estimate the temperature for each value of time.

Time of Day	1 A.M.	5 A.M.	7 A.M.	10 A.M.	12 NOON	1 P.M.
Body Temperature (°F)	98.2	96.9	96.3	97.5	97.8	98.0

b. How many temperatures (output) are assigned to any one particular time (input)?

For any given time, there is only one body temperature.

The relationship between the time of day and body temperature is an example of a function. For any specific time of day, there is one and only one corresponding temperature. In such a case, you say that the body temperature (output) is a function of the time of day (input).

Definition

> A **function** is a rule relating an input variable and an output variable in a way that assigns one and only one output value to each input value.

8. The following graph shows the distance from home over a 9-hour period as described in Problem 5.

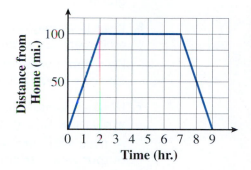

a. Use the graph to complete the following table.

Time (hr.)	1	3	4	8	9
Distance From Home (mi.)	50	100	100	50	0

b. Is the distance from home a function of the time on the trip? Explain.

Distance from home is a function of the time because for any given value of time, there is exactly one corresponding distance from home.

9. The following graph shows the height of a roller coaster as it moves away from its starting point and goes through its first loop.

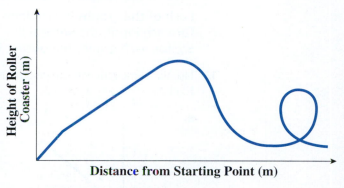

Is the height of the roller coaster a function of the distance of the coaster from the starting point?

The height is *not* a function of the roller coaster's distance from its starting point. In the loop portion, there are two heights that correspond to the same distance from the starting point. A function assigns one and only one output value to each input value.

Vertical Line Test

You can determine directly from a graph whether or not the output is a function of the input.

10. a. Refer to the graph in Problem 1, and select any specific time of day along the horizontal (input) axis. Then, move straight up or down (vertically) from that value of time to locate the corresponding point on the graph. How many points do you locate for any given time?

Every input value of time has exactly one corresponding point on the graph.

b. If you locate only one point on the graph in part a for each value of time you select, explain why this would mean that body temperature is a function of the time of day.

Every input value of time has exactly one corresponding output value for the output, body temperature.

c. If you had located more than one point on the graph in part a, explain why this would mean that body temperature is not a function of time.

A given value of the input would have more than one corresponding output value assigned to it.

The procedure in Problem 10 is referred to as the **vertical line test**.

Definition

The **vertical line test** is a visual technique used to determine whether or not a given graph represents a function. If every vertical line drawn through the graph intersects the graph at no more than one point, the graph represents a function. Equivalently, if there is a single vertical line that intersects the graph more than once, the graph does *not* represent a function.

---

11. Use the vertical line test on the graphs in Problems 2 through 4 to verify that each graph represents the output as a function of the input.

Each of the graphs in Problems 2–4 passes the vertical line test because for each input, the vertical line intersects the graph at only one point. So, for each graph, the output is a function of the input.

12. Use the vertical line test to determine which of the following graphs represent functions. Explain.

a.

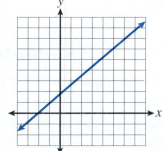

b.

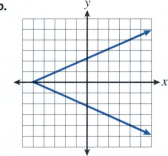

The graph in part a represents a function. It passes the vertical line test.

(Answers will vary.) The graph in part b does not represent a function. It does not pass the vertical line test. The vertical axis, for example, intersects the graph in two places.

c.

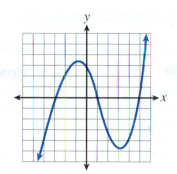

d.

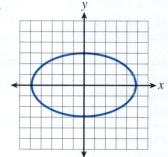

Graph c does represent a function. It passes the vertical line test.
Graph d does not represent a function. It fails the vertical line test.

SUMMARY: ACTIVITY 3.1

1. A **function** is a rule relating an input variable and an output variable in a way that assigns one and only one output value to each input value.

2. In the **vertical line test**, a graph represents a function if every vertical line drawn through the graph intersects the graph no more than once.

3. The point where a graph changes from rising to falling is called a **local maximum point**. The y-value of this point is called a **local maximum value**. The point where a graph changes from falling to rising is called a **local minimum point**. The y-value of this point is called a **local minimum value**.

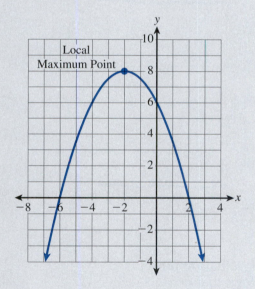

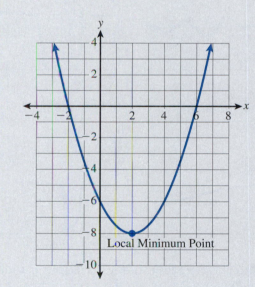

EXERCISES: ACTIVITY 3.1

1. You are a technician at the local power plant, and you have been asked to prepare a report that compares the output and efficiency of the six generators in your sector. Each generator has a graph that shows output of the generator as a function of time over the previous week, Monday through Sunday. You take all the paperwork home for the night (your supervisor wants this report on his desk at 7:00 A.M.), and to your dismay your feisty cat scatters your pile of papers out of the neat order in which you left them. Unfortunately, the graphs for generators A through F were not labeled. (You will know better next time!) You recall some information and find evidence elsewhere for the following facts.

   • Generators A and D were the only ones that maintained a fairly steady output.

   • Generator B was shut down for a little more than 2 days during midweek.

   • Generator C experienced a slow decrease in output during the entire week.

   • On Tuesday morning, there was a problem with generator E that was corrected in a few hours.

   • Generator D was the most productive over the entire week.

Match each graph with its corresponding generator. Explain in complete sentences how you arrived at your answers.

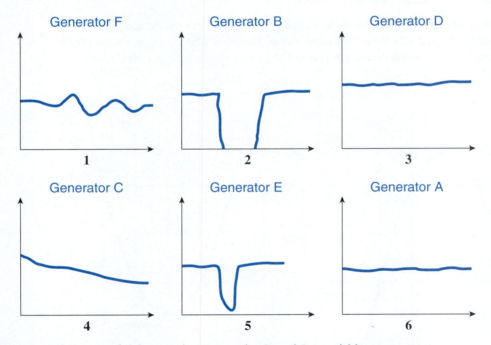

Generators A and D were fairly steady, so graphs 6 and 3 would be appropriate for them. Because D was the most productive, it follows that graph 3 represents the output of generator D and that graph 6 represents the output of generator A.

Graph 2 is the only one that shows a shutdown, so it represents the output of generator B.

The sharp drop shown in graph 5 represents the problem with generator E on Tuesday.

Generator C experienced a slow decrease in output. Graph 4 represents C as it steadily decreases.

The output of generator F is represented by graph 1 by elimination.

In Exercises 2 and 3, identify the input variable and the output variable. Then interpret the situation being represented. Indicate whether the graph rises, falls, or is constant and whether the graph reaches either a minimum (smallest) or maximum (largest) output value.

2. Time required to complete a task in relation to number of times the task is attempted

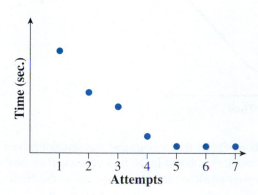

a. Input: <u>number of attempts</u>   Output: <u>time in seconds</u>

b. Interpretation:

The time required to complete a task decreases as the number of attempts increases. As a person attempts a task more times, the task takes less time to complete. After five attempts, no further improvement is made.

3. Number of units sold in relation to selling price

a. Input: <u>selling price</u>   Output: <u>number of units sold</u>

b. Interpretation:

As the selling price increases, the number of units sold increases slightly at first, reaches a maximum, and then declines until none are sold.

4. You leave home on Friday afternoon for your weekend getaway. Heavy traffic slows you down for the first half of your trip, but you make good time by the end. Express your distance from home as a function of time.

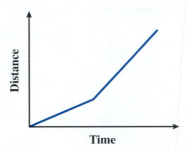

5. Your small business started slowly, losing money in its first 2 years and then breaking even in year three. By the fourth year, you made as much as you lost in the first year and then doubled your profits each of the next 2 years. Graph your profit as the output and time as the input.

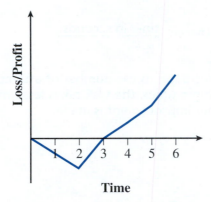

6. The following graph gives the time of day in relation to one's core body temperature.

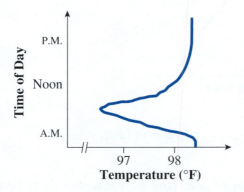

Is the time of day a function of core body temperature?

The time of day is *not* a function of a person's core body temperature. The temperature 98°F corresponds to two times of day—one in the morning and the other around noon. A function assigns one and only one output value to each input value.

7. Use the vertical line test to determine which of the following graphs represents a function. Explain your answer.

a.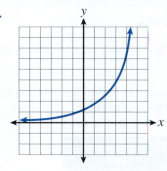

b.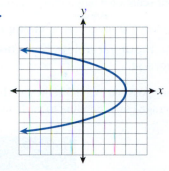

The graph in part a represents a function. It passes the vertical line test.
The graph in part b does *not* represent a function. It fails the vertical line test.

c.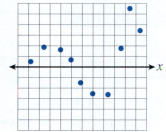

The graph in part c represents a function. It passes the vertical line test.

Activity 3.2

Course Grade

Objectives

1. Represent functions numerically, graphically, and symbolically.

2. Determine the symbolic rule that defines a function.

3. Use function notation to represent functions symbolically.

4. Identify the domain and range of a function.

5. Identify the practical domain and range of a function.

The semester is drawing to a close, and you are concerned about your grade in your anthropology course. During the semester, you have already taken four exams and scored 82, 75, 85, and 93. Your score on exam 5 will determine your final average for the anthropology course.

1. a. Identify the input and output variables.

The input variable is the score on exam 5; the output variable is the average.

b. Four possible exam 5 scores are listed in the following table. Calculate the final average corresponding to each one, and record your answers.

EXAM 5 SCORE, input	FINAL AVERAGE, output
100	87
85	84
70	81
60	79

Recall from Activity 3.1 that a function relates the input to the output in a special way. For any specific input value, there is one and only one output value.

c. Explain how the table of data in part b fits the definition of a function.

Each exam 5 score results in one and only one final average. Since each input produces exactly one output, the input/output relationship represents a function.

2. a. Plot the (input, output) pairs from the preceding table.

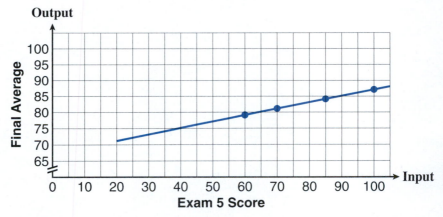

Note: The slash marks (//) indicate that the vertical axis has been cut out between 0 and 65.

b. Draw a straight line through the plotted points. Does the graph represent a function? Explain.

Yes, the graph represents a function because it passes the vertical line test.

c. Use the graph to estimate what your final average will be if you score a 95 on exam 5.

If I score a 95 on exam 5, my final average will be approximately 86.

d. Use the graph to estimate what score you will need on exam 5 to earn a final average of 85.

A score of approximately 90 will earn a final average of 85.

In the context of Problems 1 and 2, you can write that

the final average is a *function* of the score on exam 5.
                output   input

In general, you write that the *output* is a function of the *input*.

Representing Functions Verbally and Symbolically

There are several ways to represent a function. So far, you have seen that a function may be presented *numerically* by a table, as in Problem 1, and *graphically* on a grid, as in Problem 2. A function can also be given *verbally* by stating how the output value is determined for a given input value.

3. a. Describe in words how to obtain the final average for any given score on the fifth exam.

To obtain the final average, add the score on the fifth exam to the other four scores, and then divide the sum by 5.

b. Let A represent the final average and s represent the score on the fifth exam. Translate the verbal rule in part a into a symbolic rule that expresses A in terms of s.

The symbolic rule is $A = \frac{82 + 75 + 85 + 93 + s}{5}$, or $A = \frac{335 + s}{5}$.

4. a. The symbolic rule in Problem 3b represents a fourth way to represent a function. Use the symbolic rule to determine the final average for a score of 75 on exam 5.

$A = \frac{335 + 75}{5} = \frac{410}{5} = 82$

b. Use the symbolic rule to determine the score you will need on exam 5 to earn a final average of 87.

$\frac{335 + s}{5} = 87$

$335 + s = 87(5)$

$335 + s = 435$

$s = 100$

A score of 100 is needed on exam 5 to have a final average of 87.

Function Notation

Often you want to emphasize the functional relationship between input and output. You do this symbolically by writing $A(s)$, which you read as "A is a function of s." The letter within the parentheses always represents the input variable, and the letter in front of the parentheses is the function name.

The formula that defines the "final average" function is then written as

$$A(s) = \frac{335 + s}{5}.$$

To denote the final average corresponding to a specific input, you would write the desired input value in the parentheses. For example, the symbol $A(100)$ denotes the final average for a score of 100 on the fifth exam, and is pronounced "A of 100." To determine the value of $A(100)$, substitute 100 for s in the expression $\frac{335 + s}{5}$, and evaluate the expression:

$$A(100) = \frac{335 + 100}{5} = \frac{435}{5} = 87$$

You interpret the symbolic statement $A(100) = 87$ as "The final average for a fifth exam score of 100 is 87."

Notice that the symbolic way of defining a function is very useful because it clearly shows the rule for determining an output no matter what the input might be.

5. a. In the final average statement $A(75) = 82$, identify the input value and output value.

The input value is 75 and the output value is 82.

b. Interpret the practical meaning of $A(75) = 82$. Write your answer as a complete sentence.

A score of 75 on the fifth exam results in a final average of 82.

6. a. Use the symbolic formula for the final average function to determine the missing coordinate in each input/output pair.

i. (78, _82.6_)   **ii.** (_0_ ,67)

b. Write an English sentence to interpret the practical meaning of the ordered pairs.

The ordered pair (78, 82.6) indicates that a score of 78 on the fifth exam results in a final average of 82.6. The ordered pair (0, 67) indicates that a score of 0 on the fifth exam results in a final average of 67.

c. Evaluate $A(95)$ in the final average function.

$A(95) = \frac{335 + 95}{5} = \frac{430}{5} = 86$

d. Write a sentence that interprets the practical meaning of $A(95)$ in the final average function.

$A(95)$ represents the final average if the score on the fifth exam is 95.

Function Notation—A Word of Caution

In **function notation**, the parentheses *do not* indicate multiplication. If the letter f is the function name, and x represents the input variable, then the symbol $f(x)$ represents the output value corresponding to the input value x. Function notation such as $f(x)$ is used to emphasize the input/output relationship.

Example 1 *Given the symbolic rule $f(x) = x^2 + 3$, identify the name of the function and translate the rule into a verbal rule. Illustrate the rule for a given input value.*

SOLUTION

The symbolic statement $f(x) = x^2 + 3$ tells you that the function name is f and that you must square the input and then add 3 to obtain the corresponding output. For example, if $x = 4$, $f(4) = (4)^2 + 3 = 19$. This means that $(4, 19)$ is an (input, output) pair associated with the function f and $(4, 19)$ is one point on the graph of f. An (input, output) pair may also be referred to as an **ordered pair**.

7. For the function defined by $H(a) = 2a + 7$,

a. identify the function name.

The function name is H.

b. identify the input variable.

The input variable is a.

c. state the verbal rule for determining the output value for a given input value.

Multiply the input by 2 and add 7.

d. determine $H(-5)$.

$H(-5) = 2(-5) + 7 = -10 + 7 = -3$

e. write the ordered pair that represents the point on the graph of H corresponding to the fact that $H(4) = 15$.

The point on the graph corresponding to $H(4) = 15$ is $(4, 15)$.

f. determine the value of a for which $H(a) = 3$.

$2a + 7 = 3$
$2a = -4$
$a = -2$

Domain and Range

Definition

The collection of all possible replacement values of the input variable is called the **domain** of the function. The **practical domain** is the collection of replacement values of the input variable that makes practical sense in the context of a particular problem.

8. a. Determine the practical domain of the final average function. Assume that no fractional part of a point can be given and that the exam has a total of 100 points.

The practical domain is the set of integers from 0 to 100.

b. What is the domain of the function defined by $A(s) = \frac{335 + s}{5}$, where neither A nor s has any contextual significance?

The input variable can be replaced by any number.

Definition

The collection of all possible values of the output variable is the **range** of the function. The collection of all possible values of the output variable using the practical domain is the **practical range**.

9. Show or explain how you would determine the practical range of the final average function, and interpret the meaning of this range.

To calculate practical range, I would determine the output corresponding to each integer from 0 through 100 (i.e., the practical domain).
$\{67, 67.2, 67.4, \ldots, 86.4, 86.6, 86.8, 87\}$

The practical range consists of all possible final averages that I could still achieve. At worst, I could earn a final average of 67. The very best I could do is a final average of 87.

10. You are on your way to the college to take the fifth exam in the anthropology course. The gas gauge on your car indicates that you are almost out of gas. You stop to fill your car with gas.

a. Identify a reasonable input and output variable in this situation.

The input is the number of gallons pumped, and the output is the cost of a fill-up.

b. Write a verbal rule to determine the cost for any given number of gallons pumped. Assume that the price of regular-grade gasoline is $2.86 $\frac{9}{10}$ per gallon.

The cost in dollars is 2.869 times the number of gallons pumped.

c. Let C represent the cost of the gas purchased and g represent the number of gallons pumped. Translate the verbal rule in part b into a symbolic rule (equation) for C in terms of g.

$C = 2.869g$

d. What is the practical domain for the cost function? Assume that your gas tank has a maximum capacity of 17 gallons.

The practical domain is all values from 0 to 17, measured in tenths of a gallon.

e. What is the practical range for the cost function?

The practical range is $0 to $48.77.

f. Represent the output, cost, by $C(g)$. Rewrite the equation in part c using the function notation.

$C(g) = 2.869g$

g. Determine $C(15)$. Interpret the answer within the context of the situation.

$C(15) = 2.869(15) \approx 43.04$ The cost of 15 gallons of gas is $43.04.

SUMMARY: ACTIVITY 3.2

1. A **function** is a rule relating an input variable and output variable that assigns one and only one output value to each input value. Stated another way, for each input value, there is *one and only one* corresponding output value.

2. A function can be defined **numerically** as a list of **ordered pairs**, often displayed in a 2-column (vertical) table or a 2-row (horizontal) table format. A function can also be defined as a set of ordered pairs.

3. When a function is defined **graphically**, the input variable is referenced on the horizontal axis and the output variable is referenced on the vertical axis.

4. Functions can also be defined by a **verbal rule** (in words) or **symbolically** (by an equation that indicates the sequence of operations performed on the input to obtain the corresponding output value).

5. A function can be expressed symbolically using the **function notation** $f(x)$, where f is the name of the function and x is the input variable. The notation $f(x)$ represents the output value corresponding to a given x. If the output variable is denoted by y, you can write $y = f(x)$ and say that the output y is a function of x.

6. Associated with any function is a specific set of input values. The collection of all possible input values is called the **domain** of the function. In functions that result from contextual situations, the domain consists of input values that make sense within the context. Such a domain is often called a **practical domain**.

7. In a function, the collection of all possible values of the output variable is called the **range**. In functions that result from contextual situations, the **practical range** is the set of output values assigned to each element of the practical domain.

EXERCISES: ACTIVITY 3.2

1. The cost of a history club trip includes $78 for transportation rental plus a $2 admission charge for each student participating.

   a. The total cost of the trip depends on the number of students participating. Identify the input and output variables.

   The input variable is the number of students; the output variable is the total cost of the trip.

   b. Write a verbal rule to determine the total cost of the trip in terms of the number of students participating.

   The total cost of the trip will be $78 plus $2 times the number of students.

   c. Complete the following table.

NUMBER OF STUDENTS	COST OF TRIP ($)
10	98
15	108
20	118
25	128

   d. Write the symbolic rule using function notation. Use n to represent the number of students and C to represent the name of the cost function for the trip.

   $C(n) = 2n + 78$ or $C(n) = 78 + 2n$

   e. Use function notation to write a symbolic statement for "The total cost is $108 if 15 students go on the trip."

   $C(15) = 108$

   f. Choose another (input, output) pair for the cost function C. Write this pair in both ordered-pair notation and in function notation.

   (Answers will vary.) $(10, 98); C(10) = 98$

   g. What is the practical domain if the club's transportation is a bus?

   (Answers will vary.) The number of students who can sign up for the trip is restricted by the capacity of the bus. If the bus holds 45 students, then the practical domain is the set of integers from 0 to 45.

   h. Determine $C(37)$.

   $C(37) = 2(37) + 78 = 152$

   i. Describe, in words, what information about the cost of the trip is given by the statement $C(24) = 126$.

   If 24 students go on the trip, the total cost of the trip will be $126.

2. Each of the following tables defines a relationship between an input and an output. Which of the relationships represent functions? Explain your answers.

a.

Input	−8	−3	0	6	9	15	24	38	100
Output	24	4	9	72	−14	−16	53	29	7

This data set represents a function because each input value is assigned a single output value.

b.

Input	−8	−5	0	6	9	15	24	24	100
Output	24	4	9	72	14	−16	53	29	7

This data set does *not* represent a function because the input value 24 is assigned two output values, 53 and 29.

c.

Input	−8	−3	0	6	9	15	24	38	100
Output	24	4	9	72	4	−16	53	24	7

This data set represents a function because each input value is assigned a single output value.

3. Identify the input and output variables in each of the following. Then determine if the statement is true. Give a reason for your answer.

a. Your letter grade in a course is a function of your numerical grade.

The input variable is the numerical grade and the output variable is the letter grade. The statement is true. For any given numerical grade, such as 87, there is only one corresponding letter grade.

b. Your numerical grade is a function of your letter grade.

The input variable is your letter grade and the output variable is your numerical grade. The statement is not true. For example, for a letter grade of B, there can be several corresponding numerical grades, such as 82, 83.7, and 88.

4. A table is often used to define a function that has a finite number of input and output pairs. Consider the function f defined by the following table and assume that these points are the only input/output pairs that belong to the function f.

x	−3	−2	−1	0	1	2	3	4
$f(x)$	5	4	2	−1	1	3	5	6

a. What is the domain of f?

$\{-3, -2, -1, 0, 1, 2, 3, 4\}$

b. What is the range of f?

$\{-1, 1, 2, 3, 4, 5, 6\}$

c. For which set of consecutive x-values is f increasing?

f is increasing over the set $\{0, 1, 2, 3, 4\}$.

d. Determine the maximum output value of f and the input for which it occurs.

The maximum value of f is 6, and it occurs when x is 4.

e. Determine the minimum output value of f and the input for which it occurs.

The minimum value of f is -1, and it occurs when x is 0.

f. Write f as a set of ordered pairs. You should have eight ordered pairs.

$f = \{(-3, 5), (-2, 4), (-1, 2), (0, -1), (1, 1), (2, 3), (3, 5), (4, 6)\}$

g. Determine $f(3)$.

$f(3) = 5$

h. For what value of x is $f(x) = 2$?

$x = -1$

5. a. Suppose that all you know about a function f is that $f(2) = -4$. Use this information to complete the following.

i. If the input value for f is 2, then the corresponding output value is $\underline{-4}$.

ii. One point on the graph of f is $\underline{(2, -4)}$.

b. The statement in part a indicates that $f(x) = -4$ for $x = 2$. Is it possible for f to have an output of -4 for another value of x, such as $x = 5$? Explain.

Yes, the output value of a function may be the same for several inputs. However, any one input may correspond to only one output.

6. Let $f(x) = 2x - 1$. Evaluate $f(4)$.

$f(4) = 2(4) - 1 = 8 - 1 = 7$

7. Let $g(n) = 3n + 5$. Evaluate $g(-3)$.

$g(-3) = 3(-3) + 5 = -9 + 5 = -4$

8. Let $h(m) = 2m^2 + 3m - 1$. Evaluate $h(-2)$.

$h(-2) = 2(-2)^2 + 3(-2) - 1 = 2(4) - 6 - 1 = 1$

9. Let $p(x) = 3x^2 - 2x + 4$. Evaluate $p(5)$.

$p(5) = 3(5)^2 - 2(5) + 4 = 3(25) - 10 + 4 = 75 - 10 + 4 = 69$

10. Let f be a function defined by $f(x) = x + 1$. Determine the value of x for which $f(x) = 3$.

$x + 1 = 3$
$x = 2$

11. Let g be a function defined by $g(x) = 5x - 4$. Determine the value of x for which $g(x) = 11$.

$5x - 4 = 11$
$5x = 15$
$x = 3$

12. Let h be a function defined by $h(t) = \dfrac{t}{4}$. Determine the value of t for which $h(t) = 12$.

$\frac{t}{4} = 12$
$t = 48$

13. Let k be a function defined by $k(w) = 0.4w$. Determine the value of w for which $k(w) = 12$.

$0.4w = 12$

$w = 30$

14. Give at least two examples of functions from your daily life. Explain how each fits the description of a function. Be sure to identify the input variable and output variable.

(Answers will vary.)

The sales tax I pay on items I purchase is a function of the cost of the item I purchase. The tax rate in my town is 8.25%. If T represents the total cost of my purchase including tax and C represents the cost of the item before tax, the function is represented by the formula $T(C) = 1.0825C$. C is the input variable and T is the output variable. Note that for each input, C, only one output, T, is produced.

I work for a computer store and earn a base salary of $200 per week plus a 5% commission on the sales I make. My weekly gross salary is a function of the sales I make. If s represents the sales I make and g represents my weekly gross salary, the function is represented by the formula $g(s) = 0.05s + 200$. The variable s is the input and the variable g is the output.

Activity 3.3

How Fast Did You Lose?

Objective

1. Determine the average rate of change of an output variable with respect to the input variable.

You are a member of a health and fitness club. The club's registered dietitian and your personal trainer helped you develop a special 8-week diet and exercise program. The data in the following table represents your weight, w, as a function of time, t, over an 8-week period.

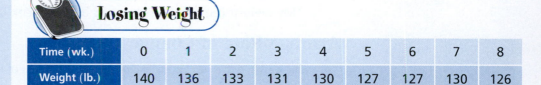

Losing Weight

Time (wk.)	0	1	2	3	4	5	6	7	8
Weight (lb.)	140	136	133	131	130	127	127	130	126

1. a. Plot the data points using ordered pairs of the form (t, w). For example, $(3, 131)$ is a data point that represents your weight at the end of the third week.

   Note to instructor: For Problem 1, just the dots are graphed. The line segments shown are answers to Problems 6, 7, and 8.

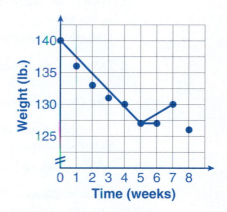

The resulting graph of points in part a is called a **scatterplot**.

   b. What is the practical domain of this function?

   The practical domain is the set {0, 1, 2, 3, 4, 5, 6, 7, 8}, which represent the weeks of your diet.

   c. What is the practical range of this function?

   (Answers will vary.) The practical range is the set of all real numbers between 126 and 140, which represent possible weights during the diet.

2. a. What is your weight at the beginning of the program?

   At the beginning of the program, my weight was 140 pounds.

   b. What is your weight at the end of the first week?

   At the end of the first week, my weight was 136 pounds.

3. To see how the health program is working for you, you analyze your weekly weight changes during the 8-week period.

   a. During which week(s) does your weight increase?

   My weight increased during the seventh week.

b. During which week(s) does your weight decrease?

My weight decreases from the beginning of the program to the end of week 5 and during the eighth week.

c. During which week(s) does your weight remain unchanged?

My weight remained unchanged during week 6.

Average Rate of Change

4. Your weight decreases during each of the first 5 weeks of the program.

a. Determine the actual change in your weight, the output value, over the first 5 weeks of the program by subtracting your initial weight from your weight at the end of the first 5 weeks.

$127 - 140 = -13$ lb.
The actual change over the first 5 weeks was -13 pounds.

b. What is the sign (positive or negative) of your answer? What is the significance of this sign?

Negative. The change in weight is negative because my weight decreases.

c. Determine the change in the input value over the first 5 weeks, that is, from $t = 0$ to $t = 5$.

The change in the input value is 5.

d. Write the quotient of the change in weight from part a divided by the change in time in part c. Interpret the meaning of this quotient.

$\frac{-13 \text{ lb.}}{5 \text{ wk.}}$; I lost 13 pounds in 5 weeks.

The change in weight describes how much weight you have gained or lost, but it does not tell how quickly you shed those pounds. That is, a loss of 3 pounds in 1 week is more impressive than a loss of 3 pounds over a month's time. Dividing the change in weight by the change in time gives an average rate at which you have lost weight over this period of time. The units for this rate are output units per input unit—in this case, pounds per week.

The rate in Problem 4d,

$$\frac{-13 \text{ lb.}}{5 \text{ wk.}} = -2.6 \text{ lb./week}$$

is called the **average rate of change** of weight with respect to time over the first 5 weeks of the program. It can be interpreted as an average loss of 2.6 pounds each week for the first 5 weeks of the program.

average rate of change $= \dfrac{\text{change in output}}{\text{change in input}}$

where

i. change in output is calculated by subtracting the first (initial) output value from the second (final) output value

ii. change in input is calculated by subtracting the first (initial) input value from the second (final) input value

Delta Notation

The rate of change in output with respect to a corresponding change in input is so important that special symbolic notation has been developed to denote it.

The uppercase Greek letter delta, Δ, is used with a variable's name to represent a change in the value of the variable from a starting point to an ending point.

For example, in Problem 4 the notation Δw represents the change in value of the output variable weight, w. It is calculated by subtracting the initial value of w, denoted by w_1, from the final value of w, denoted by w_2. Symbolically, this change is represented by

$$\Delta w = w_2 - w_1.$$

In the case of Problem 4a, the change in weight over the first 5 weeks can be calculated as

$$\Delta w = w_2 - w_1 = 127 - 140 = -13 \text{ lb.}$$

In the same way, the change in value of the input variable time, t, is written as Δt and is calculated by subtracting the initial value of t, denoted by t_1, from the final value of t, denoted by t_2. Symbolically, this change is represented by

$$\Delta t = t_2 - t_1.$$

In Problem 4c, the change in the number of weeks can be written using Δ notation as follows:

$$\Delta t = t_2 - t_1 = 5 - 0 = 5 \text{ wk.}$$

The average rate of change of weight over the first 5 weeks of the program can now be symbolically written as follows:

$$\frac{\Delta w}{\Delta t} = \frac{-13}{5} = -2.6 \text{ lb./week}$$

Note that the symbol Δ for delta is the Greek version of d, for difference, the result of a subtraction that produces the change in value.

5. Use Δ notation and determine the average rate of change of weight over the last 4 weeks of the program.

$\frac{\Delta w}{\Delta t} = \frac{126 - 130}{8 - 4} = \frac{-4}{4} = -1$ lb./week

The average rate of change over the last 4 weeks of the program was a loss of 1 pound per week.

Graphical Interpretation of the Average Rate of Change

6. **a.** On the graph in Problem 1, connect the points $(0, 140)$ and $(5, 127)$ with a line segment. Does the line segment rise, fall, or remain horizontal as you follow it from left to right?

The line segment falls.

b. Recall that the average rate of change over the first 5 weeks was -2.6 pounds per week. What does the average rate of change tell you about the line segment drawn in part a?

The line segment falls because the average rate of change is negative. The average rate also indicates that the line segment is falling 2.6 units for each 1-unit increase in time.

7. a. Determine the average rate of change of your weight over the time period from $t = 5$ to $t = 7$ weeks. Include the appropriate sign and units.

$$\frac{\text{change in weight}}{\text{change in time}} = \frac{\Delta w}{\Delta t} = \frac{130 - 127}{7 - 5} = \frac{3}{2} = 1.5 \text{ lb./week}$$

b. Interpret the rate in part a with respect to your diet.

The rate of change is positive, indicating that a weight gain occurred. Over this 2-week period, you averaged a 1.5-pound gain each week.

c. On the graph in Problem 1, connect the points $(5, 127)$ and $(7, 130)$ with a line segment. Does the line segment rise, fall, or remain horizontal as you follow it from left to right?

The line segment rises.

d. How is the average rate of change of weight over the given 2-week period related to the line segment you drew in part c?

The line segment rises because the average rate of change is positive. The line rises 1.5 units for each 1-unit increase in time.

8. a. At what average rate is your weight changing during the 6th week of your diet, that is, from $t = 5$ to $t = 6$?

$$\frac{\text{change in weight}}{\text{change in time}} = \frac{\Delta w}{\Delta t} = \frac{127 - 127}{6 - 5} = \frac{0}{1} = 0 \text{ lb./week}$$

My weight did not change.

b. Interpret the rate in part a with respect to your diet.

During the sixth week of the diet, my weight stays the same.

c. Connect the points $(5, 127)$ and $(6, 127)$ on the graph with a line segment. Does the line segment rise, fall, or remain horizontal as you follow it from left to right?

The line segment remains horizontal.

d. How is the average rate of change in part a related to the line segment drawn in part c?

An average rate of change equal to 0 corresponds to a horizontal line segment connecting the starting and ending points.

9. a. What is the average rate of change of your weight over the period from $t = 4$ to $t = 7$ weeks?

$$\frac{\Delta w}{\Delta t} = \frac{130 - 130}{7 - 4} = 0 \text{ lb./week}$$

b. Explain how the rate in part a reflects the progress of your diet over those 3 weeks.

At the end of the seventh week, I weighed the same as I did at the end of the 4th week.

SUMMARY: ACTIVITY 3.3

1. Let y_1 represent the corresponding output value for the input x_1, and y_2 represent the corresponding output value for the input x_2. As the variable x changes in value from x_1 (initial value) to x_2 (final value),

a. the change in input is represented by $\Delta x = x_2 - x_1$

b. the change in output is represented by $\Delta y = y_2 - y_1$

2. The quotient $\dfrac{\Delta y}{\Delta x} = \dfrac{y_2 - y_1}{x_2 - x_1}$ is called the **average rate of change** of y (output) with respect to x (input) over the x-interval from x_1 to x_2. The units of measurement of the quantity $\dfrac{\Delta y}{\Delta x}$ are *output units* per *input unit*.

3. The line segment connecting the points (x_1, y_1) and (x_2, y_2)

a. rises from left to right if $\dfrac{\Delta y}{\Delta x} > 0$

b. falls from left to right if $\dfrac{\Delta y}{\Delta x} < 0$

c. remains constant if $\dfrac{\Delta y}{\Delta x} = 0$

EXERCISES: ACTIVITY 3.3

1. For the years between 1900 and 2000, the following table presents the median ages (output) of U.S. men at the time when they first married.

Input, year	1900	1910	1920	1930	1940	1950	1960	1970	1980	1990	2000
Output, age	25.9	25.1	24.6	24.3	24.3	22.8	22.8	23.2	24.7	26.1	27.1

a. During which decade(s) did the median age at first marriage increase for men?

A man's age at the time of his first marriage increased during the 1960s, 1970s, 1980s, and 1990s.

b. During which decade(s) did the median age at first marriage decrease?

Male age at time of first marriage decreased during the 1900s, 1910s, 1920s, and 1940s.

c. During which decade(s) did the median age at first marriage remain unchanged?

Male age at time of first marriage remained unchanged during the 1930s and 1950s.

d. During which decade(s) was the change in median age at first marriage the greatest?

The change was greatest during the 1970s (increase) and 1940s (decrease).

2. Graph the data from Exercise 1, and connect consecutive points with line segments. Then answer the following questions using the graph.

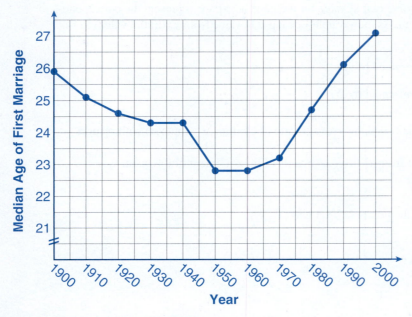

a. During which decade(s) is your graph rising?

The graph is rising during the 1960s, 1970s, 1980s, and 1990s.

b. During which decade(s) is the graph falling?

The graph is falling during the 1900s, 1910s, 1920s, and 1940s.

c. During which decade(s) is your graph horizontal?

The graph is horizontal during the 1930s and 1950s.

d. During which decade(s) is the graph the steepest?

The graph is steepest during the 1940s and 1970s.

e. Compare the answers of the corresponding parts (a–d) of Exercises 1 and 2. What is the relationship between the sign of the output change for a given input interval and the direction (rising, falling, horizontal) of the graph in that interval?

When the median age of first marriage increases, the graph rises. When the median age of first marriage decreases, the graph falls. When the median age of first marriage remains unchanged, the graph is horizontal. When the change in first-marriage median age is the greatest, the graph is steepest.

3. a. Refer to the data in Exercise 1 to determine the average rate of change per year of first-marriage age for men from 1900 through 1950.

$$\frac{\Delta \text{age}}{\Delta \text{year}} = \frac{22.8 - 25.9}{1950 - 1900} = \frac{-3.1}{50} = -0.062 \text{ yr. of age/yr.}$$

Therefore, for the years from 1900 to 1950, the average rate of change was -0.062 year of age per year.

Cluster 1   What Have I Learned?

1. Can the graph of a function intersect the vertical axis in more than one point? Explain.

No; the definition of a function requires that each input value, including $x = 0$, produce exactly one output value.

2. What is the mathematical definition of a function? Give a real-life example, and explain how this example satisfies the definition of a function.

A function is a rule that determines a correspondence between an input variable and an output variable. To each input value there corresponds a single output value. For example, sales tax is a function of the purchase price of an item, where the sales tax is a fixed percent of the purchase price, say 8%. The purchase price is the input and the sales tax is the output. For each purchase price, the rule, 8% times the purchase price, determines one and only one sales tax. In particular, if the purchase price is $50, then the sales tax has to be 8% of 50, or $0.08(50) = \$4$.

3. Describe how you can tell from its graph when a function is increasing and when it is decreasing.

If a graph is rising from left to right, the function is increasing. If the graph is falling from left to right the function is decreasing.

4. Explain the meaning of the symbolic statement $H(5) = 100$.

The statement $H(5) = 100$ means that for the function H, an input value of $x = 5$ corresponds to an output value of 100.

5. Describe how you would determine the domain and range of a function defined by a continuous graph (one with no breaks or holes). Assume that the graph you see includes all the points of the function.

I would determine the leftmost and rightmost input values. The domain would be everything between these, inclusive. To determine the range, I would determine the lowest and highest output values. The range would be all values between these, inclusive.

6. Give an example of a function from your major field of study or area of interest.

(Answers will vary.) In statistics, actual data values are very often converted to standardized scores called z-scores. For one thing, this allows comparison of data sets that have different units of measurement. We say that a z-score is a function of the data value it represents. A z-score for a data value, x_i, is determined by the formula

$$z = \frac{x_i - \bar{x}}{s_x},$$

where z is the output score, x_i is the input value, $\bar{x}$ is the average, and s_x is the standard deviation of the data set.

Exercise numbers appearing in color are answered in the Selected Answers appendix.

7. The sales tax rate in Ann Arbor, Michigan is 6%. Consider the function that associates the sales tax (output) paid on an item with the price of the item (input). Represent this function in each of the four formats discussed in this Cluster: verbally, symbolically, numerically, and graphically.

(Answers may vary.)

Verbal Definition of a Function: The sales tax on an item is calculated by multiplying the cost of the item by 0.06.

Symbolic Definition: If t represents the sales tax on an item that costs C dollars, then

$$t = 0.06C$$

Numerical Definition:

Cost of Item (dollars)	10	20	30	40	50	100	150	200	300	500	1000
Sales Tax ($)	0.60	1.20	1.80	2.40	3.00	6.00	9.00	12.00	18.00	30.00	60.00

Graphical Definition:

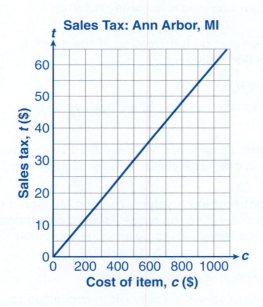

8. A function is defined by the rule $f(x) = 5x - 8$. What does $f(1)$ refer to on the graph of f?

$f(1)$ represents the output value when the input is 1. Because $f(1) = -3$, the point $(1, -3)$ is on the graph of f.

9. The notation $g(t)$ represents the weight (in grams) of a melting ice cube t minutes after being removed from the freezer. Interpret the meaning of $g(10) = 4$.

After 10 minutes, the ice cube weighs 4 grams.

10. What is true about the sign of the average rate of change between any two points on the graph of an increasing function?

If a function is increasing, the rate of change between any two points must be positive.

11. You are told that the average rate of change of a particular function is always negative. What can you conclude about the graph of that function and why?

If the rate of change of a function is negative, then the initial output value on any interval must be larger than the final value. We can conclude that the function is decreasing.

12. Describe a step-by-step procedure for determining the average rate of change between any two points on the graph of a function. Use the points represented by $(85, 350)$ and $(89, 400)$ in your explanation.

The average rate of change between any two points on the graph of a function is the quotient of the difference between the output values and the difference between the input values. Therefore, using the given points to illustrate, do the following:

   i. Subtract 350 from 400 to obtain 50, the difference between the outputs for the two points.

   ii. Subtract 85 from 89 to obtain 4, the difference between the inputs for the two points.

   iii. Write the quotient $\dfrac{\text{(difference in output)}}{\text{(difference in input)}} = \dfrac{50}{4} = 12.5$.

   iv. Interpret the quotient: The rate of change is 12.5 units of output per 1 unit of input.

Cluster 1   How Can I Practice?

1. Students at one community college in New York State pay $129 per credit hour when taking fewer than 12 credits, provided they are New York State residents. For 12 or more credit hours, they pay $1550 per semester.

   a. Determine the tuition cost for a student taking the given number of credit hours.

NUMBER OF CREDIT HOURS	TUITION COST ($)
3	387
6	774
10	1290
12	1550
16	1550
18	1550

   b. Is the tuition cost a function of the number of credit hours for the values in your completed table from part a? Explain. Be sure to identify the input and output variables in your explanation.

   Yes, for each number of credit hours (input), there is exactly one tuition cost (output).

   c. What is the practical domain of the tuition cost function? Assume there are no half-credit courses. However, there are 1- and 2-credit courses available.

   The practical domain is {0, 1, 2, 3, . . . , 19, 20, 21}.

   d. Use the table in part a to help graph the tuition cost at this college as a function of the number of credit hours taken.

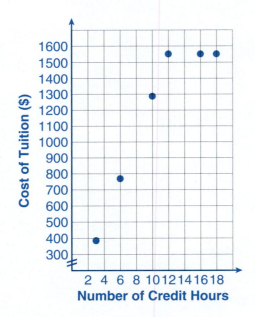

Exercise numbers appearing in color are answered in the Selected Answers appendix.

e. Suppose you have saved $700 for tuition. Use the graph to estimate the most credit hours you can take.

The most credit hours I can take for $700 is 5.

f. Does the graph of the tuition cost function pass the vertical line test?

Yes, the graph of the tuition cost function passes the vertical line test.

g. Let h represent the number of credit hours (input) and C represent the tuition cost (output). Write a symbolic rule for the cost of part-time tuition in terms of the number of credit hours taken.

$C(h) = 129h, h < 12$

h. Use your equation from part g to verify the tuition cost for the credit hours given in the table in part a.

For the table values with fewer than 12 credits:

$C(3) = 129(3) = 387; C(6) = 129(6) = 774;$ and

$C(10) = 129(10) = 1290$

2. Let f be defined by the set of two ordered pairs $\{(2, 3), (0, -5)\}$.

a. List the set of numbers that constitute the domain of f. $\{0, 2\}$

b. List the set of numbers that constitute the range of f. $\{-5, 3\}$

3. You decide to lose weight and will cut down on your calories to lose 2 pounds per week. Suppose that your present weight is 180 pounds. Sketch a graph covering 20 weeks showing your projected weight loss. Describe your graph. If you stick to your plan, how much will you weigh in 3 months (13 weeks)?

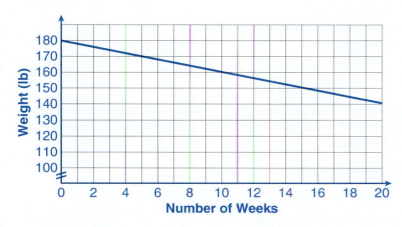

The graph is falling. In 3 months (13 weeks), I would lose 26 pounds and weigh 154 pounds.

4. A taxicab driver charges a flat rate of $2.50 plus $2.20 per mile. The fare F (in dollars) is a function of the distance driven, x (in miles). The driver wants to display a table for her customers to show approximate fares for different locations within the city.

a. Write a symbolic rule for F in terms of x.

$F(x) = 2.50 + 2.20x$

b. Use the symbolic rule to complete the following table.

x (mi)	0.25	0.5	0.75	1.0	1.5	2.0	3.0	5.0	10.0
F(x)	$3.05	$3.60	$4.15	$4.70	$5.80	$6.90	$9.10	$13.50	$24.50

5. Let $f(x) = -3x + 4$. Determine $f(-5)$.

$f(-5) = -3(-5) + 4 = 15 + 4 = 19$

6. Let $g(x) = (x + 3)(x - 2)$. Determine $g(-4)$.

$g(-4) = (-4 + 3)(-4 - 2) = (-1)(-6) = 6$

7. Let $m(x) = 2x^2 + 6x - 7$. Determine $m(3)$.

$m(3) = 2(3)^2 + 6(3) - 7 = 18 + 18 - 7 = 29$

8. Let $h(s) = (s - 1)^2$. Determine $h(-3)$.

$h(-3) = (-3 - 1)^2 = (-4)^2 = 16$

9. Let $s(x) = \sqrt{x + 3}$. Determine $s(6)$.

$s(6) = \sqrt{6 + 3} = \sqrt{9} = 3$

10. Let f be a function defined by $f(x) = x - 6$. Determine the value of x for which $f(x) = 10$.

$x - 6 = 10$

$x = 16$

11. Let g be a function defined by $g(x) = 0.8x$. Determine the value of x for which $g(x) = 16$.

$0.8x = 16$

$x = 20$

12. Let h be a function defined by $h(x) = 2x + 1$. Determine the value of x for which $h(x) = 13$.

$2x + 1 = 13$

$2x = 12$

$x = 6$

13. Let k be a function defined by $k(x) = \dfrac{x}{6}$. Determine the value of x for which $k(x) = -3$.

$\dfrac{x}{6} = -3$

$x = -18$

14. Interpret each situation represented by the following graphs. That is, describe, in words, what the graph is saying about the input/output relationship. Indicate what occurs when the function reaches either a minimum or maximum value.

a. Hours of daylight per day in relation to time of year

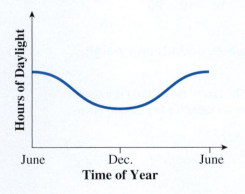

The maximum hours of daylight occur in June. The amount of daylight decreases each month until it reaches a minimum in December.

b. Population of fish in a pond in relation to the number of years since stocking

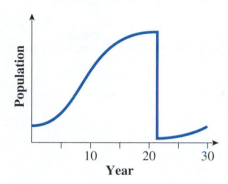

For the first few years after the pond is stocked, the population grows slowly. It increases more quickly as the fish reproduce; then it levels off. A predator (or pollutant) kills off almost all the fish 21 years after stocking. The population begins to grow when the remaining fish reproduce.

c. Distance from home (in miles) in relation to driving time (in hours)

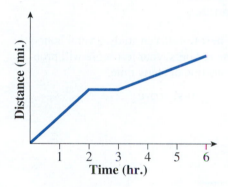

You leave home, drive for 2 hours at a constant rate, and then stop for 1 hour. Finally, you continue at a slower (but constant) speed than before.

d. Amount of money saved per year in relation to amount of money earned

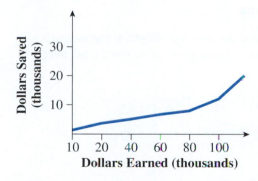

The amount saved is fairly constant when dollars earned are less than $80,000. Saving occurs at a higher rate for earnings between $80,000 and $100,000 and even more is saved when earnings exceed $100,000.

The amount needed for expenses levels off and more money is available for savings.

15. In the following situations, determine the input and output variables, and then sketch a graph that best describes the situation. Remember to label the axes with the names of the variables.

(Answers will vary.)

a. Sketch a graph that approximates average temperatures where you live as a function of the number of months since January.

Input variable: <u>number of months since January</u>

Output variable: <u>average temperature</u>

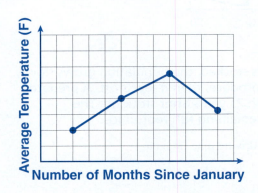

b. If you don't study, you would expect to do poorly on the next test. If you study several hours, you should do quite well, but if you study for too many more hours, your test score will probably not improve. Sketch a graph of your test score as a function of study time.

Input variable: <u>study time</u>   Output variable: <u>test score</u>

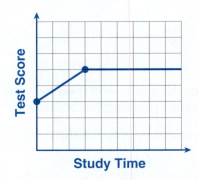

16. As part of your special diet and exercise program, you record your weight at the beginning of the program and each week thereafter. The following data gives your weight, w, over a 5-week period.

Time, t (wk.)	0	1	2	3	4	5
Weight, w (lb.)	196	183	180	177	174	171

a. Sketch a graph of the data on appropriately scaled and labeled axes.

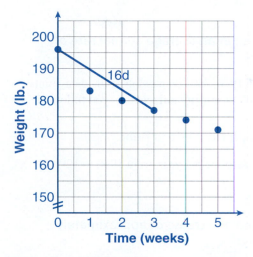

b. Determine the average rate of change of your weight during the first 3 weeks. Be sure to include the units of measurement of this rate.

$$\frac{177 - 196}{3 - 0} = -\frac{19}{3} \approx -6.3 \text{ lb./week}$$

I averaged a loss of 6.3 pounds per week during the first 3 weeks.

c. Determine the average rate of change during the 5-week period.

$$\frac{171 - 196}{5 - 0} = -\frac{25}{5} = -5 \text{ lb./week}$$

I averaged a loss of 5 pounds per week during the 5-week period.

d. On the graph in part a, connect the points $(0, 196)$ and $(3, 177)$ with a line segment. Does the line segment rise, fall, or remain horizontal as you follow the line left to right?

The line segment falls as I follow the line left to right.

e. What is the practical meaning of the average rate of change in this situation?

The average rate of change indicates how quickly I am losing weight.

f. What can you say about the average rate of change of weight during any of the time intervals in this situation?

The average rate of change is always negative. My weight is always decreasing during the 5-week program.

17. Between 1960 and 2006 automobiles in the United States changed size and shape almost annually. The amount of fuel consumed by these vehicles also changed. The following table describes the average number of gallons of gasoline consumed per year per passenger car.

Year	1960	1970	1980	1990	1995	2000	2002	2004	2005	2006
Average Gallons of Gasoline Consumed Per Passenger Car (gal.)	668	760	576	520	530	546	551	553	567	554

a. Determine the average rate of change, of gallons of gasoline per year, from 1960 to 1970.

$$\frac{760 - 668}{1970 - 1960} = \frac{92}{10} = 9.2 \text{ gal./yr.}$$

From 1960 to 1970, gas consumption increased at an average annual rate of 9.2 gallons.

b. Suppose you connected the points $(1960, 668)$ and $(1970, 760)$ on a graph of the data points with a line segment. Would the line segment rise, fall, or remain horizontal as you follow the line left to right?

The line segment would rise as I read the graph from left to right.

c. Determine the average rate of change, in gallons of gasoline per year, from 1960 to 1990.

$$\frac{520 - 668}{1990 - 1960} = \frac{-148}{30} \approx -4.93 \text{ gal./yr.}$$

From 1960 to 1990, gas consumption averaged a decrease of approximately 4.9 gallons per year.

d. Determine the average rate of change, of gallons of gasoline per year, from 2000 to 2002.

$$\frac{551 - 546}{2002 - 2000} = \frac{5}{2} = 2.5 \text{ gal./yr.}$$

From 2000 to 2002, gas consumption increased at an average annual rate of 2.5 gallons.

e. Determine the average rate of change, of gallons of gasoline per year, between 1960 and 2006.

$$\frac{554 - 668}{2006 - 1960} = \frac{-114}{46} \approx -2.48 \text{ gal./yr.}$$

From 1960 to 2006, gas consumption decreased by an average of approximately 2.48 gallons per year, or almost 2.5 gallons per year.

Cluster 2   Introduction to Linear Functions

Activity 3.4

The Snowy Tree Cricket

Objectives

1. Identify linear functions by a constant average rate of change of the output variable with respect to the input variable.

2. Determine the slope of the line drawn through two points.

3. Identify increasing linear functions using slope.

One of the more familiar late-evening sounds during the summer is the rhythmic chirping of a male cricket. Of particular interest is the snowy tree cricket, sometimes called the temperature cricket. It is very sensitive to temperature, speeding up or slowing down its chirping as the temperature rises or falls. The following data shows how the number of chirps per minute of the snowy tree cricket is related to temperature.

Chirping Crickets

t, TEMPERATURE (°F)	N(t), NUMBER OF CHIRPS/MINUTE
55	60
60	80
65	100
70	120
75	140
80	160

1. Crickets are usually silent when the temperature falls below 55°F. What is a possible practical domain for the snowy tree cricket function?

   A suitable domain is 55° to perhaps 100°F.

2. **a.** Determine the average rate of change of the number of chirps per minute with respect to temperature as the temperature increases from 55°F to 60°F.

   $$\frac{\text{change in number of chirps per minute } \Delta N}{\text{corresponding change in temperature } \Delta t} = \frac{80 - 60}{60 - 55} = \frac{20}{5} = 4$$

   The number of chirps per minute increases at a rate of 4 per degree.

   b. What are the units of measure of this rate of change?

   The units are number of chirps per minute per degree.

3. **a.** How does the average rate of change determined in Problem 2 compare with the average rate of change as the temperature increases from 65°F to 80°F?

   $$\frac{\Delta N}{\Delta t} = \frac{160 - 100}{80 - 65} = \frac{60}{15} = 4 \text{ chirps per minute per degree}$$

   The rate of increase is the same as in Problem 2.

   b. Determine the average rate of change of number of chirps per minute with respect to temperature for the temperature intervals given in the following table. The results from Problems 2 and 3 are already recorded. Add several more of your own choice. List all your results in the table.

TEMPERATURE INCREASES	AVERAGE RATE OF CHANGE (number of chirps/minute per degree F)
From 55° to 60°F	4
From 65° to 80°F	4
From 55° to 75°F	4
From 60° to 80°F	4

c. What can you conclude about the average rate of increase in the number of chirps per minute for any particular increase in temperature?

The average rate of increase in the number of chirps per minute is constant, namely, 4 per each degree increase in temperature. For example, from 60° to 70°, the average rate of change is

$\frac{\Delta N}{\Delta t} = \frac{120 - 80}{70 - 60} = \frac{40}{10} = 4$ chirps per minute per degree.

4. For any 7° increase in temperature, what is the expected increase in chirps per minute?

$(7°) \cdot (4$ chirps per minute per degree$) = 28$ chirps per minute. The increase expected for a 7° rise is 28 chirps per minute.

5. Plot the data pairs (temperature, chirps per minute) from the table preceding Problem 1. What type of graph is suggested by the pattern of points?

Note: Use slash marks (//) to indicate that the horizontal axis has been cut out between 0 and 55 and the vertical axis between 0 and 60.

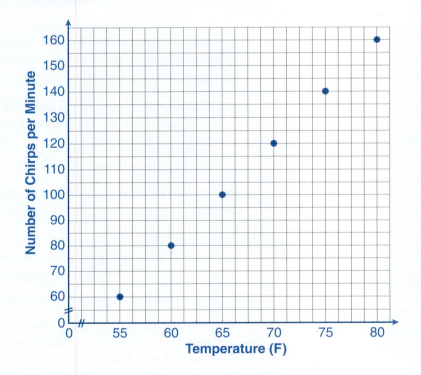

The pattern of the points suggests the graph is a line.

Linear Functions

If the average rate of change in output with respect to input remains constant (stays the same) for *any* two points in a data set, then all the points will lie on a single straight line. That is, the output is a **linear** function of the input. Conversely, if all the points of a data set lie on a straight line when graphed, the average rate of change of output with respect to input will be constant for any two data points.

6. From the graph in Problem 5, would you conclude that the number of chirps per minute is a linear function of the temperature? Explain.

Yes, the number of chirps per minute is a linear function of the temperature. Each input (temperature) results in a single output (number of chirps per minute), and the plotted data points all lie on a straight line.

As you will see in the rest of Chapter 3, many situations in the world around us can be modeled by linear functions.

Slope of a Line

The average rate of change for a linear function determines the steepness of the line and is called the *slope* of the line.

Definition

The **slope** of a line is a measure of its steepness. It is the average rate of change between any two points on the line. Symbolically, the letter m is used to denote slope:

$$\text{slope} = m = \frac{\text{change in output}}{\text{change in input}} = \frac{\Delta y}{\Delta x} = \frac{y_2 - y_1}{x_2 - x_1}$$

where (x_1, y_1) and (x_2, y_2) are any two points on the line and $x_1 \neq x_2$.

Example 1 *Determine the slope of the line containing the points $(1, -2)$ and $(3, 8)$.*

SOLUTION

Let $x_1 = 1, y_1 = -2, x_2 = 3,$ and $y_2 = 8,$ so

$$m = \frac{\Delta y}{\Delta x} = \frac{y_2 - y_1}{x_2 - x_1} = \frac{8 - (-2)}{3 - 1} = \frac{10}{2} = \frac{5}{1} = 5.$$

7. a. What is the slope of the line in the snowy tree cricket situation?

Because the average rate of change has a constant value of 4, the slope of the line is also 4.

b. Because the slope of this line is positive, what can you conclude about the direction of the line as the input variable (temperature) increases in value?

The line rises from left to right; that is, it is increasing.

c. What is the practical meaning of slope in this situation?

The number of chirps per minute increases by 4 for every 1°F increase in temperature.

On a graph, slope can be understood as the ratio of two distances. For example, in the snowy tree cricket situation, the slope of the line between the points $(55, 60)$ and $(56, 64)$ as well as between $(56, 64)$ and $(57, 68)$ is $\frac{4}{1}$, as shown on the following graph. The change in the input, 1, represents a horizontal distance (the run) in going from one point to another point on the same line. The change in the output, 4, represents a vertical distance (the rise) between the same points. The graph on the right illustrates that a horizontal distance (run) of 2 and a vertical distance (rise) of 8 from $(55, 60)$ will also locate the point $(57, 68)$ on the line.

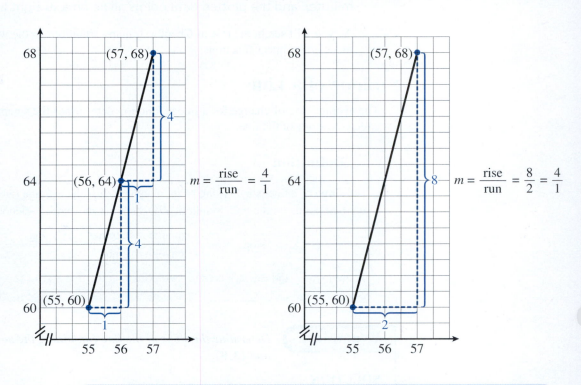

$$m = \frac{\text{rise}}{\text{run}} = \frac{4}{1} \qquad\qquad m = \frac{\text{rise}}{\text{run}} = \frac{8}{2} = \frac{4}{1}$$

Geometric Meaning of Slope

From its geometric meaning, slope can be determined using the vertical and horizontal distances between any two points on a line by the formula

$$m = \text{slope} = \frac{\text{rise}}{\text{run}} = \frac{\text{distance up } (+) \text{ or down } (-)}{\text{distance right } (+) \text{ or left } (-)}.$$

Because all slopes can be written in fraction form, including fractions having denominator 1, you can extend the geometric meaning of slope by forming equivalent fractions.

Note that a positive slope of $\frac{5}{3}$ can be interpreted as $\frac{+5}{+3} = \frac{\text{up } 5}{\text{right } 3}$ or $\frac{-5}{-3} = \frac{\text{down } 5}{\text{left } 3}$ or as any other equivalent fraction, such as $\frac{-10}{-6}$.

The slope can be used to locate additional points on a graph.

Example 2   *Use the geometric interpretation of slope to locate two additional points in the snowy cricket situation.*

a. Locate a point below $(55, 60)$.

Write the slope $m = 4$ as a fraction $\frac{4}{1}$. Since you want a point *below* $(55, 60)$, you need to form an equivalent fraction whose numerator (vertical direction) is negative. One possible choice is to multiply numerator and denominator by -3.

$$m = \frac{4 \cdot (-3)}{1 \cdot (-3)} = \frac{-12}{-3}, \text{ which indicates move } \frac{down\ 12}{left\ 3}$$

from starting point $(55, 60)$ *and places you at point* $(52, 48)$.

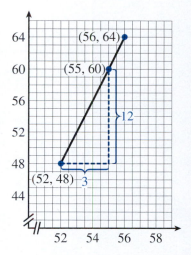

b. Locate a point between $(55, 60)$ and $(56, 64)$.

Because the *x*-coordinates are consecutive integers, you need to form an equivalent fraction whose denominator (horizontal direction) is a positive number less than one, for example, 0.5.

$$m = \frac{4 \cdot (0.5)}{1 \cdot (0.5)} = \frac{2}{0.5}, \text{ which indicates move } \frac{up\ 2}{right\ 0.5}$$

from starting point $(55, 60)$ *and places you at point* $(55.5, 62)$.

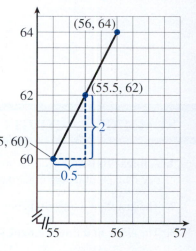

8. a. Suppose a line contains the point $(1, 2)$ and has slope $\frac{3}{1}$. Plot the point and then use the geometric interpretation of slope to determine two additional points on the line. Draw the line containing these three points.

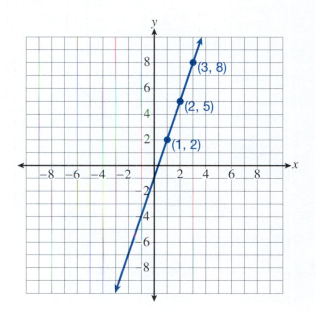

b. Consider the line containing the point $(-6, 4)$ and having slope $\frac{3}{4}$. Plot the point and then determine two additional points on the line. Draw a line through these points.

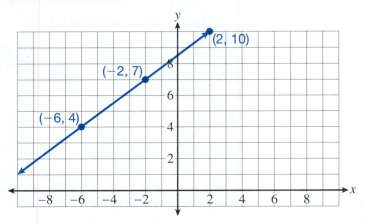

c. Consider the linear functions represented in parts a and b. Which function's output increases more rapidly? Explain.

The line with slope 3 is steeper than the line with slope $\frac{3}{4}$. Noting that $3 = \frac{12}{4}$, the output for the function in part a will increase by 12 when the input increases by 4, while the linear function output in part b will only increase by 3 for the same change in input. Therefore, the function in part a is increasing more rapidly.

SUMMARY: ACTIVITY 3.4

1. A **linear function** is one whose average rate of change of output with respect to input from any one data point to any other data point is always the same (constant) value.

2. The **graph** of a linear function is a line whose slope is the constant rate of change of the function.

3. The **slope of a line segment** joining two points (x_1, y_1) and (x_2, y_2) is denoted by m and can be calculated using the formula $m = \dfrac{\Delta y}{\Delta x} = \dfrac{y_2 - y_1}{x_2 - x_1}$, where $x_1 \neq x_2$. Geometrically, Δy represents a vertical distance (rise), and Δx represents a horizontal distance (run).

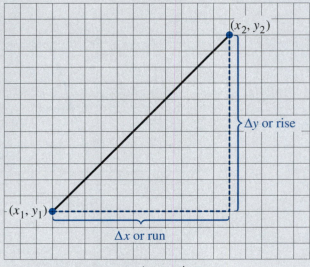

Therefore, $m = \dfrac{\Delta y}{\Delta x} = \dfrac{\text{rise}}{\text{run}}$.

4. The graph of every linear function with **positive slope** is a line rising to the right. The output values increase as the input values increase. The function is then said to be an **increasing function**.

EXERCISES: ACTIVITY 3.4

1. Consider the following data regarding the growth of the U.S. national debt from 1950 to 2008.

Number of Years Since 1950	0	10	20	30	40	50	58
National Debt (billions of dollars)	257	291	381	909	3113	5662	10,700

a. Compare the average rate of increase in the national debt from 1950 to 1960 with that from 1980 to 1990. Is the average rate of change constant? Explain, using the data.

rate of increase from 1950 to 1960: $\dfrac{291 - 257}{10 - 0} = \dfrac{34}{10} = \$3.4$ billion/yr.

rate of increase from 1980 to 1990: $\dfrac{3113 - 909}{40 - 30} = \dfrac{2204}{10} = \$220.4$ billion/yr.

The average rate of change is not constant. As the number of years since 1950 increases, the amount of debt does not change at a constant rate.

b. Plot the data points. If the points are connected to form a smooth curve, is the graph a straight line? Are the input and output variables in this problem related linearly?

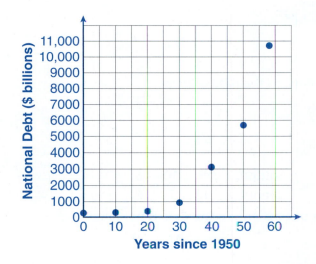

The graph is *not* a line, thus the input and output in this problem are *not* linearly related.

Exercise numbers appearing in color are answered in the Selected Answers appendix.

  c. Compare the graph in part b with the graph of the snowy tree cricket data in Problem 5. What must be true about the average rate of change between any two data points in order for all the data points to lie on a straight line?

  For all the data points to be on a line, the rate of change between any two data points must be the same, no matter which points are used in the calculation.

2. Your friend's parents began to give her a weekly allowance of $4 on her 4th birthday. On each successive birthday her allowance increased, so that her weekly allowance in dollars was equal to her age in years.

  a. List your friend's allowance for the ages in the table.

AGE (Years)	WEEKLY ALLOWANCE ($)
4	4
5	5
6	6
7	7

  b. Is her allowance a linear function of her age? Explain.

  Yes; her allowance is a linear function of her age. Each input (age) is assigned a single output (allowance amount), and the average rate of change between any two data points is a constant, $1 per year.

3. Calculate the average rate of change between consecutive data points to determine whether the output in each table is a linear function of the input.

  a.

Input	−5	0	5	8
Output	−45	−5	35	59

  $$\frac{\Delta \text{output}}{\Delta \text{input}} = \frac{40}{5} = \frac{40}{5} = \frac{24}{3} = 8$$

  The same is true for all data points, so the data is linear.

  b.

Input	2	7	12	17
Output	0	10	16	18

  $$\frac{\Delta \text{output}}{\Delta \text{input}} = \frac{10 - 0}{7 - 2} = \frac{10}{5} = \frac{2}{1}, \text{ but } \frac{\Delta \text{output}}{\Delta \text{input}} = \frac{16 - 10}{12 - 7} = \frac{6}{5}$$

  The average rate of change is not constant, so the data is not linear.

  c.

Input	−4	0	3	5
Output	−23.8	1	19.6	32

  $$\frac{\Delta \text{output}}{\Delta \text{input}} = \frac{24.8}{4} = \frac{18.6}{3} = \frac{12.4}{2} = 6.2$$

  The same is true for all data points, so the data is linear.

4. For each of the following, determine two additional points on the line. Then sketch a graph of the line.

a. A line contains the point $(-5, 10)$ and has slope $\frac{2}{3}$.

b. A line contains the point $(3, -4)$ and has slope 5.

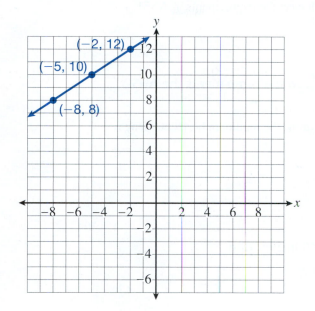

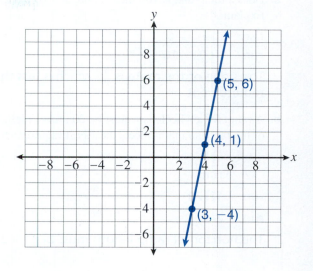

5. The concept of slope arises in many practical applications. When designing and building roads, engineers and surveyors need to be concerned about the grade of the road. The grade, usually expressed as a percent, is one way to describe the steepness of the finished surface of the road. For example, a 5% grade means that the road has a slope (rise over run) of $0.05 = \frac{5}{100}$.

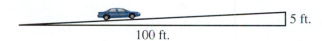

a. If a road has a 5% upward grade over a 1000-foot run, how much higher will you be at the end of that run than at the beginning?

$$\frac{\text{rise}}{\text{run}} = 0.05 = \frac{5}{100}$$

$$\frac{\text{rise}}{1000} = \frac{5}{100}$$

$$100 \text{ rise} = 5(1000)$$

$$\text{rise} = \frac{5000}{100} = 50$$

At the end of that 1000-foot run, you will be 50 feet higher than at the beginning.

b. What is the grade of a road that rises 26 feet over a distance of 500 feet?

The grade $= \dfrac{\text{rise}}{\text{run}} = \dfrac{26}{500} = 0.052 = 5.2\%$.

The road has a 5.2% grade.

6. The American National Standards Institute (ANSI) requires that the slope for a wheelchair ramp not exceed $\frac{1}{12}$.

  a. Does a ramp that is 160 inches long and 10 inches high meet the requirements of ANSI? Explain.

$$\frac{\text{rise}}{\text{run}} = \frac{10}{160} = \frac{1}{16}$$

The slope of the ramp is $\frac{1}{16}$, which is less than $\frac{1}{12}$, so the ramp meets the ANSI requirement.

  b. A ramp for a wheelchair must be 20 inches high. Determine the minimum horizontal length of the ramp so that it meets the ANSI requirement.

$$\frac{1}{12} = \frac{20}{\text{run}}$$

$$\text{run} = 12(20)$$

$$\text{run} = 240$$

The horizontal ramp length must be at least 240 inches long.

Activity 3.5

Descending in an Airplane

Objectives

1. Identify lines as having negative, zero, or undefined slopes.

2. Identify a decreasing linear function from its graph or slope.

3. Determine horizontal and vertical intercepts of a linear function from its graph.

4. Interpret the meaning of horizontal and vertical intercepts of a line.

While on a trip, you notice that the video screen on the airplane, in addition to showing movies and news, records your altitude (in kilometers) above the ground. As the plane starts its descent (at time $t = 0$), you record the following data.

 **Cleared for a Landing**

TIME, t (min)	ALTITUDE, $A(t)$ (km)
0	12
2	10
4	8
6	6
8	4
10	2

1. a. What is the average rate of change in the altitude of the plane from 2 to 6 minutes into the descent? Pay careful attention to the sign of this rate of change.

$$\frac{\Delta A}{\Delta t} = \frac{6 - 10}{6 - 2} = \frac{-4}{4} = -1$$

b. What are the units of measurement of this average rate of change?

The units are kilometers per minute.

2. a. Determine the average rate of change over several other input intervals.

(Answers will vary.)

From 6 to 10 min.: $\dfrac{\Delta A}{\Delta t} = \dfrac{2 - 6}{10 - 6} = \dfrac{-4}{4} = -1$ kpm.

From 0 to 8 min.: $\dfrac{\Delta A}{\Delta t} = \dfrac{4 - 12}{8 - 0} = \dfrac{-8}{8} = -1$ kpm.

b. What is the significance of the signs of these average rates of change?

The sign of the rate of change is negative, indicating that the altitude is decreasing as time increases.

c. Based on your calculation in Problems 1a and 2a, do you think that the data lie on a single straight line? Explain.

Yes, I think the graph is linear because the rate of change is constant.

d. What is the practical meaning of slope in this situation?

For every minute that passes, the airplane descends 1 kilometer.

3. By how much does the altitude of the plane change for each 3-minute change in time during the descent?

The altitude of the plane decreases 3 kilometers for each 3 minutes into the descent.

4. Plot the data points from the table preceding Problem 1, and verify that the points lie on a line. What is the slope of the line?

The slope of the line is − 1.

5. a. The slope of the line for the descent function in this activity is negative. What does this tell you about how the outputs change as the input variable (time) increases in value?

Since the slope of the line is negative, the outputs decrease as the input variable increases in value.

b. Is the descent function an increasing or decreasing function?

It is a decreasing function.

6. Suppose another airplane was descending at the rate of 1.5 kilometers every minute.

a. Complete the following table.

Time	0	1	2	3
Altitude	12	10.5	9	7.5

b. Plot the data points, and verify that the points lie on a line. What is the slope of the line?

The slope is $-\frac{3}{2}$.

Horizontal and Vertical Intercepts

7. a. From the time the plane (in Problems 1–5) begins its descent, how many minutes does it take to reach the ground?

The plane will reach the ground in 12 minutes.

b. Use a straightedge to connect the data points on your graph in Problem 4. Extend the line so that it crosses both axes. See graph.

Definition

A **horizontal intercept** of a graph is a point at which the graph crosses (or touches) the horizontal (input) axis.

A horizontal intercept is clearly identified by noting it is the point at which the output value is zero. The ordered pair notation for all horizontal intercepts always has the form $(a, 0)$, where a is the input value. Because it is understood that its output value is 0, the horizontal intercept is occasionally referred to simply by a.

8. a. Identify the horizontal intercept of the line in Problem 4 from the graph.

The horizontal intercept is (12, 0).

b. How is the horizontal intercept related to the answer you obtained in Problem 7a? That is, what is the practical meaning of the horizontal intercept?

The answers should be the same. The horizontal intercept is the point at which the output value is zero; that is, the plane is on the ground.

Definition

A **vertical intercept** of a graph is a point at which the graph crosses (or touches) the vertical (output) axis.

A vertical intercept is clearly identified by noting it is the point at which the input value is zero. The ordered-pair notation for all vertical intercepts always has the form $(0, b)$, where b is the output value. Because it is understood that its input value is 0, the vertical intercept is occasionally referred to simply by b.

9. a. Identify the vertical intercept of the descent function from its graph in Problem 4.

The vertical intercept is (0, 12). Its output value, 12 kilometers, is the altitude of the plane at the start of descent, when the input is zero.

b. What is the practical meaning of this intercept?

The plane has an altitude of 12 kilometers when it begins to descend.

If x represents the input variable and y represents the output variable, then the horizontal intercept is called the x-intercept, and the vertical intercept is called the y-intercept.

10. For each of the following lines, determine
  i. the slope,
  ii. the horizontal (x) intercept,
  iii. the vertical (y) intercept,

a.

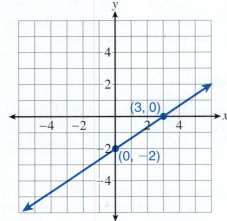

i. Slope is $\frac{2}{3}$.
ii. Horizontal intercept is $(3, 0)$.
iii. Vertical intercept is $(0, -2)$.

b.

i. Slope is $\frac{-5}{2}$.
ii. Horizontal intercept is $(2, 0)$.
iii. Vertical intercept is $(0, 5)$.

Horizontal Line

11. a. You found a promotion for unlimited access to the Internet for $20 per month. Complete the following table of values, where t is the number of hours a subscriber spends online during the month and c is the monthly access cost for that subscriber.

t, Time (hr.)	1	2	3	4	5
c, Cost ($)	20	20	20	20	20

b. Sketch a graph of the data points.

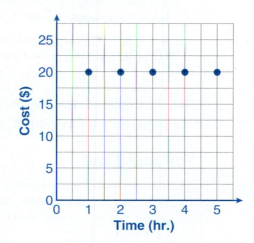

c. What is the slope of the line drawn through the points?

$$m = \frac{\text{up } 0}{\text{right } 1} = \frac{0}{1} = 0 \quad \text{The slope of the resulting line is zero.}$$

d. What single word best describes a line with zero slope?

"Horizontal" best describes a line with zero slope.

e. Determine the horizontal and vertical intercepts (if any).

The vertical intercept is $(0, 20)$. There is no horizontal intercept.

f. What can you say about the output value of every point on a horizontal line?

The output value of every point on a horizontal line is the same constant value.

> A **horizontal line** is characterized by the facts that *all* its points have identical output values and its slope is zero.

Suppose the horizontal and vertical axes are labeled, respectively, as the x- and y-axes. Then every point on a horizontal line will have the form (x, c), where the x-coordinate varies from point to point and the y-coordinate always takes the same value, c. Because of this, a horizontal line can be completely described by an equation that specifies the constant y-value: $y = c$.

Example 1   *A horizontal line contains the point $(3, 7)$. List two additional points on this line and indicate its equation.*

SOLUTION

A horizontal line is completely described by its y-coordinate, here 7. Two additional points could be $(-2, 7)$ and $(11, 7)$. Its equation is $y = 7$.

12. Use function notation to write a symbolic rule that expresses the relationship in Problem 11—the monthly cost for unlimited access to the Internet in terms of the number of hours spent online.

Let t represent the number of hours spent online during the month and $c(t)$, the monthly access cost for a subscriber.

$c(t) = 20$ The cost $c(t)$ is $20 no matter how many hours are spent online.

13. The x-axis is a horizontal line. List two points on the x-axis and indicate its equation.

(Answers my vary.) $(0, 0)$ and $(5, 0)$; equation is $y = 0$.

Vertical Line

14. To cover your weekly expenses while going to school, you work as a part-time aide in your college's health center and earn $100 each week. Complete the following table, where x represents your weekly salary and y represents your weekly expenses for a typical month.

x, Weekly Salary ($)	100	100	100	100
y, Weekly Expenses ($)	50	70	90	60

a. Sketch a graph of the data points. Do the points lie on a line? Explain.

The points lie on a vertical line.

b. Explain what happens if you use the slope formula to determine a numerical value for the slope of the line in part a.

$$m = \frac{90 - 70}{100 - 100} = \frac{20}{0} = ????$$

A numerical value for the slope cannot be determined because the denominator is zero. Division by zero is undefined.

c. Write down the slope formula. What must be true about the change in the input variable for the quotient to be defined?

$$m = \frac{\text{change in output}}{\text{change in input}}$$

For you to determine the quotient, the change in the input variable must not be zero.

d. What type of line results whenever the slope is undefined, as in this problem?

A vertical line results when the slope is undefined.

e. Determine the vertical and horizontal intercepts (if any).

The horizontal intercept is $(100, 0)$. There is no vertical intercept.

f. What can you say about the input value of every point on a vertical line?

The input value of every point on a vertical line is the same.

g. Is y a function of x? Explain.

(Reasons will vary.)

No, y is not a function of x. The graph of y with respect to x fails the vertical line test.

> A **vertical line** is characterized by the facts that *all* its points have identical input values and its slope is not defined—it has no numerical value.

When the horizontal and vertical axes are labeled, respectively, as the x- and y-axes, then every point on a vertical line will have the form (d, y), where the y-coordinate varies from point to point and the x-coordinate always takes the same value, d. Because of this, the vertical line can be completely described by an equation that specifies the constant x-value: $x = d$.

Example 2 *A vertical line contains the point $(3, 7)$. List two additional points on this line and indicate its equation.*

SOLUTION

A vertical line is completely described by its x-coordinate, here 3. Two additional points could be $(3, -5)$ and $(3, 10)$. Its equation is $x = 3$.

15. Write an equation for the situation described in Problem 13.

$x = 100$

16. The y-axis is a vertical line. List two points on the y-axis and indicate its equation.

(Answers may vary.) $(0, -2)$ and $(0, 10)$; equation is $x = 0$.

SUMMARY: ACTIVITY 3.5

1. The graph of every linear function with **negative slope** is a line falling to the right.

2. A linear function having a negative slope $(m < 0)$ is a **decreasing function**.

3. The **horizontal intercept** is the point at which the graph crosses the horizontal (input) axis. Its ordered-pair notation is $(a, 0)$; that is, the output value is equal to zero. If the input is denoted by x, then the horizontal intercept is referred to as the x-intercept.

4. The **vertical intercept** is the point at which the graph crosses the vertical (output) axis. Its ordered-pair notation is $(0, b)$; that is, the input value is equal to zero. If the output is denoted by y, then the intercept is referred to as the y-intercept.

5. The slope of a **horizontal line** is zero. Every point on a horizontal line has the same output value. The equation of a horizontal line is $y = d$, where d is a constant. For example, the graph of $y = 6$ is a horizontal line.

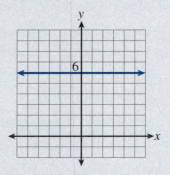

6. The slope of a **vertical line** is undefined; it has no numerical value. Every point on a vertical line has the same input value. The equation of a vertical line is $x = c$, where c is a constant. For example, the graph of $x = 5$ is a vertical line.

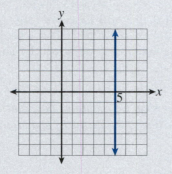

EXERCISES: ACTIVITY 3.5

1. In a science lab, you collect the following sets of data. Which of the four data sets are linear functions? If linear, determine the slope.

a.

Time (sec.)	0	10	20	30	40
Temperature (°C)	12	17	22	27	32

linear $\left(\text{constant average rate of change} = \dfrac{1}{2}\right)$

b.

Time (sec.)	0	10	20	30	40
Temperature (°C)	41	23	5	-10	-20

not linear (average rate of change not constant)

$$\frac{23 - 41}{10} = \frac{-18}{10} = -1.8; \quad \frac{-20 - (-10)}{10} = \frac{-10}{10} = -1$$

Exercise numbers appearing in color are answered in the Selected Answers appendix.

c.

Time (sec.)	3	5	8	10	15
Temperature (°C)	12	16	24	28	36

not linear (average rate of change not constant)

$$\frac{16-12}{5-3} = \frac{4}{2} = 2; \quad \frac{24-16}{8-5} = \frac{8}{3}$$

d.

Time (sec.)	3	9	12	18	21
Temperature (°C)	25	23	22	20	19

linear $\left(\text{constant average rate of change} = \dfrac{-1}{3}\right)$

2. a. You are a member of a health and fitness club. A special diet and exercise program has been developed for you by the club's registered dietitian and your personal trainer. You weigh 181 pounds and would like to lose 2 pounds every week. Complete the following table of values for your desired weight each week.

N, Number of Weeks	0	1	2	3	4
W(N), Desired Weight (lb.)	181	179	177	175	173

b. Plot the data points.

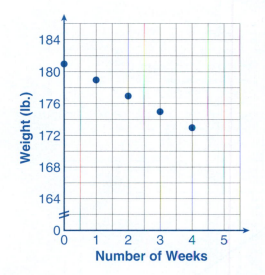

c. Explain why your desired weight is a linear function of time. What is the slope of the line containing the five data points?

My desired weight loss is a *linear* function of time because it is a constant average rate of change, −2, each week.

$$m = \frac{\Delta W}{\Delta N} = \frac{177-179}{2-1} = \frac{-2}{1} = -2$$

The slope of the line connecting the data is −2.

d. What is the practical meaning of slope in this situation?

Every week I hope to lose 2 pounds.

e. How long will it take to reach your ideal weight of 168 pounds?

It will take $6\frac{1}{2}$ weeks to reach 168 pounds.

3. Your aerobics instructor informs you that to receive full physical benefit from exercising, your heart rate must be maintained at a certain level for at least 12 minutes. The proper exercise heart rate for a healthy person, called the target heart rate, is determined by the person's age. The relationship between these two quantities is illustrated by the data in the following table.

A, Age (yr.)	20	30	40	50	60
B(A), Target Heart Rate (beats/min.)	140	133	126	119	112

a. Does the data in the table indicate that the target heart rate is a linear function of age? Explain.

(Answers may vary.) Yes, target heart rate is a linear function of age. Each input is assigned a unique output, and the average rate of change is constant.

b. What is the slope of the line for this data? What are the units?

The slope of the line for this data is $-\frac{7}{10}$ beats per minute per year of age.

c. What are suitable replacement values (practical domain) for age, A?

(Answers will vary.) Suitable replacement values for A are values greater than 0 but less than 110.

d. Plot the data points on coordinate axes where both axes start with zero, with 10 units between grid lines. Connect the points with a line.

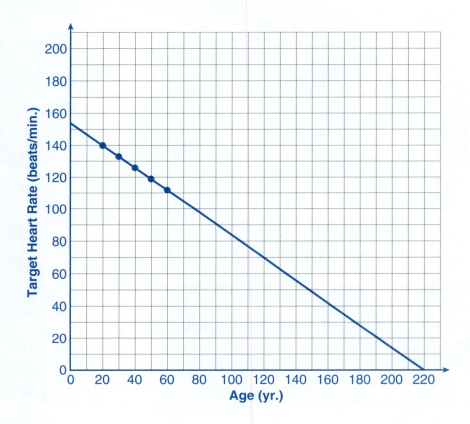

e. Extend the line to locate the horizontal and vertical intercepts. Do these intercepts have a practical meaning in the problem? Explain.

(Answers will vary.) The vertical intercept is approximately (0, 154). The horizontal intercept is approximately (220, 0). Neither intercept has a practical meaning. The vertical intercept would refer to the target heart rate of a newborn. The horizontal intercept would correspond to the target heart rate of a 220-year-old person.

4. a. Determine the slope of each of the following lines.

i.

ii.

iii.

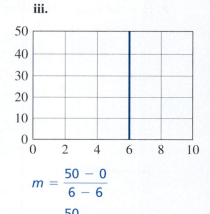

$$m = \frac{60 - 0}{20 - 0} = \frac{60}{20} = 3$$

$$m = \frac{0 - 20}{6 - 0}$$

$$= \frac{-20}{6}$$

$$= \frac{-10}{3}$$

$$m = \frac{50 - 0}{6 - 6}$$

$$= \frac{50}{0}$$

m is undefined.

b. Determine the horizontal and vertical intercept of each of the lines in part a.

INTERCEPT	GRAPH i	GRAPH ii	GRAPH iii
Horizontal	(0, 0)	(6, 0)	(6, 0)
Vertical	(0, 0)	(0, 20)	none

5. a. A horizontal line contains the point $(-3, 7)$. Determine and list three additional points that lie on the line.

(Answers may vary.) $(0, 7), (-5, 7), (10, 7)$

b. Sketch a graph of the horizontal line.

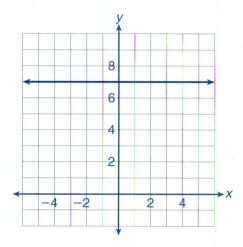

c. Identify the slope and vertical and horizontal intercepts (if they exist) of the graph.

The slope is 0; the vertical intercept is (0, 7); and there is no horizontal intercept.

d. What is the equation of this line?

$y = 7$

6. Each question refers to the graph that accompanies it. The graphed line in each grid represents the total distance a car travels as a function of time (in hours).

a. How fast is the car traveling? Explain how you obtained your result.

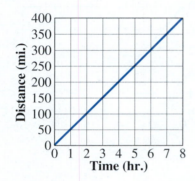

$\dfrac{400 \text{ mi.}}{8 \text{ hr.}}$ = 50 mph. The car is traveling at 50 miles per hour.

b. How can you determine visually from the following graph which car is going faster? Verify your answer by calculating the speed of each car.

The line representing the distance traveled by car B is steeper, so car B is going faster.

Car A travels 500 miles in 10 hours, a rate of 50 miles per hour.

Car B travels 520 miles in 7 hours, a rate of 74 miles per hour.

c. Describe in words the movement of the car that is represented by the following graph.

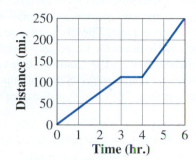

(Answers will vary.)

The car travels about 110 miles in the first 3 hours, at less than 40 miles per hour. It then stops for 1 hour before continuing. The car travels approximately 140 miles in the next 2 hours, at an approximate speed of 70 miles per hour.

7. a. Determine the slope of each of the following lines.

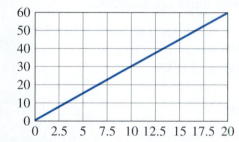
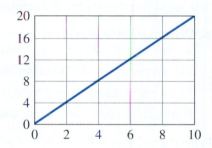
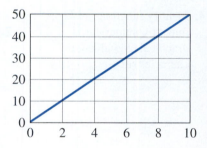

$$m = \frac{60 - 0}{20 - 0}$$

$$= \frac{60}{20}$$

$$= 3$$

$$m = \frac{20 - 0}{10 - 0}$$

$$= \frac{20}{10}$$

$$= 2$$

$$m = \frac{50 - 0}{10 - 0}$$

$$= \frac{50}{10}$$

$$= 5$$

b. At first glance, the three graphs in part a may appear to represent the same line. Do they?

No, their slopes are all different. The number of units represented by each tick mark is different on each graph, producing different slopes.

Activity 3.6

Charity Event

Objectives

1. Determine a symbolic rule for a linear function from contextual information.

2. Identify the practical meanings of the slope and intercepts of a linear function.

3. Determine the slope-intercept form of a linear function.

4. Identify functions as linear by numerical, graphical, and algebraic characteristics.

Professor Abrahamsen's social psychology class is organizing a campus entertainment night to benefit charities in the community. You are a member of the budget committee for this class project. The committee suggests an admission donation of $10 per person for food, non-alcoholic beverages, and entertainment. The committee members expect that each student in attendance will purchase a raffle ticket for $1. Faculty members volunteer to emcee the event and perform comedy sketches. Two student bands are hired at a cost of $200 each. Additional expenses include $1000 for food and drinks, $200 for paper products, $100 for posters and tickets, and $500 for raffle prizes. The college is donating the use of the gymnasium for the evening.

1. Determine the total fixed costs for the entertainment night.

   Bands $2 \cdot \$200 = \$400.00$

   Food and drink $1000.00

   Paper products $200.00

   Posters and tickets $100.00

   Raffle prizes $500.00

   Total fixed costs = $2200.00

   The total fixed costs for entertainment night are $2200.00.

2. **a.** Determine the **total revenue** (gross income *before* expenses are deducted) if 400 students attend and each buys a raffle ticket.

   $10 admission + $1.00 raffle ticket = $11.00

   $11.00 \cdot 400$ students = $4400.00

   The total revenue would be $4400.

   b. Determine the **profit** (net income *after* expenses are deducted) if 400 students attend and each buys a raffle ticket.

   $4400.00 revenue − $2200.00 costs = $2200.00 profit

   There is a $2200.00 profit if 400 students attend and each buys a raffle ticket.

3. **a.** The total revenue (gross income) for the event depends on the number, n, of students who attend. Write an expression in terms of n that represents the total revenue if n students attend and each buys a raffle ticket.

   $11n$ dollars

   b. Write a symbolic function rule defining the profit, $p(n)$, in terms of the number, n, of students in attendance. Remember that the total fixed costs for the entertainment night are $2200.

   $p(n) = 11n − 2200$ dollars

   c. List some suitable replacement values (practical domain) for the input variable n. Is it meaningful for n to have a value of $\frac{1}{2}$ or $−3$?

   Suitable replacement values would be whole numbers from 0 to the capacity of the gym. n could not have a fraction or negative value because n represents the number of students attending.

4. a. Use the symbolic rule in Problem 3b to determine the profit if 100 students attend.

$p(100) = 11(100) - 2200 = 1100 - 2200 = -1100$

If 100 students attend and each buys a raffle ticket, the profit is $-\$1100.00$.

b. What is the practical meaning of the negative value for profit in part a?

The negative value for profit indicates a *loss* of $1100.

5. If the gymnasium holds a maximum of 650 people, what is the maximum amount of money that can be donated to charity?

$p(650) = 11(650) - 2200 = 7150 - 2200 = 4950$

The maximum amount that can be given to charity is $4950.

6. a. Suppose that the members of the class want to be able to donate $1000 to community charities. Write an equation to determine how many students must attend the entertainment night for there to be a profit of $1000. Solve the equation.

$11n - 2200 = 1000$

$n = 290.\overline{90}$

291 students must attend and buy a raffle ticket for there to be a $1000 profit to donate.

b. How many students must attend for there to be $2000 to donate to the local charities?

$11n - 2200 = 2000$

$11n = 4200$

$n \approx 381.8$

382 students must attend and buy a raffle ticket for there to be $2000 to donate to local charities.

7. a. Complete the following table of values for the charity event situation.

n, Number of Students	0	50	100	200	300	400
p(n), Profit ($)	-2200	-1650	-1100	0	1100	2200

b. Determine the average rate of change in profit as the number of students in attendance increases from 300 students to 400 students.

$$\frac{\Delta \text{profit}}{\Delta \text{number of students}} = \frac{2200 - 1100}{400 - 300} = \frac{1100}{100} = \frac{11}{1} = 11 \text{ dollars per student}$$

The average rate of change is $11 per student.

c. Determine the average rate of change between consecutive data pairs in the table.

The average rate of change between any two pairs of data in the table is always 11.

d. Is profit a linear function of the number of students attending? Explain.

The profit function is linear because the average rate of change is always the same constant value, 11.

8. a. Sketch a graph of the profit function.

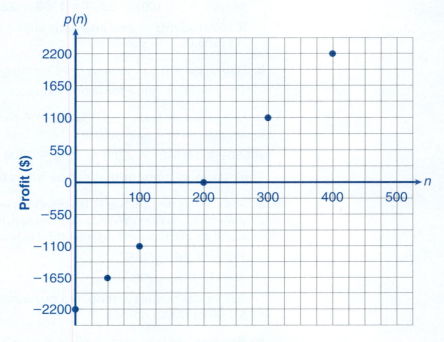

Number of Students

b. Why should you expect all the plotted points to lie on the same line?

Since the average rate of change is constant, the graph will be a straight line.

9. a. Determine the slope of the line representing the profit function in Problem 8a. What are the units of measurement of the slope? What is the practical meaning of the slope in this situation?

$$\text{slope} = \frac{2200 - 1100}{400 - 300} = \frac{1100}{100} = 11$$

The slope is 11.

The slope is measured in dollars per student.

Profit increases by $11 for each additional student attending.

b. What is the vertical intercept of the line?

The vertical intercept is $(0, -2200)$.

c. What is the practical meaning of the vertical intercept in this situation?

If no students attend, there is a loss of $2200.

d. What is the horizontal intercept of the line?

The horizontal intercept is $(200, 0)$

e. What is the practical meaning of the horizontal intercept in this situation?

Attendance of 200 students is the break-even point at which the expenses are covered but no profit is made.

Slope-Intercept Form of an Equation of a Line

The profit function defined by $p(n) = 11n - 2200$ has a symbolic form that is representative of *all* linear functions. That is, the symbolic form of a linear function consists of the sum of two terms:

- a *variable term* (the input variable multiplied by its coefficient)

- and a *constant term* (a fixed number)

10. a. Identify the variable term in the symbolic rule $p(n) = 11n - 2200$. What is its coefficient?

The variable term is 11*n*. Its coefficient is 11.

b. Identify the constant term in this symbolic rule.

The constant term is − 2200.

c. What characteristic of the linear function graph does the coefficient of the input variable *n* represent?

The coefficient represents the slope of the graph.

d. What characteristic of the linear function graph does the constant term represent?

The constant term represents the output value of the vertical intercept.

11. a. Consider a line defined by the equation $y = 2x + 7$. Use the equation to complete the following table.

x	− 2	− 1	0	1	2
y	3	5	7	9	11

b. Use the slope formula to determine the slope of the line. How does the slope compare to the coefficient of *x* in the equation?

$$\text{slope} = \frac{5 - 3}{-1 - (-2)} = \frac{2}{1} = 2$$

The slope and the coefficient are the same.

c. Determine the vertical (*y*-) intercept. How does it compare to the constant term in the equation of the line?

The vertical intercept is (0, 7). The *y*-value of the vertical intercept and the constant term are the same.

The answers to Problems 10c and d and Problem 11 generalize to all linear functions.

When the horizontal and vertical axes are labeled, respectively, as the *x*- and *y*-axes, the coordinates of every point (x, y) on a line will satisfy the equation $y = mx + b$, where m is the slope of the line and $(0, b)$ is its *y*-intercept. This equation is often called the **slope-intercept** form of the equation of a line.

Example 1

a. Given the linear function rule $y = 3x + 1$, identify the slope and y-intercept.

b. Determine which of the following points lie on the line: $(-2, -5)$, $(-1, 2)$, and $(4, 13)$.

SOLUTION

a. The slope of the line is 3, the coefficient of x; the y-intercept is $(0, 1)$.

b. Check to see which ordered pairs satisfy the equation:

Replacing x by -2: $3(-2) + 1 = -5$, which confirms that $(-2, -5)$ lies on the line.

Replacing x by -1: $3(-1) + 1 = -2$, which indicates that $(-1, 2)$ does not lie on the line.

Replacing x by 4: $3(4) + 1 = 13$, which confirms that $(4, 13)$ lies on the line.

Example 2   *Identify the slope and y-intercept of each of the following.*

LINEAR FUNCTION RULE	m, SLOPE	$(0, b)$, y-INTERCEPT
$y = 5x + 3$	5	$(0, 3)$
$y = -2x + 7$	-2	$(0, 7)$
$y = \frac{1}{2}x - 4$	$\frac{1}{2}$	$(0, -4)$
$y = 3x + 0$ or simply $y = 3x$	3	$(0, 0)$
$y = 0x + 10$ or simply $y = 10$	0	$(0, 10)$
$y = 10 + 6x$	6	$(0, 10)$
$y = 85 - 7x$	-7	$(0, 85)$

12. Identify the slope and vertical intercept of each of the following linear functions.

a. $y = -3x + 8$

slope: -3

vertical intercept: $(0, 8)$

b. $y = \frac{3}{4}x - \frac{1}{2}$

slope: $\frac{3}{4}$

vertical intercept: $\left(0, -\frac{1}{2}\right)$

c. $y = -5x$

slope: -5

vertical intercept: $(0, 0)$

d. $y = -3$

slope: 0

vertical intercept: $(0, -3)$

e. $y = 16 + 4x$

slope: 4

vertical intercept: $(0, 16)$

f. $y = 110 - 3x$

slope: -3

vertical intercept: $(0, 110)$

g. $p = -12 + 2.5n$

slope: 2.5

vertical intercept: $(0, -12)$

h. $q = -45 - 9r$

slope: -9

vertical intercept: $(0, -45)$

13. For each of the following, determine the equation of the line having the given slope and y-intercept.

 a. The slope is 3 and the y-intercept is $(0, 4)$.

   $y = 3x + 4$

 b. The slope is -1 and the y-intercept is $(0, 0)$.

   $y = -x$

 c. The slope is $\frac{2}{3}$ and the y-intercept is $(0, 6)$.

   $y = \frac{2}{3}x + 6$

 d. The slope is 0 and the y-intercept is $(0, -5)$.

   $y = -5$

SUMMARY: ACTIVITY 3.6

1. The average rate of change between any two input/output pairs of a linear function is always the same constant value.

2. The graph of every linear function is a line whose slope, m, is precisely the constant average rate of change of the function.

3. For every linear function, equally spaced input values produce equally spaced output values.

4. The symbolic rule for a linear function, also called the **slope-intercept form** of the line, is given by $y = mx + b$, where m is the slope and $(0, b)$ is the y- (vertical) intercept of the line.

EXERCISES: ACTIVITY 3.6

Exercises 1–5 refer to the charity event scenario in this activity. Suppose the budget committee decides to increase the admission fee to $12 per person. It is still expected that each student will purchase a raffle ticket for $1.

1. a. Write a new symbolic rule for profit in terms of the number of tickets sold.

   $p(n) = 13n - 2200$ dollars

 b. Complete the following table using the symbolic rule determined in part a.

Number of Tickets, n	0	50	100	150	200	300	400
Profit, $p(n)$	-2200	-1550	-900	-250	400	1700	3000

2. a. Determine the practical domain of the new profit function.

The practical domain consists of whole numbers from 0 to 650. Recall that the capacity of the gym is 650 students.

b. Sketch a graph of the new profit function.

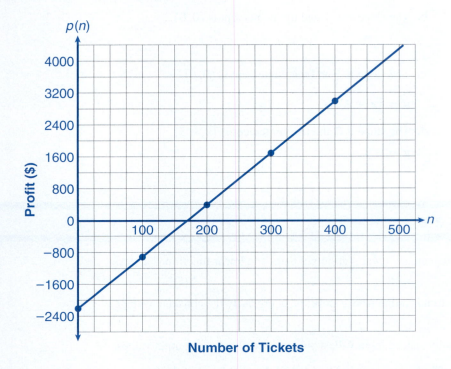

Number of Tickets

c. What is the slope of the line containing the data points? What is the practical meaning of the slope in this situation?

The slope is 13. The profit increases by $13 for each additional student attending.

d. What is the vertical intercept of the line? What is the practical meaning of the intercept in this situation?

The vertical intercept is $(0, -2200)$. The class project has a loss of $2200 if no students attend.

3. If the gymnasium could hold a maximum of 900 people instead of 650, what is the maximum amount of money that could be given to charity?

$p(900) = 13(900) - 2200 = \$11{,}700 - 2200 = \$9500$

The maximum amount that could be given to charity is $9500.

4. What attendance is needed for there to be a profit of $1000?

$13n - 2200 = 1000$

$13n = 3200$

$n \approx 246.15$

But n represents the number of people and must be a whole number. Therefore, 247 students must attend for there to be a profit of $1000.

5. a. Is there a point where the charity event will show neither a profit nor a loss? If yes, where would this point be located?

Yes, when the profit is zero, there will be neither a profit nor a loss, at the horizontal intercept of the graph.

b. The point where the profit is zero is called the **break-even point**. Use the idea of the break-even point to determine how many students must attend so that the class project will not incur a loss.

$$13n - 2200 = 0$$
$$13n = 2200$$
$$n \approx 169.23 \text{ at } p = 0.$$

But n represents the number of people and must be a whole number. Therefore, $n > 169.23$ and 170 students are needed to break even.

c. Is your answer to part b consistent with your graph in Problem 2b?

Yes, the graph seems to indicate a value just under 175 tickets.

6. Consider the line having equation $y = -3x + 1.5$.

a. Complete the following table.

x	-2	-1	0	1	2
y	7.5	4.5	1.5	-1.5	-4.5

b. Use the slope formula to determine the slope of the line. How does the slope compare to the coefficient of the input variable x in the equation of the line?

$$\text{slope} = \frac{4.5 - 7.5}{-1 - (-2)} = \frac{-3}{1} = -3$$

The slope and the coefficient are the same.

c. How does the y-intercept you determined in part a compare to the constant term in the equation of the line?

The y-intercept is $(0, 1.5)$. The y-intercept and the constant term are the same.

7. Identify the slope and vertical intercept of each of the following.

LINEAR FUNCTION RULE	m, SLOPE	$(0, b)$, VERTICAL INTERCEPT
$y = 3x - 2$	3	$(0, -2)$
$y = -2x + 5$	-2	$(0, 5)$
$y = \frac{1}{2}x + 3$	$\frac{1}{2}$	$(0, 3)$
$y = -2x$	-2	$(0, 0)$
$y = 6$	0	$(0, 6)$
$y = 9 + 3.5x$	3.5	$(0, 9)$
$w = -25 + 8x$	8	$(0, -25)$
$v = 48 - 32t$	-32	$(0, 48)$
$z = -15 - 6u$	-6	$(0, -15)$

8. Housing prices in your neighborhood have been increasing steadily since you purchased your home in 2005. The relationship between the market value, V, of your home and the length of time, x, you have owned your home is modeled by the symbolic rule

$$V(x) = 2500x + 125{,}000,$$

where $V(x)$ is measured in dollars and x in years.

   a. The graph of the symbolic rule is a line. What is the slope of this line? What is the practical meaning of slope in this situation?

   The slope of the line is 2500, indicating that the value of my home increased at a constant rate of $2500 per year since 2005.

   b. Determine the vertical (v-) intercept. What is the practical meaning of this intercept in the context of this problem?

   The v-intercept is (0, 125,000). In 2005, the value of my house was $125,000.

   c. Determine and interpret the value $V(8)$.

   $V(8) = 2500(8) + 125{,}000 = 145{,}000$ and $2005 + 8 = 2013$

   The market value of my house in 2013 will be $145,000.

9. The value of a car decreases (depreciates) immediately after it is purchased. The value of a car you recently purchased can be modeled by the symbolic rule

$$V(x) = -1350x + 18{,}500,$$

where $V(x)$ is the market value (in dollars) and x is the length of time you own your car (in years).

   a. The graph of the relationship is a line. Determine the slope of this line. What is the practical meaning of slope in this situation?

   The slope of the line is -1350, indicating that the car depreciates in value $1350 per year.

   b. Determine the vertical (V-) intercept. What is the practical meaning of this intercept?

   The v-intercept is (0, 18,500), indicating that the purchase price of the car was $18,500.

   c. Determine and interpret the value $V(3)$.

   $V(3) = -1350(3) + 18{,}500 = 14{,}450$

   In 3 years, my car will be worth $14,450.

10. **a.** Given the linear function rule $y = -x + 4$, identify the slope and y-intercept.

   The slope of the line is -1; the y-intercept is (0, 4).

   b. Determine which of the following points lie on the line: $(-1, 5)$, $(3, 1)$, and $(5, 9)$.

   Replacing x by -1: $-(-1) + 4 = 5$, so $(-1, 5)$ lies on the line.

   Replacing x by 3: $-(3) + 4 = 1$, so $(3, 1)$ lies on the line.

   Replacing x by 5: $-(5) + 1 = -4$, so $(5, 9)$ does not lie on the line.

11. **a.** Determine an equation of the line whose slope is 2 and whose y-intercept is $(0, -3)$.

   $y = 2x - 3$

   b. Determine an equation of the line whose slope is -3 and whose y-intercept is $(0, 0)$.

   $y = -3x$

   c. Determine an equation of the line whose slope is $\frac{3}{4}$ if it contains the point $(0, 1)$.

   $y = \dfrac{3}{4}x + 1$

Activity 3.7

Software Sales

Objectives

1. Identify the slope and vertical intercept from the equation of a line written in slope-intercept form.

2. Write an equation of a line in slope-intercept form.

3. Use the *y*-intercept and the slope to graph a linear function.

4. Determine horizontal intercepts of linear functions using an algebraic approach.

5. Use intercepts to graph a linear function.

You have been hired by a company that sells computer software products. In 2010, the company's total (annual) sales were $16 million. Its marketing department projects that sales will increase by $2 million per year for the next several years.

1. a. Let *t* represent the number of years since 2010. That is, $t = 0$ corresponds to 2010, $t = 1$ corresponds to 2011, and so on. Complete the following table.

t, NUMBER OF YEARS SINCE 2010	s, TOTAL SALES IN MILLIONS OF DOLLARS
0	16
1	18
2	20
5	26

b. Write a symbolic rule that would express the total sales, *s*, in terms of the number of years, *t*, since 2010.

$s = 2t + 16$

c. Does your symbolic rule define a function?

Yes, each input value has exactly one corresponding output value.

d. What is the practical domain of this total sales function?

(Answers will vary.)

The practical domain for *t* is all integers from 0 through 10 years. This would correspond to the years 2010 to 2020. It may turn out that after several years, data will show that projected values are inaccurate, and a new model may have to be determined.

e. Is the total sales function linear? Explain.

Yes, the total sales function is linear because the average rate of change is constant ($2 million per year).

2. a. Determine the slope of the total sales function. What are the units of measurement of the slope?

The slope is 2; the units are million dollars per year.

b. What is the practical meaning of the slope in this situation?

The slope indicates that the total sales are increasing by $2 million each year after 2010.

3. a. Determine the vertical (*s*-) intercept of this linear function.

The vertical intercept is (0, 16).

b. What is the practical meaning of the vertical intercept in this situation?

The vertical intercept indicates that when $t = 0$, that is, in 2010, the total sales were $16 million.

The symbolic rule for the total sales function can be written using function notation as $s(t) = 2t + 16$.

4. a. Determine $s(6)$.

$$s(6) = 2(6) + 16$$
$$s(6) = \$28 \text{ million}$$

b. Interpret the meaning of the result in part a.

In 2016 (6 years after 2010), the total sales will be $28 million.

5. Use the symbolic rule $s(t) = 2t + 16$ for the total sales function to approximate the year in which total sales will reach $32 million. What ordered pair on the graph conveys the same information?

$$32 = 2t + 16$$
$$16 = 2t$$
$$8 = t$$

Total sales will reach $32 million in 2018.

The point (8, 32) on the graph conveys this information.

6. Use the given symbolic rule $s(t) = 2t + 16$ to determine the total sales in the year 2015. What ordered pair conveys the same information?

The year 2015 corresponds to $t = 5$.

$$s(5) = 2(5) + 16 = \$26 \text{ million}$$

The ordered pair corresponding to $s(5) = 26$ is (5, 26).

The point (5, 26) conveys this information.

7. Let x represent the input and $f(x)$ represent the output. Use function notation to write the equation of each of the following lines:

a. A line having slope $\frac{2}{3}$ and vertical intercept $(0, 7)$

Substitute $\frac{2}{3}$ for m and 7 for b in $f(x) = mx + b$ to get $f(x) = \frac{2}{3}x + 7$.

b. A line having slope -3 and vertical intercept $\left(0, \frac{3}{4}\right)$

$$f(x) = -3x + \tfrac{3}{4}$$

Graphing Linear Functions Using the Vertical Intercept and Slope

> **Example 1**   *Use the vertical (s-) intercept and slope to plot the linear function $s = 2t + 16$, from Problem 1, on a rectangular grid.*

SOLUTION

Locate and mark the s-intercept, $(0, 16)$.

Write the slope $m = 2$ as a fraction $\frac{2}{1}$, which indicates a move $\dfrac{up\ 2}{right\ 1}$ from the starting point $(0, 16)$ to the point $(1, 18)$. Mark this point.

Finally, use a straightedge to draw the line through $(0, 16)$ and $(1, 18)$.

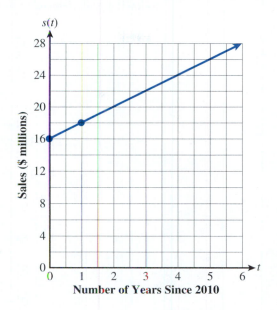

8. **a.** Start at $(0, 16)$ and explain how to use the slope, $m = 2$, to calculate the coordinates of the point $(1, 18)$ without actually moving on the graph.

   Starting at $(0, 16)$, add 1 to the horizontal coordinate, and then add 2 to the vertical coordinate.

   b. Interpret the practical meaning of the ordered pair $(1, 18)$ in terms of the total sales situation.

   $(1, 18)$ indicates that when $t = 1$, that is, in 2011, total sales are $18 million.

9. Use the slope once again to reach a third point on the line. Interpret the practical meaning of this new ordered pair in terms of the total sales situation.

   (Answers will vary.) Starting at $(1, 18)$, add 1 to the horizontal coordinate and add 2 to the vertical coordinate to reach the point $(2, 20)$. These coordinates indicate that when $t = 2$, that is, in 2012, total sales will be $20 million.

10. **a.** Use the slope to determine the change in total sales over any 6-year period.

    $\frac{2}{1} \cdot \frac{6}{6} = \frac{12}{6}$. Therefore, over any 6-year period, sales increase by $12 million.

    b. Use the result of part a to determine the coordinates of the point corresponding to the year 2016.

    2016 corresponds to $t = 6$. Starting at $(0, 16)$, add 6 to the horizontal coordinate and 12 to the vertical coordinate to obtain the point $(6, 28)$.

11. A start-up software company's sales were $6 million in 2010, and the company anticipates sales to increase by $3 million per year for the next several years.

    a. Write a symbolic rule to express the total sales, s in terms of the number of years, t, since 2010.

    $s = 3t + 6$

b. Use the vertical intercept and slope to draw the graph of the symbolic rule on the following grid. *Remember to first label the axes with appropriate scales.*

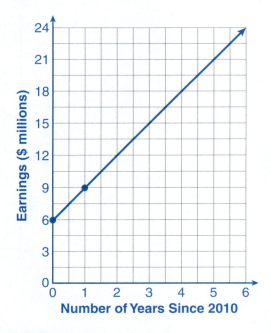

c. Use the slope to locate two more points on the line. Interpret the meanings of the two points you locate.

(Answers will vary.) Possible answers: (2, 12), In 2012, the company expects to earn $12 million. (4, 18), In 2014, the company anticipates earning $18 million.

Graphing Linear Functions Using Intercepts

Another way to graph a linear function is to plot its vertical and horizontal intercepts and then use a straightedge to draw the line containing these two points.

12. You have purchased a laptop computer so that you can use the software products you have acquired through your job. The initial cost of the computer is $1350. You expect that the computer will depreciate (lose value) at the rate of $450 per year.

a. Write a symbolic rule that will determine the value, $v(t)$, of the computer in terms of the number of years, t, that you own it.

$v(t) = -450t + 1350$

b. Write the ordered pairs that represent the vertical and horizontal intercepts.

i. To determine the vertical (v-) intercept, evaluate the expression in part a for $t = 0$ and record your result in the following table.

ii. To determine the horizontal (t-) intercept, solve the equation in part a for $v(t) = 0$, and record your result in the table.

INTERCEPTS	t, NUMBER OF YEARS	v, VALUE OF COMPUTER ($)
Vertical	0	1350
Horizontal	3	0

The vertical intercept is (0, 1350). The horizontal intercept is (3, 0).

c. On the following grid, plot the intercept points you determined in part b. Then use a straightedge to draw a straight line through the intercepts.

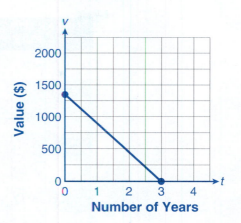

d. What is the practical meaning of the vertical (*v*-) intercept in this situation?

The vertical intercept indicates the value of the computer ($1350) at the time it was purchased.

e. What is the practical interpretation of the horizontal (*t*-) intercept in this situation?

The horizontal intercept indicates that 3 years after purchase, the resale value of your computer has decreased to zero.

f. What portion of the line in part c can be used to represent the computer value situation? (*Hint:* What is the practical domain of this function?)

The portion of the line from *t* = 0 to *t* = 3 represents the practical domain of the computer value function.

13. Determine the vertical and horizontal intercepts for each of the following. Then sketch a graph of the line using the intercepts. Use your graphing calculator to check your results.

a. $y = -3x + 6$

The vertical (*y*-) intercept is (0, 6).
The horizontal (*x*-) intercept is (2, 0).

b. $f(x) = \frac{1}{2}x - 8$

The vertical (*y*-) intercept is (0, −8).
The horizontal (*x*-) intercept is (16, 0)

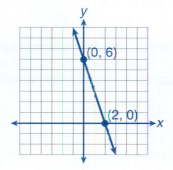

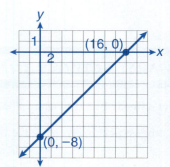

14. Identify the slope, vertical intercept, and horizontal intercept of each linear function in the following table.

LINEAR FUNCTION RULE	SLOPE	VERTICAL INTERCEPT	HORIZONTAL INTERCEPT
$y = 3x - 7$	3	$(0, -7)$	$\left(\frac{7}{3}, 0\right)$
$f(x) = -2x + 3$	-2	$(0, 3)$	$(1.5, 0)$
$y = 5x + 2$	5	$(0, 2)$	$(-0.4, 0)$
$y = 10x$	10	$(0, 0)$	$(0, 0)$
$y = 5$	0	$(0, 5)$	none
$g(x) = 12 + 4x$	4	$(0, 12)$	$(-3, 0)$
$v = 192 - 32t$	-32	$(0, 192)$	$(6, 0)$
$w = -25 + 4r$	4	$(0, -25)$	$(6.25, 0)$
$z = -200 - 8x$	-8	$(0, -200)$	$(-25, 0)$

15. Determine the intercepts and slope of the linear function having equation $3x + 4y = 12$.

$x = 0$ for the y-intercept

$$3(0) + 4y = 12$$
$$4y = 12$$
$$y = 3$$

Therefore, the y-intercept is $(0, 3)$.

$y = 0$ for the x-intercept

$$3x + 4(0) = 12$$
$$3x = 12$$
$$x = 4$$

Therefore, the x-intercept is $(4, 0)$.

The slope is $\dfrac{3 - 0}{0 - 4} = -\dfrac{3}{4}$.

SUMMARY: ACTIVITY 3.7

1. To **plot** a **linear function** using slope-intercept form:

- Plot the vertical intercept on the vertical axis.

- Write the slope in fractional form as $\dfrac{\text{change in output}}{\text{change in input}}$

- Start at the vertical intercept. Move up or down as many units as the numerator indicates, and then move to the right or left as many units as the denominator indicates. Mark the point you have reached.

- Use a straightedge to draw a line between the two points.

2. Given an equation of a line, determine its **y-intercept** by setting $x = 0$ and calculating the corresponding y-value.

3. Given an equation of a line, determine its **x-intercept** by setting $y = 0$ and calculating the corresponding x-value.

4. To plot a linear function using its intercepts:

- Determine the horizontal and vertical intercepts, and then use a straightedge to draw the line containing the two points.

EXERCISES: ACTIVITY 3.7

In Exercises 1–6, determine the slope, y-intercept, and x-intercept of each line. Then sketch each graph, labeling and verifying the coordinates of each intercept. Use your graphing calculator to check your results.

1. $y = 3x - 4$

Slope is 3.

y-intercept is $(0, -4)$.

x-intercept is $\left(\frac{4}{3}, 0\right)$.

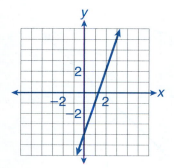

2. $y = -5x + 2$

Slope is -5.

y-intercept is $(0, 2)$.

x-intercept is $\left(\frac{2}{5}, 0\right)$ or $(0.4, 0)$.

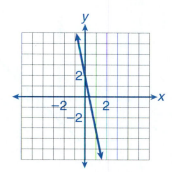

3. $y = 8$

Slope is 0.

y-intercept is $(0, 8)$.

There is no x-intercept.

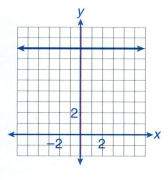

4. $y = \dfrac{x}{2} + 5$

Slope is $\frac{1}{2}$.

y-intercept is $(0, 5)$.

x-intercept is $(-10, 0)$.

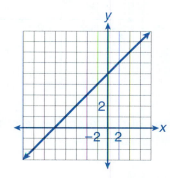

Exercise numbers appearing in color are answered in the Selected Answers appendix.

Hint: In Exercises 5 and 6, solve for *y* first.

5. $2x - y = 3$

$y = 2x - 3$

Slope is 2.

y-intercept is $(0, -3)$.

x-intercept is $\left(\frac{3}{2}, 0\right)$, or $(1.5, 0)$.

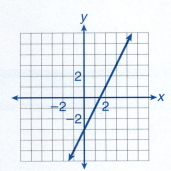

6. $3x + 2y = 1$

$y = -\frac{3}{2}x + \frac{1}{2}$

Slope is $-\frac{3}{2}$.

y-intercept is $\left(0, \frac{1}{2}\right)$.

x-intercept is $\left(\frac{1}{3}, 0\right)$.

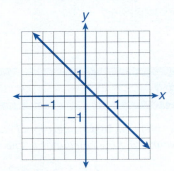

7. Graph the following linear functions in the order given. Use your graphing calculator to verify your answers. In what ways are the lines similar? In what ways are they different?

a. $y = x - 4$

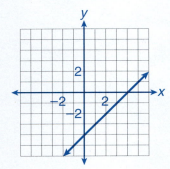

b. $f(x) = x - 2$

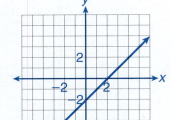

c. $g(x) = x$

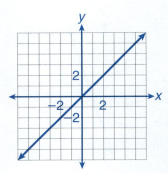

d. $y = x + 2$

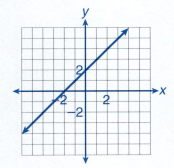

e. $y = x + 4$

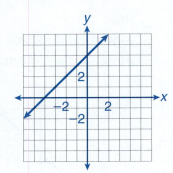

The lines all have the same slope but different *y*-intercepts.
They are parallel.

8. Graph the following linear functions in the order given. Use your graphing calculator to verify your answers. In what ways are the lines similar? In what ways are they different?

a. $y = -4x + 2$ **b.** $h(x) = -2x + 2$ **c.** $y = 2$

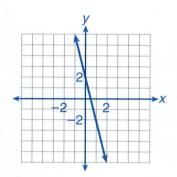   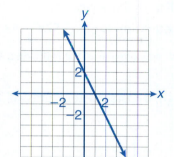   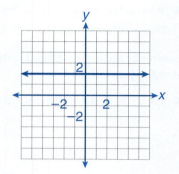

Line a has a steep negative slope.

Line b is less steep than line a but is still negative.

Line c is horizontal.

d. $g(x) = 2x + 2$ **e.** $y = 4x + 2$

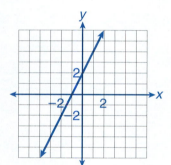   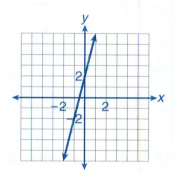

Line d has a positive slope but the same steepness as b.

The slope of line e is positive and is as steep as line a.

The lines all have the same y-intercept but different slopes.

9. What is the equation of the linear function with slope 12 and y-intercept $(0, 3)$?

$y = 12x + 3$

10. a. You start with $20 in your savings account and add $10 every week. At what rate does the amount in your account, excluding interest, change from week to week?

The amount in the account is increasing at a rate of $10 per week.

b. Write an equation that models your savings, $s(t)$, as a function of time, t (in weeks).

$s(t) = 20 + 10t$ dollars

11. a. What is the slope of the line that goes through the points $(0, 5)$ and $(2, 11)$?

$m = \dfrac{11 - 5}{2 - 0} = \dfrac{6}{2} = 3$

b. What is the equation (symbolic rule) of the line through these two points?

The y-intercept is $(0, 5)$. The equation is $y = 3x + 5$.

12. a. What is the slope of the line that goes through the points $(0, -43.5)$ and $(-1, 13.5)$?

$$m = \frac{13.5 - (-43.5)}{-1 - 0} = \frac{57}{-1} = -57$$

b. What is the equation of the line through these two points?

The y-intercept is $(0, -43.5)$. The equation is $y = -57x - 43.5$.

13. Determine the horizontal and vertical intercepts of each of the following. Use the intercepts to sketch a graph of the function.

a. $y = -3x + 12$ The x-intercept is $(4, 0)$; the y-intercept is $(0, 12)$.

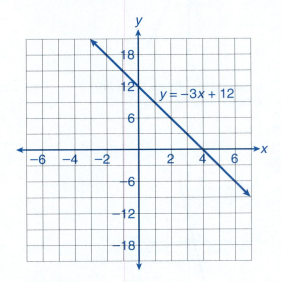

b. $y = \frac{1}{2}x + 6$ The x-intercept is $(-12, 0)$; the y-intercept is $(0, 6)$.

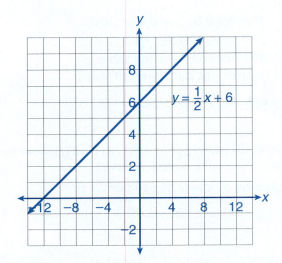

14. The average cost of a 30-second advertisement during the 1998 Super Bowl game was $1.3 million. In 2006, the average cost was $2.5 million. If x represents the number of years since 1998, the given data can be summarized as follows:

NUMBER OF YEARS SINCE 1998, x	COST OF A 30-SECOND AD, C (millions of dollars)
0	1.3
8	2.5

a. Assume that the average rate of increase in the cost of a 30-second advertisement remained constant from 1998 through 2006. Determine this rate. What characteristic of the line through $(0, 1.3)$ and $(8, 2.5)$ does this rate represent?

$$\frac{2.5 - 1.3}{8 - 0} = 0.15 \text{ million dollars per year} = \$150,000/\text{yr.}$$

The cost of a 30-second advertisement during the Super Bowl increased at a constant rate of $150,000 per year since 1998. This represents the slope of the line through $(0, 1.3)$ and $(8, 2.5)$.

b. Determine the equation of the line in part a.

$C = 0.15x + 1.3$

c. If this trend continues, what will be the cost of a 30-second ad during the 2010 Super Bowl?

For the year 2010, $x = 12$. $C = 0.15(12) + 1.3 = 3.1$

The cost of a 30-second ad during the 2010 Super Bowl will be $3,100,000 if the cost increases at the same constant rate.

15. After applying the brakes, a car traveling 60 miles per hour continues 120 feet before coming to a complete stop. This information is summarized in the following table.

Distance Traveled After Applying the Brakes, d (ft.)	0	120
Speed of the Car, v (mph)	60	0

a. Assume that the speed, v, of the car is a linear function of the distance, d, traveled after applying the brakes. Determine the slope of the line containing the points $(0, 60)$ and $(120, 0)$. What is the practical meaning of slope in this situation?

$$\frac{0 - 60}{120 - 0} = \frac{-60}{120} = -0.5$$

The slope is -0.5, indicating that the speed of the car is decreasing at a constant rate of 0.5 mph for each foot that it travels.

b. Determine the equation of the line in part a.

$v = -0.5d + 60$

c. Determine the speed of the car when it is 70 feet from where the brakes are applied.

$v = -0.5(70) + 60 = 25$ mph

The car is traveling at a speed of 25 miles per hour when it is 70 feet from where the brakes were applied.

Activity 3.8

Predicting Population

Objectives

1. Write an equation for a linear function given its slope and y-intercept.

2. Write linear functions in slope-intercept form, $y = mx + b$.

3. Interpret the slope and y-intercept of linear functions in contextual situations.

4. Use the slope-intercept form of linear equations to solve problems.

The United States Census Bureau keeps historical records on populations in the United States from the year 1790 onward. The bureau's records show that from 1940 to 1950, the yearly changes in the national population were quite close to being constant. Therefore, for the objectives of this activity, you may assume that the average rate of change of population with respect to time is a constant value in this decade. In other words, the relationship between time and population may be considered linear from 1940 to 1950.

1. a. According to the U.S. Bureau of the Census, the population of the United States was approximately 132 million in 1940 and 151 million in 1950. Write the data as ordered pairs of the form $(t, P(t))$, where t is the number of years since 1940 and $P(t)$ is the corresponding population, in millions.

(0, 132) (10, 151)

b. Plot the two data points, and draw a straight line through them. Label the horizontal axis from 0 to 25 and the vertical axis from 130 to 180, compressing the axis between 0 and 130.

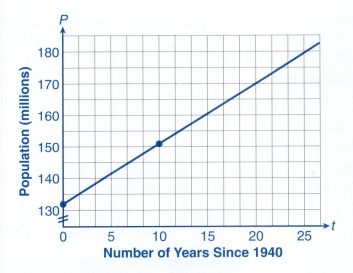

2. a. What is the average rate of change of population from $t = 0$ (1940) to $t = 10$ (1950)?

$$\frac{\Delta P}{\Delta t} = \frac{151 - 132}{10 - 0} = \frac{19 \text{ million}}{10 \text{ yr.}} = 1.9 \text{ million/yr.}$$

The average rate of change of population from 1940 to 1950 is 1.9 million people per year.

b. What is the slope of the line connecting the two points in part a? What is the practical meaning of the slope in this situation?

The slope is 1.9. Population increased at an average rate of 1.9 million people per year between 1940 and 1950.

3. What is the vertical intercept of this line? What is the practical meaning of the vertical intercept in this situation?

The vertical intercept is (0, 132). It indicates that in 1940, the population was 132 million.

4. a. Use the slope and vertical intercept from Problems 2 and 3 to write an equation for the line.

$P(t) = 1.9t + 132$

b. Assume that the average rate of change you determined in Problem 2 stays the same through 1960. Use the equation in part a to predict the U.S. population in 1960. Also estimate the population in 1960 from the graph.

$P(20) = 1.9(20) + 132 = 38 + 132 = 170$ million

The equation gives a population of 170 million in 1960, as does the graph.

5. You want to develop a population model based on more recent data. The U.S. population was approximately 249 million in 1990 and 281 million in 2000.

a. Plot these data points using ordered pairs of the form $(t, P(t))$, where t is the number of years since 1990 (now, $t = 0$ corresponds to 1990). Compress the vertical axis between 0 and 220. Draw a line through the points.

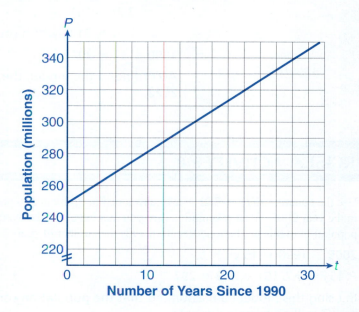

b. Determine the slope of the line in part a. What is the practical meaning of the slope in this situation? How does this slope compare with the slope in Problem 2?

$$m = \frac{281 - 249}{10 - 0} = \frac{32}{10} = 3.2 \text{ million/yr.}$$

The slope of 3.2 indicates that the population increased at an average rate of 3.2 million people per year between 1990 and 2000. The slope of this line is greater than the slope of the line in Problem 2.

c. In which decade, 1940–1950 or 1990–2000, did the U.S. population increase more rapidly? Explain your answer in terms of slope.

The U.S. population increased more rapidly between 1990 and 2000. The slope of the line segment for that period, 3.2 million per year, was greater than that between 1940 and 1950, when it was 1.9 million per year.

d. Determine the vertical intercept of the line in part a.

The vertical intercept is $(0, 249)$.

e. Write the equation of the line in part a.

$P(t) = 3.2t + 249$

6. a. Use the equation $P(t) = 3.2t + 249$, also called a linear model, developed in Problem 5 to predict the population in 2010. What assumptions do you make about the average rate of change of the population in this prediction?

$2010 - 1990 = 20$ yr.

$P(20) = 3.2(20) + 249 = 313$

The linear model yields a population of 313 million in 2010. In making this prediction, I assume that the population will grow at the same linear rate after 2000 as it did between 1990 and 2000.

b. According to the linear model $P(t) = 3.2t + 249$, in what year will the population be 350 million?

$3.2t + 249 = 350$

$3.2t = 101$

$t \approx 31.6$, which rounds up to 32

$1990 + 32 = 2022$

According to the model, the population will reach 350 million in about 2022.

EXERCISES: ACTIVITY 3.8

1. a. Use the equation $P(t) = 3.2t + 249$ developed in Problem 5 to predict the U.S. population in the year 2005. What assumptions are you making about the average rate of change of the population in this prediction? Recall that t is the number of years since 1990.

$2005 - 1990 = 15$ yr.

$P(15) = 3.2(15) + 249 = 297$

In using this model I am assuming that the population continued to grow by 3.2 million people per year.

b. The actual U.S. population in 2005 was approximately 296 million. How close was your prediction?

$297 - 296 = 1$ million

My prediction was 1 million more than the actual population.

c. What do you think was the cause of the prediction error?

I assumed that the population growth was linear, that is, it grew at a constant average rate of change.

2. a. According to the U.S. Bureau of the Census, the population of California in 2000 was approximately 34.10 million and was increasing at a rate of approximately 630,000 people per year. Let $P(t)$ represent the California population (in millions) and t represent the number of years since 2000. Complete the following table.

t	$P(t)$ (in millions)
0	34.10
1	34.73
2	35.36

Exercise numbers appearing in color are answered in the Selected Answers appendix.

b. What information in part a indicates that the California population growth is linear with respect to time? What are the slope and vertical intercept of the graph of the population data?

I can assume that the California population growth around 2000 is linear with respect to time because the U.S. Bureau of the Census indicated an approximately constant rate of growth (630,000 people per year).

630,000 = 0.630 million

The slope (rate of change) is 0.630 million per year.

The vertical intercept is (0, 34.10).

c. Write a linear function rule for $P(t)$ in terms of t.

$P(t) = 0.63t + 34.10$

d. Use the linear function in part c to estimate the population of California in 2004.

$2004 - 2000 = 4$

$P(4) = 0.63(4) + 34.10 = 36.62$ million

According to the model, the population of California was approximately 36,620,000 in 2004.

e. Use the linear model from part c to predict the population of California in 2010.

$2010 - 2000 = 10$

$P(10) = 0.63(10) + 34.10 = 40.4$ million

The model predicts the 2010 population of California will be 40,400,000.

3. a. The population of Atlanta, Georgia, was 2.96 million in 1990 and 4.11 million in 2000. This information is summarized in the accompanying table, where t is the number of years since 1990 and $P(t)$ represents the population (in millions) at a given time t.

t	$P(t)$
0	2.96
10	4.11

Assume that the average rate of change of the population over this 10-year period is constant. Determine this average rate.

$$\frac{\Delta P}{\Delta t} = \frac{4.11 - 2.96}{10 - 0} = \frac{1.15}{10} = 0.115 \text{ million/yr.}$$

b. Plot the two data points, and draw a line through them. Label the horizontal axis from 0 to 25 and the vertical axis from 0 to 8.

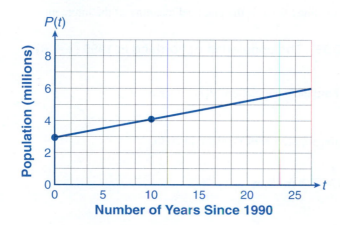

c. Determine the slope and vertical intercept of the line in part b.

The slope is 0.115. The vertical intercept is (0, 2.96).

d. Write a symbolic rule to model Atlanta's population, $P(t)$ (in millions), in terms of t.

$P(t) = 0.115t + 2.96$

e. Use this linear model to predict Atlanta's population in 2020.

$2020 - 1990 = 30$

$P(30) = 0.115(30) + 2.96 = 6.41$ million

4. a. The population of Portland, Oregon, was 2.39 million in 1990 and 2.36 million in 2000. If t represents the number of years since 1990 and $P(t)$ represents the population (in millions) at a given time, t, summarize the given information in the accompanying table.

t	$P(t)$
0	2.39
10	2.36

b. Plot the two data points on appropriately scaled and labeled coordinate axes. Draw a line connecting the points.

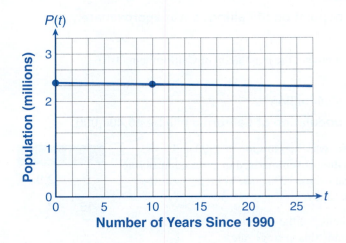

c. What is the slope of the line? What is the practical meaning of the slope in this situation?

$$m = \frac{2.36 - 2.39}{10 - 0} = \frac{-0.03}{10} = -0.003 \text{ million/yr.}$$

The population of Portland is decreasing at a rate of 0.003 million per year.

d. What is the vertical intercept of the line? What is the practical meaning of the intercept in this situation?

The vertical intercept is (0, 2.39). The population of Portland in 1990 ($t = 0$) was 2.39 million.

e. Write a symbolic rule to model Portland's population, $P(t)$, in terms of t.

$P(t) = -0.003t + 2.39$

f. Use this linear model to predict the population of Portland in 2020.

$2020 - 1990 = 30$

$P(30) = -0.003(30) + 2.39 = 2.3$ million

5. In each part, determine the equation of the line for the given information.

a. Two points on the line are $(0, 4)$ and $(7, 18)$. Use the points to first determine the slope and y-intercept. Then write the equation of the line.

$m = \dfrac{18 - 4}{7 - 0} = \dfrac{14}{7} = 2$; y-intercept is $(0, 4)$. The equation is $y = 2x + 4$.

b. The graph has y-intercept $(0, 6)$ and contains the point $(2, 1)$.

$m = \dfrac{1 - 6}{2 - 0} = \dfrac{-5}{2} = -\dfrac{5}{2}$. The equation is $y = -2.5x + 6$.

6. a. In 1990, the rate of change of the world population was approximately 0.09125 billion per year (or approximately 1 million people every 4 days). The world population was estimated to be 5.3 billion in 1990. Write a symbolic rule to model the population, P (in billions), in terms of t, where t is the number of years since 1990 ($t = 0$ corresponds to 1990).

$P(t) = 5.3 + 0.09125t$

b. Use the linear model to predict the world population in 2020.

$2020 - 1990 = 30$

$P(30) = 5.3 + 0.09125(30) = 8.0375$ billion

According to the model, in 2020 the world population will be 8,037,500,000.

c. According to the model, when will the population of the world be double the 1990 population?

$5.3 + 0.09125t = 10.6$

$0.09125t = 5.3$

$t \approx 58.08$

The population will double in approximately 58 years by 2048.

Cluster 2   **What Have I Learned?**

1. A line is given by the equation $y = -4x + 10$.

    a. Determine its x-intercept and y-intercept algebraically from the equation.

        Set $y = 0$ and solve for x.   Set $x = 0$ and solve for y.

$$-4x + 10 = 0 \qquad\qquad y = -4(0) + 10$$
$$-4x = -10 \qquad\qquad y = 0 + 10$$
$$x = 2.5 \qquad\qquad\qquad y = 10$$

        The x-intercept is $(2.5, 0)$. The y-intercept is $(0, 10)$.

     b. Use your graphing calculator to confirm these intercepts.

        From the graph to the left, you see that the x-intercept is $(2.5, 0)$ and the y-intercept is $(0, 10)$.

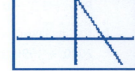

2. a. Does the slope of the line having the equation $4x + 2y = 3$ have a value of 4? Why or why not?

        No, because the equation is not in the form $y = mx + b$.

    b. Solve the equation in part a for y so that it is in the form $y = mx + b$.

$$4x + 2y = 3$$
$$2y = -4x + 3$$
$$y = -2x + 1.5$$

    c. What is the slope of the line?

        The slope is -2.

3. Explain the difference between a line with zero slope and a line with an undefined slope.

    A line with zero slope is horizontal and represents a function. A line with undefined slope is vertical and cannot represent a function.

4. Describe how you recognize that a function is linear when it is given

    a. graphically

        The graph is a straight line.

    b. symbolically

        It can be written in the form $y = mx + b$.

    c. numerically in a table

        The rate of change is constant.

5. Do vertical lines represent functions? Explain.

    No, vertical lines do not represent functions. Each point on the vertical line has the same input value but a different output value.

Cluster 2   How Can I Practice?

1. A function is linear because the average rate of change of the output with respect to the input from point to point is constant. Use this idea to determine the missing input (x) and output (y) values in each table, assuming that each table represents a linear function.

a.

x	y
1	4
2	5
3	6

$m = 1$

b.

x	y
1	4
3	8
5	12

$m = 2$

c.

x	y
0	4
5	9
10	14

$m = 1$

d.

x	y
−1	3
0	8
1	13
2	18

$m = 5$

e.

x	y
−3	11
0	8
3	5
6	2

$m = -1$

f.

x	y
−2	−5
0	−8
2	−11
4	−14

$m = -1.5$

g. Explain how you used the idea of constant average rate of change to determine the values in the tables.

(Answers will vary.)

The constant average rate of change is the amount that *y* changes for each increase of 1 in *x*. I used the first two pairs to find this rate of change and then determined how many of those to add to get to the next input.

2. The pitch of a roof is an example of slope in a practical setting. The roof slope is usually expressed as a ratio of rise over run. For example, in the building shown, the pitch is 6 to 24 or, in fraction form, $\frac{1}{4}$.

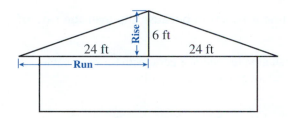

a. If a roof has a pitch of 5 to 16, how high will the roof rise over a 24-foot run?

$$\text{pitch} = \frac{\text{rise}}{\text{run}}$$

$$\frac{5}{16} = \frac{x}{24}$$

$$16x = 120$$

$$x = 7.5 \text{ ft.}$$

The roof will rise 7.5 feet.

b. If a roof's slope is 0.25, how high will the roof rise over a 16-foot run?

$$\frac{\text{rise}}{16} = \frac{25}{100}$$

$$100 \cdot \text{rise} = 400$$

$$\text{rise} = 4$$

$$= 4 \text{ ft.}$$

The roof will rise 4 feet.

c. What is the slope of a roof that rises 12 feet over a run of 30 feet?

$$\text{slope} = \frac{\text{rise}}{\text{run}} = \frac{12}{30} = \frac{2}{5}$$

The slope (pitch) of the roof is 2 to 5.

3. Determine whether any of the following tables contain input and output data that represent a linear function. In each case, give a reason for your answer.

a. You make an investment of $100 at 5% interest compounded semiannually. The following table represents the amount of money you will have at the end of each year.

TIME (yr.)	AMOUNT ($)
1	105.06
2	110.38
3	115.97
4	121.84

rate of change = 5.32, 5.59, and 5.87

No, the data does not represent a linear function because the average rate of change is not constant.

b. A cable-TV company charges a $45 installation fee and $28 per month for basic cable service. The table values represent the total usage cost since installation.

Number of Months	6	12	18	24	36
Total Cost ($)	213	381	549	717	1053

Yes, the data represents a linear function because the average rate of change is constant, $28 per month.

c. For a fee of $20 a month, you have unlimited video rental. Values in the table represent the relationship between the number of videos you rented each month and the monthly fee.

Number of Rentals	10	15	12	9	2
Cost ($)	20	20	20	20	20

Yes, the data represents a linear function because the average rate of change is constant, $0 per rental.

4. After stopping your car at a stop sign, you accelerate at a constant rate for a period of time. The speed of your car is a function of the time since you left the stop sign. The following table shows your speedometer reading each second for the next 7 seconds.

t, TIME (sec.)	s, SPEED (mph)
0	0
1	11
2	22
3	33
4	44
5	55
6	55
7	55

a. Graph the data by plotting the ordered pairs of the form (t, s) and then connecting the points.

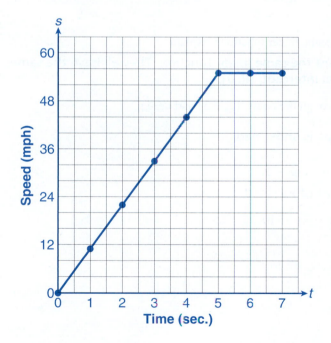

b. For what values of *t* is the graph increasing?

The graph is increasing for values of *t* from 0 through 5 seconds.

c. What is the slope of the line segment during the period of acceleration?

The slope of this line segment is 11 miles per hour per second.

d. What is the practical meaning of the slope in this situation?

For each second that passes during the first 5 seconds, speed increases 11 miles per hour.

e. For what values of *t* is the speed a constant? What is the slope of the line connecting the points of constant speed?

The speed is constant for values of *t* from 5 through 7 seconds. The slope of the line segment connecting the points of constant speed is zero.

5. a. The three lines shown in the following graphs appear to be different. Calculate the slope of each line.

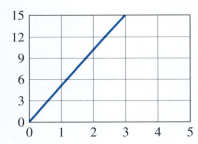
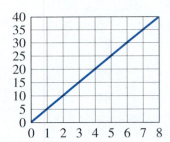
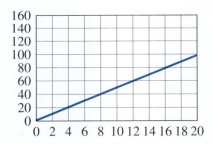

$$m = \frac{15 - 0}{3 - 0}$$

$$= \frac{15}{3}$$

$$= 5$$

$$m = \frac{40 - 0}{8 - 0}$$

$$= \frac{40}{8}$$

$$= 5$$

$$m = \frac{100 - 0}{20 - 0}$$

$$= \frac{100}{20}$$

$$= 5$$

b. Do the three graphs represent the same linear function? Explain.

Yes; the three graphs represent the same linear function. They all have the same slope, 5, and the same vertical intercept, $(0, 0)$.

6. a. Determine the slope of the line through the points $(2, -5)$ and $(2, 4)$.

$$m = \frac{4 - (-5)}{2 - 2} = \frac{9}{0}$$ The slope is undefined.

b. Determine the slope of the line $y = -3x - 2$.

The slope is -3, the coefficient of *x*.

c. Determine the slope of the line $2x - 4y = 10$.

$$-4y = -2x + 10$$

$$y = 0.5x - 2.5$$

The slope is 0.5.

d. Determine the slope of the line from the following graph.

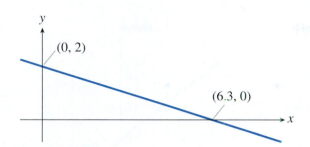

To get from $(0, 2)$ to $(6.3, 0)$, go down 2 and to the right 6.3.

$$\text{slope} = \frac{\text{rise}}{\text{run}} = \frac{-2}{+6.3} = \frac{-20}{63} \approx -0.317$$

7. Determine the vertical and horizontal intercepts for the graph of each of the following.

a. $y = 2x - 6$

$y = 2(0) - 6$

$y = 0 - 6$

The vertical intercept is $(0, -6)$.

$2x - 6 = 0$

$2x = 6$

$x = 3$

The horizontal intercept is $(3, 0)$.

b. $y = -\dfrac{3}{2}x + 10$

$y = -\frac{3}{2}(0) + 10$

$y = 10$

The vertical intercept is $(0, 10)$.

$-\frac{3}{2}x + 10 = 0$

$-\frac{3}{2}x = -10$

$x = -10\left(-\frac{2}{3}\right)$

$x = \frac{20}{3}$

The horizontal intercept is $\left(\frac{20}{3}, 0\right)$.

c. $y = 10$

This is a horizontal line through 10. The vertical intercept is $(0, 10)$. There is no horizontal intercept.

8. Determine the equation of each line.

a. The line passes through the points $(2, 0)$ and $(0, -5)$.

The vertical intercept is $(0, -5)$.

$$m = \frac{-5 - 0}{0 - 2} = \frac{-5}{-2} = 2.5$$

The equation is $y = 2.5x - 5$.

b. The slope is 7, and the line passes through the point $\left(0, \frac{1}{2}\right)$.

The equation is $y = 7x + \frac{1}{2}$.

c. The slope is 0, and the line passes through the point $(2, -4)$.

The equation is $y = -4$.

9. Sketch a graph of each of the following. Use your graphing calculator to verify your graphs.

a. $y = 3x - 6$ **b.** $f(x) = -2x + 10$

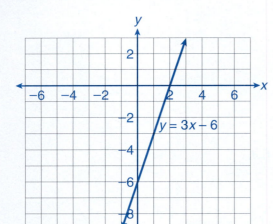   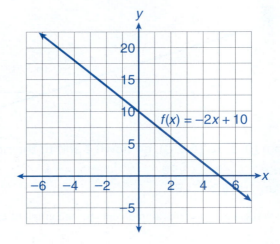

 10. Write each equation in slope-intercept form to discover what the graphs have in common. Use your graphing calculator to verify your graphs.

a. $y = 3x - 4$ **b.** $y - 3x = 6$ **c.** $y = 3x$
$y = 3x - 4$ $y = 3x + 6$

c. $3x - y = 0$
$y = 3x + 0$
The lines are parallel; they all have the same slope, 3.

 11. Write each equation in slope-intercept form to discover what the graphs have in common. Use your graphing calculator to verify your graphs.

a. $y = -2$ **b.** $y - 3x = -2$ **b.** $y = 3x - 2$ **c.** $y = x - 2$
$y = 0x - 2$ $y = 3x - 2$

c. $x = y + 2$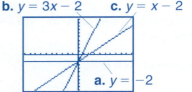
$y = x - 2$
The lines all have the same vertical intercept $(0, -2)$.

12. a. Complete the following table by listing four points that are contained on the line $x = 3$.

x	3	3	3	3
y	-4	0	2	5

(Answers will vary.)

b. What is the slope of the line in part a?
The slope is undefined.

c. Determine the vertical and horizontal intercepts, if any, of the graph of the line in part a.
The horizontal intercept is $(3, 0)$. There is no vertical intercept.

d. Does the graph of the line in part a represent a function? Explain.
The graph of the line in part a does *not* represent a function. The input 3 is assigned more than one (infinitely many) output.

Cluster 3

Problem Solving with Linear Functions

Activity 3.9

Housing Prices

Objectives

1. Determine the slope and
y-intercept of a line
algebraically and
graphically.

2. Determine the equation
for a linear function when
given two points.

3. Interpret the slope and
y-intercept of a linear
function in contextual
situations.

Despite the decrease in housing prices in recent years, there has been a steady increase in housing prices in your neighborhood since 2002. The house across the street sold for $125,000 in 2005 and then sold again in 2009 for $150,000. This data can be written in a table, where the input, x, represents the number of years since 2002 and the output, y, represents the sale price of a typical house in your neighborhood.

NUMBER OF YEARS SINCE 2002, x	HOUSING PRICE (thousands of $), y
3	125
7	150

1. Plot the two points on the grid below, and sketch the line containing them. Extend the line so that it intersects the vertical axis. Scale the input axis to include the period of years from 2002 to 2012 (from x = 0 to x = 10). Scale the output axis by increments of 25, starting at 0 and continuing through 250.

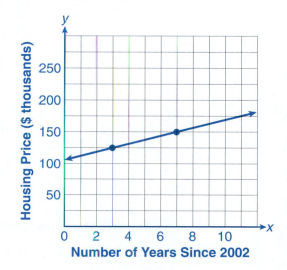

2. a. Use the points in the table to determine the slope of the line. What are its units of measurement?

$$\text{slope} = \frac{150 - 125}{7 - 3} = \frac{25}{4} = 6.25$$

It is measured in thousands of dollars per year.

b. What is the practical meaning of the slope in this situation?

A slope of 6.25 indicates that housing prices have been increasing an average of $6250 each year since 2002.

3. Estimate the y-intercept from the graph. What is the practical meaning of the y-intercept in this situation?

(Answers will vary somewhat between 100 and 115.)

The practical meaning is that in 2002, a typical house in the neighborhood sold for approximately $110,000.

4. Use the slope and your estimate of the *y*-intercept to write a linear function rule for housing price, *y*, in terms of *x*, the number of years since 2002.

(Answers will vary somewhat, depending on the estimate of the *y*-intercept.)

y = 6.25x + 110

5. Test the accuracy of the function rule you determined in Problem 4 by checking whether the coordinates of each plotted point satisfy the equation.

Check (3, 125); that is, replace *x* with 3 and *y* with 125 in the rule.

The expression on the right becomes 6.25 · 3 + 110 = 128.75. The expression on the left is *y* = 125. They are close, but not equal.

Check (7, 150): 6.25 · 7 + 110 = 153.75, which is close to but does not equal 150.

6. **a.** Starting at the point $(3, 125)$, use the slope $\dfrac{6.25}{1}$ to determine the coordinates of the vertical intercept.

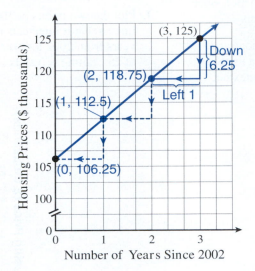

Number of Years Since 2002

To reach the *y*-intercept from the point (3, 125), I must move down and to the left. I can write the slope $\dfrac{\Delta y}{\Delta x} = \dfrac{6.25}{1}$ as $\dfrac{\Delta y}{\Delta x} = \dfrac{-6.25}{-1}$ so that each time I move down 6.25 units from a point on the line, I must move left 1 unit to get back on the line. Here I move down and to the left from (3, 125) to (2, 118.75), then in the same way to (1, 112.5), and finally, to (0, 106.25). The vertical intercept is (0, 106.25).

b. How does the result in part a compare to the estimate of the *y*-intercept you obtained in Problem 3?

(Answers will vary.)

You can also use a straightforward algebraic method to determine the exact *y*-intercept when you know two other points on the line.

Example 1 demonstrates this method.

Example 1   *Determine the y-intercept of the line containing the points (3, 125) and (7, 150) from Problem 1.*

Step 1. Determine the slope of the line. Referring to Problem 2, the slope is

$$m = \frac{150 - 125}{7 - 3} = \frac{25}{4} = 6.25.$$

Step 2. Substitute 6.25 for m in the slope-intercept form $y = mx + b$:
$y = 6.25x + b$.

Step 3. Choose one of the given points and replace the variables x and y in the rule $y = 6.25x + b$ with the coordinates of the point. Choosing the point $(3, 125)$, rewrite the equation as follows:

$$125 = 6.25 \cdot 3 + b$$

Step 4. Solve the equation for b.

$$125 = 6.25(3) + b$$
$$125 = 18.75 + b$$
$$125 - 18.75 = b$$
$$106.25 = b$$

The y-intercept is $(0, 106.25)$. The linear rule can now be written as $y = 6.25x + 106.25$.

7. a. In step 3 of the procedure outlined in Example 1, you can substitute the coordinates of either point into the equation $y = 6.25x + b$. Explain why this is possible.

Both points are on the line, so they both must satisfy the equation.

b. Use the ordered pair $(7, 150)$ in steps 3 and 4 in the method described in Example 1 to determine the value of b.

150 = 6.25(7) + b

b = 106.25

8. Summarize the algebraic procedure for determining an equation of a line from two points, neither of which is the y-intercept. Illustrate your step-by-step procedure using the points $(15, 62)$ and $(21, 80)$.

Calculate the slope: $m = \dfrac{80 - 62}{21 - 15} = \dfrac{18}{6} = 3$.

Set up the linear symbolic rule with b still unknown: $y = 3x + b$.

Substitute the coordinates of either point into the rule, and solve the resulting equation for b: $62 = 3(15) + b$; so $b = 17$.

Substitute this value for b into the linear function rule: $y = 3x + 17$.

Check (optional): Test the coordinates of the other point in the new function rule determined in the previous step: $x = 21, y = 80$. Does $80 = 3(21) + 17$? Yes.

There is one other very useful way to determine the equation of a line given its slope m and a point other than its y-intercept. This point is customarily denoted (x_1, y_1), where x_1 and y_1 should be understood to represent specific values, unlike x and y, which are understood to be variables.

Definition

The equation of the line with slope m containing point (x_1, y_1) can be written as

$$y - y_1 = m(x - x_1)$$

and is called the **point-slope** form of the equation of a line.

Example 2

a. Use the point-slope form to determine an equation of the line containing the points $(3, 8)$ and $(7, 24)$.

b. Rewrite your equation in slope-intercept form.

SOLUTION

a. The slope m of the line containing $(3, 8)$ and $(7, 24)$ is $m = \dfrac{24 - 8}{7 - 3} = \dfrac{16}{4} = 4$.

Choose either given point as (x_1, y_1). Here, $x_1 = 3$ and $y_1 = 8$. The point-slope equation then becomes

$$y - 8 = 4(x - 3).$$

b. This equation can be rewritten in slope intercept form by expanding the expression on the right side and then solving for y.

$$y - 8 = 4(x - 3)$$
$$y - 8 = 4x - 12 \quad \text{using the distributive property}$$
$$y = 4x - 4 \quad \text{adding 8 to both sides}$$

9. a. Use the point-slope form to determine an equation of the line containing the points $(15, 62)$ and $(21, 80)$.

Slope of line containing $(15, 62)$ and $(21, 80)$ is $m = 3$.

Choose either point as (x_1, y_1). The two possible equations are:

$$y - 62 = 3(x - 15) \quad \text{or} \quad y - 80 = 3(x - 21)$$

b. Rewrite your equation in part a in slope-intercept form and compare with your results from Problem 8.

$$y - 62 = 3(x - 15) \quad \text{or} \quad y - 80 = 3(x - 21)$$
$$y - 62 = 3x - 45 \qquad\qquad y - 80 = 3x - 63$$
$$y = 3x + 17 \qquad\qquad y = 3x + 17$$

10. The basal energy requirement is the daily number of calories that a person needs to maintain basic life processes. For a 20-year-old male who weighs 75 kilograms and is 190.5 centimeters tall, the basal energy requirement is 1952 calories. If his weight increases to 95 kilograms, he will require 2226 calories.

The given information is summarized in the following table.

20-YEAR-OLD MALE, 190.5 CENTIMETERS TALL		
w, Weight (kg)	75	95
B, Basal Energy Requirement (cal.)	1952	2226

a. Assume that the basal energy requirement, B, is a linear function of weight, w, for a 20-year-old male who is 190.5 centimeters tall. Determine the slope of the line containing the two points indicated in the table above.

$$\text{slope} = \frac{2226 - 1952}{95 - 75} = \frac{274}{20} = 13.7$$

b. What is the practical meaning of the slope in the context of this situation?

For each additional 1-kilogram increase in weight, a 20-year-old, 190.5-centimeter-tall male has an increase of 13.7 additional calories in his basal energy requirement.

c. Determine a symbolic rule that expresses B in terms of w for a 20-year-old, 190.5-centimeter-tall male.

$2226 = 13.7(95) + b$ $\qquad\qquad$ $B - B_1 = m(w - w_1)$

$2226 = 1301.5 + b$ $\qquad\qquad$ $B - 1952 = 13.7(w - 75)$

$924.5 = b$ $\qquad\qquad\qquad\qquad$ $B = 13.7w + 924.5$

$\qquad B = mw + b$

$\qquad\quad = 13.7w + 924.5$

The symbolic rule $B = 13.7w + 924.5$ expresses the basal energy rate, B, for a 20-year-old, 190.5-centimeter-tall male in terms of his weight, w, in kilograms.

d. Does the B-intercept have any practical meaning in this situation? Determine the practical domain of the basal energy function.

The B-intercept has no practical meaning in this situation because it would indicate a weight of 0 kilograms. A possible practical domain is a set of weights from 55 to 182 kilograms.

SUMMARY: ACTIVITY 3.9

1. To determine the equation of a line, $y = mx + b$, given two points on the line:

Step 1. Determine the slope, m.

Step 2. Substitute the value of m into $y = mx + b$, where b is still unknown.

Step 3. Substitute the coordinates of one of the known points for x and y in the equation in step 2.

Step 4. Solve the equation for b to obtain the y-intercept of the line.

Step 5. Substitute the values for m and b into $y = mx + b$.

2. To determine the point-slope equation of a line, $y - y_1 = m(x - x_1)$, given two points on the line:

Step 1. Determine the slope, m.

Step 2. Choose any one of the given points as (x_1, y_1).

Step 3. Replace m, x_1, and y_1 by their numerical values in $y - y_1 = m(x - x_1)$.

EXERCISES: ACTIVITY 3.9

1. Federal income tax paid by an individual single taxpayer is a function of taxable income. For a recent year, the federal tax for various taxable incomes is given in the following table:

i, Taxable Income ($)	15,000	16,500	18,000	19,500	21,000	22,500	24,000
t, Federal Tax ($)	1,889	2,114	2,339	2,564	2,789	3,014	3,239

a. Plot the data points, with taxable income i as input and tax t as output. Scale the input axis from $0 to $24,000 and the output axis from $0 to $4000. Explain why the relationship is linear.

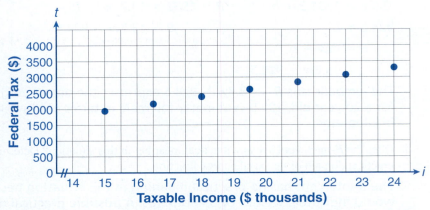

The data points on the graph lie on a straight line, which indicates that the relationship between taxable income and federal tax is linear. Therefore, the tax rate for this income range is constant.

b. Determine the slope of the line. What is the practical meaning of the slope?

$$\text{slope} = \frac{\Delta t}{\Delta i} = \frac{2,339 - 1,889}{18,000 - 15,000} = \frac{450}{3,000} = 0.15$$

The tax rate at this income level is 15%.

c. Write an equation to model this situation. Use the variable i to represent the taxable income and the variable t to represent the federal tax owed.

$t = 0.15i - 361$

d. What is the t-intercept? Does it make sense?

The t-intercept is $(0, -361)$.

The negative amount $-$361$ could mean the IRS would give $361 to an individual with no taxable income.

e. Use the equation from part c to determine the federal tax owed by a college student having a taxable income of $8600.

$t = 0.15(8600) - 361$

$t = \$929$

A college student with a taxable income of $8600 would pay $929 federal tax.

f. Use the equation from part c to determine the taxable income of a single person who paid $1686 in federal taxes.

$$1686 = 0.15i - 361$$
$$2047 = 0.15i$$
$$13{,}647 \approx i$$

A single person who paid $1686 in federal taxes had a taxable income of $13,647.

In Exercises 2–8, determine the equation of the line that has the given slope and passes through the given point. Then sketch a graph of the line.

2. $m = 3$, through the point $(2, 6)$

The equation is $y = 3x$.

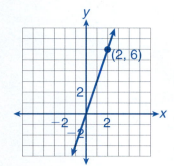

3. $m = -1$, through the point $(5, 0)$

The equation is $y = -x + 5$.

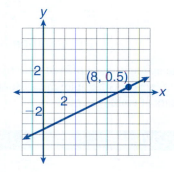

4. $m = 7$, through the point $(-3, -5)$

The equation is $y = 7x + 16$.

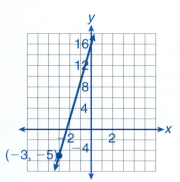

5. $m = 0.5$, through the point $(8, 0.5)$

The equation is $y = 0.5x - 3.5$.

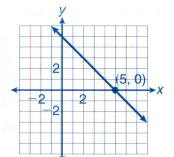

6. $m = 0$, through the point $(5, 2)$

The equation is $y = 2$.

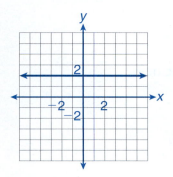

7. $m = -4.2$, through the point $(-4, 6.8)$

The equation is $y = -4.2x - 10$.

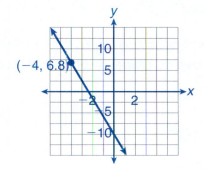

8. $m = -\dfrac{2}{7}$, through the point $\left(5, \dfrac{4}{7}\right)$

The equation is $y = -\dfrac{2}{7}x + 2$.

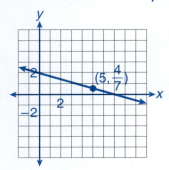

In Exercises 9–14, determine the equation of the line that passes through the given points.

9. $(2, 6)$ and $(4, 16)$

$m = \dfrac{16 - 6}{4 - 2} = \dfrac{10}{2} = 5$

The equation is $y = 5x - 4$.

10. $(-5, 10)$ and $(5, -10)$

$m = \dfrac{-10 - 10}{5 - (-5)} = \dfrac{-20}{10} = -2$

The equation is $y = -2x$.

11. $(3, 18)$ and $(8, 33)$

$m = \dfrac{33 - 18}{8 - 3} = \dfrac{15}{5} = 3$

The equation is $y = 3x + 9$.

12. $(0, 6)$ and $(-10, 0)$

$m = \dfrac{0 - 6}{-10 - 0} = \dfrac{-6}{-10} = \dfrac{3}{5}$

The equation is $y = \frac{3}{5}x + 6$.

13. $(10, 2)$ and $(-3, 2)$

$m = \dfrac{2 - 2}{-3 - 10} = \dfrac{0}{-13} = 0$

The equation is $y = 2$.

14. $(3.5, 8.2)$ and $(2, 7.3)$

$m = \dfrac{8.2 - 7.3}{3.5 - 2} = \dfrac{0.9}{1.5} = 0.6$

The equation is $y = 0.6x + 6.1$.

15. You have just graduated from college and have been offered your first job. The following table gives the salary schedule for the first few years of employment. Bonuses are not included. Let s represent your salary after x years of employment.

YEARS OF EMPLOYMENT, x	SALARY, s ($)
0	32,500
1	34,125
2	35,750
3	37,375

a. Is the salary a linear function of the years of employment? Explain.

Yes, the average rate of change (slope) is constant from one year to the next.

$m = \dfrac{34,125 - 32,500}{1 - 0} = 1625$

$m = \dfrac{35,750 - 34,125}{2 - 1} = 1625$

b. What is the practical domain for the input variable *x*?

(Answers may vary considerably.) {0, 1, 2, ..., 30}

c. Scale and label the axes appropriately and plot the points from the table on the following grid.

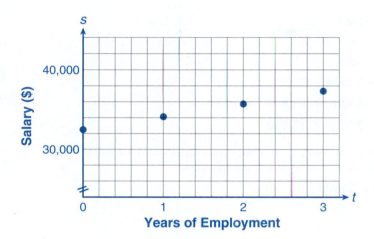

d. Determine the slope of the line containing the points.

The slope is 1625.

e. What is the practical meaning of the slope in this situation?

The slope is the rate of increase in salary, $1625 per year.

f. What is the *s*-intercept? What is the practical meaning of the intercept in this situation?

The *s*-intercept is $32,500, the starting salary after 0 years of employment.

g. Write a symbolic rule to determine the salary, *s*, after *x* years of employment.

s = 32,500 + 1625*x*

h. Assume that the rate of increase in your salary remains the same. Use the symbolic rule to determine your salary after 8 years of employment.

s = 32,500 + 1625(8) = 32,500 + 13,000 = $45,500

16. A boat departs from a marina and travels so that its distance from the marina is a linear function of time. The table below displays two ordered pairs of this function.

t (hr.)	*d* (mi.)
2	75
4	145

a. Determine the slope of the line. What is the practical meaning of slope in this situation?

$m = \dfrac{145 - 75}{4 - 2} = \dfrac{70}{2} = 35$; The boat traveled at an average speed of 35 miles per hour.

b. Write the equation of the line in slope-intercept form.

d − 75 = 35(*t* − 2), *d* − 75 = 35*t* − 70,

d − 75 + 75 = 35*t* − 70 + 75, *d* = 35*t* + 5

17. Straight-line depreciation helps spread the cost of new equipment over a number of years. The value of your company's copy machine after 1 year will be $14,700 and after 4 years will be $4800.

a. Write a linear function that will determine the value of the copy machine for any specified year.

$$m = \frac{4800 - 14{,}700}{4 - 1} = \frac{-9900}{3} = -3300$$

$$y - 14{,}700 = -3300(x - 1)$$

$$y - 14{,}700 = -3300x + 3300$$

$$y - 14{,}700 + 14{,}700 = -3300x + 3300 + 14{,}700$$

$$y = -3300x + 18{,}000$$

b. The salvage value is the value of the equipment when it gets replaced. What will be the salvage value of the copier if you plan to replace it after 5 years?

$$y = -3300(5) + 18{,}000 = \$1500$$

Project Activity 3.10

Oxygen for Fish

Objectives

1. Construct scatterplots from sets of data.

2. Recognize when patterns of points in a scatterplot are approximately linear.

3. Estimate and draw a line of best fit through a set of points in a scatterplot.

4. Use a graphing calculator to determine a line of best fit by the least-squares method.

5. Estimate the error of representing a set of data by a line of best fit.

Fish need oxygen to live, just as you do. The amount of dissolved oxygen in water is measred in parts per million (ppm). Trout need a minimum of 6 ppm to live. There are many variables that affect the amount of dissolved oxygen in a stream. One very important variable is the water temperature. To investigate the effect of temperature on dissolved oxygen, you take a water sample from a stream and measure the dissolved oxygen as you heat the water. Your results are as follows.

t, Temperature (°C)	11	16	21	26	31
d, Dissolved Oxygen (ppm)	10.2	8.6	7.7	7.0	6.4

1. Plot the data points as ordered pairs of the form (t, d). Scale your input axis from 0 to 40°C and your output axis from 0 to 15 ppm. Recall that the resulting graph of points is called a **scatterplot**.

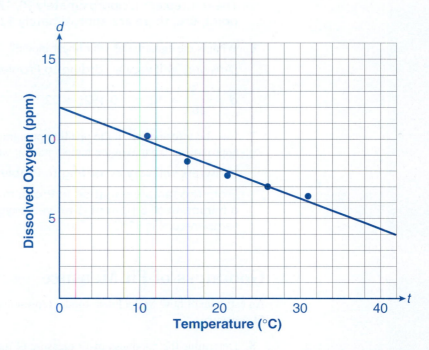

Instructor note: Just points are plotted for Problem 1; the line is added in Problem 3. Lines will vary.

2. Does there appear to be a linear relationship between the temperature, t, and the amount of dissolved oxygen, d? Is there an exact linear fit? That is, do all the points lie on the same line?

   The relationship between temperature and the amount of dissolved oxygen appears somewhat, but not exactly, linear.

3. Use a straightedge to draw a single line that you believe best represents the linear trend in the data. The resulting line is commonly called a **line of best fit**.

   (Answers will vary.)

Informally drawing a line, called the "eyeball" method, is one way to estimate a line of best fit. This line and its symbolic rule are called **linear models** for the given set of data.

4. Use two points (preferably, as far apart as possible) on the line to estimate the slope of the line. What is the practical meaning of the slope in this situation?

(Answers will vary according to lines drawn.)

Using $(0, 12)$ and $(26, 7)$,

$$m = \frac{7 - 12}{26 - 0} = \frac{-5}{26} \approx -0.192$$

The slope represents the change in ppm of dissolved oxygen for each degree increase in temperature. That is, the dissolved oxygen content decreases approximately 0.19 ppm for each $+1°C$.

5. What is the d-intercept of this line? Does this point have any practical meaning in this situation?

(Answers will vary according to lines drawn.)

The d-intercept is approximately $(0, 12)$. This indicates that at freezing point, $0°C$, there are approximately 12 ppm of dissolved oxygen.

6. What is the equation of your linear model?

(Answers will vary according to Problems 4 and 5.)

$$d = \frac{-5}{26}t + 12 \approx -0.192t + 12$$

7. Use the information in the opening paragraph of this activity and your linear model to approximate the maximum temperature at which trout can survive.

(Answers will vary according to Problem 6.)

From the sample equation in Problem 6 and the fact that trout need a minimum of 6 ppm of dissolved oxygen to live, the maximum temperature is $31°C$

Goodness-of-Fit Measure

An estimate of how well a linear model represents a given set of data is called a **goodness-of-fit measure**.

8. Determine the goodness-of-fit measure of the linear model from Problem 6.

Step 1.   Use the linear rule you derived in Problem 6 to complete the following table.

(Answers will vary according to Problem 6.)

t, INPUT	ACTUAL OUTPUT	d, MODEL'S OUTPUT	ACTUAL VALUE − MODEL VALUE	|ACTUAL VALUE − MODEL VALUE|
11	10.2	9.9	0.3	0.3
16	8.6	8.9	−0.3	0.3
21	7.7	8.0	−0.3	0.3
26	7.0	7.0	0	0
31	6.4	6.0	0.4	0.4

Step 2.   Determine the sum of the absolute values of the differences in the last column. This sum is called the **error** or **goodness-of-fit measure**. The smaller the error, the better the fit.

(Answers will vary according to Problem 6.)

$0.3 + 0.3 + 0.3 + 0 + 0.4 = 1.3$

Regression Line

Appendix

The method of least squares is a statistical procedure for determining a line of best fit from a set of data pairs. This method produces an equation of a line, called a **regression line**. Your graphing calculator uses this procedure to obtain the equation of a regression line. Appendix D shows you how to use the TI-83/TI-84 Plus to determine the equation of a regression line for a set of data pairs.

```
LinReg
y=ax+b
a=-.184
b=11.844
r²=.9566003617
r=-.9780594878
```

9. Use your graphing calculator's statistics (STAT) menu to determine the equation for the regression line in this situation.

The regression line is $y = -0.184x + 11.844$

(or $d = -0.184t + 11.844$).

10. **a.** Determine the goodness-of-fit measure for the least-squares regression line in Problem 9.

t, INPUT	ACTUAL OUTPUT	d, MODEL'S OUTPUT	ACTUAL VALUE − MODEL VALUE	|ACTUAL VALUE − MODEL VALUE|
11	10.2	9.82	0.38	0.38
16	8.6	8.9	−0.3	0.3
21	7.7	7.98	−0.28	0.28
26	7.0	7.06	−0.06	0.06
31	6.4	6.14	0.26	0.26

The goodness of fit measure is 1.28.

b. Compare the error of your line of best fit with the error of the least-squares regression line.

(Answers will vary.) My line has a larger error, 1.3, than the regression line, 1.28. The regression line is a better fit.

11. The number of Internet users in the United States increased steadily from 2000 to 2005, as indicated in the following table.

YEAR	NUMBER OF INTERNET USERS IN U.S. (Millions)
2000	121
2001	127
2002	140
2003	146
2004	156
2005	163

a. Plot the data points on an appropriately scaled and labeled coordinate axis. Let *x* represent the number of years since 2000.

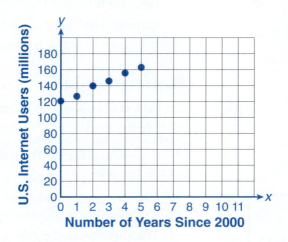

b. Use your graphing calculator's statistics menu (STAT) to determine the equation of the line that best fits the data (the regression line).

$y = 8.66x + 120.52$

c. What is the slope of the line in part b? What is the practical meaning of the slope?

The slope of the line is 8.66, indicating that every year after 2000, the number of Internet users increased by approximately 8.66 million.

d. Use the linear model from part b to predict when the number of Internet users in the United States will reach 200 million.

$200 = 8.66 \cdot x + 120.52$

$79.48 = 8.66x$

$9.18 \approx x$ Year: $2000 + 9 = 2009$

According to the model, the number of Internet users in the United States reached 200 million in 2009.

SUMMARY: ACTIVITY 3.10

1. A **line of best fit** is a line used to represent the general linear trend of a set of nonlinear data. This line and its equation form a linear model for the given set of data.

2. A **goodness-of-fit measure** is an estimate of how well a linear model represents a given set of data.

3. Graphing calculators and computer software use the **method of least squares** to determine a particular line of best fit, called a **regression line**.

EXERCISES: ACTIVITY 3.10

1. During the spring and summer, a concession stand at a community Little League baseball field sells soft drinks and other refreshments. To prepare for the season, the concession owner refers to the previous year's files, in which he had recorded the daily soft-drink sales (in gallons) and the average daily temperature (in degrees Fahrenheit). The data is shown in the table.

TEMPERATURE (°F), t	SOFT-DRINK SALES (gal.), g
52	35
55	42
61	50
66	53
72	66
75	68
77	72
84	80
90	84
94	91
97	95

a. Plot the data points as ordered pairs of the form (t, g).

Instructor's Note: For Exercise 1, plot points only. The line is drawn in Exercise 1c and will vary.

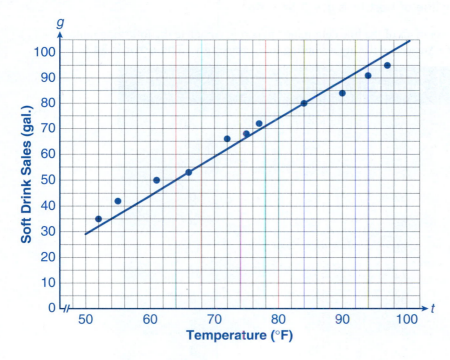

b. Does there appear to be a linear trend between the temperature, t, and the soft-drink sales, g? Is there an exact linear fit?

Yes, there appears to be a somewhat but not exactly linear relationship between the temperature and the soft-drink sales.

c. Use a straightedge to draw a line of best fit that you believe best represents the linear trend in the data.   See graph in Exercise 1.

d. Use the coordinates of two points on your line to determine the slope of your line of best fit.

(Answers will vary.)

Using $(84, 80)$ and $(74, 65)$,

$$m = \frac{65 - 80}{74 - 84} = \frac{-15}{-10} = 1.5.$$

The slope is 1.5.

e. What is the practical meaning of the slope in this situation?

(Answers will vary.) The slope indicates that the amount of soda sold increases 1.5 gallons for each increase of 1°F.

f. What is the g-intercept of your line of best fit? Does this number have any practical meaning in this situation?

(Answers will vary.)

$$80 = 1.5(84) + b$$
$$80 = 126 + b$$
$$-46 = b$$

The g-intercept is $(0, -46)$. The vertical intercept has no practical meaning; they would not play Little League baseball if the temperature were 0°F.

g. What is the equation of your line of best fit?

(Answers will vary according to Exercises 1d and 1f.)

The equation of my line of best fit is $g = 1.5t - 46$.

h. To measure the goodness-of-fit of the line determined in part g, complete the following table and compute the error. (Answers will vary according to Exercise 1g.)   The error is 34.5.

INPUT, t	ACTUAL OUTPUT	MODEL'S OUTPUT, d	|ACTUAL VALUE − MODEL VALUE|
52	35	32	3
55	42	36.5	5.5
61	50	45.5	4.5
66	53	53	0
72	66	62	4
75	68	66.5	1.5
77	72	69.5	2.5
84	80	80	0
90	84	89	5
94	91	95	4
97	95	99.5	4.5

2. a. Use your graphing calculator's statistics menu to determine the equation for the regression line in the situation in Exercise 1.

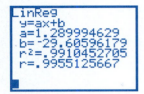

The regression line equation is approximately
$g = 1.29t - 29.6$.

b. What is the goodness-of-fit measure (the error) for the regression line in part a? Proceed as you did in Exercise 1h.

The error for the regression line is 17.35.

INPUT, t	ACTUAL OUTPUT	MODEL'S OUTPUT, $G(t)$	ACTUAL VALUE − MODEL VALUE
52	35	37.48	2.48
55	42	41.35	0.65
61	50	49.09	0.91
66	53	55.54	2.54
72	66	63.28	2.72
75	68	67.15	0.85
77	72	69.73	2.27
84	80	78.76	1.24
90	84	86.5	2.5
94	91	91.66	0.66
97	95	95.53	0.53

c. Compare the error of your line of best fit with the error of the least-squares regression line.

The error of the regression line is lower than for my line, so it is a better fit.

3. The following table shows the life expectancies at birth for men and women in the United States born in various years. For convenience, t, the number of years since 1980, has been inserted into the table. Note that the life expectancy for women has been longer than that of men for several years.

YEAR OF BIRTH	1980 $t = 0$	1985 $t = 5$	1990 $t = 10$	1995 $t = 15$	2000 $t = 20$	2005 $t = 25$
Life Expectancy For Women	77.4	78.0	78.6	79.2	79.7	80.3
Life Expectancy For Men	69.8	70.9	71.9	73.0	74.0	75.1

Source: U.S. Bureau of the Census

a. Plot the life expectancy data for women as ordered pairs of the form (t, E), where t is the number of years since 1980 and E is the life expectancy. Appropriately scale and label the coordinate axes.

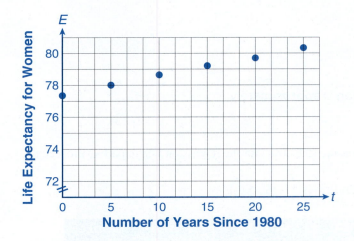

b. Use your graphing calculator to determine the equation of the regression line for the data in part a.

$E = 0.115t + 77.4$

c. According to the model relating the variables E and t, how long could a woman born in 1988 expect to live? Be careful. The input value of t is the number of years since 1980.

$t = 1988 - 1980 = 8$

$E = 0.115(8) + 77.4 \approx 78.3$

A woman born in 1988 would expect to live close to 78 years according to the model.

d. Use the regression equation for the line to predict the year of birth of a woman who can expect to live 85 years.

$85 = 0.115t + 77.4$

$7.6 = 0.115t$

$t = 66.1$

Close to 66 years after 1980; $1980 + 66 = 2046$

A woman born in 2046 can expect to live 85 years.

e. Predict the year of birth of a woman who can expect to live 100 years.

$100 = 0.115t + 77.4$

$22.6 = 0.115t$

$t \approx 197$

$1980 + 197 = 2177$

A woman born in 2177 can expect to live 100 years.

f. For which prediction (part d or e) do you have the greater confidence? Explain.

The first one; it is more reasonable. The farther the input for prediction is from the data, the less reliable the prediction.

g. Plot the life expectancy data for men on appropriately scaled and labeled coordinate axes.

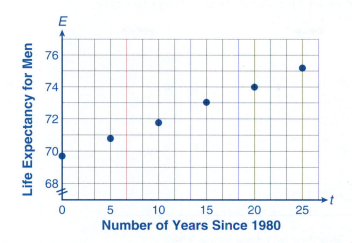

h. Determine the equation of the regression line for the data in part g.

The regression equation is $E = 0.211t + 69.8$.

i. According to this model, how long could a man born in 1988 expect to live?

$1988 - 1980 = 8$; $E = 0.211(8) + 69.8 = 71.5$.

A man born in 1988 would expect to live about 71.5 years, according to this model.

j. Predict the year of birth of a man who can expect to live 100 years.

$100 = 0.211t + 69.8$

$30.2 = 0.211t$

$t \approx 143$

$1980 + 143 = 2123$

A man born in 2123 could expect to live 100 years.

4. In 1966, the U.S. Surgeon General's health warnings began appearing on cigarette packages. At that time, approximately 43% of adults were smokers. The following data seems to demonstrate that public awareness of the health hazards of smoking has had some effect on consumption of cigarettes.

YEAR	1997	1998	1999	2000	2001	2002	2003	2004
% of Total Population 18 and Older Who Smoke	24.7	24.1	23.5	23.2	22.7	22.4	21.6	20.9

Source: U.S. National Center for Health Statistics

a. Plot the given data as ordered pairs of the form (t, P), where t is the number of years since 1997 and P is the percentage of the total population (18 and older) who smoke. Appropriately scale and label the coordinate axes.

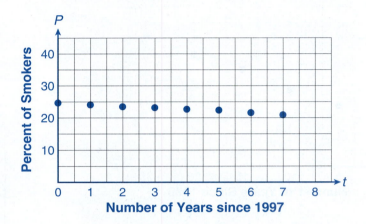

b. Determine the equation of the regression line that best represents the data.

The regression equation is $P = -0.51t + 24.675$, where P represents the percent of adult smokers in the United States t years since 1997.

c. Use the equation to predict the percent of the total population 18 and older that will smoke in 2010.

$2010 - 1997 = 13$

$P = -0.51(13) + 24.675 = 18.045\%$

According to the model, approximately 18.045% of the total population 18 and older will smoke in 2010.

Lab Activity 3.11

Body Parts

Objectives

1. Collect and organize data in a table.

2. Plot data in a scatterplot.

3. Recognize linear patterns in paired data.

Variables arise in many common measurements. Your height is one measurement that has probably been recorded frequently from the day you were born. In this Lab, you are asked to pair up and make the following body measurements: height (h); arm span (a), the distance between the tips of your two middle fingers with arms outstretched; wrist circumference (w); foot length (f); and neck circumference (n). For consistency, measure the lengths in inches.

(Answers will vary for each class.)

1. Gather the data for your entire class, and record it in the following table:

 Inch by Inch

STUDENT	HEIGHT (h)	ARM SPAN (a)	WRIST (w)	FOOT (f)	NECK (n)	FEMUR (t)

2. What are some relationships you can identify, based on a visual inspection of the data? For example, how do the heights relate to the arm spans?

   (Answers will vary.) Encourage students to discover trends in their data.

3. Construct a scatterplot for heights versus arm span on the grid below, carefully labeling the axes and marking the scales. Does the scatterplot confirm what you may have guessed in Problem 2?

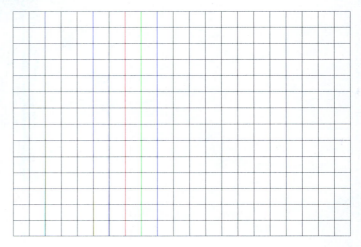

4. Use your calculator to create scatterplots for the following pairs of data, and state whether or not there appears to be a linear relationship. Comment on how the scatterplots either confirm or go against the observations you made in Problem 2.

VARIABLES	LINEAR RELATIONSHIP?
Height versus Foot Length	
Arm Span versus Wrist	
Foot Length versus Neck Circumference	

5. Determine a linear regression equation to represent the relationship between the two variables in Problems 3 and 4 that show the strongest linear pattern.

Predicting Height from Bone Length

An anthropologist studies human physical traits, place of origin, social structure, and culture. Anthropologists are often searching for the remains of people who lived many years ago. A forensic scientist studies the evidence from a crime scene in order to help solve a crime. Both of these groups of scientists use various characteristics and measurements of the human skeletal remains to help determine physical traits such as height, as well as racial and gender differences.

In the average person, there is a strong relationship between height and the length of two major arm bones (the humerous and the radius), as well as the length of the two major leg bones (the femur and the tibia).

Anthropologists and forensic scientists can closely estimate a person's height from the length of just one of these major bones.

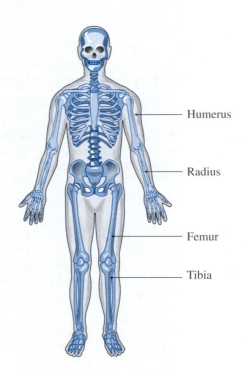
Humerus

Radius

Femur

Tibia

6. Each member of the class should measure his or her leg from the center of the kneecap to the bone on the outside of the hip. This is the length of the femur. Record the results in the appropriate place in the last column of the table in Problem 1.

a. If you want to predict height from the length of the femur, which variable should represent the independent variable? Explain.

b. Make a scatterplot of the data on a carefully scaled and labeled coordinate axes.

c. Describe any patterns you observe in the scatterplot.

d. Determine the equation of the regression line for the data.

e. Use the equation of the regression line in part d to predict the height of a person whose femur measures 17 inches.

f. Anthropologists have developed the following formula to predict the height of a male based on the length of his femur:

$$h = 1.888L + 32.010$$

where h represents the height in inches and L represents the length of the femur in inches. Use the formula to determine the height of the person whose femur measures 17 inches.

g. Compare your results from parts e and f. What might explain the difference between the height you obtained using the regression formula in part e and the height using the formula in part f?

h. Determine the regression line equation for femur length vs. height using just the male data from Problem 1. How does this new regression line equation compare with the formula $h = 1.888L + 32.010$ used by anthropologists?

7. a. Determine the linear regression equation (femur length vs. height) for the female data in the Problem 1.

b. Compare your results to the formula used by anthropologists:

$$h = 1.945x + 28.679$$

where h represents height in inches and x represents femur length in inches.

8. The work of Dr. Mildred Trotter (1899–1991) in skeletal biology led to the development of formulas used to estimate a person's height based on bone length. Her research also led to discoveries about the growth, racial and gender differences, and aging of the human skeleton. Write a brief report on the life and accomplishments of this remarkable scientist.

Cluster 3 — What Have I Learned?

1. If you know the slope and the vertical intercept of a line, how would you write the equation of the line? Use an example to demonstrate.

I would substitute the slope for m and the output value of the vertical intercept for b in $y = mx + b$. For example, if the slope were 4 and the vertical intercept were $(0, 3)$, the equation of the line would be $y = 4x + 3$.

2. Demonstrate how you would change the equation of a linear function such as $5y - 6x = 3$ into slope-intercept form. Explain your method.

$5y - 6x = 3$

$5y = 6x + 3$

$y = \frac{6}{5}x + \frac{3}{5}$ or $y = 1.2x + 0.6$

I would solve for y by first isolating the y-term on one side of the equation and then dividing by its coefficient.

3. What assumption are you making when you say that the cost, c, of a rental car (in dollars) is a linear function of the number, n, of miles driven?

I am assuming that the cost depends on the number of miles driven and that the cost per mile is constant.

4. When a scatterplot of input/output values from a data set suggests a linear relationship, you can determine a line of best fit. Why might this line be useful in your analysis of the data?

(Answers will vary.) The line of best fit gives us a symbolic rule to estimate the output values from input values other than those in the given data.

5. Explain how you would determine a line of best fit for a set of data. How would you estimate the slope and y-intercept?

(Answers will vary.) I would pick two points from the data that, when connected, yield a line that is very close to the data points. The y-intercept is the point at which the line crosses the y-axis. I would use $\dfrac{\text{rise}}{\text{run}}$ to estimate the slope.

6. Suppose a set of data pairs suggests a linear trend. The input values range from a low of 10 to a high of 40. You use your graphing calculator to calculate the regression equation in the form $y = ax + b$.

a. Do you think that the equation will provide a good prediction of the output value for an input value of $x = 20$? Explain.

(Answers will vary.) The equation should produce a decent prediction of the output value for an input value of 20 because that input is within the range of the given data.

b. Do you think that the equation will provide a good prediction of the output value for an input value of $x = 60$? Explain.

The equation might not provide a good prediction of the output value for an input of 60 because 60 is beyond the original data.

Cluster 3 How Can I Practice?

In Problems 1–8, determine the slope and the intercepts of each line.

1. $y = 2x + 1$

Slope is 2;

y-intercept is $(0, 1)$.

$0 = 2x + 1$

$-1 = 2x$

$-\frac{1}{2} = x$

x-intercept is $\left(-\frac{1}{2}, 0\right)$.

2. $y = 4 - x$

Slope is -1;

y-intercept is $(0, 4)$.

$0 = 4 - x$

$x = 4$

x-intercept is $(4, 0)$.

3. $y = -2$

Slope is 0;

y-intercept is $(0, -2)$.

There is no x-intercept.

4. $-\frac{3}{2}x - 5 = y$

Slope is $-\frac{3}{2}$;

y-intercept is $(0, -5)$.

$-\frac{3}{2}x - 5 = 0$

$-\frac{3}{2}x = 5$

$x = -\frac{10}{3}$

x-intercept is $\left(-\frac{10}{3}, 0\right)$.

5. $y = \frac{x}{5}$

Slope is $\frac{1}{5}$;

y-intercept is $(0, 0)$.

x-intercept is $(0, 0)$.

6. $y = 4x + \frac{1}{2}$

Slope is 4;

y-intercept is $\left(0, \frac{1}{2}\right)$.

$0 = 4x + \frac{1}{2}$

$-\frac{1}{2} = 4x$

$-\frac{1}{8} = x$

x-intercept is $\left(-\frac{1}{8}, 0\right)$.

7. $2x + y = 2$

$y = -2x + 2$

Slope is -2;

y-intercept is $(0, 2)$.

$2x + 0 = 2$

$x = 1$

x-intercept is $(1, 0)$.

8. $-3x + 4y = 12$

$4y = 3x + 12$

$y = \frac{3}{4}x + 3$

Slope is $\frac{3}{4}$; y-intercept is $(0, 3)$.

$-3x + 4(0) = 12$

$-3x = 12$

$x = -4$

x-intercept is $(-4, 0)$.

Determine the equation of each line described in Problems 9–12.

9. The slope is 9, and the y-intercept is $(0, -4)$.

$y = 9x - 4$

10. The line passes through the points $(0, 4)$ and $(-5, 0)$.

$m = \dfrac{0 - 4}{-5 - 0} = \dfrac{4}{5}$

$y = \dfrac{4}{5}x + 4$

11. The slope is $\dfrac{5}{3}$, and the line passes through the point $(0, -2)$.

$y = \dfrac{5}{3}x - 2$

12. The slope is zero, and the line passes through the origin.

$y = 0$

13. Identify the input and output variables, and write a linear function in symbolic form for each of the following situations. Then give the practical meaning of the slope and vertical intercept in each situation.

a. You make a down payment of $50 and pay $10 per month for your new computer.

$C(t) = 50 + 10t$, where the input variable, t, represents the number of months since computer was purchased and the output variable, C, represents the total cost, in dollars. The vertical intercept represents the $50 down payment, and the slope represents the monthly charge of $10.

b. You pay $16,000 for a new car whose value decreases by $1500 each year.

$V(x) = 16,000 - 1500x$, where the input variable, x, represents the number of years since the car was bought and the output variable, V, represents the value of the car. The vertical intercept represents the value of the car on the day of purchase. The slope represents the car's yearly depreciation.

Graph the equations in Problems 14 and 15, and determine the slope and y-intercept of each graph. Describe the similarities and differences in the graphs.

14. a. $y = -x + 2$

Slope is -1;

y-intercept is $(0, 2)$.

b. $y = -4x + 2$

Slope is -4;

y-intercept is $(0, 2)$.

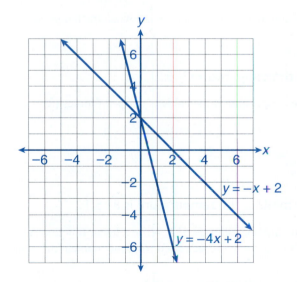

The lines in Problems 14a and b have the same y-intercept but different slopes. The graph of Problem 14b is steeper than that of 14a. Both lines are decreasing.

15. a. $g(x) = 3x - 4$

Slope is 3;

y-intercept is $(0, -4)$.

b. $y = 3x + 5$

Slope is 3;

y-intercept is $(0, 5)$.

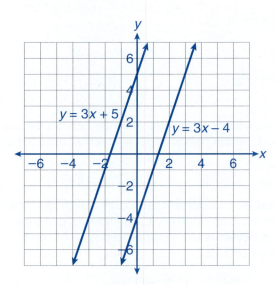

The lines in Problems 15a and b are parallel. They have the same slope but different y-intercepts.

16. a. Refer to Activity 3.4, The Snowy Tree Cricket, on page 303, to write a linear function rule that gives the number of chirps per minute, N, in terms of temperature, t.

The slope is 4. Use $(60, 80)$ for (t_1, N_1)

$N - N_1 = m(t - t_1)$

$N - 80 = 4(t - 60)$

$N - 80 = 4t - 240$

$\quad N = 4t - 160$

b. Use your linear model to determine the number of chirps per minute when the temperature is 62°F.

$N = 4(62) - 160 = 88$

There are 88 chirps per minute if the temperature is 62°F.

c. Use your linear model to determine the temperature if the crickets chirp 190 times per minute.

$190 = 4t - 160$

$350 = 4t$

$87.5 = t$

If the crickets chirp 190 times per minute, the temperature is 87.5°F.

17. a. Refer to Activity 3.5, Descending in an Airplane, on page 313. Write a linear function rule that gives the altitude, A, in terms of time, t, where t is measured from the moment the plane begins its descent.

$m = -1$ (from Problem 4a) and $b = 12$ (from Problem 9a)

$A = -t + 12$

b. Use your linear model to determine how far the plane has descended after 5 minutes.

$A = -5 + 12 = 7$

After 5 minutes, the plane has descended 5 kilometers to an altitude of 7 kilometers.

c. Use your linear model to determine how long it will take the plane to reach the ground.

When is altitude = 0?

$0 = -t + 12$

$t = 12$

It will take the plane 12 minutes to reach the ground.

18. Suppose you enter Interstate 90 in Montana and drive at a constant speed of 75 mph.

a. Write a linear function rule that represents the total distance, d, traveled on the highway as a function of time, t, in hours.

$d = 75t$

b. Sketch a graph of the function. What are the slope and d-intercept of the line? What is the practical meaning of the slope?

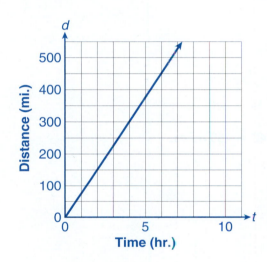

The slope is 75. The d-intercept is $(0, 0)$.

The slope, 75, represents constant speed. You travel 75 miles for every hour of travel time.

c. How long would you need to drive at 75 miles per hour to travel a total of 400 miles?

$400 = 75t$

$5\frac{1}{3} = t$

You would need to drive at 75 miles per hour for 5 hours and 20 minutes in order to travel a total of 400 miles.

d. You start out at 10:00 A.M. and drive for 3 hours at a constant speed of 75 miles per hour. You are hungry and stop for lunch. One hour later, you resume your travel, driving steadily at 60 miles per hour until 6 P.M., when you reach your destination. How far will you have traveled? Sketch a graph that shows the distance traveled as a function of time.

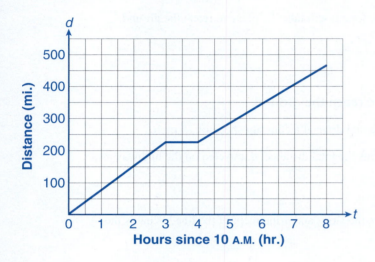

10 A.M.–1 P.M.

$75 \cdot 3 = 225$ mi.

1 P.M.–2 P.M. stopped

2 P.M.–6 P.M.

$60 \cdot 4 = 240$ mi.

The total distance traveled is $225 + 240 = 465$ miles.

19. Determine the equation of each line in parts a–f.

a. The slope is 3, and the y-intercept is 6.

$y = 3x + 6$

b. The slope is -4, and the y-intercept is -5.

$y = -4x - 5$

c. The slope is 2, and the line passes through $(0, 4)$.

$y = 2x + 4$

d. The slope is 4, and the line passes through $(6, -3)$.

$$-3 = 4(6) + b \qquad y - (-3) = 4(x - 6)$$
$$-3 = 24 + b \quad \text{or} \quad y + 3 = 4x - 24$$
$$-27 = b \qquad\qquad y = 4x - 27$$
$$y = 4x - 27$$

e. The slope is -5, and the line passes through $(4, -7)$.

$$-7 = -5(4) + b \qquad y - (-7) = -5(x - 4)$$
$$-7 = -20 + b \quad \text{or} \quad y + 7 = -5x + 20$$
$$13 = b \qquad\qquad y = -5x + 13$$
$$y = -5x + 13$$

f. The slope is 2, and the line passes through $(5, -3)$.

$$-3 = 2(5) + b \qquad y - (-3) = 2(x - 5)$$
$$-3 = 10 + b \quad \text{or} \quad y + 3 = 2x - 10$$
$$-13 = b \qquad\qquad y = 2x - 13$$
$$y = 2x - 13$$

20. Determine the equation of the line passing through each pair of points.

a. $(0, 6)$ and $(4, 14)$

$$m = \frac{14 - 6}{4 - 0} = \frac{8}{4} = 2$$

The equation is $y = 2x + 6$.

b. $(-2, -13)$ and $(0, -5)$

$$m = \frac{-5 - (-13)}{0 - (-2)} = \frac{8}{2} = 4$$

The equation is $y = 4x - 5$.

c. $(5, 3)$ and $(-1, 3)$

$$m = \frac{3 - 3}{-1 - 5} = \frac{0}{-6} = 0$$

The equation is $y = 3$.

d. $(-9, -7)$ and $(-7, -3)$

$$m = \frac{-3 - (-7)}{-7 - (-9)} = \frac{4}{2} = 2$$

$$-7 = 2(-9) + b \qquad\qquad y - (-7) = 2(x - (-9))$$
$$-7 = -18 + b \quad \text{or} \quad y + 7 = 2x + 18$$
$$11 = b \qquad\qquad y = 2x + 11$$

The equation is $y = 2x + 11$.

e. $(6, 1)$ and $(6, 7)$

$$m = \frac{7 - 1}{6 - 6} = \frac{6}{0}, \text{ undefined slope}$$

The line is vertical.

The equation is $x = 6$.

f. $(2, 3)$ and $(2, 7)$

$$m = \frac{7 - 3}{2 - 2} = \frac{4}{0}, \text{ undefined slope}$$

The line is vertical.

The equation is $x = 2$.

21. Determine the equation of the line shown on the following graph.

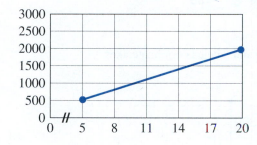

$$m = \frac{2000 - 500}{20 - 5} = \frac{1500}{15} = 100$$

$$500 = 100(5) + b \qquad\qquad y - 500 = 100(x - 5)$$
$$500 = 500 + b \quad \text{or} \quad y - 500 = 100x - 500$$
$$0 = b \qquad\qquad y = 100x$$

The equation is $y = 100x$.

22. Determine the x-intercept of each line having the given equation.

a. $y = 2x + 4$

$0 = 2x + 4$

$-4 = 2x$

$-2 = x$

x-intercept is $(-2, 0)$.

b. $y = 4x - 27$

$0 = 4x - 27$

$27 = 4x$

$\frac{27}{4} = x$

$6.75 = x$

x-intercept is $(6.75, 0)$.

c. $y = -5x + 13$

$0 = -5x + 13$

$-13 = -5x$

$\frac{-13}{-5} = x$

$2.6 = x$

x-intercept is $(2.6, 0)$.

d. $y = 2x - 13$

$0 = 2x - 13$

$13 = 2x$

$6.5 = x$

x-intercept is $(6.5, 0)$.

23. On an average winter day, the Auto Club receives 125 calls from people who need help starting their cars. The number of calls varies, however, depending on the temperature. Here is some data giving the number of calls as a function of the temperature (in degrees Celsius).

TEMPERATURE (°C)	NUMBER OF AUTO CLUB SERVICE CALLS
− 12	250
− 6	190
0	140
4	125
9	100

a. Sketch the given data on appropriately scaled and labeled coordinate axes.

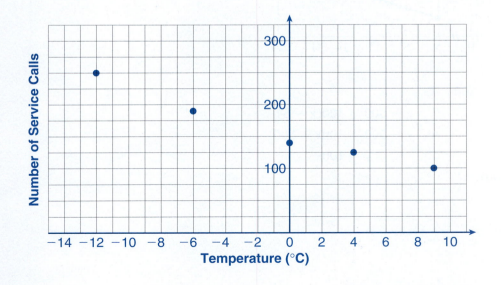

b. Use your graphing calculator to determine the equation of the regression line for the data in the preceding table.

The TI-83/TI-84 Plus calculator yields a regression line of $y = -7.11x + 153.9$, where y is the number of calls at a temperature of $x°C$.

c. Use your regression equation from part b, $y = -7.11x + 153.9$, to determine how many service calls the Auto Club can expect if the temperature drops to $-20°C$.

$y = -7.11(-20) + 153.9 = 296.10 \approx 296$ calls

The Auto Club can expect approximately 296 calls if the temperature is $-20°C$.

24. The following table contains the average tuition and required fees for full-time matriculated students at private 4-year colleges, as published by the College Board.

COLLEGE COSTS						
YEAR	1996–97	1998–99	2000–01	2002–03	2004–05	2006–07
Years Since 96–97	0	2	4	6	8	10
Cost	$12,994	$14,709	$16,072	$18,060	$20,045	$22,218

a. Let t, the number of years since 1996–97 represent the input variable and c, the average cost, the output variable. Determine an appropriate scale, and plot the data points from the accompanying table.

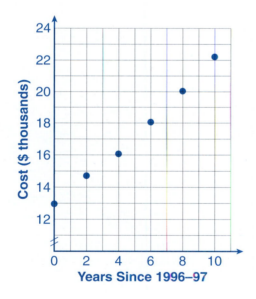

b. Does there appear to be a linear relationship between the years since 1996–97 and the cost of tuition?

The graph appears to be somewhat linear.

c. As the years since 1996–97 increases, what is the general trend in the tuition costs?

In general, the values of tuition cost increase as time increases.
The two variables are said to be positively related.

d. Determine the regression line for this data, rounding the coefficients to three decimal places.

$y = 915.943x + 12769.952$

e. Use the regression equation to predict the average tuition and fees at private 4-year colleges in 1999–2000 ($t = 3$).

$y = 915.943(3) + 12769.952 = 15,517.78$

The regression model predicts the average tuition was $15,518.

f. Use the regression equation to predict the average tuition and fees at private 4-year colleges in 2010–11 and in 2015–16.

$y = 915.943(14) + 12769.952 = 25,593.15$

$y = 915.943(19) + 12769.952 = 30,172.87$

The predicted average tuitions are $25,593 and $30,173, respectively.

g. Which prediction do you believe would be more accurate? Explain.

The 2010–11 is more believable because the input is closer to the given data.

Cluster 4 Systems of Two Linear Equations

Activity 3.12

Business Checking
Account

Objectives

1. Solve a system of two linear equations numerically.

2. Solve a system of two linear equations graphically.

3. Solve a system of two linear equations symbolically by the substitution method.

4. Recognize the connections among the three methods of solution.

5. Interpret the solution to a system of two linear equations in terms of the problem's content.

In setting up your part-time business, you have two choices for a checking account at the local bank.

	MONTHLY FEE	TRANSACTION FEE
Regular	$11.00	$0.17 for each transaction
Basic	$8.50	$0.22 for each transaction in excess of 20

1. If you anticipate making about 50 transactions each month, which checking account will be more economical?

   regular account: $11 + 0.17(50) = \$19.50$

   basic account: $8.50 + 0.22(50 - 20) = \$15.10$

   The basic account is more economical than the regular account for 50 transactions.

2. Let x represent the number of transactions. Write a function rule that expresses the total monthly cost, C, of the regular account in terms of x.

   $C = 11 + 0.17x$

3. The function for the basic account is not expressed as easily symbolically, because the transaction fee does not apply to the first 20 transactions. Assuming that you will write at least 20 checks per month, determine the equation of this function.

   $C = 8.50 + 0.22(x - 20)$, or $C = 4.10 + 0.22x$

Appendix

4. a. Complete the following table for each account, showing the monthly cost for 20, 50, 100, 150, 200, 250, and 300 transactions. Estimate the number of transactions for which the cost of the two accounts comes the closest. If you have a graphing calculator, use the table feature to complete the table. See Appendix D.

Number of Transactions	20	50	100	150	200	250	300
Cost of Regular ($)	14.40	19.50	28.00	36.50	45.00	53.50	62.00
Cost of Basic ($)	8.50	15.10	26.10	37.10	48.10	59.10	70.10

Costs come the closest for 150 transactions.

b. Use the table feature of the graphing calculator to determine the input value that produces two identical outputs. What is that value?

Using $y_1 = 0.17x + 11$ and $y_2 = 0.22(x - 20) + 8.5$,

$x = 138$ produces identical outputs.

X	Y1	Y2
135	33.95	33.8
136	34.12	34.02
137	34.29	34.24
138	34.46	34.46
139	34.63	34.68
140	34.8	34.9
141	34.97	35.12

X=138

5. a. Graph the cost equation for each account on the same coordinate axes. Plot the data points for $x \geq 20$ and then use a straightedge to draw a line connecting each set of points. Be sure to properly scale and label the axes.

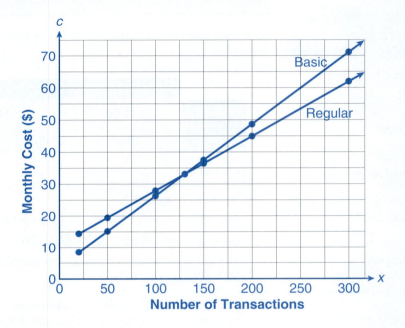

b. Estimate the coordinates of the point where the lines intersect. What is the significance of this point?

(Answers will vary.) The lines cross at approximately (135, 34). If the account holder writes 135 checks, the monthly cost will be the same for both accounts, $34.

Appendix

c. Verify your results from part b using your graphing calculator. Use the trace or intersect feature of the graphing calculator. See Appendix D for the procedure for the TI-83/TI-84 Plus. Your final screens should appear as follows.

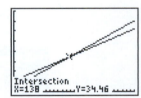

System of Two Linear Equations

Two linear equations that relate the same two variables are called a **system of linear equations**. The two cost equations from Problems 2 and 3 form a system of two linear equations,

$$C = 0.17x + 11$$
$$C = 0.22x + 4.10,$$

where $x \geq 20$.

The **solution** of a system is the set of all ordered pairs that satisfy both equations. If the system has exactly one solution, the system is called **consistent**. The solution to the cost system is (138, 34.46). This solution represents the specific number of transactions ($x = 138$) that produce identical costs in both accounts ($34.46).

In Problem 4, you solved the cost system **numerically** by completing a table and noting the value of the input that resulted in the same output. In Problem 5, you solved the cost system **graphically** by determining the coordinates of the point of intersection.

You can also determine an exact solution by solving the system of equations **algebraically**. In Problem 6, you will explore one method—the **substitution method**—for solving systems of equations algebraically. Another algebraic method for solving systems of linear equations, the **addition**, or **elimination method**, will be used in Activity 3.13.

Substitution Method for Solving a System of Two Linear Equations

Consider the following system of two linear equations.

$$y = 3x - 10$$
$$y = 5x + 14$$

To solve this system, you need to determine for what value of x are the corresponding y-values the same. The idea behind the substitution method is to replace one variable in one of the equations by an expression involving the second variable.

6. a. Use the two linear equations in the preceding system to write a single equation involving just one variable.

Substitute $3x - 10$ for y in the equation $y = 5x + 14$.
$3x - 10 = 5x + 14$

b. Solve the equation in part a for the variable.

$3x - 10 = 5x + 14$
$-10 = 2x + 14$
$-24 = 2x$
$-12 = x$

c. Use the result in part b to determine the corresponding value for the other variable in the system.

Substitute -12 for x into either equation of the system and solve for y.
$y = 3(-12) - 10 = -36 - 10 = -46$

d. Write the solution to this system as an ordered pair.

$(-12, -46)$

 e. Verify this solution numerically, by substituting the values into the original equations, as well as graphically using your graphing calculator.

 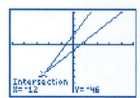

7. The algebraic process used in Problem 6 is called the substitution method for solving a system of two linear equations. Write a summary of this procedure.

Substitution Method: Replace (or *substitute*) the variable y in one equation with its algebraic expression in x from the other equation. Solve for x. Substitute this x-value into either function rule to determine the corresponding y-value.

8. a. Using the substitution method, solve the following system of checking account cost functions.

$$C = 0.17x + 11$$
$$C = 0.22x + 4.10$$

$$0.17x + 11 = 0.22x + 4.10$$
$$-0.05x = -6.9$$
$$x = 138$$
$$C = 0.17(138) + 11$$
$$C = \$34.46$$

b. Compare your result with the answers obtained using a numerical approach (Problem 4) and a graphing approach (Problem 5).

The table in Problem 4 indicates the costs are closest for 150 transactions. With some trial and error, a table can be produced on the graphing calculator that shows that 138 transactions will cost the same for both accounts, that is, $34.46. Finally, about 138 to 140 transactions are reasonable estimates from the graph in Problem 5. Therefore, all approaches yield the same answer, verifying that it is the correct answer.

c. Summarize your results by describing under what circumstances the basic account is preferable to the regular account.

For fewer than 138 transactions, the basic account is preferable; for more than 138 transactions, the regular account is more economical.

9. Your part-time business is growing to a full-time operation. You need to purchase a car for deliveries.

a. An American car costs $13,600 and depreciates $500 a year. Write an equation to determine the resale value, V, of the car after x years of use.

$$V = -500x + 13,600$$

b. A Japanese car costs $16,000 and depreciates $800 a year. Write an equation to determine the resale value, V, of this car after x years of use.

$$V = -800x + 16,000$$

c. Write a system of two linear equations that can be used to determine in how many years both cars will have the same resale value.

$$V = -500x + 13,600$$
$$V = -800x + 16,000$$

d. Solve this system numerically by completing the following table.

NUMBER OF YEARS	VALUE OF AMERICAN CAR ($)	VALUE OF JAPANESE CAR ($)
1	13,100	15,200
5	11,100	12,000
8	9,600	9,600
10	8,600	8,000
12	7,600	6,400

After 8 years, the value of the two cars will be the same, that is, $9600.

e. Solve the system graphically.

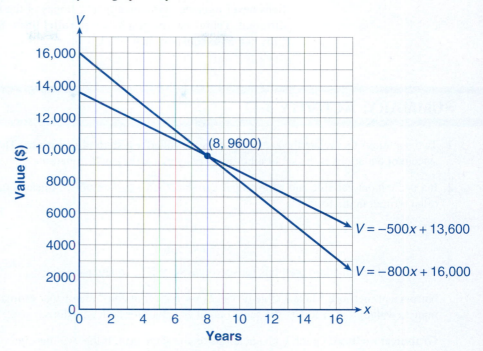

f. Solve the system algebraically using the substitution method.

$$-500x + 13{,}600 = -800x + 16{,}000$$
$$300x + 13{,}600 = 16{,}000$$
$$300x = 2400$$
$$x = 8$$
$$y = -500(8) + 13{,}600 = \$9600$$

The solution is (8, $9600).

g. Compare the results in parts d, e, and f.

The results are the same, verifying the solution is correct.

h. If you plan to keep your car for 5 years, which one would have more value and by how much?

The Japanese car would be worth $12,000 after 5 years, $900 more than the American car.

Does every system of equations have exactly one solution? Attempt to solve the system in Problem 10 algebraically. Explain your result using a graphical interpretation.

10. $y = 2x + 2$
$y = 2x - 1$
$2x + 2 = 2x - 1$
$2 = -1$

The variable drops out, leaving a false statement. Thus, there is no solution.

From the graphs, we see that the lines are parallel and thus have no point in common.

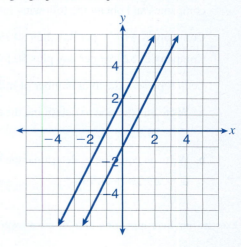

The linear system in Problem 10 is said to be **inconsistent**. There is no solution because the lines never intersect. Graphically, the slopes of the lines are equal, but the y-intercepts are different. Therefore, the graphs are **parallel lines**. Solving such a system algebraically results in a false equation such as $30 = 0$.

SUMMARY: ACTIVITY 3.12

1. Two equations that relate the same variables are called a **system of equations**. The solution of a system of equations is the set of all ordered pairs that satisfy both equations.

2. If x is the input variable and y the output variable, then a system of two linear equations is often written in the form

$$y = ax + b$$
$$y = cx + d.$$

3. There are three standard methods for solving a system of equations:

   Numerical method: Make a table of values for both equations. Identify or estimate the input (x-value) that produces the same output (y-value) for both equations.

   Graphical method: Graph both equations on the same grid. If the two lines intersect, the coordinates of the point of intersection represent the solution of the system. If the lines are parallel, the system has no solution.

   Substitution method: Replace (or *substitute*) the variable y in one equation with its algebraic expression in x from the other equation. Solve for x. Substitute this x-value into either function rule to determine the corresponding y-value. If no value is determined for x, the system has no solution.

4. A linear system is **consistent** if there is one solution, the point of intersection of the graphs.

5. A linear system is **inconsistent** if there is no solution, the lines are parallel.

EXERCISES: ACTIVITY 3.12

1. Finals are over and you are moving back home for the summer. You need to rent a truck to move your possessions from the college residence hall back to your home. You contact two local rental companies and obtain the following information for the 1-day cost of renting a truck:

   Company 1: $19.99 per day plus $0.79 per mile

   Company 2: $29.99 per day plus $0.59 per mile

   Let n represent the total number of miles driven in one day.

   a. Write an equation to determine the total cost, C, of renting a truck for 1 day from company 1.

   $C = 19.99 + 0.79n$

   b. Write an equation to determine the total cost, C, of renting a truck for 1 day from company 2.

   $C = 29.99 + 0.59n$

c. Complete the following table to compare the total cost of renting the vehicle for the day. Verify your results using the table feature of your graphing calculator.

n, NUMBER OF MILES DRIVEN	TOTAL COST, C, COMPANY 1	TOTAL COST, C, COMPANY 2
0	19.99	29.99
10	27.89	35.89
20	35.79	41.79
30	43.69	47.69
40	51.59	53.59
50	59.49	59.49
60	67.39	65.39
70	75.29	71.29
80	83.19	77.19

d. For what mileage is the 1-day rental cost the same?

$n = 50$ miles

e. Which company should you choose if you intend to use less than 50 miles?

Use company 1 if my miles will be less than 50.

f. Which company should you choose if you intend to use more than 50 miles?

Use company 2 if my miles will be greater than 50.

g. Graph the two cost functions, for n between 0 and 120 miles, on the same coordinate axes below.

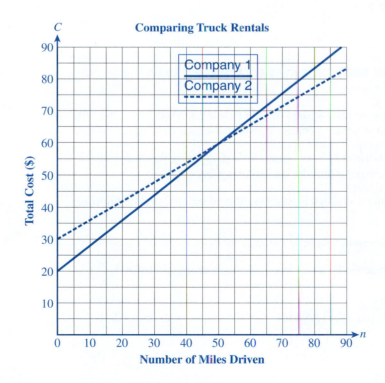

h. Use the table in part c to determine the point where the lines in part e intersect. What is the significance of the point in this situation?

The point is (50, 59.49); at 50 miles the rental cost of both companies is $59.49.

i. Determine the mileage for which the 1-day rental costs are equal by solving the following system for n using the substitution method.

$$C = 19.99 + 0.79n$$
$$C = 29.99 + 0.59n$$

$$19.99 + 0.79n = 29.99 + 0.59n$$
$$0.20n = 10$$
$$n = 50 \text{ mi.}$$

2. Two companies sell software products. In 2009, company A had total sales of $17.2 million. Its marketing department projects that sales will increase $1.5 million per year for the next several years. Company B had total sales of $9.6 million of software products in 2009 and projects that its sales will increase an average of $2.3 million each year.

Let n represent the number of years since 2009.

a. Write an equation that represents the total annual sales (in millions of dollars), s, for company A since 2009.

$$s = 1.5n + 17.2$$

b. Write an equation that represents the total annual sales (in millions of dollars), s, for company B since 2009.

$$s = 2.3n + 9.6$$

c. The two equations in parts a and b form a system. Solve this system to determine the year in which the total annual sales of both companies will be the same.

$$s = 1.5n + 17.2$$
$$s = 2.3n + 9.6$$
$$2.3n + 9.6 = 1.5n + 17.2$$
$$0.8n = 7.6$$
$$n = 9.5$$

The total annual sales of both companies are projected to be the same 9.5 years later, in 2019.

3. You are considering installing a security system in your new house. You gather the following information from two local home security dealers for similar security systems.

Dealer 1 charges $3565 to install and $15 per month for a monitoring fee, and dealer 2 charges $2850 to install and $28 per month for a monitoring fee.

Although the initial fee of dealer 1 is much higher than that of dealer 2, the monitoring fee is lower.

Let n represent the number of months you have the security system.

a. Write an equation that represents the total cost, c, of the system with dealer 1.

$$c = 15n + 3565$$

b. Write an equation that represents the total cost, c, of the system with dealer 2.

$$c = 28n + 2850$$

c. Solve the system of equations that results from parts a and b to determine in how many months the total cost of the systems will be equal.

$$c = 15n + 3565$$

$$c = 28n + 2850$$

$$28n + 2850 = 15n + 3565$$

$$13n = 715$$

$$n = 55$$

The total cost of the systems will be equal in 55 months.

d. Which system would be more economical if you plan to live in the house for more than 5 years (60 months)?

If I were going to live in the house for more than 60 months, the first system would be preferable because the installation charges would be paid off and the monthly monitoring fee is lower.

Use substitution to determine algebraically the exact solution to each system of equations in Exercises 4–7. Check your solutions numerically and by using the table feature or graphing capability of your calculator.

4. $p = q - 2$

$p = -1.5q + 3$

$q - 2 = -1.5q + 3$

$2.5q = 5$

$q = 2$

$p = q - 2$

$= 2 - 2$

$= 0$

The solution is $q = 2$, $p = 0$.

5. $n = -2m + 9$

$n = 3m - 11$

$-2m + 9 = 3m - 11$

$-5m = -20$

$m = 4$

$n = 3(4) - 11$

$= 12 - 11$

$= 1$

The solution is $m = 4$, $n = 1$.

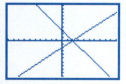

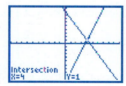

6. $y = 1.5x - 8$

$y = -0.25x + 2.5$

$1.5x - 8 = -0.25x + 2.5$

$1.75x = 10.5$

$x = 6$

$y = 1.5(6) - 8$

$= 9 - 8$

$= 1$

The solution is $x = 6$, $y = 1$.

7. $z = 3w - 1$

$z = -3w - 1$

$3w - 1 = -3w - 1$

$6w = 0$

$w = 0$

$z = 3(0) - 1$

$= -1$

The solution is $w = 0$, $z = -1$.

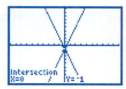

Attempt to solve the system in Exercise 8 algebraically. Explain your result using a graphical interpretation.

8. $y = -3x + 2$

$y = -3x + 3$

$-3x + 2 = -3x + 3$

$2 = 3$

The variable drops out, leaving a false statement. Thus, there is no solution.

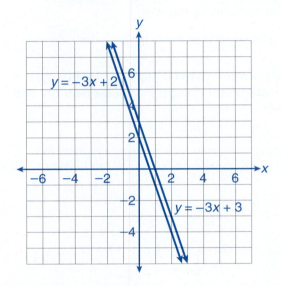

Examining the graphs, I see that the lines are parallel and thus have no points in common.

9. You want to hire someone to prune your trees and shrubs. One service you call charges a $15 consultation fee plus $8 an hour for the actual work. A neighborhood gardener says she does not include a consulting fee, but she charges $10 an hour for her work.

    a. Write an equation that describes the pruning service's charge, C, as a function of h, the number of hours worked.

    service: $C = 15 + 8h$ dollars

    b. Write an equation that describes the local gardener's charge, C, as a function of h, the number of hours worked.

    neighbor: $C = 10h$ dollars

    c. Whom would you hire for a 3-hour job?

    Three hours of the neighbor's work would cost $30. The service would charge $15 + $8(3) = $39 for a 3-hour job; so I would hire the neighbor.

    d. When, if at all, would it be more economical to hire the other service? Set up and solve a system of equations to answer this question.

    $15 + 8h = 10h$

        $15 = 2h$

    $7.5 = h$

    I would hire the service if I had any more than 7.5 hours of work.

e. Use your graphing calculator to verify your results.

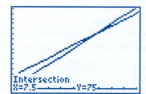

10. You and your friend are going rollerblading at a local park. There is a 5-mile path along the lake that begins at the concession stand. You rollerblade at a rate of 10 miles per hour, and your friend rollerblades at 8 miles per hour. You start rollerblading at the concession stand. Your friend starts farther down the path, 0.5 mile from the concession stand.

a. Write an equation that models your distance from the concession stand as a function of time. What are the units of the input variable? output variable? (Recall that distance = rate · time.)

t is the number of hours rollerblading, and *D* is the distance (in miles) from the concession stand.

$D = 10t$

b. Write an equation that models your friend's distance from the concession stand as a function of time.

$D = 8t + 0.5$

c. How long will it take you to catch up to your friend? In that time, how far will you have rollerbladed?

$10t = 8t + 0.5$

$2t = 0.5$

$t = 0.25$ hr

$D = 10 \cdot 0.25 = 2.5$ mi.

I would catch up with my friend in 15 minutes having rollerbladed 2.5 miles.

d. Use your graphing calculator to verify your results.

Activity 3.13

Healthy Lifestyle

Objectives

1. Solve a system of two linear equations algebraically using the substitution method.

2. Solve a system of two linear equations algebraically using the addition (or elimination) method.

You are trying to maintain a healthy lifestyle. You eat a well-balanced diet and follow a regular schedule of exercise. One of your favorite exercise activities is a combination of walking and jogging in the nearby park.

On one particular day, it takes you 1 hour and 18 minutes (1.3 hours) to walk and jog a total of 5.5 miles in the park. You are curious about the amount of time you spent walking and the amount of time you spent jogging during the workout.

Let x represent the number of hours (or part of an hour) you walked and y represent the number of hours (or part of an hour) you jogged.

1. Write an equation using x and y that expresses the total time of your workout in the park.

    $x + y = 1.3$

2. a. You walk at a steady speed of 3 miles per hour for x hours. Write an expression that represents the distance you walked.

    $3x$

   b. You jog at a constant speed of 5 miles per hour for y hours. Write an expression that represents the distance you jogged.

    $5y$

   c. Write an equation for the total distance you walked and jogged in the park.

    $3x + 5y = 5.5$

The situation just described can be represented by the system

$$x + y = 1.3$$
$$3x + 5y = 5.5.$$

Note that neither equation in this system is solved for x or y. One approach to solving this system is to solve each equation for y in terms of x.

3. a. Solve $x + y = 1.3$ for y.

    $y = -x + 1.3$

   b. Solve $3x + 5y = 5.5$ for y.

    $5y = -3x + 5.5$
    $y = -\frac{3}{5}x + 1.1$

   c. Solve the following system algebraically using the substitution method.

   $y = -x + 1.3$ $\qquad\qquad$ $y = -x + 1.3$
   $y = -\frac{3}{5}x + 1.1$ $\qquad\qquad$ $y = -0.5 + 1.3 = 0.8$
   $-x + 1.3 = -\frac{3}{5}x + 1.1$
   $-10x + 13 = -6x + 11$
   $\qquad -4x = -2$
   $\qquad\quad x = 0.5$

   The solution to the system is $x = 0.5$ and $y = 0.8$.

   I walked for a half hour and jogged for 0.8 hour (48 minutes).

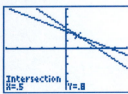

   d. Check your answer graphically using your graphing calculator. You may want to use the window Xmin $= -2.5$, Xmax $= 2.5$, Ymin $= -2.5$, Ymax $= 2.5$.

   See graphs at left.

Addition (or Elimination) Method for Solving a System of Two Linear Equations

Sometimes it is more convenient to leave each of the equations in the linear system in **standard form** $(ax + by = c)$ rather than solving for one variable in terms of the other. Look again at the original system.

> **System 1:** $x + y = 1.3$
> $3x + 5y = 5.5$

4. Your strategy in solving a linear system is to obtain a single equation involving just one variable. Apply the addition principle of equations by adding the two equations (left side to left side and right side to right side). Do you obtain a single equation containing only one variable?

$$\begin{array}{r} x + y = 1.3 \\ \underline{3x + 5y = 5.5} \\ 4x + 6y = 6.8 \end{array}$$

Adding the equations in this case will not eliminate a variable.

5. Consider a system that is similar to the one you are trying to solve.

> **System 2:** $-5x - 5y = -6.5$
> $3x + 5y = 5.5$

Apply the addition property of equations to system 2. Explain what happens.

$$\begin{array}{r} -5x - 5y = -6.5 \\ \underline{3x + 5y = 5.5} \\ -2x = -1 \end{array}$$

(Explanations may vary.)

Because the coefficients of the variable y in the system are opposites, the variable y is eliminated. I now have a single equation involving just x.

6. Compare the two systems under consideration:

> **System 1:** $x + y = 1.3$ **System 2:** $-5x - 5y = -6.5$
> $3x + 5y = 5.5$ $3x + 5y = 5.5$

Do you see any relationship between the two systems? What can you do to system 1 to make it look like system 2?

The equation $-5x - 5y = -6.5$ can be obtained from $x + y = 1.3$ by multiplying each side of the equation $x + y = 1.3$ by -5.

7. You are now ready to solve system 1 using an algebraic method called the **addition (or elimination) method**.

   a. Multiply the top equation by a factor that will produce opposite coefficients for the variable y.

$$\begin{array}{cc} -5(x + y) = -5(1.3) & \qquad -5x - 5y = -6.5 \\ 3x + 5y = 5.5 & \longrightarrow \qquad 3x + 5y = 5.5 \end{array}$$

   b. Apply the addition principle of equations.

$$\begin{array}{rl} -5x - 5y = -6.5 & \\ \underline{3x + 5y = 5.5} & \text{Add the corresponding sides of the equation.} \\ -2x + 0 = -1 & \end{array}$$

c. Solve the resulting single equation for the variable.

$$-2x = -1$$
$$x = \tfrac{1}{2}$$
$$x = 0.5$$

d. Substitute the result from part c into either one of the equations of the system to determine the corresponding value for y.

Substituting for x in $x + y = 1.3$, I have $0.5 + y = 1.3$, or $y = 0.8$.

To solve a system of two linear equations by the addition (or elimination) method,

Step 1. Line up the like terms in each equation vertically.

Step 2. If necessary, multiply one or both equations by constants so that the coefficients of one of the variables are opposites.

Step 3. Add the corresponding sides of the two equations.

8. Solve system 1 again using the addition method. This time, multiply the top equation by a number that will eliminate the variable x.

$$x + y = 1.3$$
$$3x + 5y = 5.5$$

$$-3(x + y) = -3(1.3) \quad \longrightarrow \quad -3x - 3y = -3.9$$
$$3x + 5y = 5.5 \qquad\qquad\qquad \underline{3x + 5y = 5.5}$$
$$2y = 1.6$$
$$y = 0.8$$

Substitute for y in $x + y = 1.3$, $x + 0.8 = 1.3$ or $x = 0.5$.

The solution is $(0.5, 0.8)$.

You may need to multiply one or both equations by a factor that will produce coefficients of the same variable that are additive inverses (opposites).

9. Solve the following system using the addition method.

$$2x + 3y = 2$$
$$3x + 5y = 4$$

a. Identify which variable you wish to eliminate. Multiply each equation by an appropriate factor so that the coefficients of your chosen variable become opposites. Remember to multiply *both* sides of the equation by the factor. Write the two resulting equations.

(Answer will vary depending on which variable is chosen to eliminate.)

To eliminate x, I will multiply each term in the first equation by 3 and in the second equation by -2 to obtain

$$3 \cdot 2x + 3 \cdot 3y = 3 \cdot 2 \quad \longrightarrow \quad 6x + 9y = 6$$
$$-2 \cdot 3x - 2 \cdot 5y = -2 \cdot 4 \qquad\qquad -6x - 10y = -8$$

b. Add the two equations to eliminate the chosen variable.

$$-y = -2$$

c. Solve the resulting linear equation.

$$y = 2$$

d. Determine the complete solution. Remember to check by substituting into both of the original equations.

Substitute $y = 2$ into the original first equation: $2x + 3(2) = 2$.

Solve for x: $2x + 6 = 2$, so $2x = -4$ and $x = -2$

Check: $2(-2) + 3(2) = 2$ and $3(-2) + 5(2) = 4$

The complete solution is $(-2, 2)$.

Substitution Method Revisited

When each linear equation in a system of two linear equations is written in the slope-intercept form, $y = mx + b$, the substitution method is the most efficient approach. Simply equate the two expressions from each equation and then solve for x. Sometimes, when the equations are written in standard form, $ax + by = c$, one of the equations can be easily rewritten in the form $y = mx + b$ and then the substitution method can be used. The following problem shows how.

10. Consider the system from the walking and jogging workout scenario,

$$x + y = 1.3$$
$$3x + 5y = 5.5.$$

a. First solve the equation $x + y = 1.3$ for y.

$y = 1.3 - x$

b. Then substitute this expression for y in the equation $3x + 5y = 5.5$ to obtain a single equation involving just x.

Substituting $1.3 - x$ for y, I have $3x + 5(1.3 - x) = 5.5$.

c. Solve the equation in part b.

$3x + 5(1.3 - x) = 5.5$
$3x + 6.5 - 5x = 5.5$
$-2x + 6.5 = 5.5$
$-2x = -1$
$x = \frac{1}{2}$
$x = 0.5$

d. What is the solution to the system? How does it compare to the solution determined using the addition method in Problem 7?

Substituting $x = 0.5$ in $y = 1.3 - x$, I obtain $y = 0.8$. The solution to the system is $x = 0.5$ and $y = 0.8$, the same solution as obtained in Problem 7.

11. a. Solve the following linear system using the substitution method by solving one equation for a chosen variable and then substituting in the remaining equation.

$$x - y = 5$$
$$4x + 5y = -7$$

$-y = -x + 5$
$y = x - 5$
$4x + 5(x - 5) = -7$
$4x + 5x - 25 = -7$ $x - y = 5$
$9x - 25 = -7$ $2 - y = 5$
$9x = 18$ $-y = 3$
$x = 2$ $y = -3$

The solution is $(2, -3)$.

b. Check your answer in part a by solving the system using the addition method.

$$-4(x - y) = -4(5) \longrightarrow -4x + 4y = -20$$
$$4x + 5y = -7 \longrightarrow \underline{4x + 5y = -7}$$
$$9y = -27$$
$$y = -3$$
$$x - (-3) = 5$$
$$x + 3 = 5$$
$$x = 2$$

The solution is $(2, -3)$.

SUMMARY: ACTIVITY 3.13

1. There are two common methods for solving a system of two linear equations algebraically,

   a. the **substitution** method and

   b. the **addition (or elimination)** method.

2. The procedure for solving linear systems by substitution is as follows.

   Step 1. If not already done, solve one equation for one variable (for instance, y) in terms of the other (x).

   Step 2. In the other equation, replace y by the expression in x. This equation should now contain only one variable, x.

   Step 3. Solve this equation for x.

   Step 4. Substitute the x-value you determined in part c into one of the original equations, and solve this equation for y.

3. The procedure for solving linear systems by addition is as follows.

   Step 1. Write each equation in the standard form $ax + by = c$.

   Step 2. Determine which variable you want to eliminate. Multiply one or both equations by the number(s) that will make the coefficients of this variable opposites.

   Step 3. Sum the left and right sides of the two equations, and combine like terms. This step should produce a single equation in one variable.

   Step 4. Solve the resulting equation.

   Step 5. Substitute the value from Step 4 into one of the original equations, and solve for the other variable.

4. A solution to a system of two linear equations may be expressed as an ordered pair.

EXERCISES: ACTIVITY 3.13

1. Solve the following systems algebraically using the substitution method.
Check your answers numerically and by using your graphing calculator.

a. $y = 3x + 1$

$y = 6x - 0.5$

$3x + 1 = 6x - 0.5$

$-3x = -1.5$

$x = 0.5$

$y = 3(0.5) + 1 = 2.5$

b. $y = 3x + 7$

$2x - 5y = 4$

$2x - 5(3x + 7) = 4$

$2x - 15x - 35 = 4$

$-13x - 35 = 4$

$-13x = 39$

$x = -3$

$y = 3(-3) + 7 = -2$

c. $2x + 3y = 5$

$-2x + y = -9$

$y = 2x - 9$

$2x + 3(2x - 9) = 5$

$2x + 6x - 27 = 5$

$8x - 27 = 5$

$8x = 32$

$x = 4$

$-2(4) + y = -9$

$y = -1$

d. $4x + y = 10$

$2x + 3y = -5$

$y = -4x + 10$

$2x + 3(-4x + 10) = -5$

$2x - 12x + 30 = -5$

$-10x + 30 = -5$

$-10x = -35$

$x = 3.5$

$4(3.5) + y = 10$

$y = -4$

2. Solve the following systems algebraically using the addition method.
Check your answers numerically and by using your graphing calculator.

a. $3x + y = 6$

$2x + y = 8$

$3x + y = 6$

$\underline{-2x - y = -8}$

$x = -2$

$3(-2) + y = 6$

$y = 12$

b. $-3x + 2y = 7$

$2x + 3y = 17$

$-6x + 4y = 14$

$\underline{6x + 9y = 51}$

$13y = 65$

$y = 5$

$-3x + 2(5) = 7$

$-3x + 10 = 7$

$-3x = -3$

$x = 1$

c. $7x - 5y = 1$

$3x + y = -\frac{1}{5}$

$7x - 5y = 1$ $7(0) - 5y = 1$

$\underline{15x + 5y = -1}$ $-5y = 1$

$22x = 0$ $y = -0.2$

$x = 0$

Exercise numbers appearing in color are answered in the Selected Answers appendix.

3. Solve the system both graphically and algebraically.

$3x + y = -18$

$5x - 2y = -8$

$6x + 2y = -36$

$\underline{5x - 2y = -8}$

$\quad 11x = -44$

$\quad\quad x = -4$

$5(-4) - 2y = -8$

$\quad -20 - 2y = -8$

$\quad\quad -2y = 12$

$\quad\quad\quad y = -6$

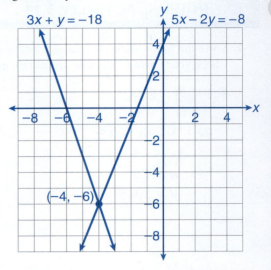

4. A catering service placed an order for eight centerpieces and five glasses, and the bill was $106. For the wedding reception it was short one centerpiece and six glasses and had to reorder. This order came to $24. Let x represent the cost of one centerpiece, and let y represent the cost of one glass.

a. Write an equation using x and y that represents the cost of the first order.

$8x + 5y = 106$

b. Write an equation using x and y that represents the cost of the second order.

$x + 6y = 24$

c. The equations in parts a and b form a system of two linear equations that can be used to determine the cost of a single centerpiece and a single glass. Write this system below.

$8x + 5y = 106$

$x + 6y = 24$

d. Solve the system using the substitution method.

$x = -6y + 24$

$8(-6y + 24) + 5y = 106$

$-48y + 192 + 5y = 106$

$\quad\quad\quad\quad -43y = -86$

$\quad\quad\quad\quad\quad y = 2$

$x + 6(2) = 24$

$\quad\quad x = 12$

$(12, 2)$

The cost of a centerpiece is $12, and the cost of a glass is $2.

e. Use the addition method to check your result in part d.

$\quad\quad 8x + 5y = 106 \quad\quad\quad\quad x + 6(2) = 24$

$-8(x + 6y) = -8 \cdot 24 \quad\quad\quad x + 12 = 24$

$\quad\quad 8x + 5y = 106 \quad\quad\quad\quad\quad x = 12$

$\underline{-8x - 48y = -192}$

$\quad\quad\quad -43y = -86$

$\quad\quad\quad\quad\quad y = 2$

f. Use the solution to the system to determine the cost of 15 centerpieces and 10 glasses.

$15(12) + 10(2) = 180 + 20 = \$200$

The cost of 15 centerpieces and 10 glasses is $200.

5. As part of a community-service project, your fraternity is asked to put up a fence in the playground area at a local day-care facility. A local fencing company will donate 500 feet of fencing. The day-care center director specifies that the length of the rectangular enclosure be 20 feet more than the width. Your task is to determine the dimensions of the enclosed region that meets the director's specifications.

a. What does the 500 represent with respect to the rectangular enclosure?

It represents the perimeter.

b. Write an equation that gives the relationship between the length, l, width, w, and 500.

$2l + 2w = 500$

c. Write an equation that expresses the length of the rectangle in terms of w.

$l = w + 20$

d. The equations in part b and c form a system of equations involving the variables l and w. Solve this system of equations to determine the dimensions of the enclosed region that satisfies the given conditions.

$$2l + 2w = 500$$
$$2(w + 20) + 2w = 500$$
$$2w + 40 + 2w = 500$$
$$4w + 40 = 500$$
$$4w = 460$$
$$w = 115$$
$$l = w + 20 = 115 + 20 = 135$$

The required width is 115 feet and the length is 135 feet.

e. Explain how you know your result in part d solves the problem you were assigned.

My results meet the director's specifications. The length 135 feet is 20 feet more than the width, 115 feet. The perimeter of the rectangle is $2(135) + 2(115) = 500$ feet.

6. The day-care center director would like the playground area in Exercise 5 to be larger than that provided by the 500 feet of fencing. The center staff also wants the length of the playground to be 30 feet more than the width. Additional donations help the center obtain a total of 620 feet of fencing.

a. Write a system of linear equations involving the length l and the width w.

$$2l + 2w = 620$$
$$l = w + 30$$

b. Solve the system in part a to determine the dimensions of the enlarged playground.

$$2l + 2w = 620$$
$$2(w + 30) + 2w = 620$$
$$2w + 60 + 2w = 620$$
$$4w + 60 = 620$$
$$4w = 560$$
$$w = 140$$
$$l = w + 30 = 140 + 30 = 170$$

The dimensions of the enlarged playground are 140 feet by 170 feet.

c. What is the area of the enlarged playground?

$$\text{Area} = l \cdot w = 140(170) = 23{,}800 \text{ square feet}$$

d. Explain how you know your result in part c solves the problem you were assigned.

My results meet the center's specifications. The length 170 feet is 30 feet more than the width, 140 feet. The perimeter of the rectangular perimeter is $2(170) + 2(140) = 620$ feet.

Modeling a Business

Objectives

1. Solve a system of two linear equations by any method.

2. Determine the break-even point of a linear system algebraically and graphically.

3. Interpret break-even points in contextual situations.

You are employed by a company that manufactures solar collector panels. To remain competitive, the company must consider many variables and make many decisions. Two major concerns are those variables and decisions that affect operating expenses (or costs) of making the product and those that affect the gross income (or revenue) from selling the product.

Costs such as rent, insurance, and utilities for the operation of the company are called *fixed costs*. These costs generally remain constant over a short period of time and must be paid whether or not any items are manufactured. Other costs, such as materials and labor, are called *variable costs*. These expenses depend directly on the number of items produced.

1. The records of the company show that fixed costs over the past year have averaged $8000 per month. In addition, each panel manufactured costs the company $95 in materials and $55 in labor. Write a symbolic rule in function notation for the total cost, $C(n)$, of producing n solar collector panels in 1 month.

$$C(n) = 150n + 8000$$

2. A marketing survey indicates that the company can sell all the panels it produces if the panels are priced at $350 each. The revenue (gross income) is the amount of money collected from the sale of the product. Write a symbolic rule in function notation for the revenue, $R(n)$, from selling n solar collector panels in 1 month.

$$R(n) = 350n$$

3. **a.** Complete the following table.

Number of Solar Panels, n	0	10	20	30	40	50	60
Total Cost ($), $C(n)$	8,000	9,500	11,000	12,500	14,000	15,500	17,000
Total Revenue ($), $R(n)$	0	3,500	7,000	10,500	14,000	17,500	21,000

b. Sketch a graph of the cost and revenue functions using the same set of coordinate axes.

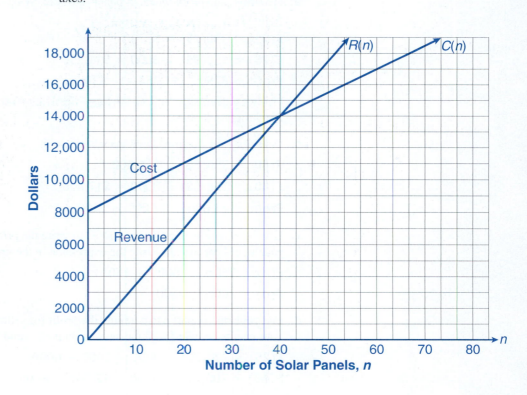

4. The point at which the cost and revenue functions are equal is called the *break-even point*.

    a. Estimate the break-even point on the graph.

    (Answers will vary.)   The break-even point is $(40, 14{,}000)$.

    b. What system of equations must be solved to determine the break-even point for your company?

    $R(n) = 350n$

    $C(n) = 8000 + 150n$

    c. Solve the system algebraically to determine the exact break-even point.

    $350n = 8000 + 150n$ $C(40) = 8000 + 150(40)$

    $200n = 8000$ $= 8000 + 6000$

    $n = 40$ $= 14{,}000$

    The break-even point is $n = 40$ and $R = C = 14{,}000$.

    d. Does your graph confirm the algebraic solution in part b?

    Yes, the lines intersect at $(40, 14{,}000)$.

5. Revenue exceeds costs when the graph of the revenue function is above the graph of the cost function. For what values of n is $R(n) > C(n)$? What do these values represent in this situation?

    By examining the graph, I see that $R(n) > C(n)$ for $n > 40$.

    There is a profit for selling more than 40 panels because revenue is greater than cost.

6. The break-even point can also be viewed with respect to profit. Profit is defined as the difference between revenue and cost. Symbolically, if $P(n)$ represents profit, then $P(n) = R(n) - C(n)$.

    a. Determine the profit when 25 panels are sold. What does the sign of your answer signify?

    $R(25) = 350(25) = \$8750$

    $C(25) = 150(25) + 8000 = \$11{,}750$

    Profit $= 8750 - 11{,}750 = -\$3000$

    The negative sign of the answer signifies that there is a loss when 25 panels are sold.

    b. Determine the profit when 60 panels are sold.

    $R(60) = 350(60) = \$21{,}000$

    $C(60) = 150(60) + 8000 = \$17{,}000$

    Profit $= 21{,}000 - 17{,}000 = \$4000$

    c. Use the equations for $C(n)$ and $R(n)$ to write the profit function in terms of n solar panels sold per month. Write the expression in the profit function in simplest terms.

    $P(n) = 350n - (8000 + 150n)$

    $P(n) = 200n - 8000$

    d. Use your rule in part c to compute the profit for selling 25 and 60 panels. How do your results compare with the answers in parts a and b?

    $P(25) = 200(25) - 8000 = 5000 - 8000 = -\$3000$

    $P(60) = 200(60) - 8000 = 12{,}000 - 8000 = \$4000$

    The results are the same.

e. What do you expect profit to be at a break-even point?

I expect the profit to be zero because at the break-even point, revenue is equal to cost.

f. Use the profit equation to confirm that the profit on n panels sold in a month is zero at the break-even point, $n = 40$.

$P(40) = 200(40) - 8000 = 0$

7. a. Graph the profit function $P(n) = R(n) - C(n)$ on the grid provided below.

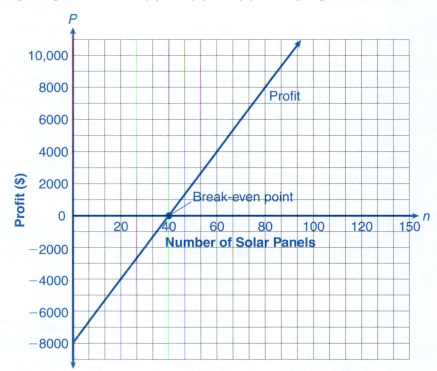

b. Locate and mark the break-even point on the graph in part a. Interpret the meaning of its coordinates.

The break-even point on this graph is given by the coordinates $(40, 0)$. The coordinates indicate that when $n = 40$ panels are sold in a month, the profit is zero.

c. Determine the vertical and horizontal intercepts of the graph. What is the meaning of each intercept in this situation?

The vertical intercept is $(0, -8000)$.

If the company sells no panels, there is a loss of $8000, due to fixed costs.

The horizontal intercept is 40 panels $(40, 0)$, the number of panels that must be sold for the company to break even.

d. What is the slope of the line? What is the practical meaning of slope in this situation?

The slope is $200. For each additional solar panel made and sold, the profit increases $200.

1. You are the sales manager of a small company that produces products for home improvement. You sell pavers (interlocking paving pieces for driveways) in bundles of 144 that cost $200 each. The equation $C(x) = 160x + 1000$ represents the cost, $C(x)$, of producing x bundles of pavers.

a. Write a symbolic rule for the revenue function $R(x)$, in dollars, from the sale of the pavers.

$R(x) = 200x$

b. Determine the slope and the C-intercept of the cost function. Explain the practical meaning of each in this situation.

The slope of the cost function is 160 dollars per bundle. This represents a cost of $160 to produce each bundle of pavers. The C-intercept is $(0, 1000)$. Fixed costs of $1000 are incurred even when no pavers are produced.

c. Determine the slope and the R-intercept of the revenue function. Explain the practical meaning of each in this situation.

The slope of the revenue function is 200 dollars per bundle. This represents the income for each bundle sold. The R-intercept is $(0, 0)$. If no pavers are sold, there is no income.

d. Graph the two functions from parts b and c on the same set of axes. Estimate the break-even point from the graph. Express your answer as an ordered pair, giving units. Check your estimate of the break-even point by graphing the two functions on your graphing calculator.

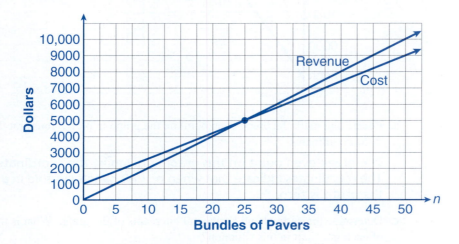

From the graph, the break-even point is (25 bundles, $5000).

e. Determine the exact break-even point algebraically. If your algebraic solution does not approximate your answer from the graph in part d, explain why.

$R(x) = C(x)$ at the break-even point

$200x = 160x + 1000$

$40x = 1000$

$x = 25$

$R(25) = 200(25)$

$= \$5000$

The break-even point is (25, 5000), which agrees with the answer obtained from the graph.

f. How many bundles of pavers does the company have to sell for it to break even?

The company would have to sell 25 bundles of pavers to break even.

g. What is the total cost to the company when you break even? Verify that the cost and revenue values are equal at the break-even point.

$C(25) = 160(25) + 1000 = \$5000$

Total cost is $5000 at the break-even point.

h. For what values of x will your revenue exceed your cost?

Revenues will exceed cost when more than 25 bundles are sold.

i. As manager, what factors do you have to consider when deciding how many pavers to make?

(Answers will vary.)

j. If you knew you could sell only 30 bundles of pavers, would you make them? Consider how much it would cost you and how much you would make. What if you could sell only 20?

The profit is only $200 on 30 bundles of pavers, so I might or might not make them.

$C(30) = 160(30) + 1000 = \$5800$

$R(30) = 200(30) = \$6000$

There is a loss of $200 if only 20 bundles are sold. I would not make them.

$C(20) = 160(20) + 1000 = \$4200$

$R(20) = 200(20) = \$4000$

Activity 3.15

How Long Can You Live?

Objective

1. Use properties of inequalities to solve linear inequalities in one variable algebraically.

We are living longer. Life expectancy in the United States is steadily increasing. The number of Americans aged 100 or older will exceed 850,000 by the middle of this century. Medical advancements have been a primary reason for Americans living longer. Another factor has been the increased awareness of a healthy lifestyle.

The life expectancy at birth of a boy born in or after 1980 can be modeled by the function

$$M(x) = 0.211x + 69.8,$$

where x represents the year of his birth given in number of years since 1980 ($x = 0$ corresponds to a 1980 birth date, $x = 5$ corresponds to a 1985 birth date, etc.), and $M(x)$ represents his life expectancy (predicted age at death).

The life expectancy at birth of a girl born in or after 1980 can be modeled by the function

$$W(x) = 0.115x + 77.4,$$

where x represents the year of her birth (given in number of years since 1980) and $W(x)$ represents her life expectancy (predicted age at death).

1. **a.** Complete the following table.

YEAR	1980	1985	1990	1995	2000	2005
x, Number of Years Since 1980	0	5	10	15	20	25
$W(x)$	77.4	78.0	78.6	79.2	79.7	80.3
$M(x)$	69.8	70.9	71.9	73.0	74.0	75.1

b. For people born between 1980 and 2005, do men or women have the greater life expectancy?

Women born between 1980 and 2005 have a greater life expectancy than men.

c. Is the life expectancy of men or women increasing more rapidly? Explain using slope.

Life expectancy for men is increasing more rapidly than for women. The slope for men is 0.211; the slope for women is 0.115.

You would like to determine when the life expectancy of men will exceed that of women. The phrase "will exceed" means "will be greater than" and indicates a mathematical relationship called an **inequality**. Symbolically, the relationship can be represented by

$$\underbrace{M(x)}_{\substack{\text{life expectancy} \\ \text{for men}}} > \underbrace{W(x)}_{\substack{\text{life expectancy} \\ \text{for women}}}$$

2. Use the life expectancy functions to write an inequality involving x to determine after which year a boy will have a greater life expectancy than a girl born the same year.

$0.211x + 69.8 > 0.115x + 77.4$

Definition

Solving an inequality in one variable is the process of determining the values of the variable that make the inequality a true statement. These values are called the **solutions** of the inequality.

Solving Inequalities in One Variable Algebraically

The process of solving an inequality in one variable algebraically is very similar to solving an equation in one variable. Your strategy is to isolate the variable on one side of the inequality symbol. You isolate the variable in an equation by performing the same operations on both sides of the equation so as not to upset the balance. In a similar manner, you isolate the variable in an inequality by performing the same operations to both sides so as not to upset the relative imbalance.

3. a. Write the statement "fifteen is greater than six" as an inequality.

$15 > 6$

b. Add 5 to each side of $15 > 6$. Is the resulting inequality a true statement? (That is, is the left side still greater than the right side?)

$20 > 11$; yes

c. Subtract 10 from each side of $15 > 6$. Is the resulting inequality a true statement?

$5 > -4$; yes

d. Multiply each side of $15 > 6$ by 4. Is the resulting inequality true?

$60 > 24$; yes

e. Multiply each side of $15 > 6$ by -2. Is the left side still greater than the right side?

$-30 > -12$; no

f. Reverse the direction of the inequality symbol in part e. Is the new inequality a true statement?

$-30 < -12$; yes

Problem 3 demonstrates two very important properties of inequalities.

Property 1. If $a < b$ represents a true inequality, and

i. if the same quantity is added to or subtracted from both sides,

or

ii. if both sides are multiplied or divided by the same *positive number*,

then the resulting inequality remains a true statement and the direction of the inequality symbol remains the same.

Example 1 *Because* $-4 < 10,$ *then*

i. $-4 + 5 < 10 + 5$, or $1 < 15$, is true and
$-4 - 3 < 10 - 3$, or $-7 < 7$, is true

ii. $-4(6) < 10(6)$, or $-24 < 60$, is true and
$\frac{-4}{2} < \frac{10}{2}$, or $-2 < 5$, is true

Property 2. If $a < b$ represents a true inequality, and if both sides are multiplied or divided by the same *negative number*, then the inequality symbol in the resulting inequality statement must be reversed ($<$ to $>$ or $>$ to $<$) for the resulting statement to be true.

Example 2   *Because* **−4 < 10,** *then*

$$-4(-5) > 10(-5), \text{ or } 20 > -50.$$

Because **−4 < 10,** *then*

$$\frac{-4}{-2} > \frac{10}{-2}, \text{ or } 2 > -5.$$

Properties 1 and 2 of inequalities are also true if $a < b$ is replaced by $a \le b$, $a > b$, or $a \ge b$. Note that $a \le b$, read a is less than or equal to b, is true if $a < b$ or $a = b$.

The following example demonstrates how properties of inequalities can be used to solve an inequality algebraically.

Example 3   *Solve* $3(x - 4) > 5(x - 2) - 8.$

SOLUTION

$$3(x - 4) > 5(x - 2) - 8$$ Apply the distributive property.

$$3x - 12 > 5x - 10 - 8$$ Combine like terms on right side.

$$3x - 12 > 5x - 18$$

$$\underline{-5x \qquad\quad -5x}$$ Subtract $5x$ from both sides; the direction of

$$-2x - 12 > -18$$ the inequality symbol remains the same.

$$\underline{+12 \quad\; +12}$$ Add 12 to both sides; the direction of the

$$\frac{-2x}{-2} < \frac{-6}{-2}$$ inequality does not change.

   Divide both sides by -2; reverse the direction!

$$x < 3$$

Therefore, any number less than 3 is a solution of the inequality $3(x - 4) > 5(x - 2) - 8$. The solution set can be represented on a number line by shading all points to the left of 3, as shown.

The open circle at 3 indicates that 3 is not a solution. A closed circle indicates that the number is a solution. The arrow shows that the solutions extend indefinitely to the left.

4. Solve the inequality $0.211x + 69.8 > 0.115x + 77.4$ from Problem 2 algebraically to determine after which year a boy will have a greater life expectancy than a girl born the same year.

$$0.211x + 69.8 > 0.115x + 77.4$$

$$0.211x - 0.115x > 77.4 - 69.8$$

$$0.096x > 7.6$$

$$x > 79.17$$

Rounding, I obtain $x > 79$. $1980 + 79 = 2059$

Therefore, if the trends given by the equations for $M(x)$ and $W(x)$ were to continue, the approximate solution to the inequality $M(x) > W(x)$ would be $x > 79$. Because x represents the number of years since 1980, $x = 79$ corresponds to the year 2059. If current trends were to continue (possible, but unlikely), men born in or after 2059 would have greater life expectancies than women.

5. a. Write an inequality to determine after which year a newborn girl is projected to live to an age of 85 or greater. $0.115x + 77.4 \ge 85$

b. Solve this inequality algebraically.

$$0.115x + 77.4 \geq 85$$
$$0.115x \geq 7.6$$
$$x \geq 66.1$$

A female born in or after 2046 (1980 + 66) is projected to have a life expectancy exceeding 85 years.

SUMMARY: ACTIVITY 3.15

1. The **solution set of an inequality** is the set of all values of the variable that satisfy the inequality.

2. The **direction of an inequality** is not changed when:

   i. The same quantity is added to or subtracted from both sides of the inequality. Stated symbolically, if $a < b$, then $a + c < b + c$ and $a - c < b - c$.

   ii. Both sides of an inequality are multiplied or divided by the same positive number.

   If $a < b$, then $ac < bc$, where $c > 0$, and $\dfrac{a}{c} < \dfrac{b}{c}$, where $c > 0$.

3. The **direction of an inequality** is **reversed** if both sides of an inequality are multiplied or divided by the same negative number. These properties can be written symbolically as

   i. if $a < b$ then $ac > bc$, where $c < 0$

   ii. if $a < b$ then $\dfrac{a}{c} > \dfrac{b}{c}$, where $c < 0$

   Properties 2 and 3 will still be true if $a < b$ is replaced by $a \leq b, a > b$, or $a \geq b$.

4. Inequalities of the form $f(x) < g(x)$ can be solved using an **algebraic approach**, in which the properties of inequalities (items 2 and 3 above) are used to isolate the variable.

   Similar statements can be made for solving inequalities of the form $f(x) \leq g(x)$, $f(x) > g(x), f(x) \geq g(x)$.

EXERCISES: ACTIVITY 3.15

In Exercises 1–3, translate the given statement into an algebraic inequality.

1. The sum of the length, l, width, w, and depth, d, of a piece of luggage to be checked on a commercial airline cannot exceed 61 inches without incurring an additional charge.

   $$l + w + d \leq 61$$

2. A PG-13 movie rating means that your age, a, must be at least 13 years for you to view the movie.

   $$a \geq 13$$

Exercise numbers appearing in color are answered in the Selected Answers appendix.

3. The cost, $C(A)$, of renting a car from company A is less expensive than the cost, $C(B)$, of renting from company B.

$C(A) < C(B)$

Solve Exercises 4–11 algebraically.

4. $3x > -6$

$\dfrac{3x}{3} > \dfrac{-6}{3}$

$x > -2$

5. $3 - 2x \leq 5$

$-2x \leq 2$

$x \geq -1$

6. $x + 2 > 3x - 8$

$x + 2 > 3x - 8$

$-2x > -10$

$x < 5$

7. $5x - 1 < 2x + 11$

$5x - 1 < 2x + 11$

$3x < 12$

$x < 4$

8. $8 - x \geq 5(8 - x)$

$8 - x \geq 5(8 - x)$

$8 - x \geq 40 - 5x$

$4x \geq 32$

$x \geq 8$

9. $5 - x < 2(x - 3) + 5$

$5 - x < 2(x - 3) + 5$

$5 - x < 2x - 6 + 5$

$5 - x < 2x - 1$

$-3x < -6$

$x > 2$

10. $\dfrac{x}{2} + 1 \leq 3x + 2$

$\dfrac{x}{2} + 1 \leq 3x + 2$

$-2.5x \leq 1$

$x \geq -0.4$

11. $0.5x + 3 \geq 2x - 1.5$

$0.5x + 3 \geq 2x - 1.5$

$-1.5x \geq -4.5$

$x \leq 3$

12. You contacted two local rental companies and obtained the following information for the 1-day cost of renting an SUV.

Company 1 charges $40.95 per day plus $0.19 per mile, and company 2 charges $19.95 per day plus $0.49 per mile.

Let n represent the total number of miles driven in 1 day.

a. Write an equation to determine the total cost, C, of renting an SUV for a day from company 1.

$C = 40.95 + 0.19n$

b. Write an equation to determine the total cost of renting an SUV for a day from company 2.

$C = 19.95 + 0.49n$

c. Use the results in parts a and b to write an inequality that can be used to determine for what number of miles it is less expensive to rent the SUV from company 2.

$19.95 + 0.49n < 40.95 + 0.19n$

d. Solve the inequality in part c.

$0.3n < 21$

$n < 70$

It is less expensive to rent from company 2 if I plan to drive fewer than 70 miles.

Cluster 4 What Have I Learned?

1. What is meant by a *solution to a system of linear equations*? How is a solution represented graphically?

   Solving a system of linear equations means finding values for the input and output variables that satisfy both equations simultaneously. That is, they are points on the graph of both equations, the point of intersection.

2. Briefly describe three different methods for solving a system of linear equations.

   1. Graphically: Graph both equations, and locate the point of intersection, if any.

   2. Algebraically:

      a. Substitution Method: Solve one equation for a given variable. Substitute the expression for that variable into the other equation. Solve the resulting equation. Substitute this value into one of the original equations to find the value for the other variable.

      b. Addition Method: With the equations in standard form, multiply one equation (or both equations) by the number that will make the coefficients of one of the variables opposites. Add the two equations to eliminate one variable, and solve the resulting equation. Substitute the resulting value into one of the original equations, and solve for the other variable.

   3. Numerically: Make a table of values and locate the input that gives a common output.

3. Describe a situation in which you would be interested in determining a break-even point. Explain how you would determine the break-even point mathematically. Interpret the break-even point with respect to the situation you describe.

   (Answers will vary.)

4. Typically, a linear system of equations has one unique solution. Under what conditions is this not the case?

   This will not be the case if the lines are parallel (no solution) or the equations represent the same line. In the latter case, all points on the line are solutions.

Cluster 4   How Can I Practice?

1. Which input value results in the same output value for $y_1 = 2x - 3$ and $y_2 = 5x + 3$?

   $2x - 3 = 5x + 3$ $y_1 = 2(-2) - 3$ $y_2 = 5(-2) + 3$

   $-3x = 6$ $y_1 = -7$ $y_2 = -7$

   $x = -2$

   An input value of -2 results in the same output, -7, for each equation.

2. Which point on the line given by $4x - 5y = 20$ is also on the line $y = x + 5$?

   $4x - 5y = 20$ $y = x + 5$

   $y = x + 5$ $y = -45 + 5$

   $4x - 5(x + 5) = 20$ $y = -40$

   $4x - 5x - 25 = 20$

   $-x = 45$

   $x = -45$

   The point $(-45, -40)$ is on both lines.

3. Solve the following system of two linear equations.

   $y = 2x + 8$

   $y = -3x + 3$

   $2x + 8 = -3x + 3$ $y = 2(-1) + 8$

   $5x = -5$ $y = 6$

   $x = -1$

   The solution is $(-1, 6)$.

4. You sell centerpieces for $19.50 each. Your fixed costs are $500 per month, and each centerpiece costs $8 to produce.

   a. Let n represent the number of centerpieces you sell. Write an equation to determine the cost, C.

   $C = 500 + 8n$

   b. Write an equation to determine the revenue, R.

   $R = 19.50n$

   c. What is your break-even point?

   cost = revenue

   $500 + 8n = 19.50n$

   $500 = 11.5n$

   $43.48 \approx n$

   I must sell approximately 44 centerpieces to break even.

5. Suppose you wish to break even (in Exercise 4) by selling only 25 centerpieces. At what price would you need to sell each centerpiece?

The cost of producing 25 centerpieces is $C(25) = 500 + 8(25) = \$700$.

Let p represent the price.

$$25p = 700$$
$$p = \$28$$

I must sell 25 centerpieces at $28 each to break even.

6. Solve the system algebraically.

$$y = -25x + 250$$
$$y = 25x + 300$$

$$-25x + 250 = 25x + 300$$
$$-50x = 50$$
$$x = -1$$
$$y = 25(-1) + 300 = 275$$

The solution is $x = -1, y = 275$.

7. Use the addition method to solve the following system.

$$4m - 3n = -7$$
$$2m + 3n = 37$$

Using the addition method,

$$6m = 30$$
$$m = 5$$
$$4m - 3n = -7$$
$$4(5) - 3n = -7$$
$$20 - 3n = -7$$
$$-3n = -27$$
$$n = 9$$

The solution is $m = 5, n = 9$.

8. Solve the following system graphically and algebraically.

$$y = 6x - 7$$
$$y = 6x + 4$$

$$6x - 7 = 6x + 4$$
$$-7 = 4$$

There is no solution. The lines are parallel.

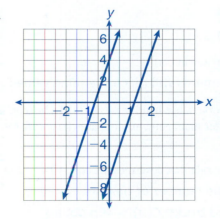

9. Solve this system algebraically and graphically

$u = 3v - 17$

$u = -4v + 11$

$3v - 17 = -4v + 11$

$7v = 28$

$v = 4$

$u = 3(4) - 17 = -5$

The solution is $v = 4$, $u = -5$.

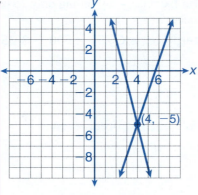

10. Use your graphing calculator to estimate the solution to the following system. (Note: Both equations first must be written in the form $y = mx + b$.)

$$342x - 167y = 418$$

$$-162x + 103y = -575$$

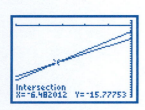

Use the window Xmin $= -10$, Xmax $= 0$, Xscl $= 2$, Ymin $= -28$, Ymax $= 0$, Yscl $= 2$.

The approximate solution is

$x \approx -6.48$,

$y \approx -15.78$.

11. The formula $A = P + Prt$ represents the value, A, of an investment of P dollars at a yearly simple interest rate, r, for t years.

a. Write an equation to determine the value, A, of an investment of \$100 at 8% for t years.

$A = 100 + 100(0.08)t = 100 + 8t$

b. Write an equation to determine the value, A, of an investment of \$120 at 5% for t years.

$A = 120 + 120(0.05)t = 120 + 6t$

c. The equations in parts a and b form a system of two linear equations. Solve this system using an algebraic approach.

$100 + 8t = 120 + 6t$

$2t = 20$

$t = 10$

d. Interpret your answer in part c.

The two investments will reach the same value in 10 years.

12. When warehouse workers use hand trucks to load a boxcar, it costs the management \$40 for labor for each boxcar. After management purchases a forklift for \$2000, it costs only \$15 for labor to load each boxcar.

a. Write a symbolic rule that describes the cost, $C(n)$, of loading n boxcars with a hand truck.

n represents the number of boxcars.

cost using hand trucks: $C(n) = 40n$ dollars

b. Write a symbolic rule that describes the total cost, $C(n)$, including the purchase price of the forklift, of loading n boxcars with the forklift.

cost using forklift: $C(n) = 2000 + 15n$

c. The equations in parts a and b form a system of linear equations. Solve the system using an algebraic approach.

$40n = 2000 + 15n$

$25n = 2000$

$n = 80$

d. Interpret your answer in part c in the context of the boxcar situation.

After 80 boxcars are loaded, the management will realize a profit on the purchase of the forklift.

13. The sign on the elevator in a seven-story building on campus states that the maximum weight the elevator can carry is 1200 pounds. As part of your work-study program, you need to move a large shipment of books to the sixth floor. Each box weighs 60 pounds.

a. Let n represent the number of boxes that you place in the elevator. If you weigh 150 pounds, write an expression that represents the total weight, w, in the elevator. Assume that only you and the boxes are in the elevator.

$w = 150 + 60n$

b. Using the expression in part a, write an inequality that can be used to determine the maximum number of boxes that you can place in the elevator at one time.

$150 + 60n \leq 1200$

c. Solve the inequality in part b.

$60n \leq 1050$

$n \leq 17.5$

The maximum number of boxes that can be placed in the elevator is 17.

Chapter 3 Summary

The bracketed numbers following each concept indicate the activity in which the concept is discussed.

CONCEPT/SKILL	DESCRIPTION	EXAMPLE
Function [3.1]	A function is a rule relating an input variable and an output variable in a way that assigns a single output value to each input value.	(see table below)

x	2	4	6	8	10
y	−1	1	2	3	4

CONCEPT/SKILL	DESCRIPTION	EXAMPLE
Vertical line test [3.1]	In the vertical line test, a graph represents a function if any vertical line drawn through the graph intersects the graph no more than once.	This graph is not a function.
Increasing function [3.1]	The graph of an increasing function rises to the right.	
Decreasing function [3.1]	The graph of a decreasing function falls to the right.	
Constant function [3.1]	The graph of a constant function is a horizontal line.	
Domain of a function [3.2]	The domain of a function is the collection of all meaningful input values. That is, those input values that produce a real-valued output.	$f(x) = \dfrac{3}{x + 1}$ The domain $= \{$real numbers, $x \neq -1\}$.

CONCEPT/SKILL	DESCRIPTION	EXAMPLE
Range of a function [3.2]	The range of a function is the collection of all possible output values.	$f(x) = x^2$ The range $= \{$real numbers $\geq 0\}$.
Practical domain and practical range [3.2]	The practical domain and range are determined by the conditions imposed on the situation being studied.	$C(x) = 35x + 15$ represents the cost to rent a car for x days. The practical domain $= \{$whole numbers $\geq 1\}$, and the practical range $= \{50, 85, 120, 155, \ldots\}$.
Function notation [3.2]	The notation $f(x)$ is a way to represent the output variable. f is the name of the function and x is the input variable. "$f(x)$" does not represent multiplication.	$f(x) = 2x - 6$
Ways to represent a function [3.2]	A function can be represented numerically by a table, graphically by a curve or scatterplot, verbally by stating how the output value is obtained for a given input, and symbolically using an algebraic rule.	
Delta notation for change [3.3]	Let y_1 and y_2 represent the output values corresponding to inputs x_1 and x_2, respectively. As the variable x changes in value from x_1 to x_2, the change in input is represented by $\Delta x = x_2 - x_1$ and the change in output is represented by $\Delta y = y_2 - y_1$.	Given the input/output pairs $(4, 14)$ and $(10, 32)$, $\Delta x = 10 - 4 = 6$ and $\Delta y = 32 - 14 = 18$.
Average rate of change over an interval [3.3]	The quotient $$\frac{\Delta y}{\Delta x} = \frac{y_2 - y_1}{x_2 - x_1}$$ is called the average rate of change of y (output) with respect to x (input) over the x-interval from x_1 to x_2.	<table><tr><td>**x**</td><td>−3</td><td>4</td><td>7</td><td>10</td></tr><tr><td>**y**</td><td>0</td><td>14</td><td>27</td><td>32</td></tr></table> The average rate of change over the interval from $x = 4$ to $x = 10$ is $$\frac{\Delta y}{\Delta x} = \frac{32 - 14}{10 - 4} = \frac{18}{6} = 3.$$

CONCEPT/SKILL	DESCRIPTION	EXAMPLE

Linear function [3.4]

A linear function is one whose average rate of change of output with respect to input from any one data point to any other data point is always the same (constant) value.

x	1	2	3	4
$f(x)$	10	15	20	25

The average rate of change between any two of these points is 5.

Graph of a linear function [3.4]

The graph of every linear function is a line.

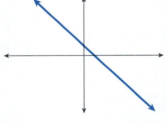

Slope of a line [3.4]

The slope of the line that contains the two points (x_1, y_1) and (x_2, y_2) is denoted by m; $m = \dfrac{\Delta y}{\Delta x} = \dfrac{y_2 - y_1}{x_2 - x_1}, (x_1 \neq x_2)$

The slope of the line containing the two points $(2, 7)$ and $(5, 11)$ is

$$\frac{11 - 7}{5 - 2} = \frac{4}{3}.$$

Positive slope [3.4]

The graph of every linear function with positive slope is a line rising to the right.

The output values increase as the input values increase.

A linear function is increasing if its slope is positive.

Negative slope [3.5]

The graph of every linear function with negative slope is a line falling to the right.

The output values decrease as the input values increase.

A linear function is decreasing if its slope is negative.

Zero slope [3.5]

The graph of every linear function with zero slope is a horizontal line.

Every point on a horizontal line has the same output value.

A linear function is constant if its slope is zero.

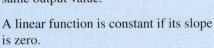

CONCEPT/SKILL	DESCRIPTION	EXAMPLE
Undefined slope [3.5]	A line whose slope is not defined (because its denominator is zero) is a vertical line. A vertical line is the only line that does not represent a function. Every point on a vertical line has the same input value.	
Horizontal intercept of a graph [3.5]	The horizontal intercept is the point at which the graph crosses the horizontal axis. Its ordered-pair notation is $(a, 0)$; that is, the second (y-) coordinate is equal to zero.	A line with horizontal intercept $(-3, 0)$ crosses the (input) x-axis 3 units to the left of the origin.
Vertical intercept of a graph [3.5]	The vertical intercept is the point at which the graph crosses the vertical axis. Its ordered-pair notation is $(0, b)$; that is, the first (x-) coordinate is equal to zero.	A line with vertical intercept $(0, 5)$ crosses the (output) y-axis 5 units above the origin.
Equation of a horizontal line [3.5]	The slope, m, of a horizontal line is 0, and the output of each of its points is the same constant value, c. Its equation is $y = c$.	An equation of the horizontal line through the point $(-2, 3)$ is $y = 3$.
Equation of a vertical line [3.5]	The input of each point on a vertical line is the same constant value, d. Its equation is $x = d$.	An equation of the vertical line through the point $(-2, 3)$ is $x = -2$.
Slope-intercept form of the equation of a line [3.6]	Represent the input variable by x, the output variable by y. Denote the slope of the line by m, the y-intercept by $(0, b)$. Then the coordinate pair, (x, y), of *every* point on the line satisfies the equation $y = mx + b$.	The line with equation $y = 3x + 4$ has a slope of 3 and y-intercept $(0, 4)$. The point $(2, 10)$ is on the line because its coordinates satisfy the equation $$10 = 3(2) + 4.$$
Identifying the slope and y-intercept [3.6]	The slope can be identified as the coefficient of the x-term. The y-intercept can be identified as the constant term.	$y = \dfrac{5}{2}x - 3$ The slope is $\dfrac{5}{2}$. The y-intercept is $(0, -3)$.
Using the slope and y-intercept to write an equation of the line [3.7]	Given a line whose slope is m and y-intercept is $(0, b)$, an equation of the line is $$y = mx + b.$$	An equation of the line with slope $\dfrac{2}{3}$ and y-intercept $(0, -6)$ is $$y = \dfrac{2}{3}x - 6.$$

CONCEPT/SKILL	DESCRIPTION	EXAMPLE
Determining the x-intercept of a line given its equation [3.7]	Because the x-intercept is the point whose y-coordinate is 0, set $y = 0$ in the equation and solve for x.	Given the line with equation $$y = 2x + 6,$$ set $y = 0$ to obtain equation $$0 = 2x + 6$$ $$-6 = 2x$$ $$-3 = x.$$ The x-intercept is $(-3, 0)$.
Determining an equation of a line given the coordinates of two points on the line [3.9]	Given (x_1, y_1) and (x_2, y_2), calculate the slope, $m = \dfrac{\Delta y}{\Delta x} = \dfrac{y_2 - y_1}{x_2 - x_1}$. The equation of the line can be written as $y = mx + b$, where m is now known, but b is not. Substitute the coordinates of either point into the equation, and then solve for b. Finally, write the equation $y = mx + b$ using the values of m and b that you have just calculated.	Given $(2, 1)$ and $(5, 10)$, the slope, m, of this line is $$\frac{10 - 1}{5 - 2} = \frac{9}{3} = 3.$$ The equation can be written as $$y = 3x + b.$$ Substitute $x = 2$ and $y = 1$: $$1 = 3 \cdot 2 + b$$ Solve for b to obtain $b = -5$. The equation of the line is $$y = 3x - 5.$$
Determining the point-slope equation of a line, $y - y_1 = m(x - x_1)$, given two points on the line. [3.9]	1. Determine the slope m 2. Choose either given point as (x_1, y_1) 3. Replace $m, x_1,$ and y_1 by their numerical values in $y - y_1 = m(x - x_1)$	Given $(2, 1)$ and $(5, 10)$, the slope of this line is $m = \dfrac{10 - 1}{5 - 2} = 3$. The equation is $$y - 1 = 3(x - 2)$$ $$y - 1 = 3x - 6$$ $$y = 3x - 5$$
Regression line [3.10]	The regression line is the line that best approximates a set of data points, even if all the points do not lie on a single line.	Input the data pairs into a graphing calculator or a spreadsheet computer application whose technology is preprogrammed to apply the "method of least squares" to determine the equation of the regression line.
System of equations [3.12]	Two equations that relate the same variables are called a system of equations. The solution of the system is the ordered pair(s) that satisfies the two equations.	For the system of equations $$2x + 3y = 1$$ $$x - y = 3,$$ the ordered pair $(2, -1)$ is a solution of the system because $x = 2, y = -1$ satisfy both equations.

CONCEPT/SKILL	DESCRIPTION	EXAMPLE

Graphical method for solving a system of linear equations [3.12]

Graph both equations on the same coordinate system. If the two lines intersect, then the solution to the system is given by the coordinates of the point of intersection.

Substitution method for solving a system of linear equations [3.12] and [3.13]

- Solve one equation for either variable, say y as an expression in x.
- In the other equation, replace y by the expression in x. (This equation should now contain only the variable x.)
- Solve this equation for x.
- Use this value of x to determine y.

For the system of equations

$$2x + 3y = 11$$
$$y = x - 3,$$

- In the second equation, y is already written as an expression in x.
- Rewrite the first equation $2x + 3(y) = 11$ as

$$2x + 3(x - 3) = 11.$$

- Solve this equation for x:

$$2x + 3x - 9 = 11$$
$$5x = 20$$
$$x = 4$$

- Solve for y:

$$y = x - 3, \text{ so } y = 4 - 3 = 1$$

Addition (or elimination) method for solving a system of linear equations [3.13]

Used when both equations are written in the form $ax + by = c$.
- Multiply each term in one equation, or both equations, by the appropriate constant that will make the coefficients of one variable, say y, opposites.
- Add the like terms in the two equations so that the y terms cancel.
- Solve the remaining equation for x.
- Use this value of x to determine y.

For the system of equations

$$3x + 2y = 5$$
$$5x + 4y = 7,$$

- Multiply *each* term in the top equation by -2, so that its y-coefficient will become -4:

$$-6x - 4y = -10$$

- Add like terms:

$$-6x - 4y = -10$$
$$\underline{5x + 4y = 7}$$
$$-x \qquad = -3$$

- Multiply each side by -1 to solve for x: $x = 3$.
- Substitute $x = 3$ into either original equation to determine y:

$$3(3) + 2y = 5$$
$$9 + 2y = 5$$
$$2y = -4; y = -2$$

CONCEPT/SKILL	DESCRIPTION	EXAMPLE
Solution set of an inequality [3.15]	The solution set of an inequality is the set of all values of the variable that satisfy the inequality.	The solution set to the inequality $2x + 3 \leq 9$ is all real numbers x which are less than or equal to 3.

Solving inequalities: direction of inequality sign [3.15]

Property 1. The direction of an inequality is not changed when

  i. The same quantity is added to or subtracted from both sides of the inequality. Stated algebraically, if $a < b$ then $a + c < b + c$ and $a - c < b - c$.

  ii. Both sides of an inequality are multiplied or divided by the same positive number. If $a < b$ then $ac < bc$, where $c > 0$ and $\frac{a}{c} < \frac{b}{c}$, where $c > 0$.

Property 2. The direction of an inequality is reversed if both sides of an inequality are multiplied by or divided by the same negative number. These properties can be written symbolically as

  i. if $a < b$, then $ac > bc$, where $c < 0$.

  ii. if $a < b$, then $\frac{a}{c} > \frac{b}{c}$, where $c < 0$.

Properties 1 and 2 will still be true if $a < b$ is replaced by $a \leq b$, $a > b$, or $a \geq b$.

For example, because $-4 < 10$, then

  i. $-4 + 5 < 10 + 5$, or
    $1 < 15$, is true;
    $-4 - 3 < 10 - 3$, or
    $-7 < 7$, is true.

  ii. $-4(6) < 10(6)$, or
    $-24 < 60$, is true;
    $\dfrac{-4}{2} < \dfrac{10}{2}$, or
    $-2 < 5$, is true.

Because $-4 < 10$, then $-4(-5) > 10(-5)$ or $20 > -50$.

Because $-4 < 10$, then $\dfrac{-4}{-2} > \dfrac{10}{-2}$ or $2 > -5$.

These properties will be true if $a < b$ is replaced by $a \leq b$, $a > b$, or $a \geq b$.

Solving inequalities of the form

$$f(x) < g(x)$$

[3.15]

Inequalities of the form $f(x) < g(x)$ can be solved using two different methods.

Method 1 is a numerical approach, in which a table of input/output pairs is used to determine values of x for which $f(x) < g(x)$.

Method 2 is an algebraic approach, in which the two properties of inequalities are used to isolate the variable.

Solve $5x + 1 \geq 3x - 7$.

 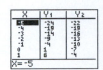

$$5x + 1 \geq 3x - 7$$
$$2x \geq -8$$
$$x \geq -4$$

1. Determine the slope and y-intercept of the line whose equation is $4y + 10x - 16 = 0$.

$4y = -10x + 16$

$y = -2.5x + 4$

The slope is -2.5.   The y-intercept is $(0, 4)$.

2. Determine the equation of the line through the points $(1, 0)$ and $(-2, 6)$.

$m = \dfrac{6 - 0}{-2 - 1} = \dfrac{6}{-3} = -2$

$0 = -2(1) + b$ $\qquad$ $y - 0 = -2(x - 1)$

$2 = b$ $\quad$ or $\quad$ $y = -2x + 2$

The equation is $y = -2x + 2$.

3. Determine the slope of the following line, and use the slope to determine an equation for the line.

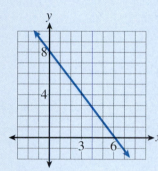

$m = \dfrac{rise}{run} = \dfrac{-8}{6}$

$m = \dfrac{-4}{3}$

The y-intercept is $(0, 8)$.

An equation for the line is

$y = \dfrac{-4}{3}x + 8$.

4. Match the graphs with the given equations. Assume that x is the input variable and y is the output variable. Each tick mark on the axes represents 1 unit.

c.
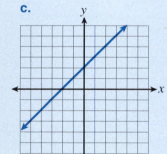

e.
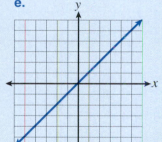

b.
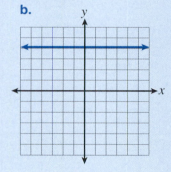

a. $y = x - 6$

d. $y = -3x - 5$

b. $y = 4$

e. $y = x$

c. $y = x + 2$

f. $y = 4 - 3x$

f.

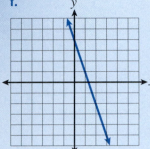

a.

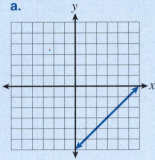

d.

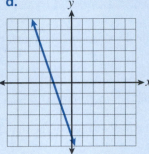

a. $y = x - 6$

b. $y = 4$

c. $y = x + 2$

d. $y = -3x - 5$

e. $y = x$

f. $y = 4 - 3x$

5. What is the equation of the line passing through the points $(0, 5)$ and $(2, 11)$? Write the final result in slope-intercept form, $y = mx + b$.

$$m = \frac{11 - 5}{2 - 0} = \frac{6}{2} = 3$$

The y-intercept is $(0, 5)$. The equation is $y = 3x + 5$.

6. The equation of a line is $5x - 10y = 20$. Determine the horizontal and vertical intercepts.

Set $x = 0$ Set $y = 0$

$5(0) - 10y = 20$ $5x - 10(0) = 20$

$-10y = 20$ $5x = 20$

$y = -2$ $x = 4$

The vertical intercept is $(0, -2)$.   The horizontal intercept is $(4, 0)$.

7. Determine the horizontal and vertical intercepts of the line whose equation is $-3x + 4y = 12$. Use the intercepts to graph the line.

$x = 0$ $-3(0) + 4y = 12$

$4y = 12$

$y = 3$

The vertical intercept is $(0, 3)$.

$y = 0$ $-3x + 4(0) = 12$

$-3x = 12$

$x = -4$

The horizontal intercept is $(-4, 0)$.

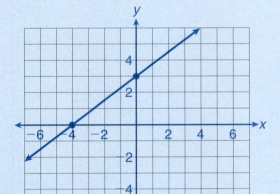

8. Determine the equation of the line that passes through the point $(0, -4)$ and has slope $\frac{5}{3}$.

$y = \frac{5}{3}x - 4$

9. Consider the lines represented by the following pair of equations: $3x + 2y = 10$ and $x - 2y = -2$.

  a. Solve the system algebraically.

  Using the addition method:

$$
\begin{array}{ll}
3x + 2y = 10 & \quad 3 \cdot 2 + 2y = 10 \\
\underline{x - 2y = -2} & \quad 6 + 2y = 10 \\
4x = 8 & \quad 2y = 4 \\
x = 2 & \quad y = 2
\end{array}
$$

  $(2, 2)$ is the solution.

  b. Determine the point of intersection of the lines by solving the system graphically.

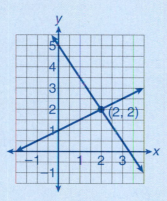

  c. Determine the point of intersection of the lines by solving the system using the tables feature of your calculator, if available.

X	Y8	Y9
0	5	1
1	3.5	1.5
2	2	2
3	.5	2.5
4	-1	3
5	-2.5	3.5
6	-4	4

X=2

The point of intersection is $(2, 2)$.

10. Solve the system of equations:

$b = 2a + 8$

$b = -3a + 3$

$2a + 8 = -3a + 3$

$5a = -5$

$a = -1$

$b = 2(-1) + 8$

$ = 6$

The solution is $a = -1, b = 6$.

11. Solve the system of equations:

$x = 10 - 4y$

$5x - 2y = 6$

$5x = 2y + 6$

$x = 0.4y + 1.2$

$10 - 4y = 0.4y + 1.2$

$-4.4y = -8.8$

$y = 2$

$x = 10 - 4(2)$

$x = 2$

The solution is $x = 2, y = 2$.

12. Corresponding values for p and q are given in the following table.

p	4	8	12	16
q	95	90	85	80

Write a symbolic rule for q as a linear function of p.

$$m = \frac{90 - 95}{8 - 4} = \frac{-5}{4} = -1.25$$

$$q = -1.25p + 100$$

13. The following table was generated by a linear function. Determine the symbolic rule that defines this function.

x	95	90	85	80	75
y	55.6	58.4	61.2	64.0	66.8

$$m = \frac{58.4 - 55.6}{90 - 95} = \frac{2.8}{-5} = -0.56$$

$64 = -0.56(80) + b$

$64 = -44.8 + b$ $y - 64 = -0.56(x - 80)$

$108.8 = b$ or $y - 64 = -0.56x + 44.8$

$y = -0.56x + 108.8$ $y = -0.56x + 108.8$

14. ABC car rental company offers midsize cars at \$24 per day and 30 cents per mile. Its competitor rents cars for \$34 a day and 25 cents per mile.

a. For each company, write a formula giving the cost, $C(x)$, of renting a car for a day as a function of the distance, x, traveled.

ABC car rental: $C(x) = 24 + 0.3x$

Competitor: $C(x) = 34 + 0.25x$

b. Sketch the graphs of both functions on the same axes.

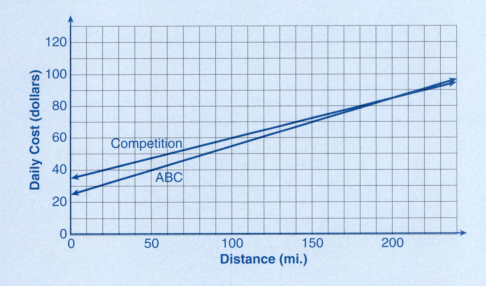

438

c. Use the graphs to determine when the two daily costs would be the same.

The lines intersect at (200, 84). The daily cost would be the same at both companies when 200 miles are traveled.

d. You intend to make a 225-mile return trip to Washington, D.C., from New York City. Which company would be cheaper to rent from?

It will be cheaper to rent from the competitor because the distance is 225 miles and the competitor's prices are cheaper if you travel over 200 miles.

15. Consider a graph of Fahrenheit temperature, °F, as a linear function of Celsius temperature, °C. You know that 212°F and 100°C represent the temperature at which water boils. You also know the input is °C; the output is °F, and that 32°F and 0°C each represent the freezing point of water.

a. What is the slope of the line through the points representing the boiling and freezing points of water?

Boiling (100°C, 212°F)

Freezing (0°C, 32°F)

$$m = \frac{32 - 212}{0 - 100} = \frac{-180}{-100} = \frac{9}{5}$$

b. What is the equation of the line?

$F = \frac{9}{5}C + 32$ or $F = 1.8C + 32$

c. Use the equation to determine the Fahrenheit temperature corresponding to 40°C.

$F = 1.8(40) + 32$

$\quad = 72 + 32$

$\quad = 104$ 104°F corresponds to 40°C.

d. Use the equation to determine the number of degrees Celsius corresponding to 170°F.

$170 = 1.8C + 32$

$138 = 1.8C$

$76.6 \approx C$ 170°F corresponds approximately to 77°c.

16. Your power company currently charges $20.63 per month as a fixed basic service charge plus a variable amount based on the number of kilowatt-hours used. The following table displays several sample monthly usage amounts and the corresponding bill for each.

a. Complete the table of electricity cost, rounding to the nearest cent.

KILOWATT-HOURS, k (kWh)	COST, C(k) ($)
600	84.35
650	89.66
700	94.97
750	100.28
800	105.59
850	110.90
900	116.21
950	121.52
1000	126.83

b. Sketch a graph of the electricity cost function, labeling the axes.

c. Write an equation for your total electric bill, $C(k)$, as a function of the number of kilowatt hours used, k.

$C(k) = 0.1062k + 20.63$

d. Determine the total cost if 875 kilowatt hours of electricity are used.

$C(875) = \$113.56$

The cost of using 875 kilowatt hours of electricity is $113.56.

e. Approximately how much electricity did you use if your total monthly bill is $150?

$150 = 0.1062k + 20.63$

$129.37 = 0.1062k$

$1218.17 = k$

I used approximately 1218 kilowatt hours of electricity.

17. Suppose that the median price, P, of a home rose from $50,000 in 1975 to $125,000 in 2005. Let t be the number of years since 1975.

a. Assume that the increase in housing prices has been linear. Determine an equation for the line representing price, P, in terms of t.

$m = \dfrac{125{,}000 - 50{,}000}{30 - 0} = \dfrac{75{,}000}{30} = 2500$

$P = 2500t + 50{,}000$ dollars

b. Use the equation from part a to complete the following table, where P represents the price in thousands of dollars.

t	0	10	20	30	40
P	50	75	100	125	150

c. Interpret the practical meaning of the values in the last line of the table.

When $t = 40$, that is, in the year 2015, the median price of the house will be $150,000.

18. For tax purposes, you may have to report the value of assets, such as a car. The value of some assets depreciates, or drops, over time. For example, a car that originally cost $20,000 may be worth only $10,000 a few years later. The simplest way to calculate the value of an asset is by using *straight-line depreciation*,

440

which assumes that the value is a linear function of time. If a \$950 refrigerator depreciates completely (to zero value) in 10 years, determine a formula for its value, $V(t)$, as a function of time, t.

$$\begin{array}{l}(0, 950) \\ (10, 0)\end{array}; \quad m = \frac{950 - 0}{0 - 10} = -95; \quad V(t) = -95t + 950$$

19. A coach of a local basketball team decides to analyze the relationship between his players' heights and their weights.

PLAYER	HEIGHT (in.)	WEIGHT (lb.)
1	68	170
2	72	210
3	78	235
4	75	220
5	71	175

a. Use your graphing calculator to determine the equation of the regression line for this data.

The regression equation is approximately
$w(h) = 6.92h - 302$.

```
LinReg
y=ax+b
a=6.921768707
b=-301.9047619
r²=.8721857164
r=.9339088373
```

b. Player 6 is 80 inches tall. Predict his weight.

$w(80) = 6.92(80) - 302 = 251.6 \approx 252$ lb.

c. If a player weighs 190 pounds, how tall would you expect him to be?

$$190 = 6.92h - 302$$
$$492 = 6.92h$$
$$71.1 \approx h$$

I would expect him to be approximately 71.1 inches tall.

20. Use the substitution method to solve the following system.

$5x + 2y = 18$

$y = x + 2$

$\begin{aligned}5x + 2(x + 2) &= 18 \\ 5x + 2x + 4 &= 18 \\ 7x &= 14 \\ x &= 2\end{aligned}$
$\qquad$
$\begin{aligned}y &= x + 2 \\ y &= 2 + 2 \\ y &= 4\end{aligned}$

The solution is $(2, 4)$.

21. Use the addition method to solve the following system.

$3x + 5y = 27$

$x - y = 1$

$\begin{aligned}3x + 5y &= 27 \\ \underline{5x - 5y} &= \underline{5} \\ 8x &= 32 \\ x &= 4\end{aligned}$
$\qquad$
$\begin{aligned}x - y &= 1 \\ 4 - y &= 1 \\ y &= 3\end{aligned}$

The solution is $(4, 3)$.

22. Solve the following inequalities algebraically.

a. $2x + 3 \geq 5$

$2x + 3 \geq 5$

$2x \geq 2$

$x \geq 1$

b. $3x - 1 \leq 4x + 5$

$3x - 1 \leq 4x + 5$

$-x \leq 6$

$x \geq -6$

23. The weekly revenue, $R(x)$, and cost, $C(x)$, generated by a product can be modeled by the following equations.

$$R(x) = 75x$$
$$C(x) = 50x + 2500$$

Determine when $R(x) \geq C(x)$ numerically, algebraically, and graphically.

Numerically:

x	R(x)	C(x)
0	0	2500
50	3750	5000
100	7500	7500
150	11,250	10,000
200	15,000	12,500

Graphically:

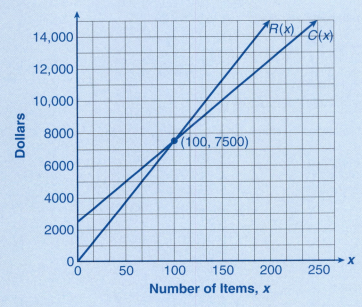

Algebraically:

$75x \geq 50x + 2500$

$25x \geq 2500$

$x \geq 100$

The revenue generated by the sale of x items exceeds the cost of producing them when 100 or more items are produced and sold.

An Introduction to Nonlinear Problem Solving

Cluster 1 — Mathematical Modeling Involving Polynomials

Activity 4.1

Fatal Crashes

Objectives

1. Identify polynomials and polynomial functions.

2. Classify a polynomial as a monomial, binomial, or trinomial.

3. Determine the degree of a polynomial.

4. Simplify a polynomial by identifying and combining like terms.

5. Add and subtract polynomials.

6. Evaluate and interpret polynomials.

In a recent year, approximately 43,000 people died in motor vehicle accidents in the United States. Does a driver's age have any relationship to his or her chances of being involved in a fatal car crash?

1. To compare the fatalities of one age group against another, you must use relative data (in the form of number of fatal crashes per 100 million miles driven). The number of fatal vehicle crashes, per 100 million miles driven, can be approximated by the function

$$F(x) = 0.013x^2 - 1.19x + 28.24,$$

where x is the age of the driver, $x \geq 16$.

a. Use the given function rule to complete the following table. Round your answers to the nearest tenth.

x, AGE OF DRIVER	F(x), FATAL CRASHES PER 100 MILLION MILES DRIVEN
16	12.5
18	11.0
20	9.6
25	6.6
35	2.5
45	1.0
55	2.1
65	5.8
75	12.1
79	15.4

b. Interpret the value $F(20)$ in the context of this problem.

According to the table value, for every 100 million miles driven by 20-year-olds, there are 9.6 (approximately 10) crashes that result in fatalities.

c. Which age group is involved in

    i. the highest number of fatal car crashes per 100 million miles driven?

        The 79-year-olds tend to be involved in the greatest number of fatal car crashes.

    ii. the lowest number of fatal crashes per 100 million miles driven?

        The 45-year-olds tend to be involved in the least number of fatal car crashes.

Polynomials

The expression $0.013x^2 - 1.19x + 28.24$ in Problem 1 of this activity is a special type of algebraic expression called a **polynomial**. Polynomials are formed by adding and subtracting terms containing positive integer powers of a variable such as x and possibly a constant term. Most of the algebraic expressions that you have encountered in this text have been polynomials.

The function $F(x) = 0.013x^2 - 1.19x + 28.24$ is an example of a **polynomial function**.

The polynomial $0.013x^2 - 1.19x + 28.24$ contains three terms: an x^2 term, with its coefficient 0.013, an x term, with its coefficient -1.19, and a constant term, 28.24. Although the second term is being subtracted, the negative sign is assigned to its coefficient. Thus it can be written as a sum: $0.013x^2 + (-1.19)x + 28.24$.

Definition

In general, a **polynomial in one variable** is an algebraic expression consisting of a sum of terms, each term being a constant or a variable raised to a positive integer power and multiplied by a coefficient. The polynomial is in **standard form** when the terms are written with the powers of x in decreasing order.

Example 1 *Consider the polynomial $2x^4 - 5x^3 + 6x + 7$. The exponent on x in each of the first three terms is a positive integer. This polynomial is in standard form, because it is written with the powers of x in decreasing order. The coefficient of x^4 is 2, the coefficient of x^3 is -5, the coefficient of x is 6, and the constant term is 7. Notice that the x^2 term is missing. However, we can think of it as being there, with coefficient 0.*

Example 2

POLYNOMIALS	EXPRESSIONS THAT ARE NOT POLYNOMIALS
$10x$	$\frac{1}{x}$
$3x^2 + 5$	$\sqrt{x}$
$2x^3 - 3x^2 + \frac{4}{3}x - \frac{1}{2}$	$10x^2 + \frac{1}{x}$
$-10x^4$	$10x\sqrt{x^2 - 1}$

Note that in a polynomial expression, the variable cannot appear in a denominator or inside a radical.

2. Write yes if the expression is a polynomial. Write no if it is not a polynomial. In each case, give a reason for your answer.

   a. $3x + \frac{2}{5}$

   Yes, it *is* a polynomial because the first term is a positive integer power of *x* and the second term is a constant.

   b. $-10x$

   Yes, it *is* a polynomial because the term contains a positive integer power of *x*.

   c. $\frac{1}{3}x^3 + 2x - 1$

   Yes, it *is* a polynomial because each term contains a positive integer power of *x* or is a constant.

   d. $5x + \dfrac{10}{x}$

   No, it is *not* a polynomial because the variable in the second term is in the denominator.

   e. $\dfrac{x + 1}{x - 2}$

   No, it is *not* a polynomial because a variable term is present in the denominator.

   f. $\sqrt{3x + 1}$

   No, it is *not* a polynomial because a variable term is present inside a radical sign.

Terminology

Polynomials with one, two, or three terms can be classified by the number of terms they contain.

Definition

1. A **monomial** is a polynomial containing a single term. It may be a number (constant) or a constant times a variable raised to a positive integer exponent.

   Examples: -3, $2x^4$, $\frac{1}{2}s^2$

2. A **binomial** is a polynomial that has exactly two terms.

   Examples: $4x^3 + 2x$, $3t - 4$, $5w^6 + 4w^2$

3. A **trinomial** is a polynomial that has exactly three terms.

   Examples: $3x^4 - 5x^2 + 10$, $5x^2 + 3x - 4$

3. Identify each of the following polynomials as a monomial, binomial, or trinomial.

    a. $15x$ __monomial__

    b. $10x^3$ __monomial__

    c. $3x^2 + 10$ __binomial__

    d. $5x^4 + 2x^3 - 1$ __trinomial__

The **degree of a monomial** is defined as the exponent on its variable. The monomial $4x^5$ has degree 5. The monomial $2x$ has degree 1. The monomial consisting of a constant, such as 8, is said to have degree 0. The **degree of a polynomial** is the highest degree among all its terms. The degree of $2x^3 + 3x^7 - 5x^2 + 10$ is 7. When a polynomial is written in standard form, its degree is the degree of its first term.

4. Determine the degree of each polynomial. Then write the polynomial in standard form.

    a. $5 + 7x$

        The degree is 1.

        Standard form is $7x + 5$.

    b. $31 - 6x^3 + 11x^2$

        The degree is 3.

        Standard form is
        $-6x^3 + 11x^2 + 31$.

    c. $-3x + 7x^2 - 8$

        The degree is 2.

        Standard form is $7x^2 - 3x - 8$.

    d. $15 + x^5 - 2x^2 + 9x$

        The degree is 5.

        Standard form is
        $x^5 - 2x^2 + 9x + 15$.

    e. 5

        The degree is 0.

        Standard form is 5.

5. A polynomial function of degree one should look familiar to you. In general, it takes the form $f(x) = ax + b$. What type of function is this?

It is a linear function, with slope $= a$ and y-intercept $(0, b)$.

The fatal-crash function defined by $F(x) = 0.013x^2 - 1.19x + 28.24$ is a second-degree polynomial function. Such functions are called **quadratic functions** and are good models for a variety of situations. The graphs of quadratic functions are U-shaped curves called **parabolas**. The properties of these graphs and the algebra of quadratic (second-degree) polynomials are explored in Cluster 2.

Simplifying, Adding, and Subtracting Polynomials

The algebra of polynomials is an extension of the procedures you already know. The basic algebraic manipulations with polynomials involve expanding using the distributive property and combining like terms.

Example 3 *Simplify the polynomial $5x^2 - 7 + 2(3x^2 - 4x + 1)$.*

SOLUTION

$5x^2 - 7 + 2(3x^2 - 4x + 1)$ Remove the parentheses by applying the distributive property.

$= 5x^2 - 7 + 6x^2 - 8x + 2$ Combine like terms.

$= 11x^2 - 8x - 5$

6. Simplify the following expressions.

   a. $-2(x + 3) + 5(3x - 1) + 3x$

   $-2x - 6 + 15x - 5 + 3x = 16x - 11$

   The simplified expression is the polynomial $16x - 11$.

   b. $(2x^2 - 7) - (4x - 3)$

   $2x^2 - 7 - 4x + 3 = 2x^2 - 4x - 4$

   The simplified expression is the polynomial $2x^2 - 4x - 4$.

   c. $2(4x^2 + 3x - 1) + 3(-2x^2 + 1) - 5x^2$

   $8x^2 + 6x - 2 - 6x^2 + 3 - 5x^2 = -3x^2 + 6x + 1$

   The simplified expression is the polynomial $-3x^2 + 6x + 1$.

7. In tennis, the length of a singles court is 3 feet less than three times its width.

   a. Let w represent the width of a singles court. Express the length of the court, l, as a function of w.

   The length, l, of the court in terms of its width, w, is given by the expression $3w - 3$, that is, $l = 3w - 3$.

   b. Express the perimeter, P, as a function of w.

   The perimeter of the rectangular court is determined by the expression $2l + 2w$, where l represents the length and w represents the width of the court. Then, in terms of w alone the perimeter is given by $2(3w - 3) + 2w$; that is, $P = 2(3w - 3) + 2w$.

   c. Simplify the function in part b.

   $P = 6w - 6 + 2w = 8w - 6$

   The simplified function is $P = 8w - 6$.

The approach used to simplify polynomials can also be used to add and subtract polynomials.

Example 4   *Add the polynomials $3x^3 - 5x^2 + 7x - 1$ and $9x^2 - 7x + 2$.*

SOLUTION

Often it is easiest to match up like terms by writing one polynomial above the other, with like terms lined up in columns.

$$\begin{array}{rl} 3x^3 - 5x^2 + 7x - 1 & \\ \underline{9x^2 - 7x + 2} & \text{Sum like terms in columns.} \\ 3x^3 + 4x^2 \ + 1 & \text{The sum is } 3x^3 + 4x^2 + 1. \end{array}$$

---

8. Add the following polynomials.

   a. $-2x^2 + 7x - 4$ and $4x^2 - 3x + 2$

   Note to instructor: The vertical method is shown for part a only.

   $$\begin{array}{r} -2x^2 + 7x - 4 \\ \underline{4x^2 - 3x + 2} \\ 2x^2 + 4x - 2 \end{array}$$

   b. $4x^3 + 15x - 4$ and $4x^2 - 3x + 20$

   $= 4x^3 + 4x^2 + 12x + 16$

   c. $125x^3 - 14x^2 - 13$ and $-100x^3 + 25x^2 - x + 11$

      $= 25x^3 + 11x^2 - x - 2$

   d. $3x^2 + 5x - 7, -7x^2 + 2x - 2,$ and $x^2 - 3x + 2$

      $= -3x^2 + 4x - 7$

Example 5   *Subtract $2x^3 - 9x^2 - 7x + 2$ from $3x^3 - 5x^2 + 7x - 1$.*

SOLUTION

Recall that subtracting $2x^3 - 9x^2 - 7x + 2$ is the same as adding its opposite. The opposite of $2x^3 - 9x^2 - 7x + 2$ is

$$-1(2x^3 - 9x^2 - 7x + 2) = -2x^3 + 9x^2 + 7x - 2.$$

Notice that the coefficient of each term in the opposite polynomial is the opposite of each coefficient in the original polynomial. Once again, you can write the polynomials in a vertical format and combine like terms:

$$\begin{array}{l} 3x^3 - 5x^2 + 7x - 1 \\ \underline{-2x^3 + 9x^2 + 7x - 2} \qquad \text{Add the opposite polynomial.} \\ x^3 + 4x^2 + 14x - 3 \end{array}$$

  9. Perform the indicated operations, and express your answer in simplest form.

    a. $(10x^2 + 2x - 4) - (6x^2 - 3x + 2)$

     Note to instructor: The vertical method is shown for part a only.

$$\begin{array}{l} 10x^2 + 2x - 4 \\ \underline{-6x^2 + 3x - 2} \\ 4x^2 + 5x - 6 \end{array}$$

    b. $(4x^3 + 13x - 4) - (4x^2 - 3x + 10)$

      $= 4x^3 + 13x - 4 - 4x^2 + 3x - 10$

      $= 4x^3 - 4x^2 + 16x - 14$

    c. $(3x^2 + 2x - 3) - (8x^2 + x - 2) + (x^2 - 3x + 2)$

      $= 3x^2 + 2x - 3 - 8x^2 - x + 2 + x^2 - 3x + 2$

      $= -4x^2 - 2x + 1$

Evaluating Polynomial Functions

The stopping distance, $D(v)$ (in feet), for a car moving at a velocity (speed) of v miles per hour is given by the following function:

$$D(v) = 0.04v^2 + 1.1v$$

Note that the function rule for $D(v)$ is a binomial in the variable v.

  10. a. Determine the value $D(55)$.

     $D(55) = 0.04(55)^2 + 1.1(55) = 181.5$ ft.

    b. Interpret the meaning of this value.

     A car traveling 55 miles per hour requires 181.5 feet to stop.

    c. Determine and interpret $D(80)$.

     $D(80) = 0.04(80)^2 + 1.1(80) = 344$ ft.

     A car traveling 80 miles per hour requires 344 feet to stop.

11. Evaluate each polynomial function for the given input value.

a. $P(a) = 15 - 2a$ Determine $P(8)$.

$P(8) = 15 - 2(8) = 15 - 16 = -1$

b. $g(x) = 2x^2 - 3x + 4$ Determine $g(-2)$.

$g(-2) = 2(-2)^2 - 3(-2) + 4 = 8 + 6 + 4 = 18$

c. $F(x) = -x^2 + 3x - 5$ Determine $F(-1)$.

$F(-1) = -(-1)^2 + 3(-1) - 5 = -1 - 3 - 5 = -9$

SUMMARY: ACTIVITY 4.1

1. A **monomial** is a single term consisting of either a number (constant) or a variable raised to a positive integer exponent along with its coefficient.

2. A **polynomial** is an algebraic expression formed by adding and subtracting monomials.

3. Polynomials can be categorized by the number of terms they contain: A **monomial** is a polynomial with one term; a **binomial** is a polynomial with exactly two terms, and a **trinomial** is a polynomial with three terms.

4. A polynomial is in **standard form** when its terms are written with the powers of x in decreasing order.

5. The **degree of a monomial** is defined as the exponent on its variable. The **degree of a polynomial** is the highest degree among all its terms.

6. A **polynomial function** is defined by $y = P(x)$, where $P(x)$ is a polynomial expression.

7. To simplify a polynomial expression, use the distributive property to expand and then combine like terms.

8. Polynomials are added and subtracted using the same techniques for simplifying polynomials.

EXERCISES: ACTIVITY 4.1

1. a. Write yes if the expression is a polynomial. Write no if the expression is not a polynomial. In each case, give a reason for your answer.

i. $4x - 10$

Yes, it *is* a polynomial because the first term is a positive integer power of x and the second term is a constant.

ii. $3x^2 + 2x - 1$

Yes, it *is* a polynomial because each term contains a positive integer power of x or is a constant.

iii. $\frac{2}{3}a$

Yes, it *is* a polynomial because the term contains a positive integer power of a.

iv. $\dfrac{1}{2}t^2 + \dfrac{4}{\sqrt{t}}$

No, it is *not* a polynomial because the variable in the second term is inside a radical sign and in the denominator.

v. $\dfrac{2}{w} - 8$

No, it is *not* a polynomial because a variable term is present in the denominator.

vi. $-x^2 + 3x^4 - 9$

Yes, it *is* a polynomial because each term contains a positive integer power of x or is a constant.

b. Classify each expression as a monomial, binomial, or trinomial.

   i. $4x - 10$ ___binomial___   **ii.** $3x^2 + 2x - 1$ ___trinomial___

   iii. $\dfrac{2}{3}a$ ___monomial___   **iv.** $-4x^3$ ___monomial___

2. a. How many terms are in the expression $2x^2 + 3x - x - 3$ as written?

There are four terms.

b. How many terms are in the simplified expression?

The simplified expression is $2x^2 + 2x - 3$.

It contains three terms.

In Exercises 3–12, perform the indicated operations, and express your answer in simplest form.

3. $3x + 2(5x - 4)$
$= 3x + 10x - 8$
$= 13x - 8$

4. $(3x - 1) + 4x$
$= 3x - 1 + 4x$
$= 7x - 1$

5. $3.1(a + b) + 8.7a$
$= 3.1a + 3.1b + 8.7a$
$= 11.8a + 3.1b$

6. $6x + 2(x - y) - 5y$
$= 6x + 2x - 2y - 5y$
$= 8x - 7y$

7. $(3 - x) + (2x - 1)$
$= 3 - x + 2x - 1$
$= x + 2$

8. $(5x^2 + 4) - (3x^2 - 2)$
$= 5x^2 + 4 - 3x^2 + 2$
$= 2x^2 + 6$

9. $9.5 - (3.5 - x)$
$= 9.5 - 3.5 + x$
$= 6 + x$

10. $4(x^3 + 2x) + 5(x^2 + 2x - 1)$
$= 4x^3 + 8x + 5x^2 + 10x - 5$
$= 4x^3 + 5x^2 + 18x - 5$

11. $3(2 - x) - 4(2x + 1)$
$= 6 - 3x - 8x - 4$
$= 2 - 11x$

12. $10(0.3x + 1) - (0.2x + 3)$
$= 3x + 10 - 0.2x - 3$
$= 2.8x + 7$

In Exercises 13–18, evaluate the polynomial function at the given value.

13. $f(x) = 3x^2 + 2x - 1$; $f(4)$

$f(4) = 3(4)^2 + 2(4) - 1$

$= 48 + 8 - 1$

$= 55$

14. $p(l) = 2(l + 2.8)$; $p(3.5)$

$p(3.5) = 2(3.5 + 2.8)$

$= 2(6.3)$

$= 12.6$

15. $V(r) = \dfrac{4\pi}{3}r^3$; $V(6)$

$V(6) = \frac{4}{3}\pi(6)^3$

$= \frac{4}{3}\pi(216) = 288\pi \approx 905$

16. $S(d) = 0.3d^2 + 4d$; $S(2.1)$

$S(2.1) = 0.3(2.1)^2 + 4(2.1)$

$= 0.3(4.41) + 8.4$

$= 1.323 + 8.4$

$= 9.723$

17. $C(x) = 4x^3 + 15$; $C(10)$

$C(10) = 4(10)^3 + 15$

$= 4000 + 15$

$= 4015$

18. $D(m) = 3(m - 20)$; $D(-5)$

$D(-5) = 3(-5 - 20)$

$= 3(-25)$

$= -75$

19. One side of a square is increased by 5 units and the other side is decreased by 3 units. If x represents a side of the original square, write a polynomial expression that represents the perimeter of the newly formed rectangle.

$2(l + w) = 2(x + 5 + x - 3) = 2(2x + 2) = 4x + 4$

The perimeter is determined by the polynomial expression $4x + 4$.

20. You want to boast to a friend about the stock that you own without telling him how much money you originally invested in the stock. Let your original stock be valued at x dollars. You watch the market once a month for 4 months and record the following.

Month	1	2	3	4
Stock Value	increased $10	doubled	decreased $4	increased by 50%

a. Use x as a variable to represent the value of your original investment, and write a polynomial expression to represent the value of your stock after the first month.

At the end of the first month my stock's value is given by $x + 10$.

b. Use the result from part a to write a polynomial expression to determine the value at the end of the second month, simplifying when possible. Continue until you determine the value of your stock at the end of the fourth month.

$2(x + 10) = 2x + 20$ gives the value of the stock at the end of the second month.

$2x + 20 - 4 = 2x + 16$ gives the value of the stock at the end of the third month.

$1.5(2x + 16) = 3x + 24$ gives the value of the stock at the end of the fourth month.

c. Do you have good news to tell your friend? Explain to him what has happened to the value of your stock over 4 months.

At the end of 4 months, I have more than 3 times the amount I started with.

d. Instead of simplifying after each step, write a single expression that represents the stock's value at the end of the 4-month period.

$1.5[2(x + 10) - 4]$

e. Simplify the expression in part d. How does the simplified expression compare with the result in part b?

$1.5(2(x + 10) - 4) = 1.5(2x + 20 - 4) = 1.5(2x + 16) = 3x + 24$

The result is the same as in part b.

21. Add the following polynomials.

a. $-5x^2 + 2x - 4$ and $4x^2 - 3x + 6$

Note to instructor: The vertical method is shown for part a only.

$$\begin{array}{r} -5x^2 + 2x - 4 \\ 4x^2 - 3x + 6 \\ \hline -x^2 - x + 2 \end{array}$$

b. $7x^3 + 12x - 9$ and $-4x^2 - 3x + 20$

$= 7x^3 - 4x^2 + 9x + 11$

c. $45x^3 - 14x^2 - 13$ and $-50x^3 + 20x^2 - x + 13$

$= -5x^3 + 6x^2 - x$

d. $x^2 + x - 7$, $-9x^2 + 2x - 3$, and $5x^2 - x + 2$

$= -3x^2 + 2x - 8$

22. Perform the indicated operations, and express your answer in simplest form.

a. $(13x^2 + 12x - 5) - (6x^2 - 3x + 2)$

Note to instructor: The vertical method is shown for part a only.

$$\begin{array}{r} 13x^2 + 12x - 5 \\ -6x^2 + 3x - 2 \\ \hline 7x^2 + 15x - 7 \end{array}$$

b. $(3x^3 + x - 4) - (3x^2 - 3x + 11)$

$= 3x^3 + x - 4 - 3x^2 + 3x - 11$

$= 3x^3 - 3x^2 + 4x - 15$

c. $(6x^2 + 2x - 3) - (7x^2 + x - 2) + (x^2 - 3x + 1)$

$= 6x^2 + 2x - 3 - 7x^2 - x + 2 + x^2 - 3x + 1$

$= -2x$

d. $(-5x^3 - 4x^2 - 7x - 8) - (-5x^3 - 4x^2 - 7x + 8)$

$= -5x^3 - 4x^2 - 7x - 8 + 5x^3 + 4x^2 + 7x - 8$

$= -16$

23. According to statistics from the U.S. Department of Commerce, the per capita personal income (i.e., the average annual income) of each resident of the United States from 1960 to 2004 can be modeled by the polynomial function

$$P(t) = 12.4584t^2 + 191.317t + 1427.7418,$$

where t equals the number of years since 1960 and $P(t)$ represents the per capita income.

a. Use $P(t)$ to estimate the per capita personal income in the year 1989.

$P(29) = 12.4584(29)^2 + 191.317(29) + 1427.7418 \approx 17{,}453$

The per capita income for 1989 is approximately $17,453.

b. Determine the value of $P(48)$. Interpret the meaning of the answer in the context of this situation.

$P(48) = 12.4584(48)^2 + 191.317(48) + 1427.7418 \approx 39{,}315$

$t = 48$ corresponds to the year 2008. If the polynomial function model continues to be valid past the year 2004, then the per capita income for U.S. residents in the year 2008 should be approximately $39,315.

24. An international rule for determining the number, n, of board feet (usable finished lumber) in a 16-foot log is modeled by the equation

$$n(d) = 0.22d^2 - 0.71d,$$

where d is the diameter of the log in inches.

a. How many board feet can be obtained from a 16-foot log with a 14-inch diameter?

$n(14) = 0.22(14)^2 - 0.71(14) \approx 33$ board ft.

Approximately 33 board feet can be obtained from a 16-foot log with a 14-inch diameter.

b. Determine and interpret $n(20)$.

$n(20) = 0.22(20)^2 - 0.71(20) = 73.8$ board ft.

Almost 74 board feet can be obtained from a 16-foot log with a 20-inch diameter.

Activity 4.2

Volume of a Storage Box

Objectives

1. Use properties of exponents to simplify expressions and combine powers that have the same base.

2. Use the distributive property and properties of exponents to write expressions in expanded form.

You are building a wooden storage box to hold your vinyl record collection. The box has three dimensions: length, l, width, w, and height, h, as shown in the following figure.

Recall that the formula for the volume V of a box (also called a *rectangular solid*) is $V = lwh$. The units of volume are cubic units, such as cubic inches, cubic feet, cubic meters, and so on.

1. Determine the volume of a storage box whose length is 3 feet, width is 2 feet, and height is 1.5 feet.

    The volume is $3 \cdot 2 \cdot 1.5 = 9$ cubic feet.

You decide to design a storage box with a square base.

2. What is the relationship between the length and width of a box with a square base?

    The length and width are equal.

3. Write the volume formula for this square-based box in terms of l and h in two ways:

    a. using three factors (including repeated factors).

        $V = llh$

    b. using two factors (including exponents).

        $V = l^2 h$

Now, suppose the box is to be a cube.

4. What is the relationship between the length, width, and height of a cube?

    A cube's length, width, and height are all identical.

5. Write the volume formula for a cube in terms of l in three ways:

    a. using three factors (including repeated factors)

        $V = lll$

    b. using two factors (including exponents)

        $V = l^2 l$

    c. using one factor (including exponents)

        $V = l^3$

6. Suppose the cube has length of 4 ft. Evaluate each of the expressions in Problem 5 to verify that all three expressions are equivalent.

    $V = 4 \cdot 4 \cdot 4 = 64\ \text{ft}^3;\quad V = 4^2 \cdot 4 = 64\ \text{ft}^3;\quad V = 4^3 = 64\ \text{ft}^3$

The three expressions lll, l^2l, and l^3 are equivalent expressions that all represent the volume of a cube with side of length l.

One of the most important skills of algebra is the ability to rewrite a given expression into an equivalent expression. In the case of the three expressions for volume of a cube, it is helpful to note that the symbol l itself can be written in exponential form as l^1. In this way you can see that

$$l^1 l^1 l^1 = l^2 l^1 = l^3$$

This equivalence illustrates the first Property of Exponents: When multiplying exponential factors with identical bases, you retain the base and add the exponents.

> **Property 1 of Exponents: Multiplying Powers with the Same Base**
>
> For any base b and positive integers m and n, $b^m b^n = b^{m+n}$.

7. Simplify the following products by rewriting each as a single power.

a. $x^2 x^5$ **b.** $ww^3 w^4$ **c.** $a^3 a^6$ **d.** $h^4 h$

x^7 w^8 a^9 h^5

When the exponential factors include coefficients, you can reorder the factors by first grouping the coefficients and then combining the exponential factors using Property 1.

8. Simplify the following products.

a. $2x^3 x^2$ **b.** $4w^3 3w^4$ **c.** $a^3 5a$ **d.** $6h^4 4h2h^2$

$= 2x^5$ $= 12w^7$ $= 5a^4$ $= 48h^7$

9. A box is constructed so that its length and height are each three times its width, w.

a. Represent the length and height in terms of w.

$l = 3w, \quad h = 3w$

b. Determine the volume of the box in terms of w.

$V = 3w \cdot 3w \cdot w = 9w^3$

10. Show how you would evaluate the expression $2^4 \cdot 3^2$. Are you able to add the exponents first?

$2^4 \cdot 3^2 = 16 \cdot 9 = 144$

No, the bases must be the same to add the exponents first.

Problem 10 illustrates that when the bases are different, the powers cannot be combined. For example, $x^4 y^2$ cannot be simplified. However, a product such as $(x^3 y^2)(x^2 y^5)$ can be rewritten as $x^3 x^2 y^2 y^5$, which can be simplified to $x^5 y^7$.

> **Procedure**
>
> **Multiplying a Series of Monomial Factors**
>
> **1.** Multiply the numerical coefficients.
>
> **2.** Use Property 1 of Exponents to simplify the product of the variable factors that have the same base. That is, add their exponents and keep the base the same.

11. Multiply the following.

a. $(-3x^2)(4x^3)$ **b.** $(5a^3)(3a^5)$

$= -12x^5$ $= 15a^8$

c. $(a^3 b^2)(ab^3)(b)$ **d.** $(3.5x)(-0.1x^4)$

$= a^4 b^6$ $= -0.35x^5$

e. $(r^2)(4.2r)$ **f.** $(4x^3 y^6)(2x^3 y)(3y^2)$

$= 4.2r^3$ $= 24x^6 y^9$

12. Use the distributive property and Property 1 of Exponents to write each of the following expressions in expanded, standard polynomial form.

a. $x^3(x^2 + 3x - 2)$ **b.** $-2x(x^2 - 3x + 4)$ **c.** $2a^3(a^3 + 2a^2 - a + 4)$

$\quad = x^3 x^2 + x^3 3x - x^3 2$ $= -2x\,x^2 - 2x(-3x) + -2x\,4$ $= 2a^6 + 4a^5 - 2a^4 + 8a^3$

$\quad = x^5 + 3x^4 - 2x^3$ $= -2x^3 + 6x^2 - 8x$

13. Rewrite each of the following in factored form.

a. $x^5 + 3x^4 - 2x^3$ **b.** $-2x^3 + 6x^2 - 8x$ **c.** $2a^6 + 4a^5 - 2a^4 + 8a^3$

$\quad x^3(x^2 + 3x - 2)$ $-2x(x^2 - 3x + 4)$ $2a^3(a^3 + 2a^2 - a + 4)$

Property 2 of Exponents: Dividing Powers Having the Same Base

Related to Property 1 is Property 2 of Exponents involving quotients, such as $\dfrac{x^5}{x^2}$, that have exponential factors with identical bases.

14. a. Complete the following table for the given values of x:

x	$\dfrac{x^5}{x^2}$	x^3
2	$\dfrac{2^5}{2^2} = \dfrac{32}{4} = 8$	$2^3 = 8$
3	$\dfrac{3^5}{3^2} = \dfrac{243}{9} = 27$	$3^3 = 27$
4	$\dfrac{4^5}{4^2} = \dfrac{1024}{16} = 64$	$4^3 = 64$

b. How does the table demonstrate that $\dfrac{x^5}{x^2}$ is equivalent to x^3?

For a given value of x, $\dfrac{x^5}{x^2}$ and x^3 produce the same value.

If you write both the numerator and denominator of $\dfrac{x^5}{x^2}$ as a chain of multiplicative factors, you obtain

$$\frac{x^5}{x^2} = \frac{x \cdot x \cdot x \cdot x \cdot x}{x \cdot x}.$$

You can then divide out the two factors of x in the denominator and the numerator to obtain x^3.

This suggests that when dividing exponential factors with identical bases, you retain the base and subtract the exponents.

Property 2 of Exponents: Dividing Powers with the Same Base

For any base $b \neq 0$ and positive integers m and n, $\dfrac{b^n}{b^m} = b^{n-m}$.

15. Simplify the following products and quotients.

a. $\dfrac{x^9}{x^2}$

x^7

b. $\dfrac{10x^8}{2x^5}$

$5x^3$

c. $\dfrac{c^3 2c^5}{3c^6}$

$\dfrac{2c^2}{3}$

d. $\dfrac{24z^{10}}{3z^3 4z}$

$2z^6$

e. $\dfrac{15d^4 2d}{3d^5}$

10

Property 3 of Exponents: Raising a Product or Quotient to a Power

Consider a cube whose side measures s inches. The surface of the cube consists of six squares—bottom and top, front, back, and left and right sides.

16. a. Determine a formula for the volume, V, of a cube whose side measures s inches.

$V = s^3$

b. Use the formula to calculate the volume of a cube whose side measures 5 inches. Include appropriate units of measure.

$V = 5^3 = 125$ in.^{3}

Suppose you construct a new cube whose side is twice as long as the cube in Problem 16a. The side of the new cube is represented by $2s$. The volume of this new cube can be determined using the formula $V = s^3$, replacing s with $2s$.

$$V = (2s)^3$$

This new formula can be simplified as follows.

$$V = (2s)^3 = 2s \cdot 2s \cdot 2s \quad = \quad 2 \cdot 2 \cdot 2 \cdot sss = 2^3 s^3$$

Therefore, the expression $(2s)^3$ is equivalent to $2^3 s^3$, or $8s^3$. In a similar fashion, the fourth power of the product xy, $(xy)^4$, is written as an equivalent expression in Problem 17.

17. a. Use the definition of an exponent to write the expression $(xy)^4$ as a repeated multiplication.

$(xy)^4 = (xy) \cdot (xy) \cdot (xy) \cdot (xy)$

b. How many factors of x are in your expanded expression?

There are four factors of x in the expanded expression.

c. How many factors of y are in your expanded expression?

There are four factors of y in the expanded expression.

d. Regroup the factors in the expanded expression so that the x factors are written first, followed by the y factors.

$x \cdot x \cdot x \cdot x \cdot y \cdot y \cdot y \cdot y$

e. Rewrite the expression in part d using exponential notation.

$x^4 \cdot y^4 = x^4 y^4$

The preceding results illustrate Property 3 of Exponents: When raising a product (or quotient) to a power, you raise each of its factors to that power.

> **Property 3 of Exponents: Raising a Product or Quotient to a Power**
>
> For any numbers a and b and positive integer n,
>
> $$(ab)^n = a^n b^n \quad \text{and} \quad \left(\frac{a}{b}\right)^n = \frac{a^n}{b^n}, \; b \neq 0.$$

18. Use Property 3 to simplify the following expressions.

   a. $(2a)^3$

   $= 2^3 a^3$

   $= 8a^3$

   b. $(3mn)^2$

   $= 3^2 m^2 n^2$

   $= 9m^2 n^2$

   c. $\left(\dfrac{10}{p}\right)^6$

   $= \dfrac{1{,}000{,}000}{p^6}$

19. a. The **surface area**, A, of a cube is the sum of the areas of its six sides (squares). Determine a formula for the surface area, A, of a cube whose side measures s inches.

   $A = 6s^2$

   b. Use your formula to calculate the surface area of a cube whose side measures 5 inches. Include appropriate units of measure.

   $A = 6 \cdot 5^2 = 6 \cdot 25 = 150 \text{ in.}^2$

   c. If you double the length of the side of the cube whose side measures s units, write a new formula, in expanded and simplified form, for the surface area.

   $A = 6(2s)^2 = 6 \cdot 2s \cdot 2s = 6 \cdot 2 \cdot 2 \cdot ss = 6 \cdot 2^2 s^2 = 24s^2$

20. Use properties of exponents to determine which of the following expressions are equivalent.

$$2(xy)^3 \qquad (2xy)^3 \qquad 2xy^3 \qquad 8x^3y^3$$

Expanding and simplifying each expression in the order given yields $2x^3y^3$, $8x^3y^3$, $2xy^3$, and $8x^3y^3$. The second and fourth expressions are the only equivalent expressions.

Property 4 of Exponents: Raising a Power to a Power

The final Property of Exponents is a combination of Property 1 and the definition of exponential notation itself. Consider the expression

$$a^2 \cdot a^2 \cdot a^2 \cdot a^2 \cdot a^2$$

Because this expression consists of a^2 written as a factor 5 times, you can use exponential notation to rewrite this product as $(a^2)^5$.

However, Property 1 indicates that $a^2 \cdot a^2 \cdot a^2 \cdot a^2 \cdot a^2$ can be rewritten as $a^{2+2+2+2+2}$, which simplifies to a^{10}.

Therefore, $(a^2)^5 = a^{10}$, illustrating Property 4 of Exponents: When raising a power to a power you multiply the exponents.

> **Property 4 of Exponents: Raising a Power to a Power**
>
> For any number b and positive integers m and n, $(b^m)^n = b^{mn}$.

21. Use the properties of exponents to simplify each of the following.

a. $(t^3)^5$

$= t^{15}$

b. $(y^2)^4$

$= y^8$

c. $(3^2)^4$

$= 3^8 = 6561$

d. $2(a^5)^3$

$= 2a^{15}$

e. $x(x^2)^3$

$= x(x^6) = x^7$

f. $-3(t^2)^4$

$= -3t^8$

g. $(5xy^2)(3x^4y^5)$

$= 15x^5y^7$

Defining Zero and Negative Integer Exponents

22. a. Determine the numerical value of the expression $\frac{2^3}{2^3}$.

$$\frac{2^3}{2^3} = \frac{8}{8} = 1$$

b. Apply Property 2 of Exponents, $\frac{b^n}{b^m} = b^{n-m}$, to simplify $\frac{2^3}{2^3}$.

$$\frac{2^3}{2^3} = 2^{3-3} = 2^0$$

c. What can you conclude about the numerical value of 2^0? Explain.

2^0 must be equal to 1 if Property 2 is true.

> **Definition**
>
> **Zero Exponent**
>
> For any nonzero number b, $b^0 = 1$.

23. a. Evaluate the expression $\frac{2^3}{2^4}$ and write your result as a fraction in simplest form.

$$\frac{2^3}{2^4} = \frac{8}{16} = \frac{1}{2}$$

b. Use Property 2 of exponents to simplify $\frac{2^3}{2^4}$.

$$\frac{2^3}{2^4} = 2^{3-4} = 2^{-1}$$

c. Use the result of part a to explain the meaning of the expression 2^{-1}.

2^{-1} is the fraction $\frac{1}{2}$ expressed in exponential form.

In Problem 23c, the fact that $2^{-1} = \frac{1}{2}$ demonstrates that meaning can be given to negative exponents.

24. a. Simplify the expression $\frac{b^3}{b^5}$ by dividing out common factors.

$$\frac{b^3}{b^5} = \frac{b^3 \cdot 1}{b^3 \cdot b^2} = \frac{b^3}{b^3} \cdot \frac{1}{b^2} = 1 \cdot \frac{1}{b^2} = \frac{1}{b^2}$$

b. Use Property 2 of exponents to simplify $\dfrac{b^3}{b^5}$.

$$\frac{b^3}{b^5} = b^{3-5} = b^{-2}$$

c. Use the result of part a to explain the meaning of the expression b^{-2}.

The expression b^{-2} is equivalent to the fraction $\dfrac{1}{b^2}$.

Problems 23 and 24 illustrate the meaning of negative exponents: Negative exponents indicate taking reciprocals.

Definition

Negative Exponents

For any nonzero number b and positive integer n, $b^{-n} = \dfrac{1}{b^n}$.

---

25. Simplify each of the following expressions. Write the result in both exponential and fraction form.

a. $\dfrac{b^0}{b^3} = \dfrac{1}{b^3}$, or b^{-3}

b. $\dfrac{x^5}{x^9} = \dfrac{1}{x^4}$, or x^{-4}

c. $\dfrac{y^{10}}{y^{12}} = \dfrac{1}{y^2}$, or y^{-2}

d. $\dfrac{12x^2}{3x^4} = \dfrac{4}{x^2}$, or $4x^{-2}$

SUMMARY: ACTIVITY 4.2

Properties of Exponents

1. **Multiplying Powers Having the Same Base**

   For any number b, if m and n represent positive integers, then $b^m \cdot b^n = b^{m+n}$.

2. **Dividing Powers Having the Same Base**

   For any nonzero number b, if m and n represent positive integers, then $\dfrac{b^m}{b^n} = b^{m-n}$.

3. **Raising a Product or a Quotient to a Power**

   For any numbers a and b, if n represents a positive integer, then

   $$(a \cdot b)^n = a^n \cdot b^n \quad \text{and} \quad \left(\frac{a}{b}\right)^n = \frac{a^n}{b^n}, \qquad b \neq 0.$$

4. **Raising a Power to a Power**

   For any number b, if m and n represent positive integers, then $(b^m)^n = b^{mn}$.

5. **Defining the Zero Exponent**

   For any nonzero number b, $b^0 = 1$.

6. **Defining Negative Exponents**

   For any nonzero number b, if n represents a positive integer, then $b^{-n} = \dfrac{1}{b^n}$.

EXERCISES: ACTIVITY 4.2

1. Refer to the properties and definitions of exponents, and explain how to simplify each of the following. (All variables are nonzero.)

a. $x \cdot x \cdot x \cdot x \cdot x = x^{1+1+1+1+1} = x^5$

The product is the base, x, raised to the sum of the exponents (Property 1).

b. $x^3 \cdot x^7 \cdot x = x^{3+7+1} = x^{11}$

The product is the base, x, raised to the sum of the exponents (Property 1).

c. $\dfrac{x^6}{x^4} = x^{6-4} = x^2$

The quotient is the base, x, raised to the difference of the exponents (Property 2).

d. $(x^4)^3 = x^{4 \cdot 3} = x^{12}$

The result is the base, x, raised to the product of the exponents (Property 4).

e. $x^{-3} = \dfrac{1}{x^3}$

The negative in the exponent means to take the reciprocal of the base x raised to the opposite of the exponent (Definition of Negative Exponents).

f. $y^0 = 1$

Any nonzero number raised to the 0 power equals 1.

g. $\dfrac{r^3}{r^8} = r^{-5} = \dfrac{1}{r^5}$

The quotient is the base, r, raised to the difference of the exponents (Property 2). Then, apply the definition of a negative exponent.

h. $(5w)^3 = 5^3 w^3 = 125w^3$

The result is obtained by raising each factor to the same power, 3 (Property 3).

In Exercises 2–22, use the properties of exponents to simplify the expressions.

2. $aa^3 = a^4$

3. $3xx^4 = 3x^5$

4. $y^2 y^3 y^4 = y^9$

5. $3t^4 5t^2 = 15t^6$

6. $-3w^2 4w^5 = -12w^7$

7. $3.4b^5 1.05b^3 = 3.57b^8$

8. $(a^5)^3 = a^{15}$

9. $4(x^2)^4 = 4x^8$

10. $(-x^{10})^5 = -x^{50}$

11. $(-3x^2)(-4x^7)(2x) = 24x^{10}$

12. $(-5x^3)(0.5x^6)(2.1y^2)$
$= -5.25x^9 y^2$

13. $(a^2 bc^3)(a^3 b^2)$
$= a^5 b^3 c^3$

14. $(-2s^2 t)(t^2)^3(s^4 t) = -2s^6 t^8$

15. $\dfrac{16x^7}{24x^3} = \dfrac{2}{3} x^4$

16. $\dfrac{10x^4 y^6}{5x^3 y^5} = 2xy$

17. $\dfrac{9a^4 b^2}{18a^3 b^2} = \dfrac{1}{2} a$

Exercise numbers appearing in color are answered in the Selected Answers appendix.

18. $\dfrac{x^4}{x^7} = x^{-3}$ or $\dfrac{1}{x^3}$

19. $\dfrac{4w^2z^2}{10w^2z^4} = \dfrac{2}{5}z^{-2}$ or $\dfrac{2}{5z^2}$

20. $11x^2(y^2)^0 = 11x^2$

21. $5a(4a^3)^2 = 80a^7$

22. $\left(\dfrac{3a^2}{2b}\right)^4 = \dfrac{81a^8}{16b^4}$

23. Use the properties of exponents and the distributive property to show how to simplify the expression $2x(x^3 - 3x^2 + 5x - 1)$.

$2x(x^3) - 2x(3x^2) + 2x(5x) - 2x(1)$

$= 2x^4 - 6x^3 + 10x^2 - 2x$

In Exercises 24–31, use the distributive property and the properties of exponents to expand the algebraic expression and write it as a polynomial in standard form.

24. $2x(x + 3)$

$= 2x^2 + 6x$

25. $y(3y - 1)$

$= 3y^2 - y$

26. $x^2(2x^2 + 3x - 1)$

$= 2x^4 + 3x^3 - x^2$

27. $2a(a^3 + 4a - 5)$

$= 2a^4 + 8a^2 - 10a$

28. $5x^3(2x - 10)$

$= 10x^4 - 50x^3$

29. $r^4(3.5r - 1.6)$

$= 3.5r^5 - 1.6r^4$

30. $3t^2(6t^4 - 2t^2 - 1.5)$

$= 18t^6 - 6t^4 - 4.5t^2$

31. $1.3x^7(-2x^3 - 6x + 1)$

$= -2.6x^{10} - 7.8x^8 + 1.3x^7$

32. Simplify the expressions $4(a + b + c)$ and $4(abc)$. Are the results the same? Explain.

$4(a + b + c) = 4a + 4b + 4c$

$4abc$ is simplified.

The expressions are not the same. The first expression can be simplified using the distributive property of multiplication over addition. It has three terms. The second expression is all one term, so the distributive property does not apply.

33. a. You are drawing up plans to enlarge your square patio. You want to triple the length of one side and double the length of the other side. If x represents a side of your square patio, write a formula for the area, A, of the new patio in terms of x.

$A = (3x)(2x) = 6x^2$

The area of the new patio is $6x^2$.

b. You discover from the plan that after doubling the one side of the patio, you must cut off 3 feet from that side to clear a bush. Write an expression in terms of x that represents the length of this side.

The length of the doubled side can be represented as $2x - 3$.

c. Use the result from part b to write a formula without parentheses to represent the area of the patio in part b. Remember that the length of the other side of the original square patio was tripled.

$A = (2x - 3)(3x) = 6x^2 - 9x$

34. A rectangular bin has the following dimensions:

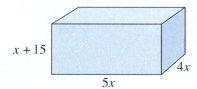

$x + 15$ $4x$ $5x$

a. Write a formula that represents the area, A, of the base of the bin.

$A = 5x(4x) = 20x^2$

b. Using the result from part a, write a formula that represents the volume, V, of the bin.

The volume of the bin is represented by $V = 20x^2(x + 15) = 20x^3 + 300x^2$.

35. A cube measures b^4 units on a side. Write a formula in terms of b that represents the volume of the cube.

The volume of the cube is represented by $V = (b^4)^3 = b^{12}$.

36. A square measures $3xy^2$ units on each side. Write an expression that represents the area of the square.

The area of the square is represented by $A = (3xy^2)^2 = 9x^2y^4$.

37. A car travels 4 hours at an average speed of $2a - 4$ miles per hour. Write an expression, without parentheses, that represents the distance traveled.

distance traveled $= 4(2a - 4) = 8a - 16$

The distance traveled is $8a - 16$ miles.

38. The radius of a cylindrical container is 3 times its height. Write a formula for the volume of the cylinder in terms of its height. Remember to indicate what each variable in your formula represents.

$r = 3h$, where r represents the radius and h represents the height

$V = \pi r^2 h$ is the formula for volume of a cylinder

$V = \pi(3h)^2 h = \pi \cdot 9h^2 h = 9\pi h^3$

The formula for the volume of this cylinder is $V = 9\pi h^3$.

39. A tuna fish company wants to make a tuna fish container whose radius is equal to its height. Write a formula for the volume of the desired cylinder in terms of its height. Be sure to indicate what each variable in your formula represents. Write the formula in its simplest form.

$r = h$; $V = \pi r^2 h$ is the formula for the volume of a cylinder with radius r and height h.

$V = \pi h^2 h = \pi h^3$

The formula for the volume of this cylinder is $V = \pi h^3$.

40. A manufacturing company says that the size of a can is the most important factor in determining profit. The company prefers the height of a can to be 4 inches greater than the radius. Write a formula for the volume of this desired cylindrical can in terms of its radius. Be sure to indicate what each variable in your formula represents. Write the formula in expanded form as a standard polynomial.

$h = r + 4$, where r represents the radius and h represents the height

$V = \pi r^2 h$ is the formula for volume of a cylinder

$V = \pi r^2 (r + 4)$

$V = \pi r^3 + 4\pi r^2$

The formula for the volume of this can is $V = \pi r^3 + 4\pi r^2$.

41. In the cylinder shown here, the relationship between the height and radius is expressed in terms of x. Write a formula, in expanded form, for the volume of the cylinder.

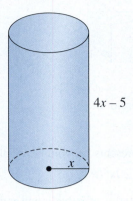

$4x - 5$

x

$V = \pi r^2 h$

$V = \pi x^2 (4x - 5) = 4\pi x^3 - 5\pi x^2$

The formula for the volume of this can is $V = 4\pi x^3 - 5\pi x^2$.

42. Determine the greatest common factor of each polynomial and then rewrite the expression in completely factored form.

a. $18x^7 + 27x^3 - 15x^2$

   $3x^2(6x^5 + 9x - 5)$

b. $14y^8 - 21y^5$

   $7y^5(2y^3 - 3)$

c. $6x^5 - 12x^4 + 9x^3$

   $3x^3(2x^2 - 4x + 3)$

d. $15t^9 - 25t^6 + 20t^2$

   $5t^2(3t^7 - 5t^4 + 4)$

Activity 4.3

Room for Work

Objectives

1. Expand and simplify the product of two binomials.

2. Expand and simplify the product of any two polynomials.

3. Recognize and expand the product of conjugate binomials: difference of squares.

4. Recognize and expand the product of identical binomials: perfect-square trinomials.

The basement of your family home is only partially finished. Currently, there is a 9-by-12-foot room that is being used for storage. You want to convert this room to a home office but want more space than is now available. You decide to knock down two walls and enlarge the room.

1. What is the current area of the storage room?

$A = 9(12) = 108$ sq. ft.

The current area of the storage room is 108 square feet.

2. If you extend the long side of the room 4 feet and the short side 2 feet, what will be the area of the new room?

$A = (9 + 2)(12 + 4) = 11(16) = 176$ sq. ft.

The area of the new room is 176 square feet.

3. Start with a diagram of the current room (9 by 12 feet), and extend the length 4 feet and the width 2 feet to obtain the following representation of the new room.

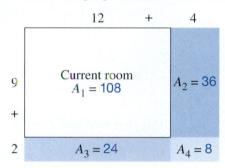

a. Calculate the area of each section, and record the results in the appropriate places on the diagram.

See preceding diagram.

$A_1 = 9(12) = 108$ sq. ft.

$A_2 = 9(4) = 36$ sq. ft.

$A_3 = 2(12) = 24$ sq. ft.

$A_4 = 2(4) = 8$ sq. ft.

b. Determine the total area of the new room by summing the areas of the four sections.

total area $= 108 + 36 + 24 + 8 = 176$ sq. ft.

c. Compare your answer from part a to the area you calculated in Problem 2.

The answers are the same.

You are not sure how much larger a room you want. However, you decide to extend the length and the width each by the same number of feet, represented by x.

4. Starting with a geometric representation of the current room (9 by 12 feet), you can extend the length and the width x feet to obtain the following diagram of the new room.

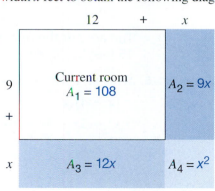

a. Determine the area of each section in your geometric model, and record your answer in the appropriate place. Note that the area in several sections will be an expression in x.

See diagram.

b. Write a formula for the total area by summing the areas of the four sections. Combine like terms, and write your answer as a polynomial in standard form.

$A = 108 + 9x + 12x + x^2 = x^2 + 21x + 108$

You can represent the area of the new room in an equivalent way by multiplying the expression for the new length by the expression for the new width. The new length of the room is represented by $12 + x$, and the new width is represented by $9 + x$, so the area of the room can be expressed as the product $(12 + x)(9 + x)$.

This product can be expanded using the distributive property twice. Each term of the first binomial is distributed over the entire second binomial. This means that each term of the first binomial multiplies every term in the second. This process results in a polynomial with four terms.

$$(12 + x)(9 + x) = \underbrace{12(9 + x)}_{\substack{\text{first column of}\\\text{geometric}\\\text{representation}}} + \underset{\text{plus}}{} \underbrace{x(9 + x)}_{\substack{\text{second column}\\\text{of geometric}\\\text{representation}}}$$

$$= 12 \cdot 9 + 12 \cdot x + x \cdot 9 + x \cdot x$$

This four-step multiplication process is often referred to as the FOIL (**F**irst + **O**uter + **I**nner + **L**ast) method, in which each letter refers to the position of the terms in the two binomials. FOIL is not a new method—it just describes a double use of the distributive property. However, it is a helpful reminder to perform all four multiplications when you are expanding the product of two binomial factors.

$$(12 + x)(9 + x) = \underbrace{12 \cdot 9}_{F} + \underbrace{12 \cdot x}_{O} + \underbrace{x \cdot 9}_{I} + \underbrace{x \cdot x}_{L}$$

Notice that $12 \cdot x$ and $x \cdot 9$ are like terms that can be combined as $21x$, so that the simplified product becomes $x^2 + 21x + 108$.

Compare the final expression with your result in Problem 4b. They should be the same.

5. The expressions $(12 + x)(9 + x)$ and $x^2 + 21x + 108$ are equivalent expressions that represent the area of a 9-by-12-foot room whose length and width have each been expanded x feet.

Use each of these expressions to determine the area of the remodeled room when $x = 3$ feet. What are the dimensions of the room? Remember to include the units.

$x^2 + 21x + 108$ $(12 + x)(9 + x)$

$3^2 + 21(3) + 108 = 180$ $12 + 3 = 15$ and $9 + 3 = 12$

          $(12 + 3)(9 + 3) = 15 \cdot 12 = 180$

The area of the remodeled room is 180 square feet and the dimensions are 15 feet by 12 feet.

6. Write $(x + 4)(x - 5)$ in expanded form, and simplify by combining like terms.

$$(x + 4)(x - 5) = x \cdot x + x(-5) + 4x + 4(-5)$$
$$= x^2 - 5x + 4x - 20$$
$$= x^2 - x - 20$$

7. Write $(3x + 1)(x + 2)$ in expanded form, and simplify by combining like terms.

$$(3x + 1)(x + 2) = 3x(x + 2) + 1(x + 2)$$
$$= 3x^2 + 6x + x + 2$$
$$= 3x^2 + 7x + 2$$

Product of Any Two Polynomials

A similar extension of the distributive property is used to expand the product of any two polynomials. Here again, each term of the first polynomial multiplies every term of the second.

Example 1 *Expand and simplify the product* $(2x + 3)(x^2 + 3x - 2)$.

SOLUTION

$$(2x + 3)(x^2 + 3x - 2)$$

$$= 2x \cdot x^2 + 2x \cdot 3x + 2x(-2) + 3 \cdot x^2 + 3 \cdot 3x + 3(-2)$$
$$= 2x^3 + 6x^2 - 4x + 3x^2 + 9x - 6$$
$$= 2x^3 + 9x^2 + 5x - 6$$

8. Expand and simplify the product $(x - 4)(3x^2 - 2x + 5)$.

$$(x - 4)(3x^2 - 2x + 5)$$
$$= x \cdot 3x^2 - x \cdot 2x + x \cdot 5 - 4 \cdot 3x^2 - 4 \cdot (-2x) + (-4) \cdot 5$$
$$= 3x^3 - 2x^2 + 5x - 12x^2 + 8x - 20$$
$$= 3x^3 - 14x^2 + 13x - 20$$

Difference of Squares

You are drawing up plans to change the dimensions of your square patio. You want to increase the length by 2 feet and reduce the width by 2 feet. This new design will change the shape of your patio, but will it change the size?

9. a. Determine the current area and new area of a square patio whose side originally measured

i. 10 ft.   **ii.** 12 ft.   **iii.** 15 ft.

Current area = 100 sq. ft.   Current area = 144 sq. ft.   Current area = 225 sq. ft.

New area = 96 sq. ft.   New area = 140 sq. ft.   New area = 221 sq. ft.

b. In all three cases, how does the proposed new area compare with the current area?

The new area seems to always be 4 square feet less than the original area.

You can examine why your observation in Problem 9b is always true regardless of the size of the original square patio. If x represents the length of a side of the square patio, the new area is represented by the product

$$(x + 2) \cdot (x - 2).$$

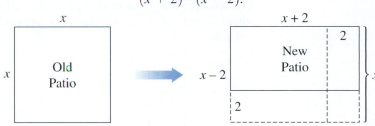

10. a. Represent the area of the old patio in terms of x.

The current area is represented by x^2.

   b. Represent the area of the new patio in terms of x. Expand and simplify this expression.

The new area would be $(x + 2)(x - 2) = x^2 - 2x + 2x - 4 = x^2 - 4$.

   c. How does the new area compare with the current area?

The new area is 4 square feet less than the original area.

11. Expand and simplify the following binomial products using the FOIL method. What do you notice about their expanded forms?

   a. $(x + 5)(x - 5)$

   $= x^2 - 5x + 5x - 25$

   $= x^2 - 25$

   b. $(3x + 4)(3x - 4)$

   $= (3x)^2 - 12x + 12x - 16$

   $= 9x^2 - 16$

The x terms have dropped out because the outer and inner products are opposites; some students may also note that the result is the difference of the squares of each of the original terms.

Look closely at the binomial factors in each product of Problem 11. In each case, the two binomial factors are **conjugate binomials**. That is, the first terms of the binomials are identical and the second terms differ only in sign (for example, $x + 5$ and $x - 5$).

When two conjugate binomials are multiplied, the like terms (corresponding to the outer and inner products) are opposites. Because opposites always sum to zero, the only terms remaining in the expanded product are the first and last products. This can be expressed symbolically as

$$(A + B)(A - B) = A^2 - B^2.$$

The expanded expression $A^2 - B^2$ is commonly called a **difference of squares**.

12. Any time two conjugate binomials are multiplied and expanded, the result is always a difference of squares. Use this pattern to write (with minimal calculation) the expanded forms of the following products.

   a. $(x + 3)(x - 3)$

   $= x^2 - 9$

   b. $(x - 10)(x + 10)$

   $= x^2 - 100$

   c. $(2x + 7)(2x - 7)$

   $= 4x^2 - 49$

   d. $(4x - 9)(4x + 9)$

   $= 16x^2 - 81$

   e. $(5x + 1)(5x - 1)$

   $= 25x^2 - 1$

   f. $(3x + 2)(3x - 2)$

   $= 9x^2 - 4$

Perfect-Square Trinomials

You rethink your plans to change the size of your square patio. You decide to extend the length of each side by 3 feet. By how many square feet will this new design enlarge your current patio?

13. a. Determine the current area and new area of a square patio if the side originally measured

i. 7 feet.   **ii.** 10 feet.   **iii.** 15 feet

Current area $= 49$ sq. ft. Current area $= 100$ sq. ft. Current area $= 225$ sq. ft.

New area $= 100$ sq. ft.   New area $= 169$ sq. ft.   New area $= 324$ sq. ft.

b. Determine the increase in patio area for each original patio dimension.

The 7-foot-sided patio is increased by 51 square feet.

The 10-foot-sided patio is increased by 69 square feet.

The 15-foot-sided patio is increased by 99 square feet.

c. Do the results in part b indicate that the new design will enlarge your current patio by a constant amount? Explain.

The increased size is not constant. It depends on the size of the original patio. The larger the original side length, the larger the increased area.

You can verify your observations in Problem 13c algebraically. If x represents the length of a side of the square patio, the new area is represented by the product

$$(x + 3)(x + 3).$$

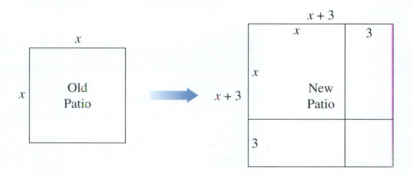

14. a. Represent the area of the current patio in terms of x.

The current area is represented by x^2.

b. Represent the area of the new patio in terms of x. Expand and simplify this expression.

The new area would be $(x + 3)(x + 3) = x^2 + 3x + 3x + 9$
$= x^2 + 6x + 9$.

c. Represent in terms of x the number of square feet by which has the patio been enlarged?

The patio has been expanded by $6x + 9$ square feet.

d. Is your result in part c consistent with the results from Problem 13c?

Yes, the amount that the patio will be enlarged is variable and depends upon the number of x feet that the patio is increased.

Yes, if $x = 7$ feet, $6x + 9 = 6 \cdot 7 + 9 = 51$ square feet.

Yes, if $x = 10$ feet, $6x + 9 = 6 \cdot 10 + 9 = 69$ square feet.

Yes, if $x = 15$ feet, $6x + 9 = 6 \cdot 15 + 9 = 99$ square feet.

15. Expand and simplify the following binomial products.

  a. $(x + 5)(x + 5)$

   $= x^2 + 5x + 5x + 25$

   $= x^2 + 10x + 25$

  b. $(x - 4)(x - 4)$

   $= x^2 - 4x - 4x + 16$

   $= x^2 - 8x + 16$

In each product of Problem 15, the binomial factors are identical. Note that the product $(x + 3)(x + 3)$ can be written as $(x + 3)^2$.

16. Rewrite each product of Problem 15 using exponential notation.

  a. $(x + 5)^2$

  b. $(x - 4)^2$

17. Look back in your solutions to Problem 15 for a relationship between the repeated binomial, $(x + a)^2$, and its expanded form.

  a. What is the relationship between the constant in the repeated binomial and the constant term in the expanded form?

   The constant term in the expanded form is the square of the constant in the binomial.

  b. What is the relationship between the coefficient of x in the repeated binomial and the coefficient of the x term in the expanded form?

   The coefficient of x is twice the constant in the binomial.

When a binomial is squared, the expanded form contains the square of each term *plus* twice the product of its terms. This is written symbolically as

$$(x + a)^2 = (x + a)(x + a) = x^2 + 2ax + a^2$$

$$\text{or } (x - a)^2 = (x - a)(x - a) = x^2 - 2ax + a^2.$$

The trinomials $x^2 + 2ax + a^2$ and $x^2 - 2ax + a^2$ are called a **perfect-square trinomial**.

Example 2   *Write the expanded form of $(x - 7)^2$.*

SOLUTION

$$\underbrace{x^2}_{\substack{\text{square} \\ \text{of the} \\ \text{first term}}} + \underbrace{2 \cdot (-7x)}_{\substack{\text{twice the} \\ \text{product of the} \\ \text{two terms}}} + \underbrace{49}_{\substack{\text{square of} \\ \text{the second} \\ \text{term}}} = x^2 - 14x + 49$$

18. Write the expanded form of each of the following using the pattern illustrated in Example 2.

  a. $(x + 8)^2$

   $= x^2 + 2(8x) + 64$

   $= x^2 + 16x + 64$

  b. $(x + 10)^2$

   $= x^2 + 2(10x) + 100$

   $= x^2 + 20x + 100$

  c. $(x - 6)^2$

   $= x^2 - 2(6x) + 36$

   $= x^2 - 12x + 36$

  d. $(x - 1)^2$

   $= x^2 - 2(1x) + 1$

   $= x^2 - 2x + 1$

e. $(x - 4)(x - 4)$
$\quad = x^2 - 2(4x) + 16$
$\quad = x^2 - 8x + 16$

f. $(x + 9)(x + 9)$
$\quad = x^2 + 2(9x) + 81$
$\quad = x^2 + 18x + 81$

19. Identify which of the following are perfect-square trinomials. Write each of the perfect-square trinomials as a square of the appropriate binomial.

a. $x^2 - 12x + 24$
no; 24 is not a square.

b. $x^2 - 4x + 4$
yes; $(x - 2)^2$

c. $x^2 - 8x - 16$
no; -16 is not a square.

d. $x^2 + 2x + 1$
yes; $(x + 1)^2$

SUMMARY: ACTIVITY 4.3

1. The product of two polynomials can be expanded using an extension of the distributive property. Each term of the first polynomial multiplies every term in the second.

2. **Conjugate binomials** are binomials whose first terms are identical and whose second terms differ only in sign: for example, $2x + 5$ and $2x - 5$.

3. When two conjugate binomials are multiplied, the expanded form is called a **difference of squares**.

$$(A + B)(A - B) = \underbrace{A^2 - B^2}_{\text{difference of squares}}$$

4. When a binomial is squared, the expanded form is called a **perfect-square trinomial**.

$$(x + a)^2 = (x + a)(x + a) = \underbrace{x^2 + 2ax + a^2}_{\text{perfect-square trinomial}}$$

$$(x - a)^2 = (x - a)(x - a) = x^2 - 2ax + a^2$$

EXERCISES: ACTIVITY 4.3

1. Multiply each of the following pairs of binomial expressions. Remember to simplify by combining like terms.

a. $(x + 1)(x + 7)$
$\quad = x^2 + 8x + 7$

b. $(w - 5)(w - 2)$
$\quad = w^2 - 7w + 10$

c. $(x + 3)(x - 2)$
$\quad = x^2 + x - 6$

d. $(x - 6)(x + 3)$
$\quad = x^2 - 3x - 18$

e. $(5 + 2c)(2 + c)$
$\quad = 10 + 9c + 2c^2$

f. $(x + 3)(x - 3)$
$\quad = x^2 - 9$

g. $(3x + 2)(2x + 1)$
$\quad = 6x^2 + 7x + 2$

h. $(5x - 1)(2x + 7)$
$\quad = 10x^2 + 33x - 7$

   i. $(6w + 5)(2w - 1)$ **j.** $(4a - 5)(8a + 3)$

       $= 12w^2 + 4w - 5$ $= 32a^2 - 28a - 15$

   k. $(x + 3y)(x - 2y)$ **l.** $(3x - 1)(x - 4)$

       $= x^2 + xy - 6y^2$ $= 3x^2 - 13x + 4$

  m. $(2a - b)(a - 2b)$ **n.** $(4c + d)(4c - d)$

       $= 2a^2 - 5ab + 2b^2$ $= 16c^2 - d^2$

2. Multiply the following. Write your answer in simplest form.

   a. $(x - 1)(3x - 2w + 5)$ **b.** $(x + 2)(x^2 - 3x + 5)$

       $= x(3x - 2w + 5) - 1(3x - 2w + 5)$ $= x(x^2 - 3x + 5) + 2(x^2 - 3x + 5)$

       $= 3x^2 - 2xw + 5x - 3x + 2w - 5$ $= x^3 - 3x^2 + 5x + 2x^2 - 6x + 10$

       $= 3x^2 - 2xw + 2x + 2w - 5$ $= x^3 - x^2 - x + 10$

   c. $(x - 4)(3x^2 + x - 2)$ **d.** $(a + b)(a^2 - 2b - 1)$

       $= x(3x^2 + x - 2) - 4(3x^2 + x - 2)$ $= a(a^2 - 2b - 1) + b(a^2 - 2b - 1)$

       $= 3x^3 + x^2 - 2x - 12x^2 - 4x + 8$ $= a^3 - 2ab - a + a^2b - 2b^2 - b$

       $= 3x^3 - 11x^2 - 6x + 8$

   e. $(x - 3)(x^2 + 3x + 9)$ **f.** $(x - 2y)(3x + xy - 2y)$

       $= x(x^2 + 3x + 9) - 3(x^2 + 3x + 9)$ $= x(3x + xy - 2y) - 2y(3x + xy - 2y)$

       $= x^3 + 3x^2 + 9x - 3x^2 - 9x - 27$ $= 3x^2 + x^2y - 2xy - 6xy - 2xy^2 + 4y^2$

       $= x^3 - 27$ $= 3x^2 + x^2y - 8xy - 2xy^2 + 4y^2$

3. Expand and simplify the following binomial products.

   a. $(x - 4)^2$ **b.** $(x + 4)^2$

       $= x^2 - 8x + 16$ $= x^2 + 8x + 16$

   c. $(x - 4)(x + 4)$ **d.** $(2x - 3)(2x + 3)$

       $= x^2 - 16$ $= 4x^2 - 9$

   e. $(x - 20)^2$ **f.** $(x + 12)^2$

       $= x^2 - 40x + 400$ $= x^2 + 24x + 144$

4. a. You have a circular patio that you wish to enlarge. The radius of the existing patio is 10 feet. Extend the radius x feet and express the area of the new patio in terms of x. Leave your answer in terms of π. Do not expand.

   $A = \pi r^2$ is the formula for the area of a circle in terms of the radius.

   $r = 10 + x$

   $A = \pi(10 + x)^2$

  b. Use the FOIL method to express the factor $(10 + x)^2$ in the formula $A = \pi(10 + x)^2$ in expanded form.

   $A = \pi(10 + x)^2$

   $A = \pi(10 + x)(10 + x)$

   $A = \pi(100 + 20x + x^2)$

  c. You decide to increase the radius 3 feet. Use the formula for area you obtained in part a to determine the area of the new patio.

   $A = \pi(10 + 3)^2 = \pi(13)^2 = 169\pi$ sq. ft.

d. Now, use the formula for area you obtained in part b to determine the area of the patio. Compare your results with your answer in part c.

$A = \pi(100 + 20(3) + 3^2) = \pi(169) = 169\pi$ sq. ft.

The results in parts c and d are the same.

5. a. You have an old rectangular birdhouse just the right size for wrens. The birdhouse measures 7 inches long, 4 inches wide, and 8 inches high. What is the volume of the birdhouse?

$V = 7(4)(8) = 224$ cu. in.

b. You want to use the old birdhouse as a model to build a new birdhouse to accommodate larger birds. You will increase the length and width the same amount, x, and leave the height unchanged. Express the volume of the birdhouse in factored form in terms of x.

$V = (7 + x)(4 + x)(8)$

c. Use the FOIL method to expand the expression you obtained in part b.

$V = 8(28 + 11x + x^2)$
$V = 224 + 88x + 8x^2$

d. If you increase the length and width 2 inches each, determine the volume of the new birdhouse. How much have you increased the volume? By what percent have you increased the volume?

$V = (7 + 2)(4 + 2)(8) = 9(6)(8) = 432$ cu. in.

The volume increased $432 - 224 = 208$ cubic inches; $\dfrac{208}{224} \approx 92.9\%$.

e. Instead of increasing both the length and width of the house, you choose to decrease the length and increase the width the same amount, x. Express the formula for the volume of the new birdhouse in factored form in terms of x.

$V = (7 - x)(4 + x)(8)$

f. Use the FOIL method to expand the expression you obtained in part e.

$V = 8(28 + 3x - x^2)$
$V = 224 + 24x - 8x^2$

g. Use the result from part f to determine the volume of the birdhouse if the length is decreased 2 inches and the width is increased 2 inches. Check your result using the volume formula you obtained in part e.

$x = 2$
$V = 224 + 24(2) - 8(2)^2 = 224 + 48 - 32 = 240$ cu. in.
Check: $V = (7 - 2)(4 + 2)(8) = 5(6)(8) = 240$ cu. in.

Cluster 1   What Have I Learned?

1. Determine numerically and algebraically which of the following three algebraic expressions are equivalent. Select any three input values, and then complete the table. Describe your findings in a few sentences.

x	$3x + 2$	$3(x + 2)$	$3x + 6$
1	5	$3(3) = 9$	9
5	17	$3(7) = 21$	21
10	32	$3(12) = 36$	36

(Answers in table will vary according to input values chosen.)

$3(x + 2)$ and $3x + 6$ are equivalent because they generate identical outputs. $3x + 2$ is not equivalent to either of the other two expressions.

2. Explain the difference between the two expressions $-x^2$ and $(-x)^2$.

The expression $-x^2$ instructs you to square the input first, and then change the sign.

The expression $(-x)^2$ instructs you to change the sign of the input and then square the result.

3. **a.** Write an algebraic expression that does not represent a polynomial, and explain why the expression is *not* a polynomial.

(Answers will vary.) $4x^2 + \frac{4}{x}$

Polynomials may not contain a variable in the denominator.

b. Write a polynomial that has three terms.

(Answers will vary.) $5x^2 + 6x - 18$

4. Does the distributive property apply when simplifying the expression $3(2xy)$? Explain.

No. For the distributive property to be applied, one factor must be a sum or difference.

5. What effect does the negative sign to the left of the parentheses have in simplifying the expression $-(x - y)$?

It reverses the signs of the terms in parentheses:
$-(x - y) = -1(x - y) = -x + y.$

6. Simplify the product $x^3 \cdot x^3$. Explain how you obtained your answer.

$x^3 \cdot x^3 = x^6$. This is so because $x^3 \cdot x^3 = (x \cdot x \cdot x)(x \cdot x \cdot x) = x^6$.

7. Is there a difference between $(3x)^2$ and $3x^2$? Explain.

Yes; in $3x^2$, only x is squared, but in $(3x)^2$, both the 3 and the x are squared: $(3x)^2 = (3x)(3x) = 9x^2$.

8. As you simplify the following expression, list each mathematical principle that you use.

$$(x - 3)(2x + 4) - 3(x + 7) - (3x - 2) + 2x^3x^5 - 7x^8 + (2x^2)^4 + (-x)^2$$

$$= \underbrace{2x^2 - 2x - 12}_{\text{FOIL}} \quad \underbrace{-3x - 21 \qquad -3x + 2}_{\text{distributive property}} \quad \underbrace{+2x^8 - 7x^8 + 16x^8 + x^2}_{\substack{\text{properties of} \\ \text{exponents}}}$$

$$(-1)^2 = 1$$

$$= \underbrace{2x^8 - 7x^8 + 16^8}_{\text{like terms}} + \underbrace{2x^2 + x^2}_{\text{like terms}} + \underbrace{-2x - 3x - 3x}_{\text{like terms}} + \underbrace{-12 - 21 + 2}_{\text{like terms}}$$

$$= 11x^8 + 3x^2 - 8x - 31$$

9. Is $(x + 3)^2$ equivalent to $x^2 + 9$? Explain why or why not.

They are not equivalent expressions. (Explanations will vary.)

$(x + 3)^2 = (x + 3)(x + 3) = x^2 + 6x + 9 \neq x^2 + 9$

Or, if $x = 1$, $(x + 3)^2 = (1 + 3)^2 = 4^2 = 16$ and $x^2 + 9 = 1^2 + 9 = 10$.
The two expressions produce different outputs for the same input.
They are not equivalent.

Cluster 1   How Can I Practice?

1. Select any three input values, and then complete the following table. Then determine numerically and algebraically which of the following expressions are equivalent.

   a. $4x^2$ **b.** $-2x^2$ **c.** $(-2x)^2$

x	$4x^2$	$-2x^2$	$(-2x)^2$
-1	4	-2	4
2	16	-8	16
3	36	-18	36

   (Table values will vary.)

   $(-2x)^2 = (-2x)(-2x) = 4x^2$

   a and *c* are equivalent numerically because they generate the same output for identical inputs and algebraically because they simplify to identical expressions.

2. Use the properties of exponents to simplify the following. Assume no variable is zero.

   a. $3x^3 \cdot x$

   $= 3x^4$

   b. $-x^2 \cdot x^5 \cdot x^7$

   $= -x^{14}$

   c. $(2x)(-6y^2)(x^3)$

   $= -12x^4y^2$

   d. $8(2x^2)(3xy^4)$

   $= 48x^3y^4$

   e. $(p^4)^5 \cdot (p^3)^2$

   $= p^{20} \cdot p^6 = p^{26}$

   f. $(3x^2y^3)(4xy^4)(x^7y)$

   $= 12x^{10}y^8$

   g. $\dfrac{w^9}{w^5}$

   $= w^4$

   h. $\dfrac{z^5}{z^5}$

   $= z^0 = 1$

   i. $\dfrac{15y^{12}}{5y^5}$

   $= 3y^7$

   j. $\dfrac{x^8y^{11}}{x^3y^5}$

   $= x^5y^6$

   k. $\dfrac{21xy^7z}{7xy^5}$

   $= 3y^2z$

   l. $(a^4)^3$

   $= a^{12}$

   m. $3(x^3)^2$

   $= 3x^6$

   n. $(-x^3)^5$

   $= -x^{15}$

   o. $(-2x^3)(-3x^5)(4x)$

   $= 24x^9$

   p. $(-5x^2)(0.3x^4)(1.2y^3)$

   $= -1.8x^6y^3$

   q. $(-3s^2t)(t^2)^3(s^3t^5)$

   $= -3s^5t^{12}$

   r. $\dfrac{a^8b^{10}}{a^9b^{12}}$

   $= a^{-1}b^{-2} = \dfrac{1}{ab^2}$

   s. $\dfrac{6x^5z}{14x^5z^2}$

   $= \dfrac{3}{7}z^{-1} = \dfrac{3}{7z}$

3. Multiply each of the following polynomials. Give the result in standard form.

a. $3(x - 5)$

$= 3x - 15$

b. $2x(x + 7)$

$= 2x^2 + 14x$

c. $-(3x - 2)$

$= -3x + 2$

d. $-3.1(x + 0.8)$

$= -3.1x - 2.48$

e. $3x(2x^4 - 5x^2 - 1)$

$= 6x^5 - 15x^3 - 3x$

f. $x^2(x^5 - x^3 + x)$

$= x^7 - x^5 + x^3$

4. Rewrite each of the following in completely factored form.

a. $7x - 28$

$= 7(x - 4)$

b. $10x^2y - 15xy^2$

$= 5xy(2x - 3y)$

c. $-7y - 49$

$= -7(y + 7)$

d. $3xy^2 - 6xy + 9y$

$= 3y(xy - 2x + 3)$

e. $8xy^3 + 16x^2y - 24x^3y$

$= 8xy(y^2 + 2x - 3x^2)$

f. $a^5b^2 + 3a^3b^5$

$= a^3b^2(a^2 + 3b^3)$

5. For each of the following expressions, list the specific operations indicated, in the order in which they are to be performed. Your sequence of operations should begin as "Start with input x, then. . . ."

a. $12 + 5(x - 2)$

Start with input x; then

1. subtract 2
2. multiply by 5
3. add 12

b. $(x + 3)^2 - 12$

Start with input x; then

1. add 3
2. square result
3. subtract 12

c. $5(3x - 2)^2 + 1$

Start with input x; then

1. multiply by 3
2. subtract 2
3. square the result
4. multiply by 5
5. add 1

d. $-x^2 + 1$

Start with input x; then

1. square x
2. multiply by -1
3. add 1

e. $(-x)^2 + 1$

Start with input x; then

1. multiply by -1
2. square the result
3. add 1

6. Simplify the following algebraic expressions.

a. $7 - (2x - 3) + 9x$

$= 7 - 2x + 3 + 9x$

$= 7x + 10$

b. $4x^2 - 3(4x^2 - 7) + 4$

$= 4x^2 - 12x^2 + 21 + 4$

$= -8x^2 + 25$

c. $x(2x - 1) + 2x(x - 3)$

$= 2x^2 - x + 2x^2 - 6x$

$= 4x^2 - 7x$

d. $3x(x^2 - 2) - 5(x^3 + x^2) + 2x^3$

$= 3x^3 - 6x - 5x^3 - 5x^2 + 2x^3$

$= -5x^2 - 6x$

e. $3[7 - 5(a - b) + 7a] - 5b$

$= 3[7 - 5a + 5b + 7a] - 5b$

$= 3[2a + 5b + 7] - 5b$

$= 6a + 15b + 21 - 5b$

$= 6a + 10b + 21$

f. $4 - [2x + 3(x + 5) - 2] + 7x$

$= 4 - [2x + 3x + 15 - 2] + 7x$

$= 4 - [5x + 13] + 7x$

$= 4 - 5x - 13 + 7x$

$= 2x - 9$

7. Determine the product for each of the following and write the final result in simplest form.

a. $(x + 2)(x - 3)$

$= x^2 - x - 6$

b. $(a - b)(a + b)$

$= a^2 - b^2$

c. $(x + 3)(x + 3)$

$= x^2 + 6x + 9$

d. $(x - 2y)(x + 4y)$

$= x^2 + 2xy - 8y^2$

e. $(2y + 1)(3y - 2)$

$= 6y^2 - y - 2$

f. $(4x - 1)(3x - 1)$

$= 12x^2 - 7x + 1$

g. $(x + 3)(x^2 - x + 3)$

$= x^3 + 2x^2 + 9$

h. $(x - 1)(x^2 + 2x - 1)$

$= x^3 + x^2 - 3x + 1$

i. $(a - b)(a^2 - 3ab + b^2)$

$= a^3 - 4a^2b + 4ab^2 - b^3$

j. $(x^2 + 3x - 1)(2x^2 - x + 3)$

$= 2x^4 + 5x^3 - 2x^2 + 10x - 3$

8. Write an expression in simplest form that represents the area of the shaded region.

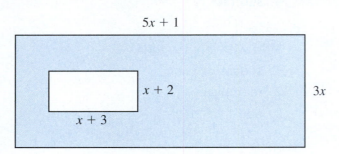

$5x + 1$

$x + 2$

$x + 3$

$3x$

$3x(5x + 1) - (x + 2)(x + 3)$

$= 15x^2 + 3x - (x^2 + 5x + 6)$

$= 15x^2 + 3x - x^2 - 5x - 6$

$= 14x^2 - 2x - 6$

9. A volatile stock began the last week of the year worth x dollars per share. The table shows the changes during that week. If you own 30 shares, express in symbolic form the total value of your stock at the end of the year.

Day	1	2	3	4	5
Change in Value/Share	doubled	lost 10	tripled	gained 12	lost half its value

$x \longrightarrow 2x \longrightarrow 2x - 10 \longrightarrow 6x - 30 \longrightarrow 6x - 18 \longrightarrow 3x - 9$ dollars per share

total value $= 30(3x - 9)$ dollars

10. a. You are drawing up plans to enlarge your square garden. You want to triple the length of one side and quadruple the length of the other side. If x represents a side of your square garden, draw a plan and write a formula for the new area in terms of x.

$A = (3x)(4x) = 12x^2$

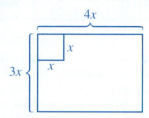

b. You discover from the plan that after tripling the one side of your garden, you must cut off 5 feet from that side to clear a bush. Draw this revised plan and express the area of your new garden in factored form in terms of x. Then, write your result as a polynomial in expanded form.

$A = (3x - 5)(4x) = 12x^2 - 20x$

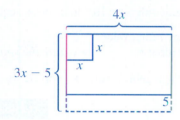

c. You also realize that if you add 5 feet to the side that you quadrupled, you could extend the garden to the walkway. What would be the new area of your garden with the changes to both sides incorporated into your plan? Draw this revision and express this area in factored form in terms of x, and expand the result.

$A = (3x - 5)(4x + 5) = 12x^2 - 5x - 25$

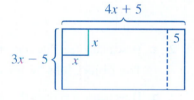

Cluster 2

Problem Solving with Quadratic Equations and Functions

Activity 4.4

The Amazing Property of Gravity

Objectives

1. Evaluate quadratic functions of the form $y = ax^2$.

2. Graph quadratic functions of the form $y = ax^2$.

3. Interpret the coordinates of points on the graph of $y = ax^2$ in context.

4. Solve a quadratic equation of the form $ax^2 = c$ graphically.

5. Solve a quadratic equation of the form $ax^2 = c$ algebraically by taking square roots.

6. Solve a quadratic equation of the form $(x \pm a)^2 = c$ algebraically by taking square roots.

Note: $a \neq 0$ in Objectives 1–5.

In the sixteenth century, scientists such as Galileo were experimenting with the physical laws of gravity. In a remarkable discovery, they learned that if the effects of air resistance are neglected, any two objects dropped from a height above Earth will fall at exactly the same speed. That is, if you drop a feather and a brick down a tube whose air has been removed, the feather and brick will fall at the same speed. Surprisingly, the function that describes the distance fallen by such an object in terms of elapsed time is a very simple one:

$$s = 16t^2,$$

where t represents the number of seconds elapsed and s represents distance (in feet) the object has fallen.

The algebraic rule $s = 16t^2$ indicates a sequence of two mathematical operations:

Start with a value
for input t $\longrightarrow$ *square the value* $\longrightarrow$ *multiply by 16* $\longrightarrow$ to obtain output s.

1. Use the algebraic rule to complete the table.

t (sec.)	s (ft.)
0	0
1	16
2	64
3	144

2. **a.** How many feet does the object fall 1 second after being dropped?

   The object falls 16 feet in 1 second.

   b. How many feet does the object fall 2 seconds after being dropped?

   The object falls 64 feet in 2 seconds.

3. **a.** Determine the average rate of change of distance fallen from time $t = 0$ to $t = 1$.

   $\begin{bmatrix} \text{average rate of change} \\ \text{from } t = 0 \text{ to } t = 1 \end{bmatrix} = \dfrac{16 - 0}{1 - 0} = 16$ ft./sec.

   b. What are the units of measurement of the average rate of change?

   The units of measurement of the average rate of change are feet per second.

   c. Explain what the average rate of change indicates about the falling object.

   The object falls at an average velocity (or speed) of 16 feet per second from $t = 0$ to $t = 1$ second.

4. Determine and interpret the average rate of change of distance fallen from time $t = 1$ to $t = 2$.

   $\begin{bmatrix} \text{average rate of change} \\ \text{from } t = 1 \text{ to } t = 2 \end{bmatrix} = \dfrac{64 - 16}{2 - 1} = 48$ ft./sec.

   The object falls at an average rate of 48 feet per second from $t = 1$ to $t = 2$ seconds.

5. Is the function $s = 16t^2$ a linear function? Explain your answer.

No; the function is not linear because the average rate of change is not constant.

6. a. If the object hits the ground after 5 seconds, determine the practical domain of the distance function.

The practical domain is the set of all numbers from 0 through 5.

b. On the following grid, sketch a curve representing the distance function over its practical domain.

c. Use your graphing calculator to verify the graph in part b in the window $0 \le x \le 5, 0 \le y \le 400$. Your graph should resemble the one below.

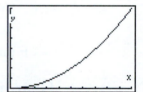

7. a. Does the point $(2.5, 100)$ lie on the graph? What do the coordinates of this point indicate about the falling object?

The graph in Problem 6b can be used to verify that $s = 100$ when $t = 2.5$.

It can also be verified algebraically or by using the table, trace, or value feature of a graphing calculator.

$$s = 16t^2$$
$$100 = 16(2.5)^2$$
$$100 = 100$$

The coordinates indicate that after 2.5 seconds, the object has fallen 100 feet.

b. Does the point $(4.5, 324)$ lie on the graph? What do the coordinates of this point indicate about the falling object?

The curve in part b should pass through $(4.5, 324)$.

It can be verified algebraically or by using the table, trace, or value feature of a graphing calculator.

$$s = 16t^2$$
$$324 = 16(4.5)^2$$
$$324 = 324$$

The coordinates indicate that after 4.5 seconds, the object has fallen 324 feet.

8. Use the graph in Problem 6b to estimate the amount of time it takes the object to fall 256 feet.

It takes the object 4 seconds to fall 256 feet.

Solving Equations of the Form $ax^2 = c$, $a \neq 0$

9. Use the algebraic rule $s = 16t^2$ to write an equation to determine the amount of time it takes the object to fall 256 feet.

Because 256 feet represents a value of s, replace s with 256 in the equation to obtain $256 = 16t^2$ or, equivalently, $16t^2 = 256$.

To solve this equation, reverse the order of operations indicated by the function rule, replacing each operation by its inverse. Here, one of the operations is "square a number." The inverse of squaring is to take a square root, denoted by the symbol $\sqrt{}$, called a **radical sign**.

Start with a value for $s \rightarrow$ *divide by 16* $\rightarrow$ *take its square root* $\rightarrow$ to obtain t.

In particular, if $s = 256$ feet:

Start with 256 $\rightarrow$ *divide by 16* $\rightarrow$ *take its square root* $\rightarrow$ to obtain t.

$$256 \div 16 = 16 \qquad \sqrt{16} = 4 \qquad t = 4$$

Therefore, it takes 4 seconds for the object to fall 256 feet.

10. Reverse the sequence of operations indicated by the algebraic rule $s = 16t^2$ to determine the amount of time it takes an object to fall 1296 feet, approximately the height of a 100-story building.

Start with 1296 $\rightarrow$ *divide by 16* $\rightarrow$ *take its square root* $\rightarrow$ to obtain t.

$$1296 \div 16 = 81 \qquad \sqrt{81} = 9 \qquad t = 9$$

It takes 9 seconds for the object to fall 1296 feet.

Graph of a Quadratic Function

Some interesting properties of the function defined by $s = 16t^2$ arise when you ignore the falling object context and consider just the algebraic rule itself.

Replace t with x and s with y in $s = 16t^2$ and consider the general equation $y = 16x^2$. First, by ignoring the context, you can allow x to take on a negative, positive, or zero value. For example, suppose $x = -5$, then

$$y = 16(-5)^2 = 16 \cdot 25 = 400.$$

11. a. Use the algebraic rule $y = 16x^2$ to complete the table. You already calculated the entries for $t = 0, 1, 2,$ and 3 when you answered Problem 1, so there is no need to recalculate those values. Just copy them into the appropriate boxes.

x	−5	−4	−3	−2.5	−2	−1.5	−1	−0.5	0	0.5	1	1.5	2	2.5	3	4	5
y	400	256	144	100	64	36	16	4	0	4	16	36	64	100	144	256	400

b. What pattern (symmetry) do you notice from the table?

For each x, $-x$ has the same output.

12. a. Sketch the graph of $y = 16x^2$, using the coordinates listed in the table in Problem 11. Scale the axes appropriately to plot the points and then draw a curve through them.

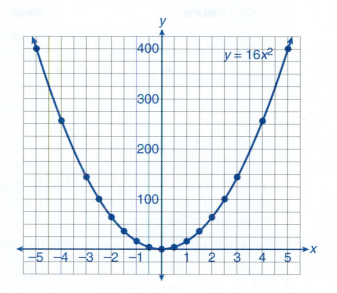

b. Use a graphing calculator to produce a graph of this function in the window $-5 \le x \le 5$, $-100 \le y \le 400$. Your graph should resemble the following one.

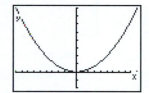

Recall from Activity 4.1 that second-degree polynomial functions are called *quadratic functions*. Their graphs are U-shaped curves called **parabolas**. The U-shaped graph of the quadratic function defined by $y = 16x^2$ is an example of a *parabola*.

Solving Equations of the Form $ax^2 = c, a \ne 0$: A Second Look

13. a. In the table in Problem 11a, how many points on the graph of $y = 16x^2$ lie 256 units above the x-axis?

Two points lie a height of 256 units above the x-axis.

b. Identify the points. What are their coordinates?

The coordinates of the points are $(-4, 256)$ and $(4, 256)$.

The x-values of the points on the graph of $y = 16x^2$ that lie 256 units above the x-axis can be determined algebraically by solving the quadratic equation $256 = 16x^2$. Example 1 demonstrates a systematic procedure to solve this equation.

Example 1   *Solve the equation* $16x^2 = 256$ *algebraically.*

SOLUTION

As discussed after Problem 9 (page 482), you can solve the equation algebraically by performing the operations in reverse order: Divide by 16 and then take the square root of both sides.

Step 1. Divide both sides by 16:

$$\frac{16x^2}{16} = \frac{256}{16} \text{ to obtain } x^2 = 16.$$

Step 2. Calculate the square root.

When x is no longer restricted to positive (or nonnegative) values only, there are *two* square roots of 16, one denoted by $\sqrt{16}$ and the other denoted by $-\sqrt{16}$. These two square roots are often written in condensed form as $\pm\sqrt{16}$. Since $\sqrt{16} = 4$ and $-\sqrt{16} = -4$, the solutions to the equation $x^2 = 16$ are $x = \pm 4$.

The solution to $16x^2 = 256$ can be written in the following systematic manner.

$$16x^2 = 256$$

$$\frac{16x^2}{16} = \frac{256}{16} \qquad \text{Divide both sides by 16.}$$

$$x^2 = 16 \qquad \text{Simplify.}$$

$$x = \pm\sqrt{16} \qquad \text{Take the positive and negative square roots.}$$

$$x = \pm 4.$$

14. a. How many points on the graph in Problem 12 lie 400 units above the x-axis?

   Two points lie a height of 400 units above the x-axis.

   b. Identify the points. What are their coordinates?

   The coordinates of the points are $(5, 400)$ and $(-5, 400)$.

   c. Set up the appropriate equation to determine the values of x for which $y = 400$, and solve it algebraically.

   $16x^2 = 400$

   $\frac{16x^2}{16} = \frac{400}{16} \qquad$ Divide both sides by 16.

   $x^2 = 25 \qquad$ Simplify.

   $x = \pm\sqrt{25} \qquad$ Take square roots.

   $x = \pm 5.$

15. Solve the following quadratic equations algebraically. Write decimal results to the nearest hundredth.

   a. $x^2 = 36$

   $x = \pm\sqrt{36}$

   $x = \pm 6$

   b. $2x^2 = 98$

   $\frac{2x^2}{2} = \frac{98}{2}$

   $x^2 = 49$

   $x = \pm\sqrt{49}$

   $x = \pm 7$

   c. $3x^2 = 375$

   $\frac{3x^2}{3} = \frac{375}{3}$

   $x^2 = 125$

   $x = \pm\sqrt{125}$

   $x \approx \pm 11.18$

   d. $5x^2 = 50$

   $\frac{5x^2}{5} = \frac{50}{5}$

   $x^2 = 10$

   $x = \pm\sqrt{10}$

   $x \approx \pm 3.16$

16. a. Refer to the graph in Problem 12a to determine how many points on the graph of $y = 16x^2$ lie 16 units *below* the x-axis.

No points lie below the x-axis.

b. Set up an equation that corresponds to the question in part a, and solve it algebraically.

$$-16 = 16x^2$$

$$\frac{-16}{16} = \frac{16x^2}{16}$$

$$-1 = x^2$$

c. How many solutions does this equation have? Explain.

There is no solution because there is no real number that, when squared, produces -1.

17. What does the graph of $y = 16x^2$ (Problem 12a) indicate about the number of solutions to the following equations? (You do not need to solve these equations.)

a. $16x^2 = 100$

There are two solutions.

b. $16x^2 = 0$

There is one solution.

c. $16x^2 = -96$

There is no solution.

18. Solve the following equations.

a. $5x^2 = 20$

$\frac{5x^2}{5} = \frac{20}{5}$

$x^2 = 4$

$x = \pm\sqrt{4}$

$x = \pm 2$

b. $4x^2 = 0$

$\frac{4x^2}{4} = \frac{0}{4}$

$x^2 = 0$

$x = \pm\sqrt{0}$

$x = 0$

c. $3x^2 = -12$

$\frac{3x^2}{3} = \frac{-12}{3}$

$x^2 = -4$

There is no solution.

Solving Equations of the Form $(x \pm a)^2 = c$

The following example demonstrates a procedure for solving a slightly more complex equation, such as $(x - 4)^2 = 9$.

Example 2 *Solve the equation $(x - 4)^2 = 9$.*

SOLUTION

The sequence of operations indicated by the expression on the left is

*Start with input $x \rightarrow$ **subtract 4** $\rightarrow$ **square** $\rightarrow$ to obtain output value* 9.

To solve the equation, reverse these steps by first taking square roots and then adding 4 as follows:

$$(x - 4)^2 = 9 \qquad \text{Take square roots}$$

$$x - 4 = \pm\sqrt{9}$$

$$x - 4 = \pm 3 \qquad \text{Rewrite as two separate equations}$$

$$x - 4 = 3 \quad \text{or} \quad x - 4 = -3 \quad \text{Solve each of these equations}$$

$$x = 7 \quad \text{or} \quad x = 1 \qquad \text{by adding 4.}$$

Therefore, $x = 7$ and $x = 1$ are solutions to the original equation $(x - 4)^2 = 9$.

19. Solve the following equations.

a. $(x + 4)^2 = 9$

$$x + 4 = \pm 3$$
$$x + 4 = -3 \quad \text{or} \quad x + 4 = 3$$
$$x = -7 \quad \text{or} \quad x = -1$$

b. $(x - 1)^2 = 25$

$$x - 1 = \pm 5$$
$$x - 1 = -5 \quad \text{or} \quad x - 1 = 5$$
$$x = -4 \quad \text{or} \quad x = 6$$

c. $(x + 5)^2 = 100$

$$x + 5 = \pm 10$$
$$x + 5 = -10 \quad \text{or} \quad x + 5 = 10$$
$$x = -15 \quad \text{or} \quad x = 5$$

d. $(x - 10)^2 = 70$

$$x - 10 = \pm\sqrt{70}$$
$$x - 10 = -\sqrt{70} \quad \text{or} \quad x - 10 = \sqrt{70}$$
$$x = 10 - \sqrt{70} \quad \text{or} \quad x = 10 + \sqrt{70}$$
$$x \approx 1.633 \quad \text{or} \quad x \approx 18.367$$

20. Suppose you have a 10-foot by 10-foot square garden plot. This year you want to increase the area of your plot and still keep its square shape.

a. Use the inner square of the sketch to represent your existing plot and the known dimensions. Use the variable x to represent the increase in length to each side. Write an expression for the length of a side of the larger square.

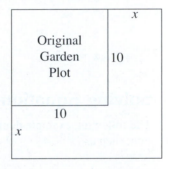

The length of each side of the larger square is 10 + x feet.

b. Express the area, A, of your new larger plot as the square of a binomial in function notation.

$A(x) = (10 + x)(10 + x) = (10 + x)^2$ sq. ft.

c. Suppose you decide to double the original area. Use the area expression in part b to write an equation to determine the required increase in the length of each side.

$$A(x) = 2 \cdot 100$$
$$(10 + x)^2 = 200$$

d. Solve the equation in part c.

$$10 + x = \pm\sqrt{200}$$
$$x = -10 \pm \sqrt{200}$$
$$x \approx 4.14, x \approx -24.14$$

Only positive values make sense here, since x represents a length.

   e. What is the length of each side of your new garden, to the nearest tenth of a foot? Check your answer by calculating the area of this new plot.

   You must expand the garden by 4.1 feet. The length of each side now is 14.1 feet.

   $A(x) = (10 + 4.1)^2 = 14.1^2 = 198.81$ square feet, which is close to 200 square feet.

SUMMARY: ACTIVITY 4.4

1. The graph of a function of the form $y = ax^2$, $a \neq 0$, is a U-shaped curve and is called a **parabola**. Such functions are called **quadratic functions**.

2. An equation of the form $ax^2 = c$, $a \neq 0$, is one form of a **quadratic equation**. Such equations can be solved algebraically by dividing both sides of the equation by a and then taking the positive and negative square roots of both sides.

3. An equation of the form $(x \pm a)^2 = c$ is solved algebraically by taking the positive and negative square roots of c and then solving the two resulting linear equations.

4. Every positive number a has two square roots, one positive and one negative. The square roots are equal in magnitude. The positive, or principal, square root of a is denoted by $\sqrt{a}$. The negative square root of a is denoted by $-\sqrt{a}$. The symbol $\sqrt{}$ is called the **radical sign**.

5. The **square root** of a number a is the number that when squared produces a. For example, the square roots of 25 are

$\sqrt{25} = 5$ because $5^2 = 25$

and

$-\sqrt{25} = -5$ because $(-5)^2 = 25$.

The square roots of 25 can be written in a condensed form as $\pm\sqrt{25} = \pm 5$.

EXERCISES: ACTIVITY 4.4

1. On Earth's Moon, gravity is only one-sixth as strong as it is on Earth, so an object on the Moon will fall one-sixth the distance it would fall on Earth in the same time. This means that the distance function for a falling object on the Moon is

$$s = \left(\frac{16}{6}\right)t^2, \text{ or } s = \frac{8}{3}t^2,$$

where t represents time since the object is released, in seconds, and s is the distance fallen, in feet.

   a. How far does an object on the Moon fall in 3 seconds?

   $s = \left(\frac{8}{3}\right)(3^2) = 24$. The object will fall 24 feet in 3 seconds.

b. How long does it take an object on the Moon to fall 96 feet?

$$96 = \frac{8}{3}t^2$$

$$36 = t^2$$

$$t = \sqrt{36} = 6$$

The object will fall 96 feet in 6 seconds.

c. Graph the Moon's gravity function on a properly scaled and labeled coordinate axis or on a graphing calculator for $t = 0$ to $t = 5$.

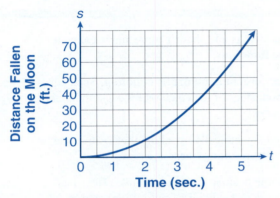

d. Use the graph to estimate how long it takes an object on the Moon to fall 35 feet.

It will take between 3.5 and 4 seconds for an object to fall 35 feet on the Moon.

e. Write an equation to determine the required time in part d. Solve the equation algebraically.

$$35 = \frac{8}{3}t^2$$

$$105 = 8t^2$$

$$t^2 = \frac{105}{8} = 13.125$$

$$t = \sqrt{13.125} \approx 3.6228.$$

The object will take approximately 3.62 seconds to fall 35 feet.

f. How does your answer in part e compare to your estimate in part d?

The answers are consistent.

2. Solve the following equations. Write decimal results to the nearest hundredth.

a. $5x^2 = 45$

$x^2 = 9$

$x = \pm\sqrt{9}$

$x = \pm 3$

b. $9x^2 = 0$

$x^2 = 0$

$x = 0$

c. $-25x^2 = 100$

$x^2 = -4$

There is no real number solution because a square of a real number cannot be negative.

d. $x^2 = 5$

$x^2 = 5$

$x = \pm\sqrt{5}$

$x \approx \pm 2.24$

e. $2x^2 = 20$

$x^2 = 10$

$x = \pm\sqrt{10}$

$x \approx \pm 3.16$

f. $\frac{x^2}{2} = 32$

$x^2 = 64$

$x = \pm\sqrt{64}$

$x = \pm 8$

3. Solve the following by first writing the equation in the form $x^2 = c$.

a. $t^2 - 49 = 0$

$t^2 = 49$

$t = \pm\sqrt{49}$

$t = \pm 7$

b. $15 + c^2 = 96$

$c^2 = 96 - 15$

$c^2 = 81$

$c = \pm\sqrt{81}$

$c = \pm 9$

c. $3a^2 - 21 = 27$

$3a^2 = 27 + 21$

$3a^2 = 48$

$a^2 = 16$

$a = \pm\sqrt{16}$

$a = \pm 4$

4. Solve the following equations.

a. $(x + 1)^2 = 16$

$x + 1 = \pm 4$

$x + 1 = 4$ or $x + 1 = -4$

$x = 3$ or $x = -5$

b. $(x - 1)^2 = 4$

$x - 1 = \pm 2$

$x - 1 = 2$ or $x - 1 = -2$

$x = 3$ or $x = -1$

c. $(x + 7)^2 = 64$

$x + 7 = \pm 8$

$x + 7 = 8$ or $x + 7 = -8$

$x = 1$ or $x = -15$

d. $(x - 9)^2 = 81$

$x - 9 = \pm 9$

$x - 9 = 9$ or $x - 9 = -9$

$x = 18$ or $x = 0$

e. $3(x - 2)^2 = 75$

$(x - 2)^2 = 25$

$x - 2 = \pm 5$

$x - 2 = 5$ or $x - 2 = -5$

$x = 7$ or $x = -3$

f. $5(x + 3)^2 = 80$

$(x + 3)^2 = 16$

$x + 3 = \pm 4$

$x + 3 = 4$ or $x + 3 = -4$

$x = 1$ or $x = -7$

g. $(2x - 5)^2 = 81$

$(2x - 5) = \pm 9$

$2x - 5 = 9$ or $2x - 5 = -9$

$2x = 14$ or $2x = -4$

$x = 7$ or $x = -2$

h. $(4 + x)^2 - 20 = 124$

$(4 + x)^2 = 124 + 20$

$(4 + x)^2 = 144$

$4 + x = \pm 12$

$4 + x = 12$ or $4 + x = -12$

$x = 8$ or $x = -16$

i. $(3x + 7)^2 = 121$

$(3x + 7) = \pm 11$

$3x + 7 = 11$ or $3x + 7 = -11$

$3x = 4$ or $3x = -18$

$x = \frac{4}{3}$ or $x = -6$

In a right triangle, as shown in the diagram to the right, the side c opposite the right angle is called the *hypotenuse,* and the other two sides a and b are called *legs.* The Pythagorean theorem states that in any right triangle, the lengths of the three sides are related by the equation $c^2 = a^2 + b^2$.

Use the Pythagorean theorem to answer Exercises 5–7.

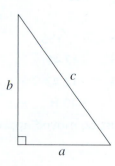

5. Determine the length of the hypotenuse in a right triangle with legs 5 inches and 12 inches.

$c^2 = 5^2 + 12^2 = 25 + 144 = 169$

$c = \sqrt{169} = 13$

The length of the hypotenuse is 13 inches.

6. One leg of a right triangle measures 8 inches and the hypotenuse 17 inches. Determine the length of the other leg.

$a^2 + 8^2 = 17^2$

$a^2 + 64 = 289$

$a^2 = 225$

$a = \sqrt{225} = 15$ The length of the other leg is 15 inches.

7. A 16-foot ladder is leaning against a wall, so that the bottom of the ladder is 5 feet from the wall (see diagram). You want to determine how much farther from the wall the bottom of the ladder must be moved so that the top of the ladder is exactly 10 feet above the floor.

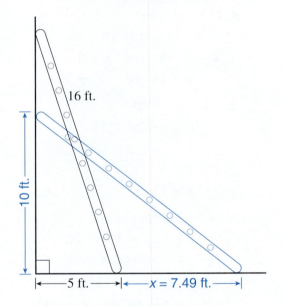

a. The new ladder position forms another right triangle in which x represents the additional distance the foot of the ladder has moved from the wall. Use the Pythagorean theorem to write an equation relating the base, height, and hypotenuse of this new right triangle.

$10^2 + (x + 5)^2 = 16^2$

b. Solve the equation from part a to determine x.

$10^2 + (x + 5)^2 = 16^2$

$100 + (x + 5)^2 = 256$

$(x + 5)^2 = 156$

$x + 5 = \pm\sqrt{156}$

$x + 5 \approx \pm 12.49$

$x + 5 \approx +12.49$ $x + 5 \approx -12.49$

$x \approx 7.49$ ft. $x \approx -17.49$ Discard; length cannot be negative.

The ladder must be moved approximately 7.5 feet so the top is 10 feet above the floor.

Activity 4.5

What Goes Up,
Comes Down

Objectives

1. Evaluate quadratic
   functions of the form
   $y = ax^2 + bx, a \neq 0$.

2. Graph quadratic functions
   of the form $y = ax^2 + bx$,
   $a \neq 0$.

3. Identify the x-intercepts
   of the graph of
   $y = ax^2 + bx$ graphically
   and algebraically.

4. Interpret the x-intercepts
   of a quadratic function in
   context.

5. Factor a binomial of the
   form $ax^2 + bx$.

6. Solve an equation of the
   form $ax^2 + bx = 0$ using
   the zero-product property.

When you kick a soccer ball up into the air, gravity acts on the ball and opposes the upward force of your kick. If the ball leaves the ground at a speed of 64 feet per second directly upward, then an algebraic rule that describes the height of the ball as a function of time is

$$h(x) = -16x^2 + 64x,$$

where x represents the number of seconds since the kick and $h(x)$ represents the ball's height (in feet) above the ground. Since the expression $-16x^2 + 64x$ is a second degree polynomial, the height function is a quadratic function.

1. Use the algebraic rule to complete the table.

x (sec.)	h(x) (ft.)
0	0
0.5	28
1	48
1.5	60
2	64
2.5	60
3	48
3.5	28
4	0

2. What does $h(x) = 0$ signify?

   $h(x) = 0$ represents a height of 0 feet. That is, the soccer ball is on the ground.

3. Explain why the heights increase and then decrease.

   The force of the kick propels the soccer ball upward. The force of gravity brings the ball back down.

4. **a.** Sketch the graph of $y = -16x^2 + 64x$ by using the table in Problem 1. Plot the points and then draw a curve through them. Scale the axes appropriately.

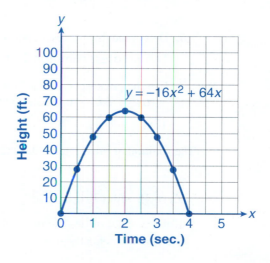

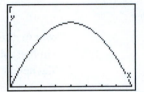

b. Use a graphing calculator to produce a graph of this function in the window $0 \le x \le 4, 0 \le y \le 80$. Your graph should resemble the one below.

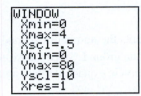

Solving Equations of the Form $ax^2 + bx = 0, a \ne 0$

The graph of $h(x) = -16x^2 + 64x$ is a parabola whose graph rises and then falls.

5. a. How many *x*-intercepts (points at which the graph intersects the *x*-axis) does this graph have?

There are two *x*-intercepts on the graph.

b. What are their coordinates?

Their coordinates are (0, 0) and (4, 0).

c. What do these *x*-intercepts signify in terms of the soccer ball?

At time 0, the ball is on the ground. After 4 seconds, the ball is back on the ground.

The **x-intercepts** have an algebraic interpretation as well. When the soccer ball is at ground level, its height, $h(x)$, is zero. Substituting 0 for $h(x)$ in the formula $h(x) = -16x^2 + 64x$, results in a quadratic equation:

$$0 = -16x^2 + 64x \text{ or, equivalently, } -16x^2 + 64x = 0.$$

Note that you *cannot* solve this equation by reversing the sequence of operations. The difficulty here is that the input variable, *x*, occurs twice, in two distinct terms, $-16x^2$ and $64x$. These two terms are not like terms, so they cannot be combined. However, both terms share common factors: Their coefficients are both multiples of 16, and they each have at least one factor of *x*.

> A common factor is called a *greatest common factor* (GCF) if there are no additional factors common to the terms in the expression.

6. Determine the greatest common factor of $-16x^2$ and $64x$.

The greatest common factor is 16x.

7. Use the greatest common factor to write the expression on the left side of the equation $-16x^2 + 64x = 0$ in factored form.

16x(−x + 4) = 0

The solution of this equation depends on a unique property of the zero on the right side.

> **Zero-Product Property**
>
> If the product of two factors is zero, then at least one of the factors must also be zero. Written symbolically, if $a \cdot b = 0$, then either $a = 0$ or $b = 0$.

Because the right side of $16x(-x + 4) = 0$ is zero, the zero-product property guarantees that either $16x = 0$ or $-x + 4 = 0$.

8. a. Solve $16x = 0$ for x.

$\frac{16x}{16} = \frac{0}{16}$

$x = 0$

b. Solve $-x + 4 = 0$ for x.

$-x = -4$

$x = 4$

c. Verify that each solution in parts a and b satisfies the original equation $-16x^2 + 64x = 0$.

$-16(0)^2 + 64(0) = 0$ and $-16(4)^2 + 64(4) = 0$

$0 = 0$ $\qquad\qquad -256 + 256 = 0$

$0 = 0$

9. Use the zero-product property to solve the following equations.

a. $x(x - 3) = 0$

$x = 0$ or $x - 3 = 0$

$x = 0$ or $x = 3$

b. $x(x + 5) = 0$

$x = 0$ or $x + 5 = 0$

$x = 0$ or $x = -5$

c. $(x - 2)(x + 2) = 0$

$x - 2 = 0$ or $x + 2 = 0$

$x = 2$ or $x = -2$

10. In each equation, use the greatest common factor to rewrite the expression in factored form, and then apply the zero-product property to determine the solutions.

a. $x^2 - 10x = 0$

$x(x - 10) = 0$

$x = 0$ or $x - 10 = 0$

$x = 0$ or $x = 10$

b. $x^2 + 7x = 0$

$x(x + 7) = 0$

$x = 0$ or $x + 7 = 0$

$x = 0$ or $x = -7$

c. $2x^2 - 12x = 0$

$2x(x - 6) = 0$

$2x = 0$ or $x - 6 = 0$

$x = 0$ or $x = 6$

d. $4x^2 - 6x = 0$

$2x(2x - 3) = 0$

$2x = 0$ or $2x - 3 = 0$

$x = 0$ or $x = \frac{3}{2}$

11. To apply the zero-product property, one side of the equation must be zero. In the following equations, first add or subtract an appropriate term to both sides of the equation to get zero on one side. Then write in factored form and apply the zero-product property.

a. $2x^2 = 8x$

$2x^2 - 8x = 0$

$2x(x - 4) = 0$

$2x = 0$ or $x - 4 = 0$

$x = 0$ or $x = 4$

b. $6x^2 = -4x$

$6x^2 + 4x = 0$

$2x(3x + 2) = 0$

$2x = 0$ or $3x + 2 = 0$

$x = 0$ or $x = -\frac{2}{3}$

In an equation such as $3x^2 - 15x = 0$, it is *always* permissible to divide *every* term of the equation (both sides) by a nonzero number. Here you see that dividing by 3 simplifies the original equation:

$$\frac{3x^2}{3} - \frac{15x}{3} = \frac{0}{3}$$ simplifies to the equivalent equation $x^2 - 5x = 0$.

Factoring and applying the zero-product property leads to $x(x - 5) = 0$, whose solutions are $x = 0$ and $x = 5$. However, note that it is *never* permissible to divide through by a variable quantity such as x. If you were to divide both sides of $3x^2 - 15x = 0$ by x, you would obtain

$$\frac{3x^2}{x} - \frac{15x}{x} = \frac{0}{x}$$

which simplifies to $3x - 15 = 0$. Solving the resulting equation, $3x - 15 = 0$, you would obtain a single solution, $x = 5$. You would have lost the other solution, $x = 0$.

12. Suppose you kick a soccer ball directly upward at a speed of 80 feet per second. An algebraic rule that describes the height of the ball as a function of time is

$$h(x) = -16x^2 + 80x,$$

where x represents the number of seconds since the kick and $h(x)$ represents the ball's height (in feet) above the ground.

a. How high is the ball 2 seconds after the kick?

$y = -16(2)^2 + 80(2) = -64 + 160 = 96$

After 2 seconds, the ball is 96 feet above the ground.

b. Write an equation that represents when the ball is on the ground.

$-16x^2 + 80x = 0$

c. Solve the equation in part b. What do the two solutions signify in terms of the soccer ball?

$-16x^2 + 80x = 0$

$16x(-x + 5) = 0$

$16x = 0$ or $-x + 5 = 0$

$x = 0$ or $x = 5$

At time 0, the ball is on the ground. After 5 seconds, the ball is back on the ground.

d. Graph this function on your graphing calculator, and identify the results of parts a and c on your graph. Use the indicated window.

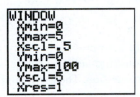 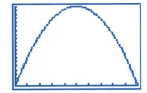

On the graph, the x-intercepts are 0 and 5, indicating the ball is on the ground.

SUMMARY: ACTIVITY 4.5

1. The **x-intercepts** on the graph of $y = ax^2 + bx$, $a \neq 0$, are the points at which the graph intersects the x-axis. (At these points, the y-coordinate is zero.) The x-intercepts can be determined algebraically by setting $y = 0$ and solving the associated equation, $ax^2 + bx = 0$.

2. **Zero-product property:** If the product of two factors is zero, then at least one of the factors must also be zero. Stated symbolically, if $a \cdot b = 0$, then either $a = 0$ or $b = 0$.

3. A quadratic equation of the form $ax^2 + bx = 0$, $a \neq 0$, is solved algebraically by factoring the equation's left side and then using the zero-product property.

$$3x^2 + 6x = 0$$
$$3x(x + 2) = 0$$
$$3x = 0 \quad \text{or} \quad x + 2 = 0$$
$$x = 0 \quad \text{or} \qquad x = -2$$

4. The solutions of the equation $ax^2 + bx = 0$ correspond precisely to the x-intercepts of the graph of $y = ax^2 + bx$. For example, the x-intercepts of the graph of $y = 3x^2 + 6x$ are $(-2, 0)$ and $(0, 0)$.

5. An equation can *always* be simplified by dividing each term on *both* sides by a nonzero number. The original and simplified equations will have identical solutions.

6. An equation may *never* be simplified by dividing both sides by a variable quantity. The resulting equation will have lost one or more solutions.

EXERCISES: ACTIVITY 4.5

1. You are relaxing on a float in your in-ground pool casually tossing a ball straight upward into the air. The height of the ball above the water is given by the formula $s = -16t^2 + 40t$, where t represents the time since the ball was tossed in the air.

a. Use the formula to determine the height of the ball for the times given in the following table.

t (sec.)	0	0.5	1	1.5	2	2.5
s (ft.)	0	16	24	24	16	0

b. Set up and solve an equation to determine when the ball will hit the water.

When the ball hits the water, the height $s = 0$. Therefore, the equation is

$$-16t^2 + 40t = 0$$
$$-8t(2t - 5) = 0$$
$$2t - 5 = 0 \quad \text{or} \quad -8t = 0$$
$$2t = 5 \quad \text{or} \qquad t = 0$$
$$t = \frac{5}{2} = 2.5$$

The ball will hit the water 2.5 seconds after it is tossed in the air.

c. Use the result in parts a and b to help you determine a practical domain.

The practical domain is the interval of time 0 to 2.5 seconds.

d. Graph the height function over its practical domain. Label and scale the axes appropriately.

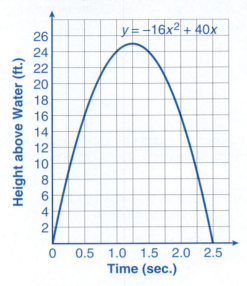

2. Use the zero-product property to solve the following equations.

a. $x(x - 2) = 0$

   $x = 0$ or $x - 2 = 0$

   $x = 0$ or $x = 2$

b. $x(x + 7) = 0$

   $x = 0$ or $x + 7 = 0$

   $x = 0$ or $x = -7$

c. $(x - 3)(x + 2) = 0$

   $x - 3 = 0$ or $x + 2 = 0$

   $x = 3$ or $x = -2$

d. $(2x - 1)(x + 1) = 0$

   $2x - 1 = 0$ or $x + 1 = 0$

   $x = \frac{1}{2}$ or $x = -1$

3. In each equation use the greatest common factor to rewrite the expression in factored form, and then apply the zero-product property to determine the solutions.

a. $x^2 - 8x = 0$

   $x(x - 8) = 0$

   $x = 0$ or $x - 8 = 0$

   $x = 0$ or $x = 8$

b. $x^2 + 5x = 0$

   $x(x + 5) = 0$

   $x = 0$ or $x + 5 = 0$

   $x = 0$ or $x = -5$

c. $3x^2 - 12x = 0$

   $3x(x - 4) = 0$

   $3x = 0$ or $x - 4 = 0$

   $x = 0$ or $x = 4$

d. $4x^2 - 2x = 0$

   $2x(2x - 1) = 0$

   $2x = 0$ or $2x - 1 = 0$

   $x = 0$ or $x = \frac{1}{2}$

e. $x^2 = 5x$

   $x^2 - 5x = 0$

   $x(x - 5) = 0$

   $x = 0$ or $x - 5 = 0$

   $x = 0$ or $x = 5$

f. $3x^2 = 48x$

   $3x^2 - 48x = 0$

   $3x(x - 16) = 0$

   $3x = 0$ or $x - 16 = 0$

   $x = 0$ or $x = 16$

4. Your model rocket is popular with the neighborhood children. Its powerful booster propels the rocket straight upward so that its height is described by the function $H(t) = -16t^2 + 480t$, where t represents the number of seconds after launch and $H(t)$ the height of the rocket in feet.

a. Write an equation that can be used to determine the t-intercepts of the graph of $H(t)$.

   $-16t^2 + 480t = 0$

b. Solve the equation obtained in part a by factoring.

$$-16t^2 + 480t = 0$$ Divide each term in this equation by -16.

$$t^2 - 30t = 0$$ Factor the left side.

$$t(t - 30) = 0$$ Set each factor equal to 0.

$t = 0$ or $t - 30 = 0$ Solve each equation.

$t = 0$ or $t = 30$

c. What do the solutions signify in terms of the rocket's flight?

The rocket is on the ground $t = 0$ seconds from launch and returns to the ground $t = 30$ seconds after launch.

d. What is the practical domain of this function?

The practical domain consists of input values from 0 through 30 seconds.

 e. Use the practical domain to help set a window for your graphing calculator. Use your graphing calculator to produce a graph of this function in the window $0 \le x \le 30, 0 \le y \le 4000$.

 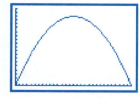

f. Use the graph to estimate how high the model rocket will go and how long after launch it will reach its maximum height.

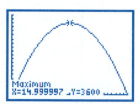

It will reach its maximum height of 3600 feet 15 seconds after launch.

5. Use the zero-product property to determine the solutions of each equation. Remember, to use the zero-product property, one side of the equation must be 0. Check your solutions by substituting back into the original equation.

Note to instructor: Checks are shown for just the first two parts of the exercise, as illustration.

a. $x^2 + 10x = 0$

$$x(x + 10) = 0$$

$x = 0$ or $x + 10 = 0$

$x = 0$ or $x = -10$

Check: $0^2 + 10(0) = 0$

and $(-10)^2 + 10(-10)$

$= 100 + (-100) = 0$

b. $2x^2 = 25x$

$$2x^2 - 25x = 0$$

$$x(2x - 25) = 0$$

$x = 0$ or $2x - 25 = 0$

$x = 0$ or $x = 12.5$

Check: $2(0)^2 = 25(0)$

$0 = 0$

$2(12.5)^2 = 25(12.5)$

$312.5 = 312.5$

c. $15y^2 + 75y = 0$

$$15y(y + 5) = 0$$

$15y = 0$ or $y + 5 = 0$

$y = 0$ or $y = -5$

d. $2(x - 5)(x + 1) = 0$

$x - 5 = 0$ or $x + 1 = 0$

$x = 5$ or $x = -1$

e. $24t^2 - 36t = 0$

$$12t(2t - 3) = 0$$

$12t = 0$ or $2t - 3 = 0$

$t = 0$ or $t = 1.5$

f. $12x = x^2$

$$12x - x^2 = 0$$

$$x(12 - x) = 0$$

$x = 0$ or $12 - x = 0$

$x = 0$ or $x = 12$

g. $2w^2 - 3w = 5w$

$$2w^2 - 8w = 0$$

$$2w(w - 4) = 0$$

$2w = 0$ or $w - 4 = 0$

$w = 0$ or $w = 4$

h. $14p = 3p - 5p^2$

$$11p + 5p^2 = 0$$

$$p(11 + 5p) = 0$$

$p = 0$ or $11 + 5p = 0$

$$5p = -11$$

$p = 0$ or $p = -\frac{11}{5}$

Activity 4.6

How High Did It Go?

Objectives

1. Recognize and write a quadratic equation in standard form, $ax^2 + bx + c = 0$, $a \neq 0$.

2. Factor trinomials of the form $x^2 + bx + c$.

3. Solve a factorable quadratic equation of the form $x^2 + bx + c = 0$ using the zero-product property.

4. Identify a quadratic function from its algebraic form.

In the previous activity, What Goes Up, Comes Down, you examined the algebraic rule that describes the height of a soccer ball kicked directly upward at 64 feet per second as a function of time,

$$h(x) = -16x^2 + 64x,$$

where x represents the number of seconds since the kick and $h(x)$ represents the ball's height (in feet) above the ground. The mathematical objective of that activity was to identify and interpret the x-intercepts of the function both graphically and algebraically.

Here is a similar situation. Suppose a soccer ball is kicked directly upward at a speed of 96 feet per second. The algebraic rule that describes the height of the soccer ball as a function of time is

$$h(x) = -16x^2 + 96x,$$

where x represents the number of seconds since the kick and $h(x)$ represents the ball's height (in feet) above the ground.

1. **a.** Determine the x-intercepts of the soccer ball function by writing and solving an appropriate equation.

$$-16x^2 + 96x = 0$$
$$16x(-x + 6) = 0$$
$$16x = 0 \quad \text{or} \quad -x + 6 = 0$$
$$x = 0 \quad \text{or} \quad x = 6$$

The x-intercepts are (0, 0) and (6, 0).

b. What do these solutions signify about the soccer ball?

At time 0, the ball is on the ground. After 6 seconds, the ball is back on the ground.

c. Complete the table, and sketch the graph of $h(x) = -16x^2 + 96x$. Scale the axes appropriately.

x	0	0.5	1	1.5	2	2.5	3	3.5	4	4.5	5	5.5	6
h(x)	0	44	80	108	128	140	144	140	128	108	80	44	0

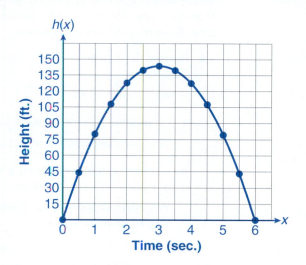

d. Use your graphing calculator to graph the function $h(x) = -16x^2 + 96x$ in the window $0 \le x \le 10$, $-10 \le y \le 150$. Your graph should resemble the one here. Identify your solutions from part a on the graph.

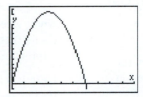

The solutions are obtained graphically by identifying the points (0, 0) and (6, 0).

Solving Equations of the Form $ax^2 + bx + c = 0$, $a \ne 0$

In this activity you will explore additional questions of interest about the flight of the soccer ball. For instance, instead of determining when the soccer ball hits the ground, you can ask, when will the soccer ball reach a height of 80 feet?

2. Use the table or the graph in Problem 1c to estimate the time(s) at which the soccer ball is 80 feet above the ground. How often during its flight does this occur?

$y = 80$ when $x = 1$ and when $x = 5$. The soccer ball is 80 feet above the ground at two different times—once on the way up, and once on the way down.

The answers to Problem 2 have an algebraic interpretation as well. When the soccer ball is at a height of 80 feet, $h(x) = 80$. Substituting 80 for $h(x)$ in the formula $h(x) = -16x^2 + 96x$, you have

$$80 = -16x^2 + 96x \quad \text{or, equivalently,} \quad -16x^2 + 96x = 80.$$

3. Explain why you *cannot* solve the equation $-16x^2 + 96x = 80$ by factoring the expression on the left side and applying the zero-product rule.

The zero-product property cannot be applied to this equation because the right side of the equation is not 0.

To use the zero-product property to solve an equation, one side of the equation *must* be zero. If need be, you can always force one side of an equation to be zero by subtracting appropriate terms from both sides of the equation.

4. Rewrite the equation $-16x^2 + 96x = 80$ as an equivalent equation whose right side is zero.

$-16x^2 + 96x - 80 = 0$

5. a. Identify the greatest common factor of the trinomial on the left side of the equation.

The greatest common factor is 16.

b. Is it permissible to divide each term of the equation by this common factor? If so, do it. If not, explain the problem.

Yes, both sides may be divided by 16, which contains no variable. The new equation is $-x^2 + 6x - 5 = 0$.

If you completed Problems 4 and 5 correctly, the original equation $-16x^2 + 96x = 80$ is most likely rewritten in equivalent form $-x^2 + 6x - 5 = 0$.

6. What is the coefficient of the x^2 term in the equation $-x^2 + 6x - 5 = 0$?

The coefficient is -1.

To simplify the equation $-x^2 + 6x - 5 = 0$ once more, multiply each term in the equation by -1. The effect of multiplication by -1 is always a reversal of the signs of each term in the equation. You should now obtain the equation

$$x^2 - 6x + 5 = 0.$$

Note here that there are no longer any common factors among the three terms on the left side. However, the expression $x^2 - 6x + 5$ can still be factored. The key to factoring $x^2 - 6x + 5$ lies in your recall of binomial multiplication from Activity 4.3.

When two binomials such as $x + a$ and $x + b$ are multiplied, the coefficients in the expanded form are closely related to the constant terms a and b in the original factored form.

$$(x + a)(x + b) = x^2 + bx + ax + ab$$
$$= x^2 + (a + b)x + ab$$

The product is a trinomial in which

i. the constant term, ab, in the expanded form on the right side is the *product* of the constants a and b,

ii. the coefficient, $a + b$, of the x term in the expanded form on the right side is the *sum* of the constants a and b.

Therefore, factoring a trinomial such as $x^2 - 6x + 5$ is accomplished by determining the two constants a and b whose product is 5 and whose sum is -6.

To determine the constants a and b, it is preferable to start by first considering the product. In this case, the product is 5, a prime number with only two possible pairs of factors, 5 and 1 or -5 and -1. It is a straightforward calculation to confirm that the second pair has the property that its sum is -6.

7. Complete the factoring of $x^2 - 6x + 5$.

$$x^2 - 6x + 5 = (x - 1)(x - 5)$$

Note that the order in which you write these factors does not matter.

$$(x - 5)(x - 1) \text{ is exactly the same product as } (x - 1)(x - 5).$$

8. a. Use the zero-product property to solve the factored form of the equation $x^2 - 6x + 5 = 0$.

$$(x - 1)(x - 5) = 0$$
$$x - 1 = 0 \quad \text{or} \quad x - 5 = 0$$
$$x = 1 \quad \text{or} \quad x = 5$$

b. How many solutions did you obtain? Interpret these solutions in the context of the soccer ball situation.

I obtained two solutions. The soccer ball is 80 feet above the ground 1 second and 5 seconds after being tossed.

The solution of $-16x^2 + 96x = 80$ that was completed in Problem 4 through Problem 8 is summarized as follows.

$$-16x^2 + 96x = 80$$

$$\underline{\qquad\qquad -80 \quad -80}$$ Subtract 80 from each side to obtain 0 on one side.

$$\frac{-16x^2}{16} + \frac{96x}{16} - \frac{80}{16} = \frac{0}{16}$$ Divide each side by the greatest common factor, 16.

$$-1(-x^2 + 6x - 5) = -1(0)$$ Multiply each side by -1.

$$x^2 - 6x + 5 = 0$$ Factor the trinomial.

$$(x - 5)(x - 1) = 0$$ Apply the zero-product property.

$$x - 5 = 0 \qquad x - 1 = 0$$

$$x = 5 \qquad\qquad x = 1$$

The equation $-16x^2 + 96x = 80$ is a **quadratic equation**. When the equation is written with zero on one side and the polynomial, written in standard form, on the other side, then the quadratic equation $-16x^2 + 96x - 80 = 0$ is said to be written in **standard form**.

Definition

Any equation that can be written in the general form $ax^2 + bx + c = 0$ is called a **quadratic equation**. Here a, b, and c are understood to be arbitrary constants, with the single restriction that $a \neq 0$.

A quadratic equation is said to be in **standard form** when the terms on the left side are written in decreasing degree order: the x^2 term first, followed by the x term, and trailed by the constant term. The right side in the standard form of the equation is *always* zero.

9. Write the quadratic equations in standard form, and determine the constants, a, b, and c.

a. $3x^2 + 5x + 8 = 0$

$a = 3, b = 5, c = 8$

b. $x^2 - 2x + 6 = 0$

$a = 1, b = -2, c = 6$

c. $x^2 + 6x = 10$

$x^2 + 6x - 10 = 0$

$a = 1, b = 6, c = -10$

d. $x^2 - 4x = 0$

$a = 1, b = -4, c = 0$

e. $x^2 = 9$

$x^2 - 9 = 0$

$a = 1, b = 0, c = -9$

f. $5x^2 = 2x + 6$

$5x^2 - 2x - 6 = 0$

$a = 5, b = -2, c = -6$

Procedure

Solving a Quadratic Equation by Factoring

1. Write the quadratic equation in standard form, $ax^2 + bx + c = 0, a \neq 0$.

2. When appropriate, divide each term in the equation by the greatest common numerical factor.

3. Factor the nonzero side.

4. Apply the zero-product property to obtain two linear equations.

5. Solve each linear equation.

6. Check your solutions in the original equation.

10. If not already done, rewrite each quadratic equation in standard form. Then factor the nonzero side and use the zero-product property to solve the equation. Check your solutions in the original equation.

a. $x^2 + 5x + 6 = 0$

$$(x + 2)(x + 3) = 0$$
$$x + 2 = 0 \quad \text{or} \quad x + 3 = 0$$
$$x = -2 \quad \text{or} \quad x = -3$$
Check: $(-2)^2 + 5(-2) + 6 = 0$
$$0 = 0$$
$$(-3)^2 + 5(-3) + 6 = 0$$
$$0 = 0$$

b. $x^2 - 8x + 12 = 0$

$$(x - 2)(x - 6) = 0$$
$$x - 2 = 0 \quad \text{or} \quad x - 6 = 0$$
$$x = 2 \quad \text{or} \quad x = 6$$
Check: $(2)^2 - 8(2) + 12 = 0$
$$0 = 0$$
$$(6)^2 - 8(6) + 12 = 0$$
$$0 = 0$$

c. $x^2 + 8 = 7x - 4$

$$x^2 - 7x + 12 = 0$$
$$(x - 3)(x - 4) = 0$$
$$x - 3 = 0 \quad \text{or} \quad x - 4 = 0$$
$$x = 3 \quad \text{or} \quad x = 4$$
Check: $(3)^2 + 8 = 7(3) - 4$
$$17 = 17$$
$$(4)^2 + 8 = 7(4) - 4$$
$$24 = 24$$

d. $x^2 + 5x - 24 = 0$

$$(x + 8)(x - 3) = 0$$
$$x + 8 = 0 \quad \text{or} \quad x - 3 = 0$$
$$x = -8 \quad \text{or} \quad x = 3$$
Check: $(-8)^2 + 5(-8) - 24 = 0$
$$0 = 0$$
$$(3)^2 + 5(3) - 24 = 0$$
$$0 = 0$$

e. $2x^2 = 6x + 20$

$$2x^2 - 6x - 20 = 0$$
$$x^2 - 3x - 10 = 0$$
$$(x + 2)(x - 5) = 0$$
$$x + 2 = 0 \quad \text{or} \quad x - 5 = 0$$
$$x = -2 \quad \text{or} \quad x = 5$$
Check: $2(-2)^2 = 6(-2) + 20$
$$8 = 8$$
$$2(5)^2 = 6(5) + 20$$
$$50 = 50$$

f. $-3x^2 + 6x = -24$

$$-3x^2 + 6x + 24 = 0$$
$$x^2 - 2x - 8 = 0$$
$$(x + 2)(x - 4) = 0$$
$$x + 2 = 0 \quad \text{or} \quad x - 4 = 0$$
$$x = -2 \quad \text{or} \quad x = 4$$
Check: $-3(-2)^2 + 6(-2) = -24$
$$-24 = -24$$
$$-3(4)^2 + 6(4) = -24$$
$$-24 = -24$$

Return, once again, to the function $h(x) = -16x^2 + 96x$ that describes the height (in feet) of a soccer ball as a function of time (in seconds).

11. a. Set up and solve the quadratic equation that describes the time when the ball is 128 feet above the ground.

$$-16x^2 + 96x = 128$$
$$-16x^2 + 96x - 128 = 0$$
$$-16(x^2 - 6x + 8) = 0$$
$$x^2 - 6x + 8 = 0$$
$$(x - 2)(x - 4) = 0$$
$$x - 2 = 0 \quad \text{or} \quad x - 4 = 0$$
$$x = 2 \quad \text{or} \quad x = 4$$

b. How many solutions does this equation have?

The equation has two solutions.

c. Interpret the solution(s) in the soccer ball context.

The soccer ball is 128 feet above the ground 2 seconds and 4 seconds after being tossed.

12. a. Set up and solve the quadratic equation that describes the time when the ball is 144 feet above the ground.

$$-16x^2 + 96x = 144$$
$$-16x^2 + 96x - 144 = 0$$
$$-16(x^2 - 6x + 9) = 0$$
$$x^2 - 6x + 9 = 0$$
$$(x - 3)(x - 3) = 0$$
$$x - 3 = 0$$
$$x = 3$$

b. How many solutions does this equation have?

The equation has one solution.

c. Interpret the solution(s) in the soccer ball context.

The soccer ball is 144 feet above the ground 3 seconds after being tossed. This is the highest the ball gets.

SUMMARY: ACTIVITY 4.6

1. Any function that is defined by an equation of the form $y = ax^2 + bx + c, a \neq 0$, is called a **quadratic function**. Here, a, b, and c are understood to be arbitrary constants, with the single restriction that $a \neq 0$.

2. Algebraically, an associated **quadratic equation** arises when y is assigned a value and you need to solve for x. A quadratic equation, written as $ax^2 + bx + c = 0$, is said to be in **standard form**.

3. If the nonzero side of a quadratic equation written in standard form can be factored, then the equation can be solved using the zero-product property.

EXERCISES: ACTIVITY 4.6

1. Solve the following quadratic equations by factoring.

a. $x^2 + 7x + 6 = 0$ **b.** $x^2 - 10x - 24 = 0$

Note to instructors: Solution checks are not shown here but it is recommended that students be encouraged to get into the habit of checking their results themselves.

$$(x + 1)(x + 6) = 0 \qquad\qquad (x - 12)(x + 2) = 0$$
$$x + 1 = 0 \quad \text{or} \quad x + 6 = 0 \qquad x - 12 = 0 \quad \text{or} \quad x + 2 = 0$$
$$x = -1 \quad \text{or} \quad x = -6 \qquad\quad x = 12 \quad \text{or} \quad x = -2$$

c. $y^2 + 11y = -28$

$\qquad y^2 + 11y + 28 = 0$

$\qquad (y + 7)(y + 4) = 0$

$\quad y + 7 = 0 \quad$ or $\quad y + 4 = 0$

$\qquad y = -7 \quad$ or $\qquad y = -4$

d. $5x^2 - 75x + 180 = 0$

$\qquad$ Divide each term by 5.

$\qquad x^2 - 15x + 36 = 0$

$\qquad (x - 12)(x - 3) = 0$

$\quad x - 12 = 0 \quad$ or $\quad x - 3 = 0$

$\qquad x = 12 \quad$ or $\qquad x = 3$

e. $x^2 + 9x + 18 = 0$

$\qquad (x + 6)(x + 3) = 0$

$\quad x + 6 = 0 \quad$ or $\quad x + 3 = 0$

$\qquad x = -6 \quad$ or $\qquad x = -3$

f. $x^2 - 9x = 36$

$\qquad x^2 - 9x - 36 = 0$

$\qquad (x - 12)(x + 3) = 0$

$\quad x - 12 = 0 \quad$ or $\quad x + 3 = 0$

$\qquad x = 12 \quad$ or $\qquad x = -3$

g. $x^2 - 9x + 20 = 0$

$\qquad (x - 5)(x - 4) = 0$

$\quad x - 5 = 0 \quad$ or $\quad x - 4 = 0$

$\qquad x = 5 \quad$ or $\qquad x = 4$

h. $2x^2 - 12 = 2x$

$\qquad 2x^2 - 2x - 12 = 0$

$\qquad x^2 - x - 6 = 0$

$\qquad (x - 3)(x + 2) = 0$

$\quad x - 3 = 0 \quad$ or $\quad x + 2 = 0$

$\qquad x = 3 \quad$ or $\qquad x = -2$

i. $x^2 - 5x = 3x - 15$

$\qquad x^2 - 8x + 15 = 0$

$\qquad (x - 5)(x - 3) = 0$

$\quad x - 5 = 0 \quad$ or $\quad x - 3 = 0$

$\qquad x = 5 \quad$ or $\qquad x = 3$

j. $2x^2 - 3x = x^2 + 10$

$\qquad x^2 - 3x - 10 = 0$

$\qquad (x - 5)(x + 2) = 0$

$\quad x - 5 = 0 \quad$ or $\quad x + 2 = 0$

$\qquad x = 5 \quad$ or $\qquad x = -2$

2. Your friend's model rocket has a powerful engine. Its height (in feet) is described by the function $H(t) = -16t^2 + 432t$, where t represents the number of seconds since launch. Set up and solve a quadratic equation to answer each of the parts a–e and record your results in the following table.

	HEIGHT H (ft.) OF ROCKET FROM GROUND	FIRST TIME t (sec.) THAT HEIGHT WAS REACHED	SECOND TIME t (sec.) THAT HEIGHT WAS REACHED	AVERAGE OF FIRST TIME AND SECOND TIME HEIGHT WAS REACHED
a.	0	0	27	$\frac{0 + 27}{2} = 13.5$
b.	800	2	25	13.5
c.	1760	5	22	13.5
d.	2240	7	20	13.5
e.	2720	10	17	13.5

a. At what times does your friend's rocket touch the ground?

$\qquad -16t^2 + 432t = 0 \quad$ Divide by -16.

$\qquad\qquad t^2 - 27t = 0$

$\qquad\qquad t(t - 27) = 0$

$\quad t = 0 \quad$ or $\quad t - 27 = 0$

$\qquad\qquad\qquad t = 27$

The rocket returns to the ground in 27 seconds.

b. When is the rocket 800 feet high?

$$-16t^2 + 432t = 800$$
$$-16t^2 + 432t - 800 = 0 \quad \text{Divide by} -16.$$
$$t^2 - 27t + 50 = 0$$
$$(t - 2)(t - 25) = 0$$
$$t - 2 = 0 \quad \text{or} \quad t - 25 = 0$$
$$t = 2 \quad \text{or} \quad t = 25$$

The rocket is 800 feet above the ground 2 seconds after launch and again, on the way down, 25 seconds after launch.

c. When is the rocket 1760 feet high?

$$-16t^2 + 432t = 1760$$
$$-16t^2 + 432t - 1760 = 0 \quad \text{Divide by} -16.$$
$$t^2 - 27t + 110 = 0$$
$$(t - 5)(t - 22) = 0$$
$$t - 5 = 0 \quad \text{or} \quad t - 22 = 0$$
$$t = 5 \quad \text{or} \quad t = 22$$

The rocket is 1760 feet above the ground 5 seconds after launch and again, on the way down, 22 seconds after launch.

d. When is the rocket 2240 feet high?

$$-16t^2 + 432t = 2240$$
$$-16t^2 + 432t - 2240 = 0 \quad \text{Divide by} -16.$$
$$t^2 - 27t + 140 = 0$$
$$(t - 7)(t - 20) = 0$$
$$t - 7 = 0 \quad \text{or} \quad t - 20 = 0$$
$$t = 7 \quad \text{or} \quad t = 20$$

The rocket is 2240 feet above the ground 7 seconds after launch and again, on the way down, 20 seconds after launch.

e. When is the rocket 2720 feet high?

$$-16t^2 + 432t = 2720$$
$$-16t^2 + 432t - 2720 = 0 \quad \text{Divide by} -16.$$
$$t^2 - 27t + 170 = 0$$
$$(t - 10)(t - 17) = 0$$
$$t - 10 = 0 \quad \text{or} \quad t - 17 = 0$$
$$t = 10 \quad \text{or} \quad t = 17$$

The rocket is 2720 feet above the ground 10 seconds after launch and again, on the way down, 17 seconds after launch.

f. For each given height in parts a–e, determine the average of the first and second times the height was reached and record in the last column of the table.

See table.

g. Use the results in the table to determine the maximum height that the rocket reached. To verify your answer, graph the solutions from parts a–e and draw a curve through the points, or use a graphing calculator.

The maximum height reached is 2916 feet, 13.5 seconds after launch.

3. Solve $(x - 1)(x + 2) = 10$. (*Hint:* First expand the expression and write the equation in standard form.)

$$x^2 + x - 2 = 10$$
$$x^2 + x - 12 = 0$$
$$(x + 4)(x - 3) = 0$$
$$x + 4 = 0 \quad \text{or} \quad x - 3 = 0$$
$$x = -4 \quad \text{or} \quad x = 3$$

4. Solve $(x - 3)^2 - 4 = 0$ in each of the following two ways.

a. Solve by taking square roots.

$$(x - 3)^2 = 4$$
$$(x - 3) = \pm\sqrt{4}$$
$$x - 3 = 2 \quad \text{or} \quad x - 3 = -2$$
$$x = 5 \quad \text{or} \quad x = 1$$

b. Write the equation in standard form and solve by factoring and the zero-product property.

$$x^2 - 6x + 9 - 4 = 0$$
$$x^2 - 6x + 5 = 0$$
$$(x - 5)(x - 1) = 0$$
$$x - 5 = 0 \quad \text{or} \quad x - 1 = 0$$
$$x = 5 \quad \text{or} \quad x = 1$$

c. Which process seems simpler to you?

(Answers will vary.)

5. You are on the roof of your apartment building with a friend. You lean over the edge of the building and toss an apple straight up into the air. The height h, of the apple above the ground can be determined by the function defined by

$$h = -16t^2 + 16t + 96,$$

where t is time measured in seconds. Note that $t = 0$ corresponds to the time the apple leaves your hand.

a. Determine the height of the apple above the ground at the time you release it.

$t = 0$ when I release the apple. Therefore, $h(0) = -16(0)^2 + 16(0) + 96 = 96$. At $t = 0$, the apple is 96 feet above the ground.

b. Write an equation that can be used to determine the time the apple strikes the ground.

$h = 0$ when the apple hits the ground, so the equation is $-16t^2 + 16t + 96 = 0$.

c. Solve the equation in part b by the factoring method.

Divide each term by -16 to obtain the equation,

$$t^2 - t - 6 = 0$$
$$(t - 3)(t + 2) = 0$$
$$t = 3 \quad \text{or} \quad t = -2$$

The apple hits the ground in 3 seconds. Discard the result $t = -2$, which is meaningless in this context.

d. What is the practical domain of the height function?

The practical domain is from $t = 0$ to $t = 3$ seconds.

e. Graph the height function on the following grid or use a graphing calculator. Use the results from part d to help decide a scale for the axes or set the proper window.

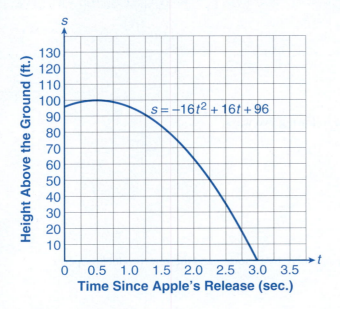

f. Use the graph to estimate the coordinates of the maximum point of the parabola.

The maximum point of the parabola is (0.5, 100).

g. What do the coordinates of the point in part f indicate about the position of the apple with respect to the ground?

The coordinates (0.5, 100) indicate that the apple at its highest is 100 feet above the ground half a second after I toss it.

Objectives

1. Use the quadratic formula to solve quadratic equations.

2. Identify the solutions of a quadratic equation with points on the corresponding graph.

In the previous activities, the soccer ball was kicked straight up into the air. Suppose the soccer goalie punted the ball in such a way as to kick the ball as far as possible down the field. The height of the ball above the field as a function of the distance downfield can be approximated by

$$h(x) = -0.017x^2 + 0.98x + 0.33,$$

where $h(x)$ represents the height of the ball (in yards) and x represents the horizontal distance (in yards) down the field from where the goalie kicked the ball.

In this situation, the graph of $h(x) = -0.017x^2 + 0.98x + 0.33$ is the actual path of the flight of the soccer ball. The graph of this quadratic function appears below.

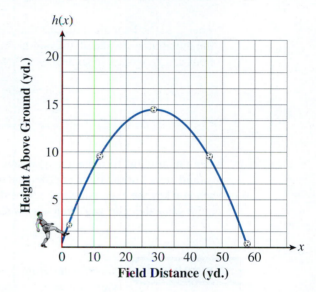

1. Use the graph to estimate how far downfield from the point of contact the soccer ball is 10 yards above the ground. How often during its flight does this occur?

   (Answers will vary.)

   The soccer ball is 10 yards above the ground twice during its flight, when it is approximately 13 yards and 45 yards from where it was kicked.

2. **a.** Use the function rule $h(x) = -0.017x^2 + 0.98x + 0.33$ to write the quadratic equation that arises in determining when the ball is 10 yards above the ground.

   $-0.017x^2 + 0.98x + 0.33 = 10$

   b. Rewrite the equation in standard form and simplify if possible.

   $-0.017x^2 + 0.98x + 0.33 = 10$
   $-0.017x^2 + 0.98x - 9.67 = 0$
   $0.017x^2 - 0.98x + 9.67 = 0$

   c. What does the graph tell you about the number of solutions to this equation?

   The graph indicates that there are two solutions.

In the previous activity, you solved the quadratic equation $-16x^2 + 96x = 80$ by simplifying its standard form to $x^2 - 6x + 5 = 0$, factoring, and applying the zero-product property. Suppose that, instead, the constant term was 6 and the equation to solve was $x^2 - 6x + 6 = 0$. The quadratic trinomial in this equation cannot be factored.

You have just encountered one of the main difficulties in solving quadratic equations algebraically—most quadratic equations cannot be solved by factoring. However, this does not imply that most quadratic equations are not solvable. By a straightforward algebraic manipulation, a quadratic equation can be turned into a formula that produces the solutions. In fact, these algebraic manipulations can even be applied to the equation $ax^2 + bx + c = 0$ to develop a general formula called the quadratic formula. Of course, the solutions will depend on the specific values of the coefficients, a, b, and c.

The Quadratic Formula

For a quadratic equation in standard form, $ax^2 + bx + c = 0$, $a \neq 0$, the solutions are

$$x = \frac{-b \pm \sqrt{b^2 - 4ac}}{2a}.$$

The $\pm$ in the formula indicates that there are two solutions, one in which the terms in the numerator are added and another in which the terms are subtracted. The two solutions can be written separately as follows:

$$x_1 = \frac{-b + \sqrt{b^2 - 4ac}}{2a} \quad \text{and} \quad x_2 = \frac{-b - \sqrt{b^2 - 4ac}}{2a}.$$

When you use this formula, the quadratic equation *must* be in standard form so that the values (and signs) of the three coefficients a, b, and c are correct. You also want to be careful when you key the expression into your calculator. The square root and quotient computations require some thought; you often must include appropriate sets of parentheses.

Example 1   *Use the quadratic formula to solve the quadratic equation* $6x^2 - x = 2$.

SOLUTION

$6x^2 - x - 2 = 0$ Write equation in standard form, $ax^2 + bx + c = 0$.

$a = 6, b = -1, c = -2$ Identify a, b, and c.

$x = \dfrac{-(-1) \pm \sqrt{(-1)^2 - 4(6)(-2)}}{2(6)}$ Substitute for a, b, and c in $x = \dfrac{-b \pm \sqrt{b^2 - 4ac}}{2a}$.

$x = \dfrac{1 \pm \sqrt{1 + 48}}{12} = \dfrac{1 \pm 7}{12}$

$x = \dfrac{1 + 7}{12} = \dfrac{8}{12} = \dfrac{2}{3}, \quad x = \dfrac{1 - 7}{12} = \dfrac{-6}{12} = -\dfrac{1}{2}$

3. Use the quadratic formula to solve the quadratic equation in Problem 2b.

$0.017x^2 - 0.98x + 9.67 = 0; \ a = 0.017, \ b = -0.98, \ c = 9.67$

$x = \dfrac{-(-0.98) \pm \sqrt{(-0.98)^2 - 4(0.017)(9.67)}}{2(0.017)}$

$= \dfrac{0.98 \pm \sqrt{0.9604 - 0.65756}}{0.034}$

$= \dfrac{0.98 \pm \sqrt{0.30284}}{0.034}$

$x = \dfrac{0.98 + \sqrt{0.30284}}{0.034} \approx 45 \quad \text{or} \quad x = \dfrac{0.98 - \sqrt{0.30284}}{0.034} \approx 12.64 \approx 13$

4. Solve $3x^2 + 20x + 7 = 0$ using the quadratic formula.

$$3x^2 + 20x + 7 = 0$$

$$a = 3, b = 20, c = 7$$

$$x = \frac{-(20) \pm \sqrt{(20)^2 - 4(3)(7)}}{2(3)}$$

$$x = \frac{-20 \pm \sqrt{400 - 84}}{6} = \frac{-20 \pm \sqrt{316}}{6}$$

$$x = \frac{-20 + \sqrt{316}}{6} \approx -0.4, \qquad x = \frac{-20 - \sqrt{316}}{6} \approx -6.3$$

SUMMARY: ACTIVITY 4.7

The solutions of a quadratic equation in standard form, $ax^2 + bx + c = 0, a \neq 0$, are

$$x = \frac{-b \pm \sqrt{b^2 - 4ac}}{2a}.$$

The $\pm$ symbol in the formula indicates that there are two solutions, one in which the terms in the numerator are added and one in which the terms are subtracted.

EXERCISES: ACTIVITY 4.7

Use the quadratic formula to solve the equations in Exercises 1–5.

1. $x^2 + 2x - 15 = 0$

$$a = 1, b = 2, c = -15$$

$$x = \frac{-(2) \pm \sqrt{(2)^2 - 4(1)(-15)}}{2(1)}$$

$$= \frac{-2 \pm 8}{2}$$

$$x = -5 \quad \text{or} \quad x = 3$$

2. $4x^2 + 32x + 15 = 0$

$$a = 4, b = 32, c = 15$$

$$x = \frac{-(32) \pm \sqrt{(32)^2 - 4(4)(15)}}{2(4)}$$

$$= \frac{-32 \pm 28}{8}$$

$$x = -0.5 \quad \text{or} \quad x = -7.5$$

3. $-2x^2 + x + 1 = 0$

$$a = -2, b = 1, c = 1$$

$$x = \frac{-(1) \pm \sqrt{(1)^2 - 4(-2)(1)}}{2(-2)}$$

$$= \frac{-1 \pm 3}{-4}$$

$$x = 1 \quad \text{or} \quad x = -\frac{1}{2}$$

4. $2x^2 + 7x = 0$

$a = 2, b = 7, c = 0$

$x = \dfrac{-7 \pm \sqrt{(7)^2 - 4(2)(0)}}{2(2)}$

$= \dfrac{-7 \pm \sqrt{49}}{4}$

$= \dfrac{-7 \pm 7}{4}$

$x = 0$ or $x = -3.5$

5. $-x^2 + 10x + 9 = 0$

$a = -1, b = 10, c = 9$

$x = \dfrac{-10 \pm \sqrt{(10)^2 - 4(-1)(9)}}{2(-1)}$

$\approx \dfrac{-10 \pm 11.66}{-2}$

$x \approx 10.83$ or $x \approx -0.83$

6. Your friend's model rocket has a powerful engine. Its height (in feet) is described by the function $H(t) = -16t^2 + 432t$, where t represents the number of seconds since launch. Set up and solve a quadratic equation to answer parts a–c.

a. When is the rocket 400 feet high?

$-16t^2 + 432t = 400$

$-16t^2 + 432t - 400 = 0$ Divide by -16.

$t^2 - 27t + 25 = 0$

Here, $a = 1$, $b = -27$, and $c = 25$.

Apply the quadratic formula.

$t = \dfrac{27 + \sqrt{(-27)^2 - 4 \cdot 1 \cdot 25}}{2 \cdot 1}$ or $t = \dfrac{27 - \sqrt{(-27)^2 - 4 \cdot 1 \cdot 25}}{2 \cdot 1}$

$t = \dfrac{27 + \sqrt{629}}{2} \approx \dfrac{52.08}{2} \approx 26$ sec. or $t = \dfrac{27 - \sqrt{629}}{2} \approx \dfrac{1.92}{2} \approx 1$ sec.

The rocket is 400 feet above the ground approximately 1 second and approximately 26 seconds after launch.

b. When is the rocket 2000 feet high?

$-16t^2 + 432t = 2000$

$-16t^2 + 432t - 2000 = 0$ Divide by -16.

$t^2 - 27t + 125 = 0$

Here, $a = 1$, $b = -27$, and $c = 125$.

Apply the quadratic formula.

$t = \dfrac{27 + \sqrt{(-27)^2 - 4 \cdot 1 \cdot 125}}{2 \cdot 1}$ or $t = \dfrac{27 - \sqrt{(-27)^2 - 4 \cdot 1 \cdot 125}}{2 \cdot 1}$

$t = \dfrac{27 + \sqrt{229}}{2} \approx \dfrac{42.13}{2} \approx 21.1$ sec. or $t = \dfrac{27 - \sqrt{229}}{2} \approx \dfrac{11.87}{2} \approx 5.9$ sec.

The rocket is 2000 feet above the ground approximately 5.9 seconds and approximately 21.1 seconds after launch.

c. When is the rocket 2800 feet high?

$$-16t^2 + 432t = 2800$$

$-16t^2 + 432t - 2800 = 0$ Divide by -16.

$t^2 - 27t + 175 = 0$

Here, $a = 1$, $b = -27$, and $c = 175$.

Apply the quadratic formula.

$$t = \frac{27 + \sqrt{(-27)^2 - 4 \cdot 1 \cdot 175}}{2 \cdot 1} \quad \text{or} \quad t = \frac{27 - \sqrt{(-27)^2 - 4 \cdot 1 \cdot 175}}{2 \cdot 1}$$

$$t = \frac{27 + \sqrt{29}}{2} \approx \frac{32.4}{2} \approx 16.2 \text{ sec.} \quad \text{or} \quad t = \frac{27 - \sqrt{29}}{2} \approx \frac{21.6}{2} \approx 10.8 \text{ sec.}$$

The rocket is 2800 feet above the ground approximately 10.8 seconds and approximately 16.2 seconds after launch.

7. Solve $x^2 - 2x - 15 = 0$ in each of the following two ways.

a. Factor and then use the zero-product property.

$$x^2 - 2x - 15 = 0$$

$$(x - 5)(x + 3) = 0$$

$x - 5 = 0 \quad \text{or} \quad x + 3 = 0$

$\quad\quad x = 5 \quad \text{or} \quad\quad x = -3$

b. Use the quadratic formula.

$a = 1$, $b = -2$, $c = -15$

$$x = \frac{-(-2) \pm \sqrt{(-2)^2 - 4(1)(-15)}}{2(1)}$$

$$= \frac{2 \pm \sqrt{4 + 60}}{2}$$

$$= \frac{2 \pm \sqrt{64}}{2}$$

$$= \frac{2 \pm 8}{2}$$

$$= \frac{2 + 8}{2} = \frac{10}{2} = 5 \quad \text{or} \quad x = \frac{2 - 8}{2} = \frac{-6}{2} = -3$$

$x = 5 \quad\quad\quad\quad\quad\quad \text{or} \quad x = -3$

c. Which process seems more efficient to you?

(Answers will vary.)

8. Solve $(2x - 1)^2 = 15$ in each of the following two ways.

a. Solve by taking square roots.

$$(2x - 1)^2 = 15$$

$$2x - 1 = \pm\sqrt{15}$$

$2x - 1 = \sqrt{15} \quad\quad \text{or} \quad 2x - 1 = -\sqrt{15}$

$\quad\quad 2x = \sqrt{15} + 1 \quad \text{or} \quad\quad 2x = -\sqrt{15} + 1$

$$x = \frac{\sqrt{15} + 1}{2} \approx \frac{4.87}{2} \approx 2.44 \quad \text{or} \quad x = \frac{-\sqrt{15} + 1}{2} \approx \frac{-2.87}{2} \approx -1.44$$

b. Solve using the quadratic formula.

$$(2x - 1)^2 = 15$$
$$4x^2 - 4x + 1 = 15$$
$$4x^2 - 4x - 14 = 0$$

Here, $a = 4$, $b = -4$, and $c = -14$.
Apply the quadratic formula.

$$x = \frac{4 + \sqrt{(-4)^2 - 4 \cdot 4 \cdot (-14)}}{2 \cdot 4} \quad \text{or} \quad x = \frac{4 - \sqrt{(-4)^2 - 4 \cdot 4 \cdot (-14)}}{2 \cdot 4}$$

$$x = \frac{4 + \sqrt{240}}{8} \approx \frac{19.49}{8} \approx 2.44 \quad \text{or} \quad x = \frac{4 - \sqrt{240}}{8} \approx \frac{-11.49}{8} \approx -1.44$$

c. Which process seems more efficient to you?

(Answers will vary.)

9. a. Determine the x-intercepts, if any, of the graph of $y = x^2 - 2x + 5$.

The x-intercepts occur when $y = 0$ so I need the solution to $x^2 - 2x + 5 = 0$.

Using the quadratic formula, with $a = 1$, $b = -2$, $c = 5$,

$$x = \frac{-(-2) \pm \sqrt{(-2)^2 - 4(1)(5)}}{2(1)}$$

$$= \frac{2 \pm \sqrt{4 - 20}}{2}$$

$$= \frac{2 \pm \sqrt{-16}}{2}$$

There are no solutions because $\sqrt{-16}$ is not a real number. Thus, there are no x-intercepts.

b. Sketch the graph of $y = x^2 - 2x + 5$ using your graphing calculator and verify your answer to part a.

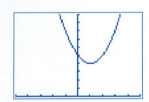

The graph of $y = x^2 - 2x + 5$ does not cross the x-axis so there are no x-intercepts.

10. The following data from the National Health and Nutrition Examination Survey indicates that the number of American adults who are overweight or obese is increasing.

	1961	1972	1978	1991	1999	2004	2006
Years Since 1960, t	1	12	18	31	39	44	46
Percentage of Americans Who are Overweight or Obese, $P(t)$	45	47	47	56	64.5	66.3	66.9

The quadratic function $P(t) = 0.01t^2 + 0.053t + 44.537$ can be used to model this data.

a. Determine the percentage of overweight or obese Americans in the year 2010.

$t = 2010 - 1960 = 50$

$P(50) = 0.01(50)^2 + 0.053(50) + 44.537 \approx 72.2\%$

According to the model, almost 72% of American adults were overweight or obese in the year 2010.

b. Determine the year after 1960 when the model predicts that the percentage of overweight or obese Americans will first exceed 75%.

$0.01t^2 + 0.053t + 44.537 = 75$

$0.01t^2 + 0.053t - 30.463 = 0$

$t = \dfrac{-0.053 \pm \sqrt{(0.053)^2 - 4(0.01)(-30.463)}}{2(0.01)}$

$t \approx 52.6, \; -57.9$

$t = 52.6$ corresponds to the year 2013, which means that the model predicts that the percentage of overweight or obese Americans in 2013 will exceed 75%. The negative value $t = -57.9$ does not apply.

Cluster 2   What Have I Learned?

1. Write (but do not solve) a quadratic equation that arises in each of the following situations.

  a. In determining the x-intercepts of the graph of $y = x^2 - 4x - 12$.

  $x^2 - 4x - 12 = 0$

  b. In determining the point(s) on the graph of $y = 100 - x^2$ whose y-coordinate is 40.

  $100 - x^2 = 40$

  c. In determining the point(s) at which the graph of $y = 3x^2 - 75$ crosses the x-axis.

  $3x^2 - 75 = 0$

2. Describe a reasonable process for solving each of the following quadratic equations.

  a. $2x^2 = 50$

  Divide each term by 2 and then take square roots.

  b. $3x^2 - 12x = 0$

  Determine the GCF, factor it out, and then use the zero-product property.

  c. $x^2 + 2x - 15 = 0$

  Factor the left side of the equation and then apply the zero-product property.

  d. $(x - 4)^2 = 9$

  Take the square root of each side of the equation and then solve the resulting linear equations.

  e. $x^2 + 3x - 1 = 0$

  Use the quadratic formula.

3. Solve each of the quadratic equations in Exercise 2 using the process you described. Remember to check your answers in the original equations.

  a. $2x^2 = 50$

  $\dfrac{2x^2}{2} = \dfrac{50}{2}$

  $x^2 = 25$

  $x = \pm 5$

  b. $3x^2 - 12x = 0$

  $3x(x - 4) = 0$

  $3x = 0 \quad \text{or} \quad x - 4 = 0$

  $x = 0 \quad \text{or} \qquad x = 4$

  c. $x^2 + 2x - 15 = 0$

  $(x + 5)(x - 3) = 0$

  $x + 5 = 0 \quad \text{or} \quad x - 3 = 0$

  $x = -5 \quad \text{or} \qquad x = 3$

  d. $(x - 4)^2 = 9$

  $x - 4 = \pm 3$

  $x - 4 = 3 \quad \text{or} \quad x - 4 = -3$

  $x = 7 \quad \text{or} \qquad x = 1$

  e. $x^2 + 3x - 1 = 0$

  $x = \dfrac{-3 \pm \sqrt{9 - 4 \cdot 1 \cdot (-1)}}{2 \cdot 1}$

  $x = \dfrac{-3 + \sqrt{13}}{2} \approx 0.30 \quad \text{or} \quad x = \dfrac{-3 - \sqrt{13}}{2} \approx -3.30$

4. Which of the following equations are equivalent (have identical solutions) to $4x^2 - 12x = 0$? Explain how to rewrite $4x^2 - 12x = 0$ into each equivalent equation.

$$2x^2 - 6x = 0 \qquad 4x - 12 = 0 \qquad x^2 - 3x = 0$$

$2x^2 - 6x = 0$ *is* equivalent. Divide each term of $4x^2 - 12x = 0$ by 2.

$4x - 12 = 0$ is *not* equivalent. It was obtained by dividing each term of $4x^2 - 12x = 0$ by x.

$x^2 - 3x = 0$ *is* equivalent. Divide each term of $4x^2 - 12x = 0$ by 4.

5. What is the graph of a quadratic function called?

The graph of a quadratic function is called a *parabola*.

6. Does the graph of every parabola have x-intercepts? Use the graph of $y = x^2 + 5$ and $y = -x^2 - 4$ to explain.

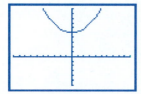 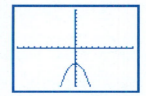

No, a parabola can open up and have a turning point above the x-axis, or it can open down and have a turning point below the x-axis.

7. How many y-intercepts are possible for a quadratic function, $y = ax^2 + bx + c$? Explain why.

There will always be *one* y-intercept.

The x-coordinate of the y-intercept is always zero. Therefore, $y(0) = a(0)^2 + b(0) + c = c$.

This shows that the y-intercept is always $(0, c)$.

Cluster 2   How Can I Practice?

1. a. Use at least two algebraic methods to solve $(x - 3)^2 = 4$.

Take roots of each side, then solve for x.

$x - 3 = \pm\sqrt{4}$

$x - 3 = \pm 2$

$x - 3 = 2$ or $x - 3 = -2$

$x = 5$ or $x = 1$

Put equation in standard form and then solve by factoring and the zero-product property.

$x^2 - 6x + 9 = 4$

$x^2 - 6x + 5 = 0$

$(x - 5)(x - 1) = 0$

$x - 5 = 0$ or $x - 1 = 0$

$x = 5$ or $x = 1$

b. Which method do you prefer? Explain.

(Answers will vary.)

2. Given $f(x) = 2x^2 - 8x$.

a. Complete the table of values.

x	−4	−2	−1	0	2	3	5	6
f(x)	64	24	10	0	−8	−6	10	24

b. Use the points in the table in part a to sketch a graph of $f(x)$. Scale the axes appropriately.

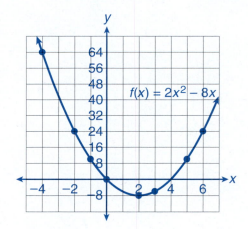

c. Verify the graph using a graphing calculator, if available.

d. Determine the x-intercepts algebraically, and verify your answer graphically.

$$2x^2 - 8x = 0$$
$$x^2 - 4x = 0$$
$$x(x - 4) = 0$$
$$x = 0 \quad \text{or} \quad x = 4$$

The x-intercepts are (0, 0) and (4, 0). The graph crosses the x-axis at 0 and 4, so the intercepts are correct.

3. Solve the following equations *by factoring*, if possible. Check your solutions in the original equation.

a. $x^2 - 9x + 20 = 0$

$$(x - 4)(x - 5) = 0$$
$$x - 4 = 0 \quad \text{or} \quad x - 5 = 0$$
$$x = 4 \quad \text{or} \quad x = 5$$

b. $x^2 - 9x + 14 = 0$

$$(x - 2)(x - 7) = 0$$
$$x - 2 = 0 \quad \text{or} \quad x - 7 = 0$$
$$x = 2 \quad \text{or} \quad x = 7$$

c. $x^2 - 11x = -24$

$$x^2 - 11x + 24 = 0$$
$$(x - 8)(x - 3) = 0$$
$$x - 8 = 0 \quad \text{or} \quad x - 3 = 0$$
$$x = 8 \quad \text{or} \quad x = 3$$

d. $m^2 + m = 6$

$$m^2 + m - 6 = 0$$
$$(m + 3)(m - 2) = 0$$
$$m + 3 = 0 \quad \text{or} \quad m - 2 = 0$$
$$m = -3 \quad \text{or} \quad m = 2$$

e. $2x^2 - 4x - 6 = 0$

Divide both sides by 2.

$$x^2 - 2x - 3 = 0$$
$$(x - 3)(x + 1) = 0$$
$$x - 3 = 0 \quad \text{or} \quad x + 1 = 0$$
$$x = 3 \quad \text{or} \quad x = -1$$

f. $3t^2 + 21t - 18 = 0$

Divide both sides by 3.

$$t^2 + 7t - 6 = 0$$

The equation is not solvable by factoring because the trinomial is not factorable.

g. $-7x + 10 = -x^2$

Rewrite in standard form.

$$x^2 - 7x + 10 = 0$$
$$(x - 5)(x - 2) = 0$$
$$x - 5 = 0 \quad \text{or} \quad x - 2 = 0$$
$$x = 5 \quad \text{or} \quad x = 2$$

h. $-9a - 12 = -3a^2$

Rewrite in standard form.

$$3a^2 - 9a - 12 = 0$$

Divide both sides by 3.

$$a^2 - 3a - 4 = 0$$
$$(a - 4)(a + 1) = 0$$
$$a - 4 = 0 \quad \text{or} \quad a + 1 = 0$$
$$a = 4 \quad \text{or} \quad a = -1$$

i. $2x^2 + 3x = 0$

$$x(2x + 3) = 0$$
$$x = 0 \quad \text{or} \quad 2x + 3 = 0$$
$$x = 0 \quad \text{or} \quad x = -\frac{3}{2}$$

j. $2x^2 = 12x$

Rewrite in standard form.

$$2x^2 - 12x = 0$$

Divide both sides by 2.

$$x^2 - 6x = 0$$
$$x(x - 6) = 0$$
$$x = 0 \quad \text{or} \quad x - 6 = 0$$
$$x = 0 \quad \text{or} \quad x = 6$$

k. $v^2 + 8v = 0$

$v(v + 8) = 0$

$v = 0$ or $v + 8 = 0$

$v = 0$ or $v = -8$

4. a. Determine the x-intercepts of $f(x) = x^2 + 5x - 6$.

$x^2 + 5x - 6 = 0$

$(x + 6)(x - 1) = 0$

$x + 6 = 0$ or $x - 1 = 0$

$x = -6$ or $x = 1$

The x-intercepts are $(-6, 0)$ and $(1, 0)$.

b. Verify your answer to part a using your graphing calculator.

The graph crosses the x-axis at -6 and 1, so the intercepts are correct.

5. Use the quadratic formula to solve the following equations.

a. $x^2 = -5x + 6$

Rewrite in standard form first.

$x^2 + 5x - 6 = 0$

$x = \dfrac{-(5) \pm \sqrt{(5)^2 - 4(1)(-6)}}{2(1)}$

$x = \dfrac{-5 \pm \sqrt{49}}{2} = \dfrac{-5 \pm 7}{2}$

$x = -6$ or $x = 1$

b. $6x^2 + x = 15$

Rewrite in standard form first.

$6x^2 + x - 15 = 0$

$x = \dfrac{-(1) \pm \sqrt{(1)^2 - 4(6)(-15)}}{2(6)}$

$x = \dfrac{-1 \pm \sqrt{361}}{12} = \dfrac{-1 \pm 19}{12}$

$x = \dfrac{3}{2}$ or $x = -\dfrac{5}{3}$

c. $x^2 - 7x - 18 = 0$

$x = \dfrac{-(-7) \pm \sqrt{(-7)^2 - 4(1)(-18)}}{2(1)}$

$x = \dfrac{7 \pm \sqrt{121}}{2} = \dfrac{7 \pm 11}{2}$

$x = 9$ or $x = -2$

d. $3x^2 - 8x + 4 = 0$

$x = \dfrac{-(-8) \pm \sqrt{(-8)^2 - 4(3)(4)}}{2(3)}$

$x = \dfrac{8 \pm \sqrt{16}}{6} = \dfrac{8 \pm 4}{6}$

$x = 2$ or $x = \dfrac{2}{3}$

e. $4x^2 + 28x - 32 = 0$

Divide both sides by 4.

$x^2 + 7x - 8 = 0$

$x = \dfrac{-(7) \pm \sqrt{(7)^2 - 4(1)(-8)}}{2(1)}$

$x = \dfrac{-7 \pm \sqrt{81}}{2} = \dfrac{-7 \pm 9}{2}$

$x = -8$ or $x = 1$

f. $2x^2 + 4x - 3 = 0$

$x = \dfrac{-(4) \pm \sqrt{(4)^2 - 4(2)(-3)}}{2(2)}$

$x = \dfrac{-4 \pm \sqrt{40}}{4} \approx \dfrac{-4 \pm 6.324}{4}$

$x \approx 0.581$ or $x \approx -2.581$

g. $x^2 + 7 = 6x$

Rewrite in standard form first.

$x^2 - 6x + 7 = 0$

$x = \dfrac{-(-6) \pm \sqrt{(-6)^2 - 4(1)(7)}}{2(1)}$

$x = \dfrac{6 \pm \sqrt{8}}{2} \approx \dfrac{6 \pm 2.828}{2}$

$x \approx 4.414$ or $x \approx 1.586$

h. $x^2 + 6x = -6$

Rewrite in standard form first.

$x^2 + 6x + 6 = 0$

$x = \dfrac{-(6) \pm \sqrt{(6)^2 - 4(1)(6)}}{2(1)}$

$x = \dfrac{-6 \pm \sqrt{12}}{2} \approx \dfrac{-6 \pm 3.464}{2}$

$x \approx -1.268$ or $x \approx -4.732$

i. $x^2 + 2 = 4x$

Rewrite in standard form first.

$x^2 - 4x + 2 = 0$

$x = \dfrac{-(-4) \pm \sqrt{(-4)^2 - 4(1)(2)}}{2(1)}$

$x = \dfrac{4 \pm \sqrt{8}}{2} \approx \dfrac{4 \pm 2.828}{2}$

$x \approx 3.414$ or $x \approx 0.586$

j. $2x - 6 = -x^2$

Rewrite in standard form first.

$x^2 + 2x - 6 = 0$

$x = \dfrac{-(2) \pm \sqrt{(2)^2 - 4(1)(-6)}}{2(1)}$

$x = \dfrac{-2 \pm \sqrt{28}}{2} \approx \dfrac{-2 \pm 5.292}{2}$

$x \approx 1.646$ or $x \approx -3.646$

6. The stopping distance of a moving vehicle is the distance it travels *after* the brakes are applied. The stopping distance, d (in feet), of a car moving at a speed of v miles per hour is a quadratic function defined by the equation

$$d = 0.04v^2 + 1.1v.$$

a. Determine the stopping distance of a car traveling at 40 miles per hour.

$d = 0.04(40)^2 + 1.1(40) = 108$ ft.

b. How fast is a car traveling before the brakes are applied if it leaves skid marks that measure 307.5 feet?

$0.04v^2 + 1.1v = 307.5$

$0.04v^2 + 1.1v - 307.5 = 0$

Use the quadratic formula with $a = 0.04$, $b = 1.1$, $c = -307.5$.

$v = \dfrac{-1.1 \pm \sqrt{1.1^2 - 4 \cdot (0.04) \cdot (-307.5)}}{2 \cdot (0.04)}$;

$v = \dfrac{-1.1 \pm \sqrt{50.41}}{0.08}$;

$v = \dfrac{-1.1 \pm 7.1}{0.08}$

$v = 75$ or $v = -102.5$. The negative speed is not a meaningful value. The car was traveling 75 miles per hour before the brakes were applied.

7. The area of a circular region is given by the formula $A = \dfrac{\pi \cdot d^2}{4}$, where d is its diameter.

a. Determine the area of a circular region whose diameter is 8 feet.

$A = \dfrac{\pi(8)^2}{4} = 16\pi \approx 50.3$

The area of the circular region is a little over 50 square feet.

b. Determine the area of a circular region whose radius is 15 feet.

If $r = 15$, diameter $= 30$; $A = \dfrac{\pi(30)^2}{4} = 225\pi \approx 706.9$.

The area of the circular region is approximately 707 square feet.

c. Determine the diameter of a circular region whose area is 25π square feet.

$\dfrac{\pi \cdot (d)^2}{4} = 25\pi$; multiply both sides by $\dfrac{4}{\pi}$.

$\dfrac{4}{\pi} \cdot \dfrac{\pi \cdot (d)^2}{4} = 25\pi \cdot \dfrac{4}{\pi}$

$d^2 = 100$; so $d = 10$ feet. Because d is a length, only the positive square root makes sense here.

The diameter is 10 feet.

d. Determine, to the nearest inch, the diameter of a circle whose area is 300 square inches.

$\dfrac{\pi \cdot (d)^2}{4} = 300$; multiply both sides by $\dfrac{4}{\pi}$.

$\dfrac{4}{\pi} \cdot \dfrac{\pi \cdot (d)^2}{4} = 300 \cdot \dfrac{4}{\pi}$

$d^2 = \dfrac{1200}{\pi}$; so $d = \sqrt{\dfrac{1200}{\pi}} \approx 19.54$

Because d is a length, only the positive square root makes sense here.
The diameter is approximately 20 inches.

8. An object is thrown upward with an initial velocity, v (measured in feet per second) from an initial height above the ground, h (in feet). The object's height above the ground, s (in feet), in t seconds after it is thrown, is given by the general formula

$$s = h + vt - 16t^2.$$

A ball is tossed upward with an initial velocity of 20 feet per second from a height of 4 feet. So the general formula becomes $s = 4 + 20t - 16t^2$.

a. How high is the ball after half a second?

$t = 0.5$

$s = 4 + 20(0.5) - 16(0.5)^2 = 10$

The ball is 10 feet high after half a second.

b. How high is it after 1 second?

$t = 1$

$s = 4 + 20(1) - 16(1)^2 = 8$

The ball is 8 feet high after 1 second.

c. How high is the ball after 3 seconds?

$t = 3$

$s = 4 + 20(3) - 16(3)^2 = -80$

This result is possible only if there is a hole in the ground.

By the time 3 seconds pass, the ball is already on the ground.

d. Complete the following table to determine the height of the ball for the given number
of seconds. Use the data in the table to determine the maximum height attained by the ball.
After how many seconds does the ball reach this height? Explain your reasoning.

NUMBER OF SECONDS	HEIGHT (ft.)
0.6	10.24
0.61	10.2464
0.62	10.2496
0.625	10.25
0.63	10.2496
0.64	10.2464
0.65	10.24

Both before and after 0.625 second, the height is less than 10.25 feet.
Therefore, the ball reaches its maximum height in 0.625 second.

Cluster 3   Other Nonlinear Functions

Activity 4.8

Inflation

Objectives

1. Recognize an exponential function as a rule for applying a growth factor or a decay factor.

2. Graph exponential functions from numerical data.

3. Recognize exponential functions from symbolic rules.

4. Graph exponential functions from symbolic rules.

Exponential Growth

Inflation means that a current dollar will buy less in the future. According to the U.S. Consumer Price Index, the inflation rate for 2005 was 3.4%. This means that a 1-pound loaf of white bread that cost a dollar in January 2005 cost $1.034 in January 2006. The change in price is usually expressed as an annual percentage rate, known as the inflation rate.

1. **a.** At the 2005 inflation rate of 3.4%, how much would a $60 pair of shoes cost next year?

   3.4% of $60 is 0.034(60) = $2.04

   $60 + $2.04 = $62.04

   The pair of shoes will cost $62.04 next year.

   b. Assume that the rate of inflation remains at 3.4% next year. How much will the shoes cost in the year following next year?

   3.4% of $62.04 is approximately $2.11.

   The pair of shoes will cost $64.15 in the year following next year.

2. **a.** Assume that the inflation rate increases somewhat and remains 4% per year for the next decade. Calculate the cost of a currently priced $10 pizza for each of the next 10 years and record the results in the following table. Use the variable t, years from now, as the input and $c(t)$, the cost of pizza, as the output. Round to the nearest cent.

YEARS FROM NOW, t	COST OF PIZZA, $c(t)$, $
0	10.00
1	10.40
2	10.82
3	11.25
4	11.70
5	12.17
6	12.65
7	13.16
8	13.69
9	14.23
10	14.80

   b. Is the cost a function of time? Explain.

   Yes. For each input value of t, there is only one corresponding output value of c.

3. a. Determine the average rate of change of the cost of the pizza in the first year (from $t = 0$ to $t = 1$), the fifth year (from $t = 4$ to $t = 5$), and the tenth year (from $t = 9$ to $t = 10$). Describe what these results mean for the cost of pizza over the next 10 years.

The average rate of change in the first year is $\dfrac{10.40 - 10.00}{1 - 0} = \dfrac{0.4}{1} = 0.4,$

that is, \$0.40 per year.

The average rate of change in the fifth year is $\dfrac{12.17 - 11.70}{5 - 4} = \dfrac{0.47}{1} = 0.47,$

that is, \$0.47 per year.

The average rate of change in the tenth year is $\dfrac{14.80 - 14.23}{10 - 9} = \dfrac{0.57}{1} = 0.57,$

that is, \$0.57 per year.

The more years go by, the faster the cost increases. (Answers will vary.)

b. Is the cost function linear? Explain.

No. The average rate of change is not constant.

To complete the table in Problem 2, you could calculate each output value by multiplying the previous output by 1.04, the inflation's growth factor (see Activities 1.9 and 1.10). Thus, to obtain the cost after 10 years, you would multiply the original cost, \$10, by 1.04, a total of 10 times. Symbolically, you write $10(1.04)^{10}$. Therefore, you can model the cost of pizza algebraically as

$$c(t) = 10(1.04)^t,$$

where $c(t)$ represents the cost and t represents the number of years from now.

Definition

A quantity y that increases by the same percent each year is said to increase exponentially. Such a quantity can be expressed algebraically as a function of time t by the rule

$$y = ab^t,$$

where a is its starting value (at $t = 0$), b is the fixed growth factor, and t represents the number of years that have elapsed. The function defined by $y = ab^t$ is called an **exponential function**. Note that the input variable t occurs as the exponent of the growth factor.

---

4. Under a constant annual inflation rate of 4%, what will a pizza cost after 20 years?

At a 4% inflation rate, the pizza will cost $10(1.04)^{20} = \$21.91$ in 20 years.

5. a. Complete the following table. Results from Problems 2 and 4 are already entered.

t	0	10	20	30	40	50
$c(t)$	10.00	14.80	21.91	32.43	48.01	71.07

b. Plot the data on an appropriately scaled and labeled axis.

c. As the input t increases, how does the corresponding cost change?

The cost of the pizzas increases more rapidly.

d. Using the graph in part b, determine how many years it will take for the price of a $10 pizza to double.

It will take more than 17 years for the price to double.

6. a. A graph of the function $c(t) = 10(1.04)^t$ is shown here in a window for t between -30 and 30 and $c(t)$ between 0 and 30. Use the same window to graph this function on a graphing calculator.

b. Use the trace or table feature of the graphing calculator to examine the coordinates of some points on the graph. Does your plot of the numerical data in Problem 5 agree with the graph shown here?

Yes, all the points that I plotted on the graph in Problem 3 appear in the first quadrant on the graph here.

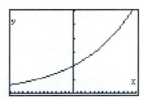

c. Is the entire graph in part a really relevant to the original problem? Explain.

The entire graph is not relevant because the practical domain is only nonnegative values.

d. Resize your window to include only the first quadrant from $x = 0$ to $x = 30$ and $y = 0$ to $y = 30$. Regraph the function.

 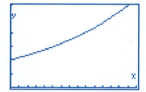

e. Use the calculator to determine how many years it will take for the price of a pizza to double? Explain how you determined your answer.

It will take more than 17 years for the price to double.

(Explanations will vary.) To determine my answer I looked at the output values using my calculator's table feature to see when the output came close to $20. Then I refined the table to use increments of 0.1. Alternatively, I could use the trace feature on my calculator and follow the output values until I reach a point where the output is 20 or very close to 20.

Exponential Decay

You have just purchased a new automobile for $21,000. Much to your dismay, you have just learned that you should expect the value of your car to depreciate by 15% per year!

7. What is the decay factor (see page 56, Activity 1.9) for the yearly depreciation rate 15%?

The decay factor associated with a 15% depreciation rate is $1.00 - 0.15 = 0.85$.

8. Use the decay factor from Problem 7 to determine the car's values for the years given in the table. Record your results in the table.

DEPRECIATION: TAKING ITS TOLL

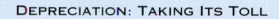

t (NUMBER OF YEARS FROM PURCHASE YEAR)	0	1	2	3	4	5
V (DOLLAR VALUE OF CAR IN THOUSANDS OF DOLLARS)	21	17.9	15.2	12.9	11	9.3

In Problem 8, you could calculate the retail value of the car in any given year by multiplying the previous year's value by 0.85, the depreciation decay factor. Thus, to obtain the retail value after 5 years, you could multiply the original value by 0.85 a total of 5 times, as follows:

$$21(0.85)(0.85)(0.85)(0.85)(0.85) \approx 9.3 \text{ thousand, or } \$9300$$

or equivalently, in exponential form,

$$21(0.85)^5 \approx 9.3 \text{ thousand, or } \$9300$$

Therefore, you can model the value of the car, V, algebraically by the formula

$$V(t) = 21(0.85)^t,$$

where t represents the number of years you own the car and V represents the car's value in thousands of dollars.

9. Determine the value of the car in 6 years.

The value of the car in 6 years will be $V = 21(0.85)^6 \approx 7.9$ thousand, or $7900.

10. Graph the depreciation formula for the car's value, $V(t) = 21(0.85)^t$, as a function of the time from the year of purchase. Use the following grid or a graphing calculator. Extend your graph to include 10 years from the date of purchase.

11. How long will it take for the value of the car to decrease to half its original value? Explain how you determined your answer.

It will take a little over 4 years for the car to be worth half its value.

(Explanations will vary.)

I determined the result by graphing $y = 10{,}500$ on the same axes and found the intersection.

SUMMARY: ACTIVITY 4.8

An **exponential function** is a function in which the input variable appears as the exponent of the growth (or decay) factor, as in

$$y = ab^t,$$

where a is the starting output value (at $t = 0$), b is the fixed growth or decay factor, and t represents the time that has elapsed.

EXERCISES: ACTIVITY 4.8

1. a. Complete the following table, in which $f(x) = \left(\frac{1}{2}\right)^x$ and $g(x) = 2^x$.

x	−2	−1	0	1	2
f(x)	4	2	1	$\frac{1}{2}$	$\frac{1}{4}$
g(x)	$\frac{1}{4}$	$\frac{1}{2}$	1	2	4

b. Use the points in part a to sketch a graph of the given functions f and g on the same coordinate axes. Compare the graphs. List the similarities and differences.

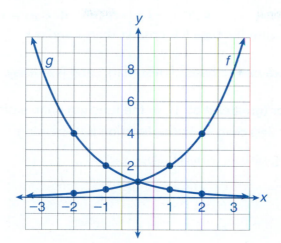

The graphs are mirror images of each other across the y-axis. Both graphs intersect the y-axis at the same place. Neither graph crosses the x-axis; that is, the output is never negative. $f(x) = \left(\frac{1}{2}\right)^x$ is a decreasing function, and $f(x) = 2^x$ is an increasing function.

 c. Use a graphing calculator to graph the functions $f(x) = \left(\frac{1}{2}\right)^x$ and $g(x) = 2^x$ given in part a in the window Xmin $= -2$, Xmax $= 2$, Ymin $= -2$, and Ymax $= 5$. Compare to the graphs you obtained in part b.

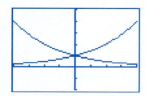

The graphs are the same.

2. Suppose the inflation rate is 5% per year and remains the same for the next 6 years.

a. Determine the growth factor for a 5% inflation rate.

The growth factor is $1 + 0.05 = 1.05$.

b. A pair of athletic shoes costs $110. If the inflation rate remains constant at 5% per year, write an algebraic rule to determine the cost, $c(t)$, of the shoes after t years.

$c(t) = 110(1.05)^t$

c. Complete the following table for the cost of a pair of athletic shoes that costs $110 now. Round to the nearest cent.

t, Years From Now	0	1	2	3	4	5	6
$c(t)$, Cost of Athletic Shoes ($)	110	115.50	121.28	127.34	133.71	140.39	147.41

d. What would be the cost of the athletic shoes in 10 years if the inflation rate stayed at 5%?

$c(10) = 110(1.05)^{10} \approx 179.18$

At an inflation rate of 5%, the $110 pair of athletic shoes would cost $179.18 in 10 years.

3. An exponential function may be increasing or decreasing. Determine which is the case for each of the following functions. Explain how you determined each answer.

a. $f(x) = 5^x$ **b.** $g(x) = \left(\frac{1}{2}\right)^x$ **c.** $h(t) = 1.5^t$ **d.** $k(p) = 0.2^P$

a. The function is increasing because the base 5 is greater than 1 and is a growth factor.

b. The function is decreasing because the base $\frac{1}{2}$ is less than 1 and is a decay factor.

c. The base 1.5 is a growth factor (greater than 1), so the function is increasing.

d. The base 0.2 is a decay factor (less than 1), so the function is decreasing.

4. a. Evaluate the functions in the following table for the input values, x.

Input x	0	1	2	3	4	5
g(x) = 3x	0	3	6	9	12	15
f(x) = 3ˣ	1	3	9	27	81	243

b. Compare the rate of increase of the functions $f(x) = 3^x$ and $g(x) = 3x$ from $x = 0$ to $x = 5$ by calculating the average rate of change for each function from $x = 0$ to $x = 5$. Determine which function grows faster on average in the given interval.

$$\frac{\Delta f}{\Delta x} = \frac{f(5) - f(0)}{5 - 0} = \frac{3^5 - 3^0}{5 - 0} = \frac{243 - 1}{5} = \frac{242}{5} = 48.4$$

$$\frac{\Delta g}{\Delta x} = \frac{g(5) - g(0)}{5 - 0} = \frac{3(5) - 3(0)}{5 - 0} = \frac{15 - 0}{5} = \frac{15}{5} = 3$$

$f(x) = 3^x$ increases at a faster rate on average than does $g(x) = 3x$. In this interval, $f(x)$ is increasing about 16 times faster.

5. a. For investment purposes, you recently bought a house for $150,000. You expect that the price of the house will increase $18,000 a year. How much will your investment be worth in 1 year? in 2 years?

My investment will be worth 150,000 + (18,000) = $168,000 in 1 year and 150,000 + 2(18,000) = $186,000 in 2 years.

b. Suppose the $18,000 increase in your house continues for many years. What type of function characterizes this growth? If P represents the housing price (in $1000s) as a function of t (in elapsed years), write this function rule.

This growth is characterized by a linear function. Its function rule is $P = 150 + 18t$.

c. You decided to buy a second investment house, again for $150,000, but in another area. Here the prices have been rising at a rate of 12% a year. If this rate of increase continues, how much will your investment be worth in 1 year? in 2 years?

My second investment will be worth 150,000(1.12) = $168,000 in 1 year and $168,000(1.12) = $188,160 in 2 years.

d. Suppose the 12% increase in housing prices continues for many years. What type of function characterizes this growth? If P represents the housing price (in $1000s) as a function of t (in elapsed years), write this function rule.

This growth is characterized by an exponential function.

Its function rule is $P = 150(1.12)^t$.

e. Which investment will give you the better return?

In 1 year, each investment gives the same return, $18,000. However, if I can wait another year, I can make $2160 more on the second investment. So after 1 year, the second investment will give the better return.

6. As a radiology specialist you use the radioactive substance iodine-131 to diagnose conditions of the thyroid gland. Iodine-131 decays at the rate of 8.3% per day. Your hospital currently has a 20-gram supply.

a. What is the decay factor for the decay of the iodine?

The decay factor associated with a 0.083 decay rate is $1.00 - 0.083 = 0.917$.

b. Write an exponential decay formula for N, the number of grams of iodine-131 remaining, in terms of t, the number of days from the current supply of 20 grams.

The formula is $N = 20(0.917)^t$.

c. Use the function rule from part b to determine the number of grams remaining for the days listed in the following table. Record your results in the table to the nearest hundredth.

t, Number of Days Starting From a 20-Gram Supply of Iodine-131	0	4	8	12	16	20	24
N, Number of Grams of Iodine-131 Remaining From a 20-Gram Supply	20.00	14.14	10.00	7.07	5.00	3.54	2.50

d. Determine the number of grams of iodine-131 remaining from a 20-gram supply after 2 months (60 days).

$N = 20(0.917)^{60} \approx 0.11$. After 2 months, approximately a tenth of a gram of iodine-131 remains from a 20-gram supply.

e. Graph the decay formula for iodine-131, $N = 20(0.917)^t$, as a function of the time t (days). Use appropriate scales and labels on the following grid or an appropriate window on a graphing calculator.

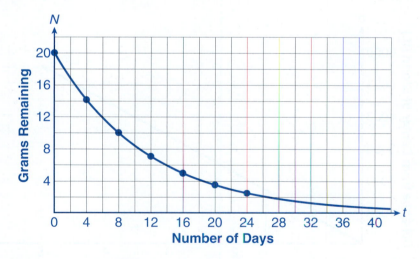

f. Use the graph or table to determine how long will it take for iodine-131 to decrease to half its original value. Explain how you determined your answer.

It will take about 8 days for iodine-131 to decay from 20 grams to 10 grams.

(Explanations will vary.)

I determined the result by graphing $y = 10$, which is half the amount, on the same axes and found the intersection.

Activity 4.9

A Thunderstorm

Objectives

1. Recognize the equivalent forms of the direct variation statement.

2. Determine the constant of proportionality in a direct variation problem.

3. Solve direct variation problems.

One of nature's more spectacular events is a thunderstorm. The sky lights up, delighting your eyes, and seconds later your ears are bombarded with the boom of thunder. Because light travels faster than sound, during a thunderstorm you see the lightning before you hear the thunder. The formula

$$d = 1080t$$

describes the distance, d in feet, you are from the storm's center if it takes t seconds for you to hear the thunder.

1. Complete the following table for the model $d = 1080t$.

t, in Seconds	1	2	3	4
d, in Feet	1080	2160	3240	4320

2. What does the ordered pair (3, 3240) from the above table mean in a practical sense?

If you hear thunder 3 seconds after you see lightning, you are 3240 feet from the center of the storm.

3. As the value of t increases, what happens to the value of d?

d increases as well.

The relationship between time t and distance d in this situation is an example of **direct variation**. As t increases, d also increases.

Definition

Two quantities are said to **vary directly** if whenever one quantity increases, the other quantity increases by the same multiplicative factor. The ratio of the two quantities is always constant. For instance, when one quantity doubles, so does the other.

4. Graph the distance, d, as a function of time, t, using the values in Problem 1.

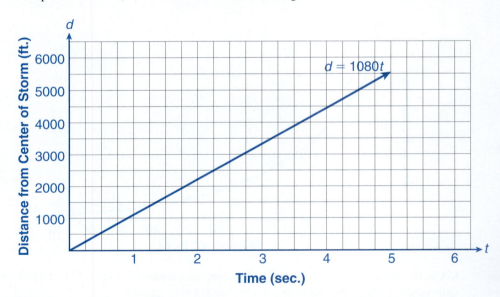

5. If the time it takes for you to hear the thunder after you see the lightning decreases, what is happening to the distance between you and the center of the storm?

The distance decreases by the same factor. For example, if the time interval between seeing the lightning and hearing the thunder is halved, then the distance of the storm's center from you is halved as well.

If x represents the input variable and y represents the output variable, then the following statements are equivalent.

a. y varies directly as x

b. y is directly proportional to x

c. $y = kx$ for some constant k.

d. The ratio $\dfrac{y}{x}$ always has the same constant value, k.

The number represented by k is called the **constant of proportionality** or **constant of variation**.

6. The amount of garbage, G, varies directly with the population, P. The population of Coral Springs, Florida, is 0.13 million and creates 2.6 million pounds of garbage each week. Determine the amount of garbage produced by Jacksonville with a population of 0.81 million.

    a. Write an equation relating G and P and the constant of variation k.

    $G = kP$

    b. Determine the value of k.

    $2.6 = k(.13)$ $k = 20$

    c. Rewrite the equation in part a using the value of k from part b.

    $G = 20P$

    d. Use the equation in part c and the population of Jacksonville to determine the weekly amount of garbage produced.

    $G = 20(0.81) = 16.2$ million pounds of garbage each week.

Procedure

To solve direct variation problems, the procedure of Problem 6 is summarized, with x representing the input variable and y representing the output variable.

   i. Write an equation of the form $y = kx$ relating the variables and the constant of proportionality, k.

  ii. Determine a value for k, using a known relationship between x and y.

 iii. Rewrite the equation in part i, using the value of k found in part ii.

 iv. Substitute a known value for x or y and determine the unknown value.

---

7. A worker's gross wages, w, vary directly as the number of hours the worker works, h. The following table shows the relationship between the wages and the hours worked.

Hours Worked, h	15	20	25	30	35
Wages, w	$172.50	$230.00	$287.50	$345.00	$402.50

a. Graph the gross wages, w, as a function of hours worked, h, using the values in the preceding table

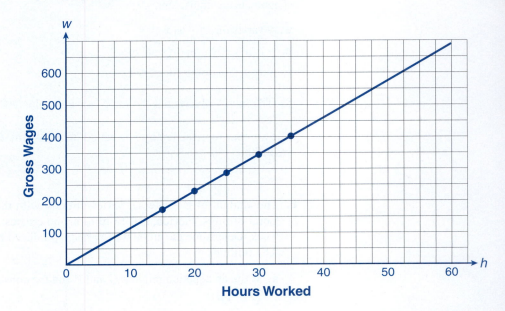

b. The relationship is defined by $w = kh$. Pick one ordered pair from the table and use it to determine the value of k.

$230 = k(20)$ $k = 11.5$

c. What does k represent in this situation?

k represents the hourly wage, $11.50 per hour.

d. Use the formula to determine the gross wages for a worker who works 40 hours.

$w = 11.50(40) = \$460.00$

Direct variation can involve higher powers of x, like x^2, x^3, or in general, x^n.

Example 1 *Let s vary directly as the square of t. If s = 64 when t = 2, determine the direct variation equation.*

SOLUTION

Because s varies directly as the square of t, you have

$$s = kt^2,$$

where k is the constant of variation. Substituting 64 for s and 2 for t, you have

$$64 = k(2)^2 \quad \text{or} \quad 64 = 4k \quad \text{or} \quad k = 16.$$

Therefore, the direct variation equation is

$$s = 16t^2.$$

8. The power P generated by a certain wind turbine varies directly as the square of the wind speed w. The turbine generates 750 watts of power in a 25-mile-per-hour wind. Determine the power it generates in a 45-mile-per-hour wind.

$P = kw^2 \quad 750 = k(25)^2 \quad k = 750/25^2 \quad k = 1.2$

$P = 1.2(45)^2 = 2430$ watts

SUMMARY: ACTIVITY 4.9

1. Two variables are said to **vary directly** if whenever one variable increases, the other variable increases by the same multiplicative factor. The ratio of the two variables is always constant.

2. The following statements are equivalent. The number represented by k is called the **constant of proportionality or constant of variation**.

   a. y varies directly as x

   b. y is directly proportional to x

   c. $y = kx$ for some constant x

   d. The ratio $\dfrac{y}{x}$ always has the same constant value, k.

3. The equation $y = kx^n$, where $k \neq 0$ and n is a positive integer, defines a **direct variation function** in which y varies directly as x^n. The constant, k, is called **constant of proportionality** or the **constant of variation**.

EXERCISES: ACTIVITY 4.9

1. The amount of sales tax, s, on any item varies directly as the list price of an item, p. The sales tax in your area is 8%.

   a. Write a formula that relates the amount of sales tax, s, to the list price, p, in this situation.

   $s = 0.08p$

   b. Use the formula in part a to complete the following table:

LIST PRICE, p ($)	SALES TAX, s ($)
10	0.80
20	1.60
30	2.40
50	4.00
100	8.00

c. Graph the sales tax, s, as a function of list price, p, using the values in the preceding table.

d. Determine the list price of an item for which the sales tax is $3.60.

$3.60 = 0.08p,$ $p = 3.60/0.08,$ $p = \$45.00$

2. The amount you tip a waiter, t, varies directly as the amount of the check, c.

a. Assuming you tip your waiter 15% of the amount of the check, write a formula that relates the amount of the tip, t, to the amount of the check, c.

$t = 0.15c$

b. Use the formula in part a to complete the following table:

Amount of Check, c ($)	15	25	35	50	100
Amount of Tip, t ($)	2.25	3.75	5.25	7.50	15.00

c. Graph the amount of the tip, t, as a function of the amount of the check, c, using the values in the preceding table.

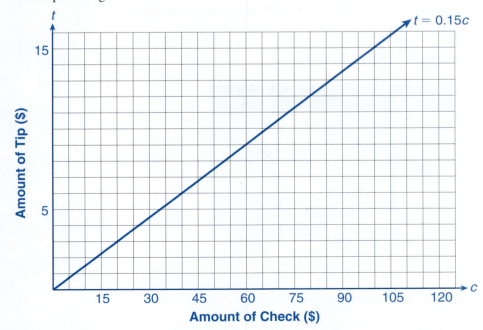

d. A company has a policy that no tip can exceed $12.00 and still be reimbursed by the company. What is the maximum cost of a meal for which all of the tip will be reimbursed if the tip is 15% of the check?

$12 = 0.15c$ $c = 12/0.15$ $c = \$80.00$

3. The distance, s, that an object falls from rest varies directly as the square of the time, t, of the fall. A ball dropped from the top of a building falls 144 feet in 3 seconds.

a. Write a formula that relates the distance the ball has fallen, s, in feet to the time, t, in seconds since it has been released.

$s = kt^2$ $144 = k(3)^2$ $k = 16$ $s = 16t^2$

b. Use the formula in part a to complete the following table:

TIME SINCE THE BALL WAS RELEASED, t (sec)	DISTANCE FALLEN, s (ft)
0.5	4
1	16
1.5	36
2	64
2.5	100

c. Graph the distance fallen, s, as a function of the time since the ball was released, x, using the values in the preceding table.

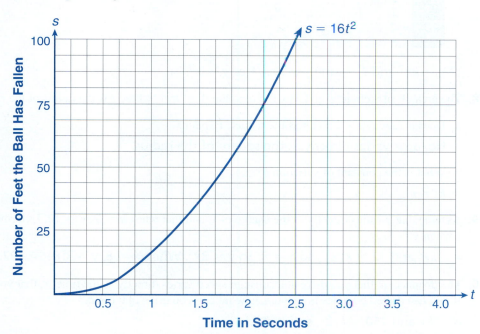

d. Estimate the height of the building if it takes approximately 2.3 seconds for the ball to hit the ground.

$S = 16(2.3)^2$ $S \approx 85$ feet

4. y varies directly as x, $y = 25$ when $x = 5$. Determine y when $x = 13$.

$y = kx$ $25 = k(5)$ $k = 5$ $y = 5x$

$y = 5(13)$ $y = 65$

5. y varies directly as x, $y = 7$ when $x = 21$. Determine x when $y = 5$.

$y = kx$ $\quad$ $7 = k(21)$ $\quad$ $k = \frac{1}{3}$ $\quad$ $y = \frac{1}{3}x$

$5 = \frac{1}{3}x$ $\quad$ $x = 15$

6. Given that y varies directly as x, consider the following table:

x	2	4	7	10	12
y	4	8	14	20	24

  a. Determine a formula that relates x and y.

    $y = kx$ $\quad$ $8 = k(4)$ $\quad$ $k = 2$ $\quad$ $y = 2x$

  b. Use the formula in part a to complete the preceding table.

7. For each table, determine the pattern and complete the table. Then write a direct variation equation for each table.

  a. y varies directly as x.

x	$\frac{1}{4}$	1	4	8
y	2	8	32	64

    $y = kx$ and $8 = k(1)$, so $k = 8$ or $y = 8x$.

  b. y varies directly as x^3.

x	$\frac{1}{2}$	1	3	6
y	$\frac{1}{8}$	1	27	216

    $y = kx^3$ and $1 = k(1)^3$, so $k = 1$ or $y = x^3$.

8. The area, A, of a circle is given by the function $A = \pi r^2$, where r is the radius of the circle.

  a. Does the area vary directly as the radius? Explain.

    No, area varies directly as the square of the radius.

  b. What is the constant of variation k?

    $k = \pi$

9. Assume that y varies directly as the square of x, and that when $x = 2$, $y = 12$. Determine y when $x = 8$.

$y = kx^2$ and $12 = k(2)^2$, so $k = 3$. So $y = 3x^2$. When $x = 8$, $y = 3(8)^2 = 192$.

10. The distance, d, that you drive at a constant speed varies directly as the time, x, that you drive. If you can drive 150 miles in 3 hours, how far can you drive in 6 hours?

$d = kx$ and $150 = k(3)$, so $k = 50$.

Now $d = 50x$, so in 6 hours I drive $d = 50(6) = 300$ miles.

11. The number of meters, d, that a skydiver falls before her parachute opens varies directly as the square of the time, t, that she is in the air. A skydiver falls 20 meters in 2 seconds. How far will she fall in 2.5 seconds?

$d = kt^2$, and $20 = k(2)^2$, so $k = 5$.

Now $d = 5t^2$, so in 2.5 seconds the skydiver travels $d = 5(2.5)^2 = 31.25$ meters.

Activity 4.10

Diving Under Pressure, or Don't Hold Your Breath

Objectives

1. Recognize functions of the form $y = \dfrac{k}{x}, x \neq 0$, as nonlinear.

2. Recognize equations of the form $xy = k$ as inverse variation.

3. Graph an inverse variation relationship from symbolic rules.

4. Solve equations of the form $\dfrac{a}{x} = b, x \neq 0$.

Do you know why you shouldn't hold your breath when scuba diving? Safe diving depends on a fundamental law of physics that states that the volume of a given mass of gas (air in your lungs) will decrease as the pressure increases (when the temperature remains constant). This law is known as Boyle's Law, named after its discoverer, the seventeenth-century scientist Sir Robert Boyle.

In this activity, you will discover the answer to the opening question.

1. You have a balloon filled with air that has a volume of 10 liters at sea level. The balloon is under a pressure of 1 atmosphere (atm) at sea level. Physicists define 1 atmosphere to be equal to 14.7 pounds of pressure per square inch.

    a. For every 33-foot increase in depth, the pressure will increase by 1 atmosphere. Therefore, if you take the balloon underwater to a depth of 33 feet below sea level, it will be under 2 atmospheres of pressure. At a depth of 66 feet, the balloon will be under 3 atmospheres of pressure. Complete column 2 (pressure) in the following table.

    b. At sea level, the balloon is under 1 atmosphere of pressure. At a depth of 33 feet, the pressure is 2 atmospheres. Since the pressure is now twice as much as it was at sea level, Boyle's Law states that the volume of the balloon decreases to one-half of its original volume, or 5 liters. Taking the balloon down to 66 feet, the pressure compresses the balloon to one-third of its original volume, or 3.33 liters; at 99 feet down the volume is one-fourth the original, or 2.5 liters, etc. Complete the third column (volume) in the following table.

DEPTH OF WATER (ft.) BELOW SEA LEVEL	PRESSURE (atm)	VOLUME OF AIR IN BALLOON (L)	PRODUCT OF PRESSURE AND VOLUME
Sea level	1	10	$1 \cdot 10 = 10$
33	2	5	$2 \cdot 5 = 10$
66	3	$3.3\overline{3}$	$3 \cdot 3.3\overline{3} = 10$
99	4	2.5	$4 \cdot 2.5 = 10$
132	5	2	$5 \cdot 2 = 10$
297	10	1	$10 \cdot 1 = 10$

    c. Note, in column 4 of the table, that the product of the volume and pressure at sea level is 10. Compute the product of the volume and pressure for the other depths in the table, and record them in the last column.

    d. What is the result of your computations in part c?

    All the products are the same, namely, 10.

2. Let p represent the pressure of a gas and v the volume at a given (constant) temperature. Use what you observed in Problem 1 to write an equation for Boyle's Law for this experiment.

    $pv = 10$

Boyle's Law is an example of **inverse variation** between two variables. In the case of Boyle's Law, when one variable increases by a multiplicative factor p, the other decreases by the multiplicative factor $\dfrac{1}{p}$, so that their product is always the same positive constant. In general, an inverse variation is represented by an equation of the form $xy = k$, where x and y represent the variables and k represents the constant.

Sometimes when you are investigating an inverse variation, you may want to consider one variable as a function of the other variable. In that case, you can solve the equation $xy = k$ for y as a function of x. By dividing each side of the equation by x, you obtain $y = \dfrac{k}{x}$.

3. Write an equation for Boyle's Law where the volume, v, is expressed as a function of the pressure, p.

The equation for Boyle's Law takes the form $v = \frac{10}{p}$.

4. a. Graph the volume, v, as a function of pressure, p, using the values in the preceding table.

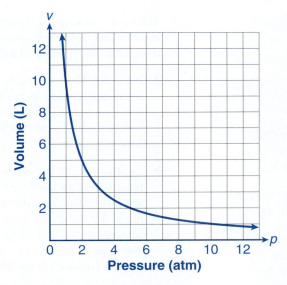

b. If you decrease the pressure (by rising to the surface of the water), what will happen to the volume? (Use the graph in part a to help answer this question.)

As the pressure decreases, the volume increases.

5. a. Suppose you are 99 feet below sea level, where the pressure is 4 atmospheres and where the volume is one-fourth of the sea-level volume. You use a scuba tank to fill the balloon back up to a volume of 10 liters. What is the product of the pressure and volume in this case?

The product of the pressure and volume is $4 \cdot 10 = 40$.

b. Now suppose you take the balloon up to 66 feet where the pressure is 3 atmospheres. What is the volume now?

The volume is $\frac{40}{3} \approx 13.33$ liters.

c. You continue to take the balloon up to 33 feet where the pressure is 2 atmospheres. What is the volume?

The volume will be $\frac{40}{2} = 20$ liters.

d. Finally you are at sea level. What is the volume of the balloon?

The volume will be $\frac{40}{1} = 40$ liters.

e. Suppose the balloon can expand only to 30 liters. What will happen before you reach the surface?

The balloon will burst.

6. Explain why the previous problem shows that, when scuba diving, you should not hold your breath if you use a scuba tank to fill your lungs.

My lungs can act very much like a pair of balloons. If I were to descend to 33 feet or lower and breathe compressed air from a scuba tank, I would increase the amount of air and volume in my lungs. If I were to hold my breath while rising to the surface, my lungs could overexpand and I could seriously injure myself.

7. a. Use the graph in Problem 4 to determine the volume of the balloon when the pressure is 12 atmospheres.

(Answers will vary.)

From the graph, I estimate the volume to be 0.8 liter.

b. From the graph, what do you estimate the pressure to be when the volume is 11 liters?

The pressure would be about 0.9 atmospheres.

c. Use the equation $v = \dfrac{10}{p}$ to determine the volume when the pressure is 12 atmospheres. How does your answer compare to the result in part a?

The volume at 12 atmospheres is $0.8\overline{3}$ liters. The answers agree.

d. Use the equation $v = \dfrac{10}{p}$ to determine the pressure when the volume is 11 liters. How does your answer compare to the result in part b?

$$11 = \frac{10}{p}$$

$$11p = 10$$

$$p = \frac{10}{11} \approx 0.91$$

The pressure is 0.91 atmosphere when the volume is 11 liters. This answer agrees with that from part b, as it should.

8. Solve each of the following.

a. $\dfrac{20}{x} = 10$

$\dfrac{20}{x} \cdot x = 10x$

$20 = 10x$

$2 = x$

b. $\dfrac{150}{x} = 3$

$\dfrac{150}{x} \cdot x = 3 \cdot x$

$150 = 3x$

$50 = x$

9. Suppose another balloon contains 12 liters of air at sea level. The inverse variation between pressure and volume is represented by the equation

$$pv = 12.$$

a. Use the equation to write pressure as a function of the volume for this balloon.

The function is $p = \dfrac{12}{v}$.

b. Choose five values for *v* between 0 and 12. For each value, determine the corresponding value of *p*. Record your results in the following table.

v	2	3	4	6	8
p	6	4	3	2	1.5

(Answers will vary.)

c. On the following grid, use the points you obtained in part b to graph the pressure as a function of the volume.

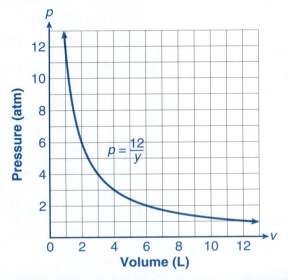

SUMMARY: ACTIVITY 4.10

1. For a function defined by $y = \dfrac{k}{x}$, where $k > 0$, as the input *x* increases, the output *y* decreases.

2. The function rule $y = \dfrac{k}{x}$, $k > 0$, can also be expressed in the form $xy = k$. This form shows explicitly that as one variable increases by a multiplicative factor *p*, the other decreases by a multiplicative factor $\dfrac{1}{p}$ so that the product of the two variables is always the constant value, *k*. The relationship between the variables defined by this function is known as **inverse variation**.

3. The input value $x = 0$ is not in the domain of the function defined by $y = \dfrac{k}{x}$.

EXERCISES: ACTIVITY 4.10

1. A guitar string has a number of frequencies at which it will naturally vibrate when it is plucked. These natural frequencies are known as the harmonics of the guitar string and are related to the length of the guitar string in an inverse relationship. For the first harmonic (also known as the fundamental frequency) the relationship can be expressed by the formula $v = 2f \cdot L$, where *v* represents the speed of the wave produced by the guitar string in meters per second, *f* represents the frequency in cycles per second (hertz), and *L* represents the length of the guitar string in meters. The speed of the wave does not depend on the frequency or length of the string, but on the medium (such as air) through which it travels. So for a given medium, the speed is a constant value.

a. The speed of waves in a particular guitar string is found to be 420 meters per second. Determine the fundamental frequency in hertz (first harmonic) of the string if its length is 70 centimeters.

$$L \text{ in meters} = \frac{70}{100} = 0.70 \text{ m}$$

$$420 = 2 \cdot 0.7 \cdot f$$

$$f = \frac{420}{2 \cdot 0.7} = 300$$

The fundamental frequency is 300 cycles per second (hertz).

b. To observe the effect that changing the length of the string has on the first harmonic, write the frequency, f, as a function of L in the formula $420 = 2f \cdot L$.

$$f = \frac{420}{2L} \quad \text{or} \quad f = \frac{210}{L}$$

c. Use the formula you wrote in part b to determine the frequencies for the string lengths given in the following table. The wave speed is 420 meters per second.

LENGTH OF GUITAR STRING (m)	FREQUENCY (Hz)
0.2	1050
0.3	700
0.4	525
0.5	420
0.6	350

d. Graph the points in the table in part c on the following grid and connect them with a smooth curve.

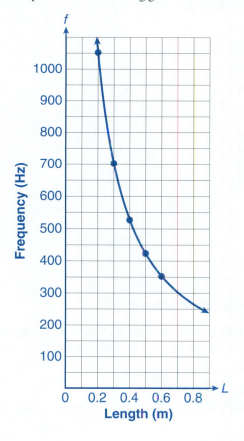

e. From your graph, estimate the frequency for a string length of 35 centimeters.

The frequency would be approximately 600 hertz.

f. What string length would produce a frequency of 840 hertz?

The string length would have to be 0.25 meter, or 25 centimeters.

2. A bicyclist, runner, jogger, and walker each completed a journey of 30 miles. Their results are recorded in the table below.

	CYCLIST	RUNNER	JOGGER	WALKER
v, Average Speed (mph)	30	15	10	5
t, Time Taken (hr.)	1	2	3	6
$d = vt$	30	30	30	30

a. Compute the product vt for each person, and record the products in the last row of the table.

b. Write the time to complete the journey as a function of average speed.

The time to complete the journey as a function of average speed is given by $t = \dfrac{30}{v}$, where t represents the time to complete the journey and v represents the average speed.

c. On the following grid, graph the function you wrote in part b for a journey of 30 miles.

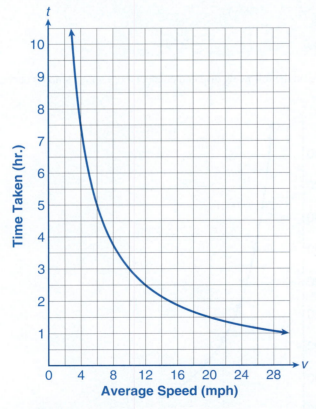

d. From the graph, determine how long the cyclist would take if she decreased her average speed to 20 miles per hour. Verify your answer from the graph by using the equation you wrote in part b.

$$t = \frac{30}{v} = \frac{30}{20} = 1.5 \text{ hr.}$$

The equation and graph both show that she would take 1.5 hours to complete her journey.

e. From the graph, determine how long it would take the runner to complete her run if she slowed to 12 miles per hour. Verify your answer from the graph by using your equation from part b.

$$t = \frac{30}{v} = \frac{30}{12} = 2.5 \text{ hr.}$$

The equation and graph both show that she would take 2.5 hours to complete her run.

3. The weight, w, of a body varies inversely as the square of its distance, d, from the center of Earth. The equation $wd^2 = 3.2 \times 10^9$ represents this relationship for a 200-pound person.

a. Rewrite the equation to express the weight of a body, w, as a function of the distance, d, from the center of Earth.

$$w = \frac{3.2 \times 10^9}{d^2}$$

b. How much would a 200-pound man weigh if he were 1000 miles above the surface of Earth? The radius of Earth is approximately 4000 miles. Explain how you solve this problem.

If the man is 1000 miles above Earth's surface, he is 5000 miles from the center of Earth. Therefore,

$$w = \frac{3.2 \times 10^9}{5000^2} = \frac{3.2 \times 10^9}{(5 \times 10^3)(5 \times 10^3)} = \frac{3.2 \times 10^9}{(25 \times 10^6)} = 0.128 \times 10^3 = 128.$$

The man weighs 128 pounds when he is 1000 miles above the surface of Earth.

4. a. Solve $xy = 120$ for y. Is y a function of x? What is its domain?

$$y = \frac{120}{x}$$

Yes, y is a function of x. The domain is all real numbers except $x = 0$.

b. Solve $xy = 120$ for x. Is x a function of y? What is its domain?

$$x = \frac{120}{y}$$

Yes, x is a function of y. The domain is all real numbers except $y = 0$.

5. Solve the following equations.

a. $30 = \dfrac{120}{x}$

$30x = 120$

$x = \dfrac{120}{30}$

$x = 4$

b. $2 = \dfrac{250}{x}$

$2x = 250$

$x = 125$

c. $-8 = \dfrac{44}{x}$

$-8x = 44$

$x = \dfrac{44}{-8}$

$x = -5.5$

d. $25 = \dfrac{750}{x + 2}$

$x + 2 = \dfrac{750}{25}$

$x + 2 = 30$

$x = 28$

Hang Time

Objectives

1. Recognize functions of the form $y = a\sqrt{x}$ as nonlinear.

2. Evaluate and solve equations that involve square roots.

3. Graph square root functions from numerical data.

4. Graph square root functions from symbolic rules.

The time, t (output, in seconds), that a basketball player spends in the air during a jump is called hang time and depends on the height, s (input, in feet), of the jump according to the formula

$$t = \tfrac{1}{2}\sqrt{s}.$$

To determine the hang time of a jump 1 foot above the court floor, you perform a sequence of two operations:

Start with s (1 ft.) $\rightarrow$ take square root $\rightarrow$ multiply by $\tfrac{1}{2}$ $\rightarrow$ to obtain t ($\tfrac{1}{2}$ second in the air).

A function such as $f(s) = \tfrac{1}{2}\sqrt{s}$ is known as a square root function because determining the output involves taking the square root of an expression involving the input.

1. A good college athlete can usually jump about 2 feet above the court floor. Determine his hang time.

$$\tfrac{1}{2}\sqrt{2} \approx 0.707 \text{ sec.}$$

2. Michael Jordan is known to have jumped as high as 39 inches above the court floor. What was his hang time on that jump?

$$39 \text{ in.} \left(\tfrac{1 \text{ ft.}}{12 \text{ in.}}\right) = 3.25 \text{ ft.}$$

$$\tfrac{1}{2}\sqrt{3.25} \approx 0.901 \text{ sec.}$$

> **Example 1** *How high would a jump have to be to have a hang time of 0.5 second?*
>
> **SOLUTION**
>
> You can determine the height from the hang time formula. You can do this by "reversing direction," or replacing each operation with its inverse operation in a sequence of algebraic steps. Note that the inverse of the square root operation is squaring.
>
> Start with 0.5 sec. $\rightarrow$ multiply by 2 $\rightarrow$ square $\rightarrow$ to obtain 1 ft.
>
> You can also solve the problem in a formal algebraic way by applying these inverse operations to each side of the equation.
>
> Substitute 0.5 second for t in the formula $t = \tfrac{1}{2}\sqrt{s}$ to obtain the equation $0.5 = \tfrac{1}{2}\sqrt{s}$ and proceed as follows:
>
> $2(0.5) = 2 \cdot \tfrac{1}{2}\sqrt{s}$ **Multiply by 2.**
>
> $1 = \sqrt{s}$ **Simplify.**
>
> $(1)^2 = \left(\sqrt{s}\right)^2$ **Square both sides.**
>
> $1 \text{ ft.} = s$ **Simplify.**

3. Use the hang time formula and the data regarding Michael Jordan's hang time on a 39-inch jump to determine whether it is reasonable for a professional basketball player to have a hang time of 3 seconds. Explain your answer.

$$\tfrac{1}{2}\sqrt{s} = 3$$

$$\sqrt{s} = 6$$

$$s = 36 \text{ ft.}$$

A hang time of 3 seconds requires a jump of 36 feet, almost 12 times as high as Michael Jordan's record height of 3.25 feet. So a hang time of 3 seconds is unreasonable.

4. a. What would be a reasonable domain and range for the hang time function having equation $t = \frac{1}{2}\sqrt{s}$?

(Answers will vary.) For example, domain: $0 \leq s \leq 4$, and range: $0 \leq t \leq 2$.

b. Complete the following table for the given jump heights.

HEIGHT OF JUMP, s (ft.)	HANG TIME, t (sec.)
0	0
1	0.5
2	0.71
3	0.87
4	1.0

c. Plot the points (t, s) using the table you completed in part b and connect them with a smooth curve.

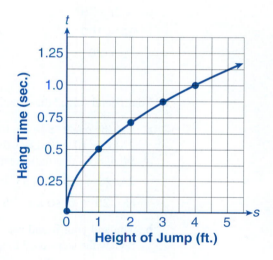

5. a. Is the hang time function $t = \frac{1}{2}\sqrt{s}$ a linear function? Use the graph in problem 4c to explain.

No, the graph of the data in Problem 4c is not a straight line.

Also, for each unit change in the input t, the corresponding change in the output s is not a constant.

b. Solve the hang time formula $t = \frac{1}{2}\sqrt{s}$ for s in terms of t.

Start with $t \rightarrow$ multiply by 2 $\rightarrow$ square each term on both sides to obtain $s = 4t^2$. Symbolically, the steps are

$$t = \frac{1}{2}\sqrt{s}$$
$$2t = \sqrt{s}$$
$$(2t)^2 = s$$
$$s = 4t^2$$

    c. Use the formula you obtained in part b to explain why the relationship between s and t is not linear.

    The t^2 in the formula $s = 4t^2$ that I obtained in part b shows that the relationship between hang time and jump height is not linear.

6. Solve each of the following equations. Check your result in the original equation.

    Note to instructor: Students should show checks. Examples are shown in parts a and b.

    a. $\sqrt{x} = 9$

        $x = 81$

        Check: $\sqrt{81} = 9$

    b. $\sqrt{t} + 3 = 10$

        $\sqrt{t} = 7$

        $t = 49$

        Check: $\sqrt{49} + 3 = 10$

        $7 + 3 = 10$

        $10 = 10$

    c. $4\sqrt{x} = 12$

        $\sqrt{x} = 3$

        $x = 9$

    d. $\sqrt{t} + 1 = 5$

        $t + 1 = 25$

        $t = 24$

    e. $\sqrt{5t} = 10$

        $5t = 100$

        $t = 20$

    f. $\sqrt{2t} - 3 = 0$

        $\sqrt{2t} = 3$

        $2t = 9$

        $t = 4.5$

7. Your friend solved the equation $2\sqrt{x} + 10 = 0$ for x and obtained $x = 25$ as the result.

    a. Use the result, $x = 25$, to evaluate the left side of the original equation, $2\sqrt{x} + 10 = 0$. Does the value of the left side equal the value of the right side? What does that indicate about the result, $x = 25$, that your friend obtained?

    $2\sqrt{25} + 10 = 2(5) + 10 = 20$

    No, $20 \neq 0$, so $x = 25$ is *not* a solution to the equation.

    b. You and your friend want to determine why $x = 25$ is not a solution. One way is to review the steps used to solve the equation. Do you see where the error is?

    1. $2\sqrt{x} + 10 = 0$

    2.   **$-10 = -10$**

    3.   **$\dfrac{2\sqrt{x}}{2} = \dfrac{-10}{2}$**

    4.   **$\sqrt{x} = -5$**

    5.   **$(\sqrt{x})^2 = (-5)^2$**

    6.   **$x = 25$**

    In step 4, the statement is $\sqrt{x} = -5$. This statement cannot be true for any real number x, because the square root symbol, $\sqrt{}$, always denotes a *positive* square root.

c. Another way to investigate why $x = 25$ is not a solution to the equation $2\sqrt{x} + 10 = 0$ is to consider the graphs of $y_1 = 2\sqrt{x} + 10$ and $y_2 = 0$. Do the graphs intersect? What does that indicate about the solution(s) to the equation $2\sqrt{x} + 10 = 0$?

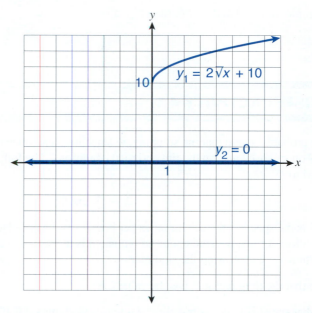

The graphs do not intersect, which means that the left side of the equation $2\sqrt{x} + 10 = 0$ is never equal to 0, no matter what value x has. Therefore, the equation has no solution.

SUMMARY: ACTIVITY 4.11

1. The function defined by $y = a\sqrt{x}$, where $a \neq 0$ and x is nonnegative, is called a **square root function** because determining the output involves the square root of an expression containing the input.

2. For $a > 0$, the graph of a square root function has the following general shape.

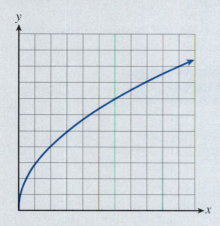

3. Squaring and taking square roots are inverse operations.

4. To evaluate an output of a square root function $y = a\sqrt{x}$ for a specified input, do the following.

Start with $x \rightarrow$ take square root $\rightarrow$ multiply by $a \rightarrow$ to obtain y.

For example, if $a = 3$ and $x = 4$, then $y = 3\sqrt{4} \rightarrow y = 3 \cdot 2 \rightarrow y = 6$.

5. To solve a square root equation $y = a\sqrt{x}$ for a specified output, do the following:

Start with $y \rightarrow$ divide by $a \rightarrow$ square $\rightarrow$ to obtain x.

For example, if $a = 3$ and $y = 6$, then $6 = 3\sqrt{x} \rightarrow 2 = \sqrt{x} \rightarrow 2^2 = x \rightarrow x = 4$.

EXERCISES: ACTIVITY 4.11

1. You are probably aware that the stopping distance for a car, once the brakes are firmly applied, depends on its speed and on the tire and road conditions. The following formula expresses this relationship:

$$S = \sqrt{30fd},$$

where S is the speed of the car (in miles per hour); f is the drag factor, which takes into account the condition of tires, pavement, and so on; and d is the stopping distance (in feet), which can be determined by measuring the skid marks.

Assume that the drag factor, f, has the value 0.83.

a. At what speed is your car traveling if it leaves skid marks of 100 feet?

$S = \sqrt{30(0.83)(100)} = \sqrt{2490} \approx 49.9$ mph

My car will leave skid marks 100 feet long if it is traveling almost 50 miles per hour.

b. Your friend stops short at an intersection when she notices a police car nearby. She is pulled over and claims that she was doing 25 miles per hour in a 30-miles-per-hour zone. The skid marks measure 60 feet. Should she be issued a speeding ticket?

$S = \sqrt{30(0.83)(60)} = \sqrt{1494} \approx 38.65$ mph

She was exceeding the posted speed limit and should be issued a speeding ticket.

c. Solve the formula $S = \sqrt{30fd}$ for d.

$d = \dfrac{S^2}{30f}$

d. Determine the stopping distances (length of the skids) for the speeds given in the following table. Round your answers to the nearest whole foot.

STOPPING DISTANCE, (ft.)	SPEED, S (mph)
100	50
121	55
145	60
170	65
197	70
226	75

e. Use the data in the preceding table to graph speed as a function of stopping distance.

2. Solve each of the following equations (if possible) using an algebraic approach. Check your answers.

Note to instructor: Students should check these solutions in the original equation.

a. $3x^2 = 27$
$x^2 = 9$
$x = \pm 3$

b. $5x^2 + 1 = 16$
$5x^2 = 15$
$x^2 = 3$
$x = \pm\sqrt{3}$

c. $t^2 + 9 = 0$
$t^2 = -9$
There is no real number solution.

There is no real number whose square is a negative number.

d. $3r^2 + 10 = 2r^2 + 14$
$r^2 = 4$
$r = \pm 2$

e. $\sqrt{x} - 4 = 3$
$\sqrt{x} = 7$
$x = 49$

f. $5\sqrt{r} = 10$
$\sqrt{r} = 2$
$r = 4$

g. $\sqrt{x + 4} - 3 = 0$
$\sqrt{x + 4} = 3$
$x + 4 = 9$
$x = 5$

h. $4 = \sqrt{2w}$
$16 = 2w$
$w = 8$

3. a. Complete the following table. Round answers to the nearest tenth.

x	0	1	2	3	4	5
$f(x) = 3\sqrt{x}$	0	3	4.2	5.2	6	6.7
$g(x) = \sqrt{3x}$	0	1.7	2.4	3	3.5	3.9

b. Sketch a graph of each function on the same coordinate axes.

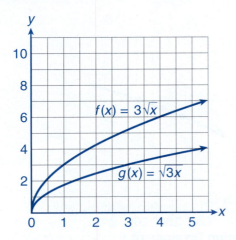

c. Both of the functions are increasing. Which function is increasing more rapidly?

Function *f* increases more rapidly.

4. In your science class, you are learning about pendulums. The period of a pendulum is the time it takes the pendulum to make one complete swing back and forth. The relationship between the period and the pendulum length is described by the formula $T = \dfrac{\pi}{4}\sqrt{\dfrac{l}{6}}$, where l is the length of the pendulum in inches and T is the period in seconds.

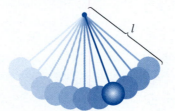

a. The length of the pendulum inside your grandfather clock is 24 inches. What is its period?

$$T = \frac{\pi}{4}\sqrt{\frac{24}{6}} = \frac{\pi}{2} \approx 1.57 \text{ sec.}$$

The period is approximately 1.57 seconds.

b. A metal sculpture hanging from the ceiling of an art museum contains a large pendulum. In science class, you are asked to determine the length of the pendulum without measuring it. You use a stopwatch and determine that its period is approximately 3 seconds. What is the length of the pendulum in feet?

$$3 = \frac{\pi}{4}\sqrt{\frac{l}{6}}$$

$$3 \cdot \frac{4}{\pi} = \frac{4}{\pi} \cdot \frac{\pi}{4}\sqrt{\frac{l}{6}}$$

$$\frac{12}{\pi} = \sqrt{\frac{l}{6}}$$

$$\left(\frac{12}{\pi}\right)^2 = \frac{l}{6}$$

$$l = \left(\frac{12}{\pi}\right)^2 (6) \approx 87.54$$

The length of the pendulum is approximately 87.54 inches, or 7.3 feet.

Cluster 3 What Have I Learned?

1. The diameter, d, of a sphere having volume V is given by the cube root formula,

$$d = \sqrt[3]{\frac{6V}{\pi}}$$

a. Use what you have learned in this cluster to explain the steps you would use to calculate a value for d, given a value for V.

The order of operations to evaluate d given a value of V are:

Start with $V \rightarrow$ multiply by 6 $\rightarrow$ divide by $\pi \rightarrow$ take the cube root to obtain the value for d.

b. Use what you learned to explain the steps you would take to solve the formula for V in terms of d.

To solve for V in terms of d, you reverse the process performed in

part a: Start with $d \rightarrow$ cube $\rightarrow$ multiply by $\pi \rightarrow$ divide by 6 to obtain the formula for V in terms of d.

Symbolically, the steps are

$$d = \sqrt[3]{\frac{6V}{\pi}} \rightarrow d^3 = \frac{6V}{\pi} \rightarrow \pi d^3 = 6V \rightarrow V = \frac{\pi d^3}{6}$$

2. The number of cellular phone subscribers in the U.S. has been growing dramatically since 1988. The following table lists some data about the number of subscribers in terms of years since 1988.

YEAR	1988	1990	1992	1994	1996	1998	2000
Number of Years, t, Since 1988	0	2	4	6	8	10	12
Number of Cell Phone Subscribers, S (in millions)	2	5	11	24	44	69	119

a. Graph the data in the table. Is the graph linear or nonlinear?   The graph is nonlinear.

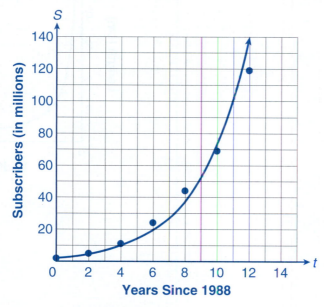

Years Since 1988

Exercise numbers appearing in color are answered in the Selected Answers appendix.

b. The function $S(t) = 2.5(1.4)^t$ closely models the data between 1988 and 2000. What type of function is it?

The formula represents an exponential function because the input variable is an exponent in the expression defining the function.

c. Graph the function in part b on the grid in part a. Describe what you observe, and interpret what it means regarding the cell phone subscriber data.

I observe that the graph of the function is very close to the data points, except that the graph is too high by about 20 million for the last data point (for 2000). It suggests that the growth of U.S. cell phone subscribers has been exponential during this period.

d. What does this function predict for the number of cell phone subscribers in 2008? Is this prediction reasonable?

In 2008, $t = 2008 - 1988 = 20$.

$S(t) = 2.5(1.4)^{20} \approx 2092$

The function predicts 2092 million (2.092 billion) U.S. cell phone subscribers in 2008. This is way more than the entire U.S. population of around 300 plus million, so it is totally unreasonable.

e. The actual number of U.S. cell phone subscribers in 2008 was estimated at 263 million. What do you conclude about the exponential function used in this exercise?

While the exponential growth seemed reasonable over the years 1988 to 2000, when the growth in usage was rapid, the population of the U.S. provides a limit as to how many subscribers there can be. So the exponential function used here is not appropriate outside the time interval of the original data.

3. The equation $xy = 4$ describes the relationship between the width x and length y of a rectangle with a fixed area of 4 square centimeters. Answer the following questions based on what you learned in this cluster.

a. Write the length as a function of the width for this rectangle, and graph the function from $x = 0$ to $x = 10$ and $y = 0$ to $y = 10$.

$$y = \frac{4}{x}$$

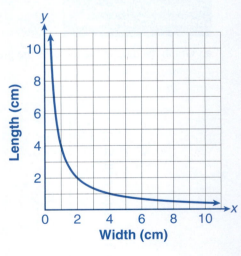

b. Describe the graph in a few sentences, noting its domain and range and whether it is an increasing or decreasing function.

The domain is the set of all positive real numbers, $x > 0$. Note that $x = 0$ is not in the domain. Similarly, the range is the set of all positive real numbers, excluding 0. The curve is a decreasing function. As the input values of x approach 0, the output values y continue increasing, and the graph approaches the y-axis but never crosses it. Similarly, as the input values increase indefinitely, the output values decrease towards 0, and the graph approaches the x-axis but never crosses it.

c. Answer the following questions regarding the relationship between the length and width of the rectangle.

i. For what value will the width and length be the same? What shape will the rectangle take for this value?

Length = width implies that $y = x$.

$$x = \frac{4}{x}$$

$$x^2 = 4$$

$x = 2$ When the length and width are 2 centimeters, the rectangle is a square.

ii. When will the rectangle be twice as long as its width? When will it be half as long?

If the length is twice the width, then $y = 2x$.

$$2x = \frac{4}{x}$$

$$2x^2 = 4$$

$$x^2 = 2$$

$$x = \sqrt{2} \approx 1.4$$

The length is twice the width when the width is $\sqrt{2}$, or approximately 1.4 centimeters.

If the length is half the width, then $y = \frac{1}{2}x$.

$$\frac{1}{2}x = \frac{4}{x}$$

$$\frac{1}{2}x^2 = 4$$

$$x^2 = 8$$

$$x = \sqrt{8} \approx 2.8$$

The length is half the width when the width is $\sqrt{8}$, or approximately 2.8 centimeters.

d. Determine several other possible dimensions for the rectangle. Use a ruler to draw those rectangles and the rectangles from parts i and ii above. Is there any one shape whose proportions are most pleasing to your eye? Explain your answer.

(Answers will vary.)

Note to instructor: Have students compare their answers and perhaps vote for the most pleasing dimensions.

Cluster 3   How Can I Practice?

1. Solve each of the following formulas for the specified variable.

   a. $V = \pi r^2 h$, solve for r

   $r = \sqrt{\dfrac{V}{\pi h}}$

   b. $V = \frac{4}{3}\pi r^3$, solve for r

   $r = \sqrt[3]{\dfrac{3V}{4\pi}}$

   c. $a^2 + b^2 = c^2$, solve for b

   $b = \sqrt{c^2 - a^2}$

2. The formula $V = \sqrt{\dfrac{1000p}{3}}$ shows the relationship between the velocity of the wind, V (in miles per hour), and wind pressure, p (in pounds per square foot).

   a. The pressure gauge on a bridge indicates a wind pressure, p, of 10 pounds per square foot. What is the velocity, V, of the wind?

   $V = \sqrt{\dfrac{1000(10)}{3}} \approx 57.7$

   The velocity of the wind is approximately 58 miles per hour.

   b. If the velocity of the wind is 50 miles per hour, what is the wind pressure?

   $50 = \sqrt{\dfrac{1000p}{3}}$

   $p = 7.5$

   The wind pressure is 7.5 pounds per square foot.

3. Your union contract provides that your salary will increase 3% in each of the next 5 years. Your current salary is $32,000.

   a. Represent your salary symbolically as a function of years, and state the practical domain.

   $s(y) = 32{,}000(1.03)^y$, where s represents salary and y represents year.

   The practical domain covers the length of the contract, 0 to 5 years.

   b. Use the formula from part a to determine your salary in each of the next 5 years.

YEARS, y	SALARY, $s(y)$
0	$32,000
1	$32,960
2	$33,949
3	$34,967
4	$36,016
5	$37,097

   c. What is the percent increase in your salary over the first 2 years of the contract?

   My salary will increase $33,949 − $32,000 = $1949 in 2 years. Relative to my $32,000 salary at the start of the contract, my salary will increase $\dfrac{1949}{32{,}000} = 0.0609$, or 6.09%.

d. What is the total percent increase in your salary over the life of the contract?

$$\frac{37{,}097 - 32{,}000}{32{,}000} = \frac{5097}{32{,}000} = 0.15928, \text{ or } 15.928\%, \text{ is the total percent increase}$$

over the life of the contract.

e. If the next 5-year contract also includes a 3% raise each year, how much will you be earning in 10 years?

$$s(10) = 32{,}000(1.03)^{10}$$
$$= \$43{,}005$$

In 10 years my yearly salary will be $43,005.

4. Your candy company has found that the price-demand function, $N(p) = \dfrac{5700}{p}$, models quite well the relationship between the price, p (in dollars), of a candy bar that you make and the number, N, of bars that you will sell.

a. How many bars does your company expect to sell if the price is $1.50?

$$N(s) = \frac{5700}{1.50} = 3800$$

My company expects to sell 3800 candy bars if the price is $1.50.

b. What price should you charge if you wish to sell 6000 bars?

$$6000 = \frac{5700}{p}$$

$$6000p = 5700$$

$$p = \frac{5700}{6000} = 0.95 = 95$$

My company should charge $0.95 each if we wish to sell 6000 candy bars.

c. Use your graphing calculator to verify your answers to parts a and b graphically. Use the window: Xmin = 0, Xmax = 4, Xscl = 0.25, Xmin = 0, Xmax = 7000, and Xscl = 200.

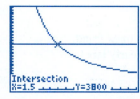 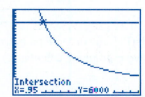 

d. What is the domain of this price-demand function, based on its algebraic definition?

The domain is all real numbers except 0.

e. Determine a reasonable practical domain for the price-demand function?

(Answers will vary, and students can compare their answers.)

The practical domain is all real numbers greater than 0 and less than $2 (or so).

5. **a.** Solve $xy = 25$ for y. Is y a function of x? What is its domain?

$$y = \frac{25}{x}$$

Yes, y is a function of x. The domain is all real numbers except 0.

b. Solve $xy = 25$ for x. Is x a function of y? What is its domain?

$$x = \frac{25}{y}$$

Yes, x is a function of y. The domain is all real numbers except 0.

Solve the following equations.

6. $24 = \dfrac{36}{x}$

$24x = 36$

$x = 1.5$

7. $2500 = \dfrac{50}{x}$

$2500x = 50$

$x = 0.02$

8. $18 = \dfrac{1530}{x + 2}$

$18(x + 2) = 1530$

$18x + 36 = 1530$

$18x = 1494$

$x = \dfrac{1494}{18} = 83$

9. $-5 = \dfrac{120}{x - 5}$

$-5(x - 5) = 120$

$-5x + 25 = 120$

$-5x = 95$

$x = -19$

Note to instructor: Also have students show a numerical check.

10. $3\sqrt{x} = 3$

$\sqrt{x} = 1$

$x = 1$

11. $\sqrt{x} - 9 = 0$

$\sqrt{x} = 9$

$x = 81$

12. $\sqrt{x + 1} = 7$

$x + 1 = 49$

$x = 48$

13. The ordered pairs $(x, f(x))$ in the following table represent a function.

x	5	8	12	15	20	30
$f(x)$	288	180	120	96	72	48
$x \cdot f(x)$	1440	1440	1440	1440	1440	1440

a. Calculate the product $x \cdot f(x)$ for each ordered pair $(x, f(x))$ in the table.

b. Use the information you obtained in part a to write a formula for $f(x)$ in terms of x.

$f(x) = \dfrac{1440}{x}$

c. Graph the function from part b on the following grid. (Plot the ordered pairs from lines 1 and 2 of the table, and draw a curve through them or use your graphing calculator.) Clearly label your axes with an appropriate scale.

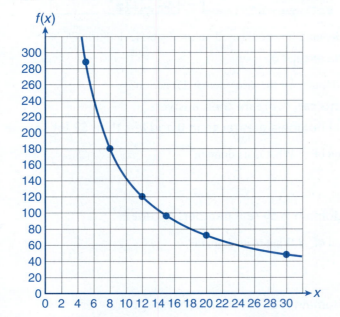

d. Use your graph to estimate the value of x for which $f(x) = 200$.

From the graph, $f(x) = 200$ when x is approximately 7.

e. Use the formula from part b to calculate the value of x for which $f(x) = 200$. How does your result compare to your result in part d?

$$200 = \frac{1440}{x}$$

$$200x = 1440$$

$$x = 7.2$$

The estimate in part d is close to the value calculated from the formula.

14. When you take medicine, your body metabolizes and eliminates the medication until there is none left in your body. The half-life of a medication is the time is takes for your body to eliminate one-half of the amount present. For many people the half-life of the medicine Prozac is 1 day.

a. What fraction of a dose is left in your body after 1 day?

$\frac{1}{2}$ of the dose will remain after 1 day.

b. What fraction of the dose is left in your body after 2 days? (This is one-half of the result from part a.)

$\frac{1}{4}$ of the dose will remain after 2 days.

c. Complete the following table. Let t represent the number of days after a dose of Prozac is taken and let Q represent the fraction of the dose of Prozac remaining in your body.

t (days)	Q (mg)
0	1
1	$\frac{1}{2}$
2	$\frac{1}{4}$
3	$\frac{1}{8}$
4	$\frac{1}{16}$

d. The values of the fraction of the dosage, Q, can be written as powers of $\frac{1}{2}$. For example, $1 = \left(\frac{1}{2}\right)^0, \frac{1}{2} = \left(\frac{1}{2}\right)^1$, and so on. Complete the following table by writing each value of Q in the preceding table of as a power of $\frac{1}{2}$. The values for 1 and $\frac{1}{2}$ have already been entered.

t (days)	Q (mg)
0	$\left(\frac{1}{2}\right)^0$
1	$\left(\frac{1}{2}\right)^1$
2	$\left(\frac{1}{2}\right)^2$
3	$\left(\frac{1}{2}\right)^3$
4	$\left(\frac{1}{2}\right)^4$

e. Use the result of part d to write an equation for Q in terms of t. Is this function linear? Explain.

$Q = \left(\frac{1}{2}\right)^t$ or $Q = 0.5^t$; no, this function is exponential.

f. Sketch a graph of the data in part a and the function in part e on an appropriately scaled and labeled coordinate axis.

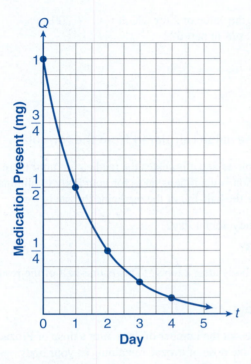

15. The length, L (in feet), of skid distance left by a car varies directly as the square of the initial velocity, v (in miles per hour), of the car.

a. Write a general equation for L as a function of v. Let k represent the constant of variation.

$L = kv^2$

b. Suppose a car traveling at 40 miles per hour leaves a skid distance of 60 feet. Use this information to determine the value of k.

$60 = k(40)^2, k = 0.0375$

c. Use the function to determine the length of the skid distance left by the car traveling at 60 miles per hour.

$L = 0.0375(60)^2 = 135$ ft

The bracketed numbers following each concept indicate the activity in which the concept is discussed.

CONCEPT/SKILL	DESCRIPTION	EXAMPLE
Polynomial [4.1]	Polynomials are formed by adding or subtracting terms containing positive integer powers of a variable, such as x, and possibly a constant term.	$25x^2 - 10x + 1$
Degree of a monomial [4.1]	Degree of a monomial is defined as the exponent on its variable.	Degree of $11x^5$ is 5.
Degree of a polynomial [4.1]	Degree of a polynomial is the highest degree among all its terms.	Degree of $6x^3 - 7x^2 - 5x + 4$ is 3.
Classifying polynomials by the number of terms they contain [4.1]	A monomial has one term, a binomial has two terms, and a trinomial has three terms.	monomial: $6x$ binomial: $2x^3 - 54$ trinomial: $25x^2 - 10x + 1$
Simplifying expressions containing polynomials [4.1]	1. If necessary, apply the distributive property to remove parentheses. 2. Combine like terms.	$5x^2 - 3x(x + 2) + 2x$ $= 5x^2 - 3x^2 - 6x + 2x$ $= 2x^2 - 4x$
Adding or subtracting polynomials [4.1]	Apply the same process used to simplify polynomials.	
Properties of Exponents [4.2]	Property 1: $b^m \cdot b^n = b^{m+n}$ Property 2: $\dfrac{b^m}{b^n} = b^{m-n}, b \neq 0$ Property 3: $(a \cdot b)^n = a^n b^n$ and $\left(\dfrac{a}{b}\right)^n = \dfrac{a^n}{b^n}$ Property 4: $(b^m)^n = b^{m \cdot n}$ Definition: $b^0 = 1, b \neq 0$ Definition: $b^{-n} = \dfrac{1}{b^n}, b \neq 0$	1. $5^7 \cdot 5^6 = 5^{13}$ 2. $\dfrac{7^8}{7^5} = 7^3$ 3. $(2 \cdot 3)^4 = 2^4 \cdot 3^4$ and $\left(\dfrac{3}{4}\right)^2 = \dfrac{3^2}{4^2}$ 4. $(3^2)^4 = 3^8$ 5. $10^0 = 1$ 6. $2^{-3} = \dfrac{1}{2^3}$
Multiplying a series of monomial factors [4.2]	1. If present, remove parentheses using Property of Exponents. 2. Multiply the numerical coefficients. 3. Apply Property 1 of Exponents to variable factors that have the same base.	$3x^4(x^3)^2 \cdot 5x^3$ $= 3x^4 x^6 \cdot 5x^3$ $= 15x^{13}$
Writing an algebraic expression in expanded form [4.2]	Use the distributive property to expand, applying the Properties of Exponents.	$-2x^2(x^3 - 5x + 3)$ $= -2x^5 + 10x^3 - 6x^2$
Factoring completely [4.2]	Determine the GCF, and then use the properties of exponents to write the polynomial in factored form.	$24x^5 - 16x^3 + 32x^2$ $= 8x^2(3x^3 - 2x + 4)$

CONCEPT/SKILL	DESCRIPTION	EXAMPLE
Multiplying a monomial times a polynomial [4.2]	Multiply each term of the polynomial by the monomial.	$2x(x^2 - 5x + 3)$ $= 2x \cdot x^2 - 2x \cdot 5x + 2x \cdot 3$ $= 2x^3 - 10x^2 + 6x$
Multiplying two binomials (also known as the FOIL process) [4.3]	Multiply every term in the second binomial by each term in the first. The FOIL method (**F**irst + **O**uter + **I**nner + **L**ast) is a multiplication process in which each of the letters F, O, I, L refers to the position of the factors in the two binomials.	$(12 + x) \cdot (9 + x)$ $(12 + x)(9 + x)$ $= 108 + 12x + 9x + x^2$ $= 108 + 21x + x^2$
Multiplying two polynomials [4.3]	Multiply every term in the second polynomial by each term in the first.	$(x - 3)(x^2 - 2x + 1)$ $= x(x^2 - 2x + 1) - 3(x^2 - 2x + 1)$ $= x^3 - 2x^2 + x - 3x^2 + 6x - 3$ $= x^3 - 5x^2 + 7x - 3$
Parabola [4.4]	A parabola is the U-shaped graph described by a function of the form $y = ax^2 + bx + c, a \neq 0$.	
Solving an equation of the form $ax^2 = c, a \neq 0$ [4.4]	Solve an equation of this form algebraically by dividing both sides of the equation by a and then taking the square root of both sides.	$2x^2 = 8$ $\dfrac{2x^2}{2} = \dfrac{8}{2}$ $x^2 = 4, x = \pm 2$
Solving an equation of the form $(x \pm a)^2 = c$ [4.4]	Solve an equation of this form algebraically by taking the square root of both sides of the equation and then solving each of the resulting linear equations.	$(x + 3)^2 = 16$ $x + 3 = \pm 4$ $x = -3 + 4 = 1$ $x = -3 - 4 = -7$
Zero-product property [4.5]	If the product of two factors is zero, then at least one of the factors must also be zero.	$(x + 6)(x - 1) = 0$ $x + 6 = 0, \quad x - 1 = 0$ $x = -6 \quad \text{or} \quad x = 1$

CONCEPT/SKILL	DESCRIPTION	EXAMPLE
Solving an equation of the form $ax^2 + bx = 0, a \neq 0$ [4.5]	Solve an equation of this form algebraically by factoring the left side and then using the zero-product property.	$x^2 - 3x = 0$ $x(x - 3) = 0$ $x = 0, \quad x - 3 = 0$ $x = 0 \quad \text{or} \quad x = 3$
x-intercepts of the graph of $y = ax^2 + bx, a \neq 0$ [4.5]	The x-intercepts of the graph of $y = ax^2 + bx$ are precisely the solutions of the equation $ax^2 + bx = 0$.	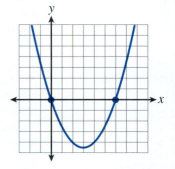
Quadratic function [4.6]	A quadratic function is a polynomial of degree 2 that can be written in the general form $y = ax^2 + bx + c$, where a, b, and c are arbitrary constants and $a \neq 0$.	$y = 5x^2 - 6x + 1$
Quadratic equation [4.6]	A quadratic equation is any equation that can be written in the general form $ax^2 + bx + c = 0, a \neq 0$.	$5x^2 - 6x + 1 = 0$
Quadratic equation in standard form [4.6]	$ax^2 + bx + c = 0, a \neq 0$	The equation $5x^2 - 6x + 1 = 0$ is written in standard form; $5x^2 - 6x = -1$ is not.
Solving quadratic equations by factoring [4.6]	1. Write the equation in standard form (i.e., with one side equal to zero). 2. Divide each term in the equation by the greatest common numerical factor. 3. Factor the nonzero side. 4. Apply the zero-product property to obtain two linear equations. 5. Solve each linear equation. 6. Check your solutions in the original equation.	$5x^2 - 10x = 15$ $5x^2 - 10x - 15 = 0$ $x^2 - 2x - 3 = 0$ $(x - 3)(x + 1) = 0$ $x - 3 = 0 \quad \text{or} \quad x + 1 = 0$ $x = 3 \quad \text{or} \quad x = -1$
Quadratic formula [4.7]	The solutions of the quadratic equation $ax^2 + bx + c = 0$ are $$x = \frac{-b \pm \sqrt{b^2 - 4ac}}{2a}, a \neq 0.$$	$5x^2 - 6x + 1 = 0$ $a = 5, b = -6, c = 1$ $$x = \frac{-b \pm \sqrt{b^2 - 4ac}}{2a}$$ $$x = \frac{-(-6) \pm \sqrt{(-6)^2 - 4 \cdot 5 \cdot 1}}{2 \cdot 5}$$ $$= \frac{6 \pm \sqrt{16}}{10}$$ $$x = \frac{1}{5} \quad \text{or} \quad x = 1$$

CONCEPT/SKILL	DESCRIPTION	EXAMPLE
Exponential function [4.8]	An **exponential function** is a function in which the input variable appears as the exponent of the growth (or decay) factor, as in $$y = ab^t,$$ where a is the starting output value (at $t = 0$), b is the fixed growth or decay factor, and t represents the time that has elapsed.	$y = 16(2.1)^x$
Direct variation [4.9]	Two variables are said to **vary directly** if whenever one variable increases, the other variable increases by the same multiplicative factor. The ratio of the two variables is always constant.	$y = 12x$ When $x = 3$, $y = 12(3) = 36$. If x doubles to 6, y also doubles, becoming 72.
Inverse variation function [4.10]	An inverse variation function is a function defined by the equation $xy = k$ (or $y = \dfrac{k}{x}$). When $k > 0$ and the input x increases by a constant multiplicative factor p, the output y decreases by the multiplicative factor $\dfrac{1}{p}$. The input value $x = 0$ is not in the domain of this function.	$y = \dfrac{360}{x}$ When $x = 4$, $y = \dfrac{360}{4} = 90$. If x doubles to 8, y is halved and becomes 45.
Square root function [4.11]	A square root function is a function of the form $y = a\sqrt{x}$, where $a \neq 0$. The domain of the function is $x \geq 0$.	$t = \frac{1}{2}\sqrt{s}$
Graph of a square root function [4.11]	The graph of $y = a\sqrt{x}$, where $a > 0$, increases and is not linear.	 $y = a\sqrt{x}, a > 0$

In Exercises 1–11, perform the indicated operations and simplify as much as possible.

1. $-3(2x - 5) + 7(3x - 8)$

$= -6x + 15 + 21x - 56$

$= 15x - 41$

2. $-4(5x^2 - 7x + 8) + 2(3x^2 - 4x - 2)$

$= -20x^2 + 28x - 32 + 6x^2 - 8x - 4$

$= -14x^2 + 20x - 36$

3. Add $5x^3 - 12x^2 - 7x + 4$ and $-3x^3 + 3x^2 + 6x - 9$.

$\quad 5x^3 - 12x^2 - 7x + 4 + (-3x^3 + 3x^2 + 6x - 9)$

$= 5x^3 - 12x^2 - 7x + 4 - 3x^3 + 3x^2 + 6x - 9$

$= 2x^3 - 9x^2 - x - 5$

4. Subtract $3x^3 + 5x^2 - 12x - 17$ from $2x^3 - 8x^2 + 4x - 19$.

$\quad 2x^3 - 8x^2 + 4x - 19 - (3x^3 + 5x^2 - 12x - 17)$

$= 2x^3 - 8x^2 + 4x - 19 - 3x^3 - 5x^2 + 12x + 17$

$= -x^3 - 13x^2 + 16x - 2$

5. $x^4(3x^3 + 2x^2 - 5x - 20)$

$= 3x^7 + 2x^6 - 5x^5 - 20x^4$

6. $-2x(3x^3 - 4x^2 + 8x - 5) + 3x(5x^3 - 8x^2 - 7x + 2)$

$= -6x^4 + 8x^3 - 16x^2 + 10x + 15x^4 - 24x^3 - 21x^2 + 6x$

$= 9x^4 - 16x^3 - 37x^2 + 16x$

7. $(2x^3)^4$

$= 2^4(x^3)^4$

$= 16x^{12}$

8. $(5xy^3)^3$

$= 5^3 x^3 (y^3)^3$

$= 125x^3 y^9$

9. $(-8a^2b^3)(-2a^3b^2)$

$= 16a^5 b^5$

10. $\dfrac{y^{11}}{y^3}$

$= y^{11-3} = y^8$

11. $\dfrac{45t^8}{9t^5}$

$= 5t^3$

In Exercises 12 and 13, factor the given expression completely.

12. $27x^5 - 9x^3 - 12x^2$

$= 3x^2(9x^3 - 3x - 4)$

13. $8x^2 + 4x$

$= 4x(2x + 1)$

14. A car travels at the average speed of $5a - 13$ miles per hour for 2 hours. Write a polynomial expression in expanded form for the distance traveled.

$(5a - 13) \cdot 2 = 10a - 26$

15. Write each of the following products in expanded form.

  a. $(x - 4)(x - 6)$

$\quad = x^2 - 10x + 24$

  b. $(x + 3)(x - 8)$

$\quad = x^2 - 5x - 24$

Answers to all Gateway exercises are included in the Selected Answers appendix.

c. $(x + 3)(x - 3)$
$= x^2 - 9$

d. $(x - 6)^2$
$= (x - 6)(x - 6)$
$= x^2 - 12x + 36$

e. $(2x - 3)(x - 5)$
$= 2x^2 - 10x - 3x + 15$
$= 2x^2 - 13x + 15$

f. $(3x + 4)(2x - 7)$
$= 6x^2 - 21x + 8x - 28$
$= 6x^2 - 13x - 28$

g. $(2x + 5)(2x - 5)$
$= 4x^2 - 25$

h. $(x + 4)(x + 4)$
$= x^2 + 8x + 16$

i. $(x + 3)(2x^2 + x - 3)$
$= 2x^3 + x^2 - 3x + 6x^2 + 3x - 9$
$= 2x^3 + 7x^2 - 9$

j. $(x + 5)(3x^2 - 2x - 4)$
$= 3x^3 - 2x^2 - 4x + 15x^2 - 10x - 20$
$= 3x^3 + 13x^2 - 14x - 20$

16. Your circular patio has radius 11 feet. To enlarge the patio you increase the radius by y feet. Write a formula for the area of the new patio in terms of y, in both factored and expanded form.

$A = \pi(11 + y)^2$ (factored form)
$A = \pi(y^2 + 22y + 121)$
$A = \pi y^2 + 22\pi y + 121\pi$ (expanded form)

17. You have a square garden. Let x represent one side of your garden. If you increase one side 7 feet and decrease the other side 2 feet, write a formula for the area of the new garden in terms of x, in both factored and expanded form.

$A = (x + 7)(x - 2)$ (factored form)
$A = x^2 + 5x - 14$ (expanded form)

18. Solve each of the following equations.

a. $75x^2 = 300$
$x^2 = 4$
$x = \pm 2$

b. $9x^2 = 81$
$x^2 = 9$
$x = \pm 3$

c. $4x^2 = -16$
$x^2 = -4$
no solution

d. $(x + 2)^2 = 64$
$x + 2 = \pm 8$
$x + 2 = 8$ or $x + 2 = -8$
$x = 6$ or $x = -10$

e. $(x - 3)^2 = 25$
$x - 3 = \pm 5$
$x - 3 = 5$ or $x - 3 = -5$
$x = 8$ or $x = -2$

f. $(x + 8)^2 = 49$
$x + 8 = \pm 7$
$x + 8 = 7$ or $x + 8 = -7$
$x = -1$ or $x = -15$

19. Solve each of the following equations.

a. $x(x + 8) = 0$
$x = 0$ or $x = -8$

b. $4x^2 - 12x = 0$
$4x(x - 3) = 0$
$x = 0$ or $x = 3$

c. $5x^2 - 25x = 0$
$5x(x - 5) = 0$
$x = 0$ or $x = 5$

d. $3x(x - 5) = 0$
$x = 0$ or $x = 5$

e. $(4x - 1)(x - 6) = 0$

$4x - 1 = 0 \qquad x - 6 = 0$

$\quad 4x = 1 \qquad\qquad x = 6$

$\qquad x = \frac{1}{4}$

$x = \frac{1}{4} \quad$ or $\quad x = 6$

20. You are sitting on the front lawn and toss a ball straight up into the air. The height, s (in feet), of the ball above the ground t seconds later is described by the function $s = -16t^2 + 48t$.

a. Determine the height of the ball after 0.5 second.

$s = -16(0.5)^2 + 48(0.5)$

$s = 20$

The ball is 20 feet high after half a second.

b. After how many seconds does the ball reach a height of 32 feet?

Replace s by 32 and solve the quadratic equation $-16t^2 + 48t = 32$.

$\qquad -16t^2 + 48t - 32 = 0$

$\qquad\qquad t^2 - 3t + 2 = 0$

$\qquad\qquad (t - 2)(t - 1) = 0$

$t - 2 = 0 \quad$ or $\quad t - 1 = 0$

$\qquad t = 2 \quad$ or $\qquad t = 1$

The ball reaches a height of 32 feet after 1 second and then again on its way down after 2 seconds.

c. When does the ball reach the ground?

Solve the quadratic equation:

$\qquad -16t^2 + 48t = 0$

$\qquad\qquad t^2 - 3t = 0$

$\qquad\qquad t(t - 3) = 0$

$t = 0 \quad$ or $\quad t - 3 = 0$

$t = 0 \quad$ or $\qquad t = 3$

The ball reaches the ground after 3 seconds.

21. Solve the following equations by factoring and the zero-product property.

a. $x^2 + 4x - 12 = 0$

$(x + 6)(x - 2) = 0$

$x + 6 = 0 \quad$ or $\quad x - 2 = 0$

$\qquad x = -6 \quad$ or $\qquad x = 2$

$\qquad x = -6 \quad$ or $\qquad x = 2$

b. $x^2 = 5x + 14$

$x^2 - 5x - 14 = 0$

$(x - 7)(x + 2) = 0$

$x - 7 = 0 \quad$ or $\quad x + 2 = 0$

$\qquad x = 7 \quad$ or $\qquad x = -2$

c. $x^2 - 5x = 24$

$x^2 - 5x - 24 = 0$

$(x - 8)(x + 3) = 0$

$x - 8 = 0 \quad$ or $\quad x + 3 = 0$

$\qquad x = 8 \quad$ or $\qquad x = -3$

d. $x^2 - 10x + 21 = 0$

$(x - 3)(x - 7) = 0$

$x - 3 = 0 \quad$ or $\quad x - 7 = 0$

$\qquad x = 3 \quad$ or $\qquad x = 7$

e. $x^2 + 2x - 24 = 0$

$(x - 4)(x + 6) = 0$

$x - 4 = 0$ or $x + 6 = 0$

$x = 4$ or $x = -6$

f. $y^2 + 3y = 40$

$y^2 + 3y - 40 = 0$

$(y + 8)(y - 5) = 0$

$y + 8 = 0$ or $y - 5 = 0$

$y = -8$ or $y = 5$

22. Solve the following equations using the quadratic formula.

a. $3x^2 - 2x - 6 = 0$

$a = 3, b = -2, c = -6$

$$x = \frac{-(-2) \pm \sqrt{(-2)^2 - 4(3)(-6)}}{2(3)}$$

$$= \frac{2 \pm \sqrt{4 - (-72)}}{6} = \frac{2 \pm \sqrt{76}}{6}$$

$$x = \frac{2 + \sqrt{76}}{6} \approx 1.786$$

$$x = \frac{2 - \sqrt{76}}{6} \approx -1.120$$

$x \approx 1.786$ or $x \approx -1.120$

b. $5x^2 + 7x = 12$

$5x^2 + 7x - 12 = 0$

$a = 5, b = 7, c = -12$

$$x = \frac{-(7) \pm \sqrt{(7)^2 - 4(5)(-12)}}{2(5)}$$

$$= \frac{-7 \pm \sqrt{49 - (-240)}}{10}$$

$$= \frac{-7 \pm \sqrt{289}}{10}$$

$$= \frac{-7 \pm 17}{10}$$

$$x = \frac{-7 + 17}{10} = \frac{10}{10} = 1$$

$$x = \frac{-7 - 17}{10} = \frac{-24}{10} = -2.4$$

$x = 1$ or $x = -2.4$

c. $-2x^2 + 4x = -9$

$-2x^2 + 4x + 9 = 0$

$a = -2, b = 4, c = 9$

$$x = \frac{-(4) \pm \sqrt{(4)^2 - 4(-2)(9)}}{2(-2)}$$

$$= \frac{-4 \pm \sqrt{16 - (-72)}}{-4}$$

$$= \frac{-4 \pm \sqrt{88}}{-4}$$

$$x = \frac{-4 + \sqrt{88}}{-4} \approx -1.345$$

$$x = \frac{-4 - \sqrt{88}}{-4} \approx 3.345$$

$x \approx -1.345$ or $x \approx 3.345$

d. $2x^2 - 9x = -12$

$2x^2 - 9x + 12 = 0$

$a = 2, b = -9, c = 12$

$$x = \frac{-(-9) \pm \sqrt{(-9)^2 - 4(2)(12)}}{2(2)}$$

$$= \frac{9 \pm \sqrt{81 - 96}}{4}$$

$$= \frac{9 \pm \sqrt{-15}}{4}$$

There is no real solution because $\sqrt{-15}$ is not a real number.

23. Solve the following equations.

a. $\dfrac{90}{x} = 15$

$\dfrac{90}{x} \cdot x = 15x$

$90 = 15x$

$6 = x$

b. $\dfrac{56}{x} = -8$

$\dfrac{56}{x} \cdot x = -8 \cdot x$

$56 = -8x$

$-7 = x$

c. $3\sqrt{x} = 21$

$\sqrt{x} = 7$

$x = 49$

d. $\sqrt{x} - 25 = 0$

$\sqrt{x} = 25$

$x = 625$

e. $\sqrt{x + 3} = 6$

$x + 3 = 36$

$x = 33$

24. Due to inflation your college's tuition is predicted to increase by 6% each year.

a. If tuition is $300 per credit now, determine how much it will be in 5 years and in 10 years.

The growth factor is 100% + 6% = 106%, or 1.06. Let T represent the tuition.

$T = 300(1.06)^t$

$T = 300(1.06)^5 = 401.47$

In 5 years the tuition will be $401.47 per credit.

$T = 300(1.06)^t$

$T = 300(1.06)^{10} \approx 537.25$

In 10 years a credit will cost $537.25.

b. Calculate the average rate of change in tuition over the next 5 years.

$\dfrac{401.47 - 300}{5} = 20.294$

Tuition increases approximately $20.29 per credit per year over the next 5 years.

c. Calculate the average rate of change in tuition over the next 10 years.

$\dfrac{537.25 - 300}{10} = 23.725$

Tuition increases approximately $23.73 per credit per year.

d. If the inflation stays at 6%, approximately when will the tuition double? Explain how you determined your answer.

Tuition will double in about 12 years if inflation stays at 6%. (Answers will vary.)

25. Aircraft design engineers use the following formula to determine the proper landing speed, V (in feet per second):

$$V = \sqrt{\dfrac{841L}{cs}}$$

where L is the gross weight of the aircraft in pounds, c is the coefficient of lift, and s is the wing surface area in square feet.

Determine L if $V = 115$, $c = 2.81$, and $s = 200$.

$$115 = \sqrt{\frac{841L}{2.81(200)}}$$

$$115^2 = \frac{841L}{2.81(200)}$$

$$2.81(200)(115^2) = 841L$$

$$\frac{2.81(200)(115^2)}{841} = L$$

$$L \approx 8837.6$$

The aircraft weighs approximately 8840 pounds.

26. If an object travels at a constant velocity, v (miles per hour), for t hours, the distance traveled (miles) is given by the formula $d = vt$.

   a. If you are driving a distance of 20 miles at a speed of 50 miles per hour, how long will it take you to arrive at your destination?

$$20 = 50 \cdot t, \text{ so } t = \frac{20}{50} = 0.4 \text{ hr., or 24 min.}$$

It will take me 24 minutes to arrive at my destination.

   b. If it takes you 40 minutes to bicycle a distance of 20 miles, what is your speed in miles per hour? (*Hint:* Use correct units for time.)

$$40 \text{ min.} = 40 \text{ min.} \cdot \frac{1 \text{ hr.}}{60 \text{ min.}} = \frac{2}{3} \text{ hr.}$$

$$20 = v \cdot \frac{2}{3}, \text{ so } v = 20 \cdot \frac{3}{2} = 30$$

I am bicycling at a rate of 30 miles per hour.

27. An object is dropped from the roof of a five-story building whose floors are vertically 12 feet apart. The function that relates the distance fallen, s (in feet), to the time, t (in seconds), that it falls is $s = 16t^2$.

To answer the questions in parts a–c, do the following:

* Determine the distance that the object falls.

* Substitute the distance in the equation $s = 16t^2$.

* Solve the resulting equation.

   a. How long will it take the object to reach the floor level of the fifth floor?

The object will have fallen 12 feet. Therefore,

$$12 = 16t^2$$

$$0.75 = t^2$$

$$t = \sqrt{0.75} \approx 0.866 \text{ sec.}$$

It will take the object approximately 0.87 second to reach the floor level of the fifth floor.

b. How long will it take the object to reach the floor level of the third floor?

The object will have fallen 36 feet:

$36 = 16t^2$

$2.25 = t^2$

$t = \sqrt{2.25} = 1.5$ sec.

It will take 1.5 seconds for the object to reach the floor level of the third floor.

c. How long will it take the object to reach ground level?

The object will have fallen 60 feet.

$60 = 16t^2$

$3.75 = t^2$

$t = \sqrt{3.75} \approx 1.94$ sec.

It will take the object approximately 1.94 seconds to reach ground level.

28. In recent years there has been a great deal of discussion about global warming and the possibility of developing alternative energy sources that are not carbon-based such as coal, oil, and natural gas. One of the fastest growing sources has been wind energy.

The following table shows the rapid increase in installed capacity of wind power in the U.S. from 2001 to 2008. The wind power capacity is measure in megawatts, MW. Note the input variable (year) increases in steps of one unit (year).

YEAR	2001	2002	2003	2004	2005	2006	2007	2008
Installed Capacity, MW	4261	4685	6374	6740	9149	11,575	16,596	25,176

a. Is this a linear function? Explain.

This is not a linear function; the rate of change is not constant.

b. The wind data in the first table can be modeled by an exponential function defined by $N(t) = 3716(1.282)^t$, where $N(t)$ represents the generating capacity of windmills in the U.S. (in MW) and t represents the number of years since 2001. Note that $t = 0$ corresponds to 2001, $t = 1$ to 2002 and so on. Complete the following table.

t	CALCULATION FOR THE GENERATING CAPACITY OF WINDMILLS IN THE U.S.	EXPONENTIAL FORM	GENERATING CAPACITY IN MW
0	3716	$3716(1.282)^0$	3716
1	3716(1.282)	$3716(1.282)^1$	4763
2	3716(1.282)(1.282)	$3716(1.282)^2$	6105
3	3716(1.282)(1.282)(1.282)	$3716(1.282)^3$	7825

c. How well do the data determined by the exponential model fit the actual data in part a?

It is close, but not exact.

d. Use the model to estimate the generating capacity of windmills in the year 2012.

The year 2012 corresponds to $t = 11$; $N(11) = 3716(1.282)^{11} \approx 57,022$ MW

Fractions

Proper and Improper Fractions

A fraction in the form $\frac{a}{b}$ is called **proper** if a and b are counting numbers and a is less than b ($a < b$). If a is greater than or equal to b ($a \geq b$), then $\frac{a}{b}$ is called **improper**.

Note: a can be zero but b cannot, since you may not divide by zero.

Example 1: $\frac{2}{3}$ is proper since $2 < 3$.

Example 2: $\frac{7}{5}$ is improper since $7 > 5$.

Example 3: $\frac{0}{4} = 0$

Example 4: $\frac{8}{0}$ has no numerical value and, therefore, is said to be **undefined**.

Reducing a Fraction

A fraction, $\frac{a}{b}$, is in **lowest terms** if a and b have no common divisor other than 1.

Example 1: $\frac{3}{4}$ is in lowest terms. The only common divisor is 1.

Example 2: $\frac{9}{15}$ is not in lowest terms. 3 is a common divisor of 9 and 15.

To **reduce a fraction $\frac{a}{b}$ to lowest terms**, divide both a and b by a common divisor until the numerator and denominator no longer have any common divisors.

Example 1: Reduce $\frac{4}{8}$ to lowest terms.

Solution: $\dfrac{4}{8} = \dfrac{4 \div 4}{8 \div 4} = \dfrac{1}{2}$

Example 2: Reduce $\dfrac{54}{42}$ to lowest terms.

Solution: $\dfrac{54}{42} = \dfrac{54 \div 2}{42 \div 2} = \dfrac{27}{21} = \dfrac{27 \div 3}{21 \div 3} = \dfrac{9}{7}$

Mixed Numbers

A **mixed number** is the sum of a whole number plus a proper fraction.

Example 1: $3\frac{2}{5} = 3 + \frac{2}{5}$

Changing an Improper Fraction, $\frac{a}{b}$, to a Mixed Number

1. Divide the numerator, a, by the denominator, b.

2. The quotient becomes the whole-number part of the mixed number.

3. The remainder becomes the numerator, and b remains the denominator, of the fractional part of the mixed number.

Example 1: Change $\frac{7}{5}$ to a mixed number.

Solution: $7 \div 5 = 1$ with a remainder of 2.

So $\frac{7}{5} = 1 + \frac{2}{5} = 1\frac{2}{5}$.

Example 2: Change $\frac{22}{6}$ to a mixed number.

Solution: First reduce $\frac{22}{6}$ to lowest terms.

$$\frac{22}{6} = \frac{22 \div 2}{6 \div 2} = \frac{11}{3}$$

$11 \div 3 = 3$ with a remainder of 2.

So $\frac{11}{3} = 3 + \frac{2}{3} = 3\frac{2}{3}$.

Changing a Mixed Number to an Improper Fraction

1. To obtain the numerator of the improper fraction, multiply the denominator by the whole number and add the original numerator to this product.

2. Place this sum over the original denominator.

Example 1: Change $4\frac{5}{6}$ to an improper fraction.

Solution: $6 \cdot 4 + 5 = 29$

So $4\frac{5}{6} = \frac{29}{6}$.

Exercises

1. Reduce these fractions, if possible.

   a. $\frac{12}{16}$ b. $\frac{33}{11}$ c. $\frac{21}{49}$ d. $\frac{13}{31}$

   $\frac{3}{4}$ 3 $\frac{3}{7}$ $\frac{13}{31}$

2. Change each improper fraction to a mixed number. Reduce, if possible.

   a. $\frac{20}{16}$ b. $\frac{34}{8}$ c. $\frac{48}{12}$ d. $\frac{33}{5}$

   $1\frac{1}{4}$ $4\frac{1}{4}$ 4 $6\frac{3}{5}$

3. Change each mixed number to an improper fraction. Reduce, if possible.

   a. $7\frac{2}{5}$ b. $8\frac{6}{10}$ c. $9\frac{6}{7}$ d. $11\frac{3}{4}$

   $\frac{37}{5}$ $\frac{86}{10} = \frac{43}{5}$ $\frac{69}{7}$ $\frac{47}{4}$

Finding a Common Denominator of Two Fractions

A **common denominator** is a number that is divisible by each of the original denominators.

The **least common denominator** is the smallest possible common denominator.

A common denominator can be obtained by multiplying the original denominators. Note that this product may not necessarily be the *least* common denominator.

Example 1: Determine a common denominator for $\frac{5}{6}$ and $\frac{7}{9}$.

Solution: $6 \cdot 9 = 54$, so 54 is a common denominator.
However, 18 is the least common denominator.

Equivalent Fractions

Equivalent fractions are fractions with the same numerical value.

To obtain an equivalent fraction, multiply or divide both numerator and denominator by the same nonzero number.

Note: Adding or subtracting the same number to the original numerator and denominator does *not* yield an equivalent fraction.

Example 1: Determine a fraction that is equivalent to $\frac{3}{5}$ and has denominator 20.

Solution: $\dfrac{3}{5} = \dfrac{3 \cdot 4}{5 \cdot 4} = \dfrac{12}{20}$ **Note:** $\dfrac{3 + 15}{5 + 15} = \dfrac{18}{20} = \dfrac{9}{10} \neq \dfrac{3}{5}$

Comparing Fractions: Determining Which Is Greater, Is $\frac{a}{b} < \frac{c}{d}$ or is $\frac{a}{b} > \frac{c}{d}$?

1. Obtain a common denominator by multiplying b and d.

2. Write $\frac{a}{b}$ and $\frac{c}{d}$ as equivalent fractions, each with denominator $b \cdot d$.

3. Compare numerators. The fraction with the greater (or lesser) numerator is the greater (or lesser) fraction.

Example 1: Determine whether $\frac{3}{5}$ is less than or greater than $\frac{7}{12}$.

Solution: A common denominator is $5 \cdot 12 = 60$.

$$\frac{3}{5} = \frac{3 \cdot 12}{5 \cdot 12} = \frac{36}{60}, \quad \frac{7}{12} = \frac{7 \cdot 5}{12 \cdot 5} = \frac{35}{60}$$

Since $36 > 35$, $\frac{3}{5} > \frac{7}{12}$.

Exercises

Compare the two fractions and indicate which one is larger.

1. $\frac{4}{7}$ and $\frac{5}{8}$ **2.** $\frac{11}{13}$ and $\frac{22}{39}$ **3.** $\frac{5}{12}$ and $\frac{7}{16}$ **4.** $\frac{3}{5}$ and $\frac{14}{20}$

    $\frac{5}{8}$ $\frac{11}{13}$ $\frac{7}{16}$ $\frac{14}{20}$

Addition and Subtraction of Fractions with the Same Denominators

To add (or subtract) $\frac{a}{c}$ and $\frac{b}{c}$:

1. Add (or subtract) the numerators, a and b.

2. Place the sum (or difference) over the common denominator, c.

3. Reduce to lowest terms.

Example 1: Add $\frac{5}{16} + \frac{7}{16}$.

Solution: $\dfrac{5}{16} + \dfrac{7}{16} = \dfrac{5+7}{16} = \dfrac{12}{16} \qquad \dfrac{12}{16} = \dfrac{12 \div 4}{16 \div 4} = \dfrac{3}{4}$

Example 2: Subtract $\frac{19}{24} - \frac{7}{24}$.

Solution: $\dfrac{19}{24} - \dfrac{7}{24} = \dfrac{19-7}{24} = \dfrac{12}{24} \qquad \dfrac{12}{24} = \dfrac{12 \div 12}{24 \div 12} = \dfrac{1}{2}$

Addition and Subtraction of Fractions with Different Denominators

To add (or subtract) $\frac{a}{b}$ and $\frac{c}{d}$, where $b \neq d$:

1. Obtain a common denominator by multiplying b and d.

2. Write $\frac{a}{b}$ and $\frac{c}{d}$ as equivalent fractions with denominator $b \cdot d$.

3. Add (or subtract) the numerators of the equivalent fractions and place the sum (or difference) over the common denominator, $b \cdot d$.

4. Reduce to lowest terms.

Example 1: Add $\frac{3}{5} + \frac{2}{7}$.

Solution: A common denominator is $5 \cdot 7 = 35$.

$$\frac{3}{5} = \frac{3 \cdot 7}{5 \cdot 7} = \frac{21}{35}, \qquad \frac{2}{7} = \frac{2 \cdot 5}{7 \cdot 5} = \frac{10}{35},$$

$$\frac{3}{5} + \frac{2}{7} = \frac{21}{35} + \frac{10}{35} = \frac{31}{35}$$

$\frac{31}{35}$ is already in lowest terms.

Example 2: Subtract $\frac{5}{12} - \frac{2}{9}$.

Solution: Using a common denominator, $12 \cdot 9 = 108$:

$$\frac{5}{12} = \frac{5 \cdot 9}{12 \cdot 9} = \frac{45}{108}$$

$$\frac{2}{9} = \frac{2 \cdot 12}{9 \cdot 12} = \frac{24}{108}$$

$$\frac{5}{12} - \frac{2}{9} = \frac{45}{108} - \frac{24}{108} = \frac{21}{108}$$

$$\frac{21}{108} = \frac{21 \div 3}{108 \div 3} = \frac{7}{36}$$

Using the least common denominator, 36:

$$\frac{5}{12} = \frac{5 \cdot 3}{12 \cdot 3} = \frac{15}{36}$$

$$\frac{2}{9} = \frac{2 \cdot 4}{9 \cdot 4} = \frac{8}{36}$$

$$\frac{5}{12} - \frac{2}{9} = \frac{15}{36} - \frac{8}{36} = \frac{7}{36}$$

Addition and Subtraction of Mixed Numbers

To add (or subtract) mixed numbers:

1. Add (or subtract) the whole-number parts.

2. Add (or subtract) the fractional parts. In subtraction, this may require borrowing.

3. Add the resulting whole-number and fractional parts to form the mixed number sum (or difference).

Example 1:   Add $3\frac{2}{3} + 5\frac{3}{4}$.

Solution: Whole-number sum: $3 + 5 = 8$

Fractional sum: $\dfrac{2}{3} + \dfrac{3}{4} = \dfrac{2 \cdot 4}{3 \cdot 4} + \dfrac{3 \cdot 3}{4 \cdot 3} = \dfrac{8}{12} + \dfrac{9}{12} = \dfrac{17}{12} = 1\frac{5}{12}$

Final result: $8 + 1\frac{5}{12} = 8 + 1 + \frac{5}{12} = 9\frac{5}{12}$

Example 2:   Subtract $4\frac{1}{3} - 1\frac{7}{8}$.

Solution: Because $\frac{1}{3}$ is smaller than $\frac{7}{8}$, you must borrow as follows:

$4\frac{1}{3} = 3 + 1 + \frac{1}{3} = 3 + 1\frac{1}{3} = 3\frac{4}{3}$

The original subtraction now becomes $3\frac{4}{3} - 1\frac{7}{8}$.

Subtracting the whole-number parts: $3 - 1 = 2$.

Subtracting the fractional parts:

$\dfrac{4}{3} - \dfrac{7}{8} = \dfrac{4 \cdot 8}{3 \cdot 8} - \dfrac{7 \cdot 3}{8 \cdot 3} = \dfrac{32}{24} - \dfrac{21}{24} = \dfrac{11}{24}$

Final result is $2 + \frac{11}{24} = 2\frac{11}{24}$.

Alternatively, to add (or subtract) mixed numbers:

1. Convert each mixed number to an improper fraction.

2. Add (or subtract) the fractions.

3. Rewrite the result as a mixed number.

Example 3:   Subtract $4\frac{1}{3} - 1\frac{7}{8}$.

Solution: $4\frac{1}{3} = \dfrac{13}{3}$ and $1\frac{7}{8} = \dfrac{15}{8}$

$\dfrac{13}{3} - \dfrac{15}{8} = \dfrac{13 \cdot 8}{3 \cdot 8} - \dfrac{15 \cdot 3}{8 \cdot 3} = \dfrac{104}{24} - \dfrac{45}{24} = \dfrac{59}{24} = 2\frac{11}{24}$

Exercises

Add or subtract as indicated.

1. $1\frac{3}{4} + 3\frac{1}{8}$
   $4\frac{7}{8}$

2. $8\frac{2}{3} - 7\frac{1}{4}$
   $1\frac{5}{12}$

3. $6\frac{5}{6} + 3\frac{11}{18}$
   $10\frac{4}{9}$

4. $2\frac{1}{3} - \frac{4}{5}$
   $1\frac{8}{15}$

5. $4\frac{3}{5} + 2\frac{3}{8}$
   $6\frac{39}{40}$

6. $12\frac{1}{3} + 8\frac{7}{10}$
   $21\frac{1}{30}$

7. $14\frac{3}{4} - 5\frac{7}{8}$
   $8\frac{7}{8}$

8. $6\frac{5}{8} + 9\frac{7}{12}$
   $16\frac{5}{24}$

Multiplying Fractions

To multiply fractions, $\frac{a}{b} \cdot \frac{c}{d}$:

1. Multiply the numerators, $a \cdot c$, to form the numerator of the product fraction.

2. Multiply the denominators, $b \cdot d$, to form the denominator of the product fraction.

3. Write the product fraction by placing $a \cdot c$ over $b \cdot d$.

4. Reduce to lowest terms.

Example 1: Multiply $\frac{3}{4} \cdot \frac{8}{15}$.

Solution: $\quad \frac{3}{4} \cdot \frac{8}{15} = \frac{3 \cdot 8}{4 \cdot 15} = \frac{24}{60}, \quad \frac{24}{60} = \frac{24 \div 12}{60 \div 12} = \frac{2}{5}$

Note: It is often simpler and more efficient to cancel any common factors of the numerators and denominators *before* multiplying.

Dividing by a Fraction

Dividing *by* a fraction, $\frac{c}{d}$, is equivalent to multiplying by its reciprocal, $\frac{d}{c}$.

To divide fraction $\frac{a}{b}$ by $\frac{c}{d}$, written $\frac{a}{b} \div \frac{c}{d}$:

1. Rewrite the division as an equivalent multiplication, $\frac{a}{b} \cdot \frac{d}{c}$.

2. Proceed by multiplying as described above.

Example 1: Divide $\frac{2}{3}$ by $\frac{1}{2}$.

Solution: $\quad \frac{2}{3} \div \frac{1}{2} = \frac{2}{3} \cdot \frac{2}{1} = \frac{2 \cdot 2}{3 \cdot 1} = \frac{4}{3}$

Note: $\frac{4}{3}$ is already in lowest terms.

Exercises

Multiply or divide as indicated.

1. $\frac{11}{14} \cdot \frac{4}{5}$ $\quad$ **2.** $\frac{3}{7} \div \frac{3}{5}$ $\quad$ **3.** $\frac{8}{15} \cdot \frac{3}{4}$ $\quad$ **4.** $6 \div \frac{2}{5}$

$\frac{22}{35}$ $\qquad\qquad$ $\frac{5}{7}$ $\qquad\qquad$ $\frac{2}{5}$ $\qquad\qquad$ 15

Multiplying or Dividing Mixed Numbers

To multiply (or divide) mixed numbers:

1. Change the mixed numbers to improper fractions.

2. Multiply (or divide) the fractions as described earlier.

3. If the result is an improper fraction, change to a mixed number.

Example 1: Divide $8\frac{2}{5}$ by 3.

Solution: $\quad 8\frac{2}{5} = \frac{42}{5}, \quad 3 = \frac{3}{1}$

$\frac{42}{5} \div \frac{3}{1} = \frac{42}{5} \cdot \frac{1}{3} = \frac{42}{15}$

$\frac{42}{15} = \frac{42 \div 3}{15 \div 3} = \frac{14}{5}$

$\frac{14}{5} = 2\frac{4}{5}$

Exercises

Multiply or divide as indicated.

1. $3\frac{1}{2} \cdot 4\frac{3}{4}$

$\frac{133}{8} = 16\frac{5}{8}$

2. $10\frac{3}{5} \div 4\frac{2}{3}$

$\frac{159}{70} = 2\frac{19}{70}$

3. $5\frac{4}{5} \cdot 6$

$\frac{174}{5} = 34\frac{4}{5}$

4. $4\frac{3}{10} \div \frac{2}{5}$

$\frac{215}{20} = 10\frac{3}{4}$

Fractions and Percents

To convert a fraction or mixed number to a percent:

1. If present, convert the mixed number to an improper fraction.

2. Multiply the fraction by 100 and attach the % symbol.

3. If the result is an improper fraction, change it to a mixed number.

Example 1: Write $\frac{1}{12}$ as a percent.

Solution: $\frac{1}{12} \cdot 100\% = \frac{100}{12}\% = 8\frac{1}{3}\%$

Example 2: Write $2\frac{1}{4}$ as a percent.

Solution: $2\frac{1}{4} = \frac{9}{4} \cdot 100\% = 225\%$

Exercises

Convert each fraction to a percent.

1. $\frac{3}{4}$

75%

2. $2\frac{3}{5}$

260%

3. $\frac{2}{3}$

$66\frac{2}{3}\%$

4. $\frac{5}{9}$

$55\frac{5}{9}\%$

To convert a percent to a fraction:

1. Remove the % symbol and multiply the number by $\frac{1}{100}$.

Example 1: Write $8\frac{1}{3}\%$ as a fraction.

Solution: $8\frac{1}{3}\% = 8\frac{1}{3} \cdot \frac{1}{100} = \frac{25}{3} \cdot \frac{1}{100} = \frac{1}{12}$

Exercises

Convert each percent to a fraction.

1. 75%

$\frac{3}{4}$

2. 60%

$\frac{3}{5}$

3. $66\frac{2}{3}\%$

$\frac{2}{3}$

4. $55\frac{5}{9}\%$

$\frac{5}{9}$

Decimals

Reading and Writing Decimal Numbers

Decimal numbers are written numerically according to a place value system. The following table lists the place values of digits to the *left* of the decimal point.

hundred million	ten million	million,	hundred thousand	ten thousand	thousand,	hundred	ten	one	*decimal point*

The following table lists the place values of digits to the *right* of the decimal point.

decimal point	tenths	hundredths	thousandths	ten-thousandths	hundred-thousandths	millionths

To read or write a decimal number in words:

1. Use the first place value table to read the digits to the left of the decimal point, in groups of three from the decimal point.

2. Insert the word *and*.

3. Read the digits to the right of the decimal point as though they were not preceded by a decimal point, and then attach the place value of its rightmost digit:

Example 1: Read and write the number 37,568.0218 in words.

				3	7	5	6	8	.
hundred million	ten million	million,	hundred thousand	ten thousand	thousand,	hundred	ten	one	*decimal point*

and

.	0	2	1	8		
decimal point	tenths	hundredths	thousandths	ten-thousandths	hundred-thousandths	millionths

Therefore, 37,568.0218 is read thirty-seven thousand, five hundred sixty-eight and two hundred eighteen ten-thousandths.

Example 2: Write the number seven hundred eighty-two million, ninety-three thousand, five hundred ninety-four and two thousand four hundred three millionths numerically in standard form.

7	8	2,	0	9	3,	5	9	4	.
hundred million	ten million	million,	hundred thousand	ten thousand	thousand,	hundred	ten	one	decimal point

"and"

.	0	0	2	4	0	3
decimal point	tenths	hundredths	thousandths	ten-thousandths	hundred-thousandths	millionths

That is, 782,093,594.002403

Exercises

1. Write the number 9467.00624 in words.

Nine thousand, four hundred sixty-seven and six hundred twenty-four hundred-thousandths.

2. Write the number 35,454,666.007 in words.

Thirty-five million, four hundred fifty-four thousand, six hundred sixty-six and seven thousandths.

3. Write the number numerically in standard form: four million, sixty-four and seventy-two ten-thousandths.

4,000,064.0072

4. Write the number numerically in standard form: seven and forty-three thousand fifty-two millionths.

7.043052

Rounding a Number to a Specified Place Value

1. Locate the digit with the specified place value (target digit).

2. If the digit directly to its right is less than 5, keep the target digit. If it is 5 or greater, increase the target digit by 1.

3. If the target digit is to the right of the decimal point, delete all digits to its right.

4. If the target digit is to the left of the decimal point, replace any digits between the target digit and the decimal point with zeros as placeholders. Delete the decimal point and all digits that follow it.

Example 1: Round 35,178.2649 to the nearest hundredth.

The digit in the hundredths place is 6. The digit to its right is 4. Therefore, keep the 6 and delete the digits to its right. The rounded value is 35,178.26.

Example 2: Round 35,178.2649 to the nearest tenth.

The digit in the tenths place is 2. The digit to its right is 6. Therefore, increase the 2 to 3 and delete the digits to its right. The rounded value is 35,178.3.

Example 3: Round 35,178.2649 to the nearest ten thousand.

The digit in the ten thousands place is 3. The digit to its right is 5. Therefore, increase the 3 to 4 and insert four zeros to its right as placeholders. The rounded value is 40,000, where the decimal point is not written.

Exercises

1. Round 7456.975 to the nearest hundredth.

   7456.98

2. Round 55,568.2 to the nearest hundred.

   55,600

3. Round 34.6378 to the nearest tenth.

   34.6

Converting a Fraction to a Decimal

To convert a fraction to a decimal, divide the numerator by the denominator.

Example 1: Convert $\frac{4}{5}$ to a decimal.

Solution: On a calculator:

     Key in $\boxed{4}$ $\boxed{\div}$ $\boxed{5}$ $\boxed{=}$
     to obtain 0.8.

Using long division:
$$\begin{array}{r} 0.8 \\ 5\overline{)4.0} \\ \underline{4.0} \\ 0 \end{array}$$

Example 2: Convert $\frac{1}{3}$ to a decimal.

Solution: On a calculator:

     Key in $\boxed{1}$ $\boxed{\div}$ $\boxed{3}$ $\boxed{=}$
     to obtain 0.3333333.

Using long division:
$$\begin{array}{r} 0.333 \\ 3\overline{)1.000} \\ \underline{-9} \\ 10 \\ \underline{-9} \\ 1 \end{array}$$

Since a calculator's display is limited to a specified number of digits, it will cut off the decimal's trailing right digits.

Since this long division process will continue indefinitely, the quotient is a repeating decimal, 0.33333 . . . and is instead denoted by $0.\overline{3}$. The bar is placed above the repeating digit or above a repeating sequence of digits.

Exercises

Convert the given fractions into decimals. Use a repeating bar, if necessary.

1. $\frac{3}{5}$	**2.** $\frac{2}{3}$	**3.** $\frac{7}{8}$	**4.** $\frac{1}{7}$	**5.** $\frac{4}{9}$
0.6	$0.\overline{6}$	0.875	$0.\overline{142857}$	$0.\overline{4}$

Converting a Terminating Decimal to a Fraction

1. Read the decimal.

2. The place value of the rightmost nonzero digit becomes the denominator of the fraction.

3. The original numeral, with the decimal point removed, becomes the numerator. Drop all leading zeros.

Example 1: Convert 0.025 to a fraction.

Solution: The rightmost digit, 5, is in the thousandths place.

So, as a fraction, $0.025 = \frac{25}{1000}$, which reduces to $\frac{1}{40}$.

Example 2: Convert 0.0034 to a fraction.

Solution: The rightmost digit, 4, is in the ten-thousandths place.

So, as a fraction, $0.0034 = \frac{34}{10,000}$, which reduces to $\frac{17}{5000}$.

Exercises

Convert the given decimals to fractions.

1. 0.4	2. 0.125	3. 0.64	4. 0.05
$\frac{4}{10} = \frac{2}{5}$	$\frac{125}{1000} = \frac{1}{8}$	$\frac{64}{100} = \frac{16}{25}$	$\frac{5}{100} = \frac{1}{20}$

Converting a Decimal to a Percent

Multiply the decimal by 100 (that is, move the decimal point two places to the right, inserting placeholding zeros when necessary) and attach the % symbol.

Example 1: 0.78 written as a percent is 78%.

Example 2: 3 written as a percent is 300%.

Example 3: 0.045 written as a percent is 4.5%.

Exercises

Write the following decimals as equivalent percents.

1. 0.35	2. 0.076	3. 0.0089	4. 6.0
35%	7.6%	0.89%	600%

Converting a Percent to a Decimal

Divide the percent by 100 (that is, move the decimal point two places to the left, inserting placeholding zeros when necessary) and drop the % symbol.

Example 1: 5% written as a decimal is 0.05.

Example 2: 625% written as a decimal is 6.25.

Example 3: 0.0005% written as a decimal is 0.000005.

Exercises

Write the following percents as decimals.

1. 45%	2. 0.0987%	3. 3.45%	4. 2000%
0.45	0.000987	0.0345	20.00

Comparing Decimals

1. Write the decimals one below the next, lining up their respective decimal points.

2. Read the decimals from left to right, comparing corresponding place values. The decimal with the first and largest nonzero digit is the greatest number.

Example 1: Order from greatest to least: 0.097, 0.48, 0.0356.

Solution: Align by decimal point: 0.097

0.48

0.0356

Since 4 (in the second decimal) is the first nonzero digit, 0.48 is the greatest number. Next, since 9 is greater than 3, 0.097 is next greatest; and finally, 0.0356 is the least number.

Example 2: Order from greatest to least: 0.043, 0.0043, 0.43, 0.00043.

Solution: Align by decimal point: 0.043

0.0043

0.43

0.00043

Since 4 (in the third decimal), is the first nonzero digit, then 0.43 is the greatest number. Similarly, next is 0.043; followed by 0.0043; and finally, 0.00043 is the least number.

Exercises

Place each group of decimals in order, from greatest to least.

1. 0.058   0.0099   0.105   0.02999
   0.105   0.058   0.02999   0.0099

2. 0.75   1.23   1.2323   0.9   0.999
   1.2323   1.23   0.999   0.9   0.75

3. 13.56   13.568   13.5068   13.56666
   13.568   13.56666   13.56   13.5068

Adding and Subtracting Decimals

1. Write the decimals one below the next, lining up their respective decimal points. If the decimals have differing numbers of digits to the right of the decimal point, place trailing zeros to the right in the shorter decimals.

2. Place the decimal point in the answer, lined up with the decimal points in the problem.

3. Add or subtract the numbers as usual.

Example 1: Add: 23.5 + 37.098 + 432.17.

Solution:

$$\begin{array}{r} 23.500 \\ 37.098 \\ +\ 432.170 \\ \hline 492.768 \end{array}$$

Example 2: Subtract 72.082 from 103.07.

Solution:
$$\begin{array}{r} 103.070 \\ -\ \ 72.082 \\ \hline 30.988 \end{array}$$

Exercises

1. Calculate: $543.785 + 43.12 + 3200.0043$.

   3786.9093

2. Calculate: $679.05 - 54.9973$.

   624.0527

Multiplying Decimals

1. Multiply the numbers as usual, ignoring the decimal points.

2. Sum the number of decimal places in each number to determine the number of decimal places in the product.

Example 1: Multiply 32.89 by 0.021.

Solution:
$$\begin{array}{r} 32.89 \longrightarrow \textbf{2 decimal places} \\ \times\ 0.021 \longrightarrow \textbf{3 decimal places} \\ \hline 3289\ \ \ \\ 6578\ \ \ \ \\ \hline 0.69069 \longrightarrow \textbf{5 decimal places} \end{array}$$

Example 2: Multiply 64.05 by 7.3.

Solution:
$$\begin{array}{r} 64.05 \longrightarrow \textbf{2 decimal places} \\ \times\ \ \ \ 7.3 \longrightarrow \textbf{1 decimal places} \\ \hline 19215\ \ \\ 44835\ \ \ \\ \hline 467.565 \longrightarrow \textbf{3 decimal places} \end{array}$$

Exercises

Multiply the following decimals.

1. 12.53×8.2 **2.** 115.3×0.003 **3.** 14.62×0.75

   102.746   0.3459   10.965

Dividing Decimals

1. Write the division in long division format.

2. Move the decimal point the same number of places to the right in both divisor and dividend so that the divisor becomes a whole number. Insert zeros if necessary.

3. Place the decimal point in the quotient directly above the decimal point in the dividend and divide as usual.

Example 1: Divide   92.4   by   0.25.

    (*dividend*)   (*divisor*)

Solution: $0.25\overline{)92.4}$ becomes $25\overline{)9240}$:

$$
\begin{array}{r}
369.6 \\
25\overline{)9240.0} \\
-75 \\
\hline
174 \\
-150 \\
\hline
240 \\
-225 \\
\hline
150 \\
-150 \\
\hline
0
\end{array}
$$

So, $92.4 \div 0.25 = 369.6$.

Example 2: 0.00052 ÷ 0.004

    (*dividend*)   (*divisor*)

Solution: $0.004\overline{)0.00052}$ becomes $4\overline{)0.52}$:

$$
\begin{array}{r}
0.13 \\
4\overline{)0.52} \\
-4 \\
\hline
12 \\
-12 \\
\hline
0
\end{array}
$$

So, $0.00052 \div 0.004 = 0.13$.

Exercises

1. Divide 12.05 by 2.5.

4.82

2. Divide 18.9973 by 78.

≈ 0.243555

3. Divide 14.05 by 0.0002.

70,250

4. Calculate 150 ÷ 0.03.

5000

5. Calculate 0.00442 ÷ 0.017.

0.26

6. Calculate 69.115 ÷ 0.0023.

30,050

Algebraic Extensions

Properties of Exponents

The basic properties of exponents are summarized as follows:

> If a and b are both positive real numbers and n and m are any real numbers, then
>
> **1.** $a^n a^m = a^{n+m}$ **2.** $\dfrac{a^n}{a^m} = a^{n-m}$ **3.** $(a^n)^m = a^{nm}$
>
> **4.** $a^{-n} = \dfrac{1}{a^n}$ **5.** $(ab)^n = a^n b^n$ **6.** $a^0 = 1$

Note: The properties of exponents are covered in detail in Activity 4.2.

Property 1: $a^n a^m = a^{n+m}$ If you are multiplying two powers of the same base, add the exponents.

Example 1: $x^4 \cdot x^7 = x^{4+7} = x^{11}$

Note: The exponents were added and the base did not change.

Property 2: $\dfrac{a^n}{a^m} = a^{n-m}$ If you are dividing two powers of the same base, subtract the exponents.

Example 2: $\dfrac{6^6}{6^4} = 6^{6-4} = 6^2 = 36$

Note: The exponents were subtracted and the base did not change.

Property 3: $(a^n)^m = a^{nm}$ If a power is raised to a power, multiply the exponents.

Example 3: $(y^3)^4 = y^{12}$

Note: The exponents were multiplied. The base did not change.

Property 4: $a^{-n} = \dfrac{1}{a^n}$ Sometimes presented as a definition, Property 4 states that any base raised to a negative power is equivalent to the reciprocal of the base raised to the positive power. Note that the negative exponent does not have any effect on the sign of the base. This property could also be viewed as a result of the second property of exponents as follows:

Consider $\dfrac{x^3}{x^5}$. Using Property 2, $x^{3-5} = x^{-2}$. If you view this expression algebraically, you have three factors of x in the numerator and five in the denominator. If you divide out the three common factors, you are left with $\dfrac{1}{x^2}$. Therefore, if Property 2 is true, then $x^{-2} = \dfrac{1}{x^2}$.

Example 4: Write each of the following without negative exponents.

$$\text{a. } 3^{-2} \qquad\qquad \text{b. } \frac{2}{x^{-3}}$$

Solution: a. $3^{-2} = \dfrac{1}{3^2} = \dfrac{1}{9}$ \qquad b. $\dfrac{2}{x^{-3}} = \dfrac{2}{\dfrac{1}{x^3}} = 2 \div \dfrac{1}{x^3} = 2 \cdot x^3 = 2x^3$

Property 5: $(ab)^n = a^n b^n$ **If a product is raised to a power, each factor is raised to that power.**

Example 5: $(2x^2y^3)^3 = 2^3 \cdot (x^2)^3 \cdot (y^3)^3 = 8x^6y^9$

Note: Because the base contained three factors each of those was raised to the third power. The common mistake in an expansion such as this is not to raise the coefficient to the power.

Property 6: $a^0 = 1, a \neq 0$. Often presented as a definition, Property 6 states that any nonzero base raised to the zero power is 1. This property or definition is a result of Property 2 of exponents as follows:

Consider $\dfrac{x^5}{x^5}$. Using Property 2, $x^{5-5} = x^0$. However, you know that any fraction in which the numerator and the denominator are equal is equivalent to 1. Therefore, $x^0 = 1$.

Example 6: $\left(\dfrac{2x^3}{3yz^5}\right)^0 = 1$

Given a nonzero base, if the exponent is zero, the value is 1.

A factor can be moved from a numerator to a denominator or from a denominator to a numerator by changing the *sign of the exponent.*

Example 7: Simplify and express your result with positive exponents only.

$$\frac{x^3 y^{-4}}{2x^{-3}y^{-2}z}$$

Solution: Move the x^{-3} and the y^{-2} factors from the denominator to the numerator making sure to change the sign of the exponents. This results in:

$$\frac{x^3 y^{-4} x^3 y^2}{2z}$$

Simplify the numerator using Property 1 and then use Property 4:

$$\frac{x^{3+3} y^{-4+2}}{2z} = \frac{x^6 y^{-2}}{2z} = \frac{x^6}{2y^2 z}$$

Exercises

Simplify and express your results with positive exponents only. Assume that all variables represent only nonzero values.

1. 5^{-3}

$\frac{1}{5^3} = \frac{1}{125}$

2. $\frac{1}{x^{-5}}$

x^5

3. $\frac{3x}{y^{-2}}$

$3xy^2$

4. $\frac{10x^2y^5}{2x^{-3}}$

$5x^5y^5$

5. $\frac{5^{-1}z}{x^{-1}z^{-2}}$

$\frac{xz^3}{5}$

6. $5x^0$

5

7. $(a + b)^0$

1

8. $-3(x^0 - 4y^0)$

9

9. $x^6 \cdot x^{-3}$

x^3

10. $\frac{4^{-2}}{4^{-3}}$

$4^1 = 4$

11. $(4x^2y^3) \cdot (3x^{-3}y^{-2})$

$\frac{12y}{x}$

12. $\frac{24x^{-2}y^3}{6x^3y^{-1}}$

$\frac{4y^4}{x^5}$

13. $\frac{(14x^{-2}y^{-3}) \cdot (5x^3y^{-2})}{6x^2y^{-3}z^{-3}}$

$\frac{35z^3}{3xy^2}$

14. $\left(\frac{2x^{-2}y^{-3}}{z^2}\right) \cdot \left(\frac{x^5y^3}{z^{-3}}\right)$

$2x^3z$

15. $\frac{(16x^4y^{-3}z^{-2})(3x^{-3}y^4)}{15x^{-3}y^{-3}z^2}$

$\frac{16x^4y^4}{5z^4}$

Addition Method for Solving a System of Two Linear Equations

The basic strategy for the addition method is to reduce a system of two linear equations to a single linear equation by eliminating a variable.

For example, consider the x-coefficients of the linear system

$$2x + 3y = 1$$
$$4x - y = 9$$

The coefficients are 2 and 4. The LCM of 2 and 4 is 4. Use the multiplication principle to multiply each side of the first equation by -2. The resulting system is

$$\begin{array}{l} -2(2x + 3y = 1) \\ 4x - y = 9 \end{array} \quad \text{or, equivalently,} \quad \begin{array}{l} -4x - 6y = -2 \\ 4x - y = 9 \end{array}$$

Multiplying by -2 produces x-coefficients that are additive inverses or opposites. Now add the two equations together to eliminate the variable x.

$$-7y = 7$$

Solving for y, $y = -1$ is the y-value of our solution. To find the value of x, substitute -1 for y and solve for x in any equation that involves x and y. For an alternative to determining the value of x, consider the original system and the coefficients of y, 3 and -1 in the original system. The LCM is 3. Since the signs are already opposites, multiply the second equation by 3.

$$\begin{array}{l} 2x + 3y = 1 \\ 3(4x - y = 9) \end{array} \quad \text{or, equivalently,} \quad \begin{array}{l} 2x + 3y = 1 \\ 12x - 3y = 27 \end{array}$$

Adding the two equations will eliminate the y-variable.

$$14x = 28$$

Solving for x, $x = 2$. Therefore, the solution is $(2, -1)$. This should be checked to make certain that it satisfies both equations.

Depending on the coefficients of the system, you may need to change both equations when using the addition method. For example, the coefficients of x in the linear system

$$5x - 2y = 11$$
$$3x + 5y = -12$$

are 5 and 3. The LCM is 15. Multiply the first equation by 3 and the second by -5 as follows:

$$3(5x - 2y = 11)$$
$$-5(3x + 5y = -12)$$
or, equivalently,
$$15x - 6y = 33$$
$$-15x - 25y = 60$$

Add the two equations to eliminate the x terms from the system.

$$-31y = 93$$

Solving for y, $y = -3$. Substituting this value for y in the first equation of the original system yields

$$5x - 2(-3) = 11$$
$$5x + 6 = 11$$
$$5x = 5$$
$$x = 1$$

Therefore, $(1, -3)$ is the solution of the system.

Exercises

Solve the following systems using the addition method. If the system has no solution or both equations represent the same line, state this as your answer.

1. $x - y = 3$
$x + y = -7$
$x = -2, y = -5$

2. $x + 4y = 10$
$x + 2y = 4$
$x = -2, y = 3$

3. $-5x - y = 4$
$-5x + 2y = 7$
$x = -1, y = 1$

4. $4x + y = 7$
$2x + 3y = 6$
$x = 1.5, y = 1$

5. $3x - y = 1$
$6x - 2y = 5$
no solution

6. $4x - 2y = 0$
$3x + 3y = 5$
$x = \frac{5}{9}, y = \frac{10}{9}$

7. $x - y = 9$
$-4x - 4y = -36$
$x = 9, y = 0$

8. $-2x + y = 6$
$4x + y = 1$
$x = -\frac{5}{6}, y = \frac{13}{3}$

9. $\frac{3}{2}x + \frac{2}{5}y = \frac{9}{10}$
$\frac{1}{2}x + \frac{6}{5}y = \frac{3}{10}$
$x = \frac{3}{5}, y = 0$

10. $0.3x - 0.8y = 1.6$
$0.1x + 0.4y = 1.2$
$x = 8, y = 1$

Factoring Trinomials with Leading Coefficient ≠ 1

With patience, it is possible to factor many trinomials by trial and error, using the FOIL method in reverse.

Factoring Trinomials by Trial and Error

1. Factor out the greatest common factor.

2. Try combinations of factors for the first and last terms in the two binomials.

3. Check the outer and inner products to match the middle term of the original trinomial.

4. If the check fails, then repeat steps 2 and 3.

Example 1: Factor $6x^2 - 7x - 3$.

Solution:

Step 1. There is no common factor, so go to step 2.

Step 2. You could factor the first term as $6x(x)$ or as $2x(3x)$. The last term has factors of 3 and 1, disregarding signs. Suppose you try $(2x + 1)(3x - 3)$.

Step 3. The outer product is $-6x$. The inner product is $3x$. The sum is $-3x$. The check fails.

Step 4. Suppose you try $(2x - 3)(3x + 1)$. The outer product is $2x$. The inner product is $-9x$. The sum is $-7x$. It checks. Therefore, $6x^2 - 7x - 3 = (2x - 3)(3x + 1)$.

Exercises

Factor each of the following completely.

1. $x^2 + 7x + 12$

   $(x + 3)(x + 4)$

2. $6x^2 - 13x + 6$

   $(3x - 2)(2x - 3)$

3. $3x^2 + 7x - 6$

   $(3x - 2)(x + 3)$

4. $6x^2 + 21x + 18$

   $3(2x + 3)(x + 2)$

5. $9x^2 - 6x + 1$

   $(3x - 1)(3x - 1)$

6. $2x^2 + 6x - 20$

   $2(x + 5)(x - 2)$

7. $15x^2 + 2x - 1$

   $(5x - 1)(3x + 1)$

8. $4x^3 + 10x^2 + 4x$

   $2x(2x + 1)(x + 2)$

Solving Equations by Factoring

Many quadratic and higher-order polynomial equations can be solved by factoring using the zero-product property.

Solving an Equation by Factoring

1. Use the addition principle to move all terms to one side, so that the other side of the equation is zero.

2. Simplify and factor the nonzero side.

3. Use the zero-product property to set each factor equal to zero and then solve the resulting equations.

4. Check your solutions in the original equation.

Example 1: Solve the equation $3x^2 - 2 = -x$.

Solution: Adding x to both sides, you obtain $3x^2 + x - 2 = 0$. Since there are no like terms, factor the trinomial.

$$(3x - 2)(x + 1) = 0$$

Using the zero-product property,

$$3x - 2 = 0 \quad \text{or} \quad x + 1 = 0$$
$$3x = 2 \quad \text{or} \quad x = -1$$
$$x = \tfrac{2}{3}$$

The two solutions are $x = \tfrac{2}{3}$ and $x = -1$. The check is left for the reader to do.

Example 2: Solve the equation $3x^3 - 8x^2 = 3x$.

Solution: Subtracting $3x$ from both sides, you obtain $3x^3 - 8x^2 - 3x = 0$. Since there are no like terms, factor the trinomial.

$$x(3x^2 - 8x - 3) = 0$$
$$x(3x + 1)(x - 3) = 0$$

Using the zero-product property,

$$x = 0 \quad \text{or} \quad 3x + 1 = 0 \quad \text{or} \quad x - 3 = 0$$
$$3x = -1 \quad \text{or} \quad x = 3$$
$$x = -\tfrac{1}{3}$$

The three solutions are $x = 0$, $x = -\tfrac{1}{3}$, and $x = 3$. The check is left for the reader to do.

Exercises

Solve each of the following equations by factoring, if possible.

1. $x^2 - x - 63 = 0$

not solvable by factoring

2. $3x^2 - 9x - 30 = 0$

$x = 5, x = -2$

3. $-7x + 6x^2 = 10$

$x = 2, x = -\tfrac{5}{6}$

4. $3y^2 = 2 - y$

$y = -1, y = \tfrac{2}{3}$

5. $-28x^2 + 15x - 2 = 0$

$x = \tfrac{2}{7}, x = \tfrac{1}{4}$

6. $4x^2 - 25 = 0$

$x = 2.5, x = -2.5$

7. $(x + 4)^2 - 16 = 0$

$x = 0, x = -8$

8. $(x + 1)^2 - 3x = 7$

$x = 3, x = -2$

9. $2(x + 2)(x - 2) = (x - 2)(x + 3) - 2$

$x = 0, x = 1$

10. $18x^3 = 15x^2 + 12x$

$x = 0, x = -\tfrac{1}{2}, x = \tfrac{4}{3}$

Getting Started with the TI-83/TI-84 Plus Family of Calculators

ON-OFF

To turn on the calculator, press the (ON) key. To turn off the calculator, press (2nd) and then (ON).

Most keys on the calculator have multiple purposes. The number or symbolic function/command written directly on the key is accessed by simply pressing the key. The symbolic function/commands written above each key are accessed with the aid of the (2nd) and (ALPHA) keys. The command above and to the left is color coded to match the (2nd) key. That command is accessed by first pressing the (2nd) key and then pressing the key itself. Similarly, the command above and to the right is color coded to match the (ALPHA) key and is accessed by first pressing the (ALPHA) key and then pressing the key itself.

Contrast

To adjust the contrast on your screen, press and release the (2nd) key and hold (▲) to darken and (▼) to lighten.

Mode

The (MODE) key controls many calculator settings. The activated settings are highlighted. For most of your work in this course, the settings in the left-hand column should be highlighted.

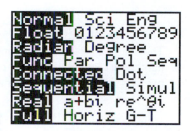

To change a setting, move the cursor to the desired setting and press (ENTER).

The Home Screen

The home screen is used for calculations.

You may return to the home screen at any time by using the QUIT command. This command is accessed by pressing. All calculations in the home screen are subject to the order of operations convention.

Enter all expressions as you would write them. Always observe the order of operations. Once you have typed the expression, press (ENTER) to obtain the simplified result. Before you press (ENTER), you may edit your expression by using the arrow keys, the delete command (DEL), and the insert command (2nd) (DEL).

Three keys of special note are the reciprocal key (X⁻¹), the caret (^) key, and the negative key ((-)).

Typing a number and then pressing the reciprocal command key (X⁻¹) will give the reciprocal of the number. The reciprocal of a nonzero number, n, is $\frac{1}{n}$. As noted in the screen below, when performing an operation on a fraction, the fraction MUST be enclosed in parentheses before accessing this command.

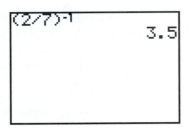

The caret key (^) is used to raise numbers to powers

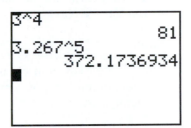

The negative key ((-)) on the bottom of the keyboard is different from the subtraction key (-). They cannot be used interchangeably. The negative key is used to change the sign of a single number or symbol; it will not perform a subtraction operation. If you mistakenly use the negative key in attempting to subtract, you will likely obtain an ERROR message.

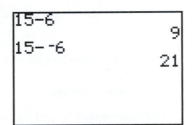

A table of some frequently used keys and their functions follows.

KEY	FUNCTION DESCRIPTION
ON	Turns calculator on or off.
CLEAR	Clears the line you are currently typing. If cursor is on a blank line when CLEAR is pressed, it clears the entire home screen.
ENTER	Executes a command.
(−)	Calculates the additive inverse.
MODE	Displays current operating settings.
DEL	Deletes the character at the cursor.
^	Symbol used for exponentiation.
ANS	Storage location of the result of the most recent calculation.
ENTRY	Retrieves the previously executed expression so that you may edit it.

ANS and ENTRY

The last two commands in the table can be real time savers. The result of your last calculation is always stored in a memory location known as ANS. It is accessed by pressing (2nd) ((−)) or it can be automatically accessed by pressing any operation button.

Suppose you want to evaluate $12.5\sqrt{1 + 0.5 \cdot (0.55)^2}$. It could be evaluated in one expression and checked with a series of calculations using ANS.

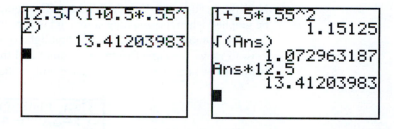

After you have keyed in an expression and pressed (ENTER), you cannot move the cursor back up to edit or recalculate this expression. This is where the ENTRY ((2nd) (ENTER)) command is used. The ENTRY command retrieves the previous expression and places the cursor at the end of the expression. You can use the left and right arrow keys to move the cursor to any location in the expression that you wish to modify.

Suppose you want to evaluate the compound interest expression $P\left(1 + \frac{r}{n}\right)^{nt}$, where P is the principal, r is the interest rate, n is the number of compounding periods annually, and t is the number of years, when $P = \$1000$, $r = 6.5\%$, $n = 1$, and $t = 2, 5,$ and 15 years.

Using the ENTRY command, this expression would be entered once and edited twice.

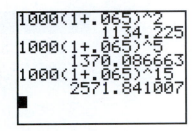

```
1000(1+.065)^2
              1134.225
1000(1+.065)^5
           1370.086663
1000(1+.065)^15
           2571.841007
■
```

Note that there are many last expressions stored in the ENTRY memory location. You can repeat the ENTRY command as many times as you want to retrieve a previously entered expression.

Functions and Graphing with the TI-83/TI-84 Plus Family of Calculators

"Y =" Menu

Functions of the form $y = f(x)$ can be entered into the TI-83/TI-84 Plus using the "Y = " menu. To access the "Y = " menu, press the $\boxed{Y=}$ key. Type the expression $f(x)$ after Y_1 using the $\boxed{X,T,\theta,n}$ key for the variable x and press $\boxed{\text{ENTER}}$.

For example, enter the function $f(x) = 3x^5 - 4x + 1$.

```
Plot1 Plot2 Plot3
\Y1■3X^5-4X+1
\Y2=
\Y3=
\Y4=
\Y5=
\Y6=
\Y7=
```

Note the = sign after Y_1 is highlighted. This indicates that the function Y_1 is active and will be graphed when the graphing command is executed and will be included in your table when the table command is executed. The highlighting may be turned on or off by using the arrow keys to move the cursor to the = symbol and then pressing $\boxed{\text{ENTER}}$. Notice in the screen below that Y_1 has been deactivated and will not be graphed nor appear in a table.

```
Plot1 Plot2 Plot3
\Y1=3X^5-4X+1
\Y2=
\Y3=
\Y4=
\Y5=
\Y6=
\Y7=
```

Once the function is entered in the "Y = " menu, function values may be evaluated in the home screen.

For example, given $f(x) = 3x^5 - 4x + 1$, evaluate $f(4)$. In the home screen, press $\boxed{\text{VARS}}$.

Move the cursor to Y-VARS and press $\boxed{\text{ENTER}}$.

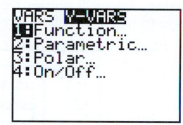 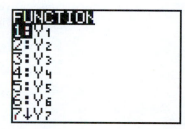

Press $\boxed{\text{ENTER}}$ again to select Y_1. Y_1 now appears in the home screen.

To evaluate $f(4)$, press $\boxed{\text{(}}$ $\boxed{4}$ $\boxed{\text{)}}$ after Y_1 and press $\boxed{\text{ENTER}}$.

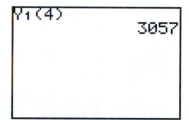

Tables of Values

If you are interested in viewing several function values for the same function, you may want to construct a table.

Before constructing the table, make sure the function appears in the "Y = " menu with its "=" highlighted. You may also want to deactivate or clear any functions that you do not need to see in your table. Next, you will need to check the settings in the Table Setup menu. To do this, use the TBLSET command ($\boxed{\text{2nd}}$ $\boxed{\text{WINDOW}}$).

As shown in the preceding screen, the default setting for the table highlights the Auto options for both the independent (*x*) and dependent (*y*) variables. Choosing this option will display ordered pairs of the function with equally spaced *x*-values. TblStart is the first *x*-value to be displayed, and here is assigned the value -2. ΔTbl represents the equal spacing between consecutive *x*-values, and here is assigned the value 0.5. The TABLE command (2nd GRAPH)) brings up the table displayed in the screen below.

Use the (▲) and (▼) keys to view additional ordered pairs of the function.

If the input values of interest are not evenly spaced, you may want to choose the Ask mode for the independent variable from the Table Setup menu.

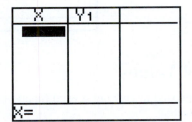

The resulting table is blank, but you can fill it by choosing any values you like for *x* and pressing (ENTER) after each.

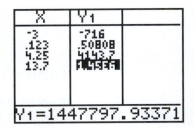

Note that the number of digits shown in the output is limited by the table width, but if you want more digits, move the cursor to the desired output and more digits appear at the bottom of the screen.

Graphing a Function

Once a function is entered in the "Y = " menu and activated, it can be displayed and analyzed. For this discussion we will use the function $f(x) = -x^2 + 10x + 12$. Enter this as Y_1 making sure to use the negation key ⊝ and not the subtraction key ⊟.

The Viewing Window

The viewing window is the portion of the rectangular coordinate system that is displayed when you graph a function.

Xmin defines the left edge of the window.

Xmax defines the right edge of the window.

Xscl defines the distance between horizontal tick marks.

Ymin defines the bottom edge of the window.

Ymax defines the top edge of the window.

Yscl defines the distance between vertical tick marks.

In the standard viewing window, Xmin = -10, Xmax = 10, Xscl = 1, Ymin = -10, Ymax = 10, and Yscl = 1.

To select the standard viewing window, press ⬚ZOOM⬚ ⬚6⬚.

You will view the following:

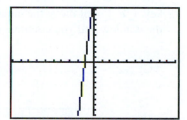

Is this an accurate and/or complete picture of your function, or is the window giving you a misleading impression? You may want to use your table function to view the output values that correspond to the input values from -10 to 10.

X	Y₁		
-10	-188		
-9	-159		
-8	-132		
-7	-107		
-6	-84		
-5	-63		
-4	-44		
X= -10			

X	Y₁		
-3	-27		
-2	-12		
-1	1		
0	12		
1	21		
2	28		
3	33		
X=3			

X	Y₁		
4	36		
5	37		
6	36		
7	33		
8	28		
9	21		
10	12		
X=10			

The table indicates that the minimum output value on the interval from $x = -10$ to $x = 10$ is -188, occurring at $x = -10$, and the maximum output value is 37, occurring at $x = 5$. Press WINDOW and reset the settings to the following:

$$\text{Xmin} = -10, \text{Xmax} = 10, \text{Xscl} = 1,$$
$$\text{Ymin} = -190, \text{Ymax} = 40, \text{Yscl} = 10$$

Press GRAPH to view the graph with these new settings.

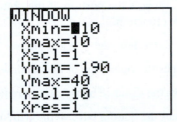 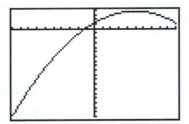

The new graph gives us a much more complete picture of the behavior of the function on the interval $[-10, 10]$.

The coordinates of specific points on the curve can be viewed by activating the trace feature. While in the graph window, press TRACE. The function equation will be displayed at the top of the screen, a flashing cursor will appear on the curve at the middle of the screen, and the coordinates of the cursor location will be displayed at the bottom of the screen.

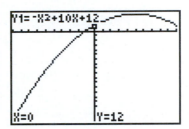

The left arrow key, ◀, will move the cursor toward smaller input values. The right arrow key, ▶, will move the cursor toward larger input values. If the cursor reaches the edge of the window and you continue to move the cursor, the window will adjust automatically.

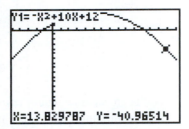

Zoom Menu

The Zoom menu offers several options for changing the window very quickly.

The features of each of the commands are summarized in the following table.

ZOOM COMMAND	DESCRIPTION
1: ZBox	Draws a box to define the viewing window.
2: Zoom In	Magnifies the graph near the cursor.
3: Zoom Out	Increases the viewing window around the cursor.
4: ZDecimal	Sets a window so that Xscl and Yscl are 0.1.
5: ZSquare	Sets equal size pixels on the x- and y-axes.
6: ZStandard	Sets the window to standard settings.
7: ZTrig	Sets built-in trig window variables.
8: ZInteger	Sets integer values on the x- and y-axes.
9: ZoomStat	Sets window based on the current values in the stat lists.
0: ZoomFit	Replots graph to include the max and min output values for the current Xmin and Xmax.

Solving Equations Graphically Using the TI-83/TI-84 Plus Family of Calculators

The Intersection Method

This method is based on the fact that solutions to the equation $f(x) = g(x)$ are input values of x that produce the same output for the functions f and g. Graphically, these are the x-coordinates of the intersection points of $y = f(x)$ and $y = g(x)$.

The following procedure illustrates how to use the intersection method to solve $x^3 + 3 = 3x$ graphically.

Step 1. Enter the left-hand side of the equation as Y_1 in the "Y = " editor and the right-hand side as Y_2. Select the standard viewing window.

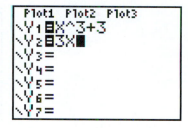

Step 2. Examine the graphs to determine the number of intersection points.

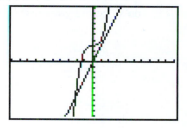

You may need to examine several windows to be certain of the number of intersection points.

Step 3.   Access the Calculate menu by pushing $\boxed{\text{2nd}}$ $\boxed{\text{TRACE}}$; then choose option 5: intersect.

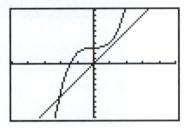   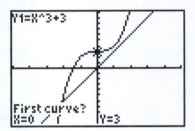

The cursor will appear on the first curve in the center of the window.

Step 4.   Move the cursor close to the desired intersection point and press $\boxed{\text{ENTER}}$.

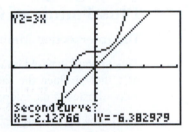

The cursor will now jump vertically to the other curve.

Step 5.   Repeat step 4 for the second curve.

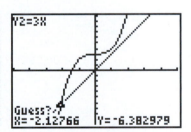

Step 6.   To use the cursor's current location as your guess, press $\boxed{\text{ENTER}}$ in response to the question on the screen that asks Guess? If you want to move to a better guess value, do so before you press $\boxed{\text{ENTER}}$.

The coordinates of the intersection point appear below the word Intersection.

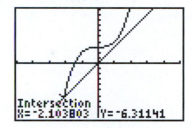

The *x*-coordinate is a solution to the equation.

If there are other intersection points, repeat the process as necessary.

Using the TI-83/TI-84 Plus Family of Calculators to Determine the Linear Regression Equation for a Set of Paired Data Values

Example 1:

INPUT	OUTPUT
2	2
3	5
4	3
5	7
6	9

Enter the data into the calculator as follows:

1. Press (STAT) and choose EDIT.

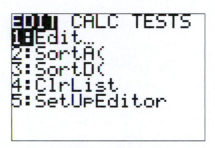

2. The calculator has six built-in lists, L1, L2, . . . , L6. If there is data in L1, clear the list as follows:

   a. Use the arrows to place the cursor on L1 at the top of the list. Press (CLEAR), followed by (ENTER), followed by the down arrow.

   b. Follow the same procedure to clear L2 if necessary.

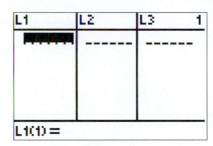

c. Enter the input values into L1 and the corresponding output values into L2.

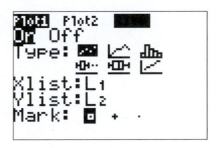

To see a scatterplot of the data proceed as follows.

1. STAT PLOT is the 2nd function of the (Y=) key. You must press (2nd) before pressing (Y=) to access the STAT PLOT menu.

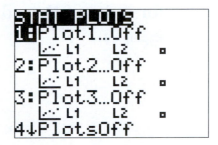

2. Select Plot 1 and make sure that Plots 2 and 3 are Off. The screen shown below will appear. Select On and then choose the scatterplot option (first icon) on the Type line. Confirm that your x and y values are stored, respectively, in L_1 and L_2. The symbols L_1 and L_2 are 2nd functions of the (1) and (2) keys, respectively. Finally, select the small square as the mark that will be used to plot each point.

3. Press (Y=) and clear or deselect any functions currently stored.

4. To display the scatterplot, have the calculator determine an appropriate window by pressing (ZOOM) and then (9) (ZoomStat).

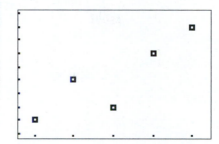

The following instructions will calculate the linear regression equation and store it in Y_1.

1. Press (STAT) and right arrow to highlight CALC.

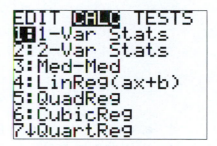

2. Choose 4: LinReg (ax + b). LinReg (ax + b) will be pasted to the home screen. To tell the calculator where the data is, press (2nd) and (1) (for L1), then (,), then (2nd) and (2) (for L2) because the Xlist and Ylist are stored in L_1 and L_2, respectively. The display should look like this:

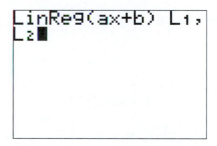

3. Press (,) and then press (VARS).

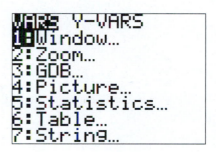

4. Right arrow to highlight Y-VARS.

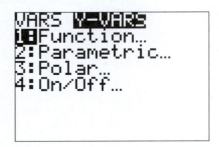

5. Choose 1, FUNCTION.

6. Choose 1 for Y_1 (or 2 for Y_2, etc., if you prefer to store the regression equation in another location).

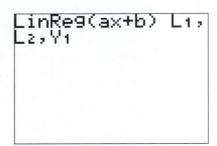

7. Press (ENTER).

LinReg
y=ax+b
a=1.6
b=-1.2

The linear regression equation for this data is $y = 1.6x - 1.2$.

8. To display the regression line on the scatterplot screen, press (GRAPH).

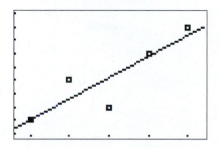

9. Press the (Y=) key to view the equation.

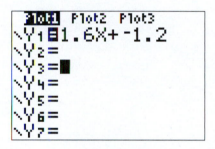

Learning Math Opens Doors: Twelve Keys to Success

1. Are you in the right math course for your skill level?

Students are sometimes placed in the wrong math course for a variety of reasons. Your answers to the following questions will help you and your instructor determine if you are in the correct math course for your skill level. This information will also help your instructor understand your background more quickly, thus providing you with a better learning experience.

a. Did you register for this math course because you took a placement exam and, as a result of your exam score, you selected or were placed in this course?

b. If you were not placed in this class because of a placement exam, please list reasons you are taking this course.

c. Name _____

d. Phone and/or e-mail address _____

Course Title	Grade
1. _____	_____
2. _____	_____

e. List the mathematics courses you have taken at the *college level*.

Course Title	Grade
1. _____	_____
2. _____	_____
3. _____	_____

f. List the mathematics courses you have taken at the *high school level*.

g. Have you taken this course before?   Yes _____ No _____

h. When did you take your last mathematics course? _____

i. Are you a full-time or part-time student? _____

j. Do you have a job? _____ If yes, how many hours per week do you work? _____

k. Do you take care of children or relatives at home? _____

You should now share your information with your instructor to make sure you are in the correct class. It is a waste of a semester of time and money if the course is too easy for you or too difficult for you. Take control of and responsibility for your learning!

2. What is your attitude about mathematics?

a. Write one word that describes how you feel about learning mathematics.
_____

b. Was the word that you wrote a positive word, a negative word, or a neutral word? _____

If your word was a positive word, you are on your way to success in this course. If your word was negative, then before progressing any further, you may want to determine why you have negative feelings toward mathematics.

c. Write about a positive or negative experience that you have had related to mathematics.

_____

_____

_____

_____

If you have not had success in the past, try something different. Here are a few suggestions:

d. Make a list of all the materials (pencils, notebook, calculator) that you might need for the course. Make sure you have all the materials required for the course.

_____

_____

e. Find a new location to study mathematics. List two or three good places you can study.

1. _____ 2. _____ 3. _____

f. Study with a classmate. Help each other organize the material. Ask each other questions about assignments. Write down names, phone numbers/e-mail addresses of two or three other students in your class to study with.

1. _____

2. _____

3. _____

g. How did you study mathematics in the past? Write a few sentences about what you did outside of class to learn mathematics.

_____

_____

_____

h. What will you change from your past to help you become more successful? Write down your strategy for success.

_____

_____

_____

3. Do you attend all classes on time and are you organized?

Class work is vital to your success. Make it a priority to attend every class. Arrive in sufficient time and be ready to start when class begins. Start this exercise by recording the months and dates for the entire semester in each box. Then write in each box when and where your class meets corresponding to your schedule. Keep track of exam and quiz dates, deadlines, study sessions, homework, and anything else that can help you succeed in the course.

Schedule routine medical and other appointments so you don't miss class. Allow time for traffic jams, finding a parking space, bus delays, and other emergencies that may occur. Arriving late interferes with your learning and the learning of your fellow students. Make sure assignments are completed and handed in on time. Attending class and being on time and organized is in your control.

✔ Place a check in each day that you are in class. Circle the check to indicate that you were on time. Place an *a* in each day you miss class.

Sunday	Monday	Tuesday	Wednesday	Thursday	Friday	Saturday

4. When is that assignment due?

In mathematics, assignments are usually given each class time. Assignments are meant to reinforce what you learned in class. If you have trouble completing your assignment, take charge and get help immediately. The following table will help you organize your assignments and the dates they are due. Keeping a record of your assignments in one location will help you know at a glance what and when your assignment is due.

Date Assigned	Assignment	Specific Instructions	Dates	
			Due	Completed

Date Assigned	Assignment	Specific Instructions	Dates	
			Due	Completed

5. Do you keep track of your progress?

Keeping track of your own progress is a great way to monitor the steps you are taking toward success in this course. Different instructors may use different methods to determine your grade.

a. Explain how your instructor will determine your final grade for this course.

_____

_____

_____

_____

b. Use the following table to keep track of your grades in this course. (You may want to change the headings to reflect your instructor's grading system; an Excel spreadsheet may also be helpful for this exercise.)

Date	Type of Assessment	Topic(s) and/ or Chapters	Number of Points Correct	Total Number of Points	Comments

c. Determine your final average using your instructor's grading system.

6. How well do you know your textbook?

Knowing the structure of your textbook helps you use the book more effectively to reach your learning goals. The Preface and "To the Student" sections at the beginning of the book provide guidance and list other valuable resources. These include CDs/DVDs, Web sites, and computer software with tutorials and/or skill practice for added help in the course. Use the following guidelines to learn the structure and goals of the textbook.

a. According to your syllabus, which chapters will you be studying this semester?

b. Find and underline the titles of the chapters in the table of contents.

c. Read the Preface and summarize the key points.

d. In the Preface, you should have found the student supplements for the course. List them.

e. What are the key points the authors make to you in the "To the Student" section?

f. Each chapter is divided into smaller parts called clusters. Name clusters in chapter 4.

g. Each cluster is made up of smaller sections called activities. What is the title of the activity in chapter 3, cluster 1, activity 1?

h. Within an activity you will see information that has a box around it. Why is this information important? Look at chapter 2, cluster 1, activity 1 before you respond.

i. Look at the end of each activity. There you will find the Exercises section. Doing these exercises should help you better understand the concepts and skills in the activity. Why are some of the numbers of the exercises in color?

j. Look at the end of cluster 1 in chapter 4. You will see a section entitled "What Have I Learned?" This section will help you review the key concepts in the cluster. What are the concepts taught in this cluster?

k. Look at the end of cluster 1 in chapter 4. What is the section called right after "What Have I Learned?" This section will help you practice all of the skills in the cluster. What are the skills taught in this cluster?

l. At the end of each chapter you will see a Gateway and a Summary. Look at the end of chapter 2 and write how each of these sections will help you in this course.

m. Locate and briefly review the glossary. Select a mathematical term from the glossary and write the term and its definition here.

7. Where does your time go?

Managing your time is often difficult. A good rule to follow when studying is to allow 2 hours of study time for each hour of class time. It is best not to have marathon study sessions, but, rather, to break up your study time into small intervals. Studying a little every day rather than "cramming" once a week helps to put the information into your long-term memory. Ask yourself this question: Is going to school a top priority? If you answered yes, then your study time must be a priority if you are going to be successful.

a. In the past, approximately how many hours per week did you study mathematics? Reflect on whether it was enough time to be successful.

b. The commitments in your life can be categorized. Estimate the number of hours that you spend each week on each category. (There are 168 hours in a week.)

Class Hours (Class) —————————   Chore Hours (C) —————————

Study Hours (ST) —————————   Sleep Hours (Sl) —————————

Work Hours (W) —————————   Personal Hours (P) —————————

Commute Hours (T) —————————   Leisure Hours (L) —————————

Other (O) —————————

c. Use the information from part b to fill in the grid with a workable schedule using the appropriate letter to represent each category.

Time	Mon.	Tues.	Wed.	Thurs.	Fri.	Sat.	Sun.
12 midnight							
1:00 A.M.							
2:00							
3:00							
4:00							
5:00							
6:00							
7:00							
8:00							
9:00							
10:00							
11:00							
12:00 noon							
1:00 P.M.							
2:00							
3:00							
4:00							
5:00							
6:00							
7:00							
8:00							
9:00							
10:00							
11:00							

d. Is your schedule realistic? Does it represent a typical week in *your* life? If not, make adjustments to it now.

e. Circle the hours in your schedule that you will devote to studying.

f. Shade or highlight the hours that you will use for studying mathematics.

g. Do you have some times scheduled each day to study mathematics? How much time is devoted to mathematics each day?

h. Studying math soon after class time will help you review what you learned in class. Did you place any hours to study math close to your actual class time?

i. Scheduling time before class can help you prepare for the class. Did you schedule any study time just before class time?

j. Review your schedule once more and make any changes.

k. Periodically review your schedule to see if it is working and whether you are following it. How are you doing?

8. Where will you use mathematics?

Students often ask math instructors, "Where will I ever use this?" Here is an opportunity to show where you have used mathematics, including what you have learned from this course.

a. Describe one way in which you use math in your everyday life.

_____

b. Find an article in a newspaper that contains a graph or some quantitative information.

c. Describe ways in which you use math in your job.

_____

_____

d. Describe a problem in a course you are taking or took (not a mathematics course) that involved a math concept or skill. Include a description of the math concept or skill that is involved.

_____

_____

After you have been in this course for a few weeks, answer parts e, f, and g.

e. Write a problem from one of your courses (not a mathematics course) or from your work experience where you used concepts or skills that you have learned in *this* mathematics class.

_____

_____

f. Show how you solved the problem.

_____

_____

g. Write the math skill(s) or concept(s) that you learned in this math course that helped you solve this problem.

_____

_____

9. Do you need extra help?

Receiving extra help when you need it may mean the difference between success and failure in your math course. If you are struggling, don't wait, seek help immediately. When you go for extra help, identify specific areas in which you need help. Don't just say "I'm lost."

a. Your instructor usually will note office hours right on or near the office door or they may be listed in the syllabus. Take a few extra minutes and locate the office. List your instructor's office hours below. Write the office room number next to the hours. Circle the office hours that work with your schedule.

b. Does your college have a center where you can go for tutoring or extra help? If there is one, what is the room number? Take a minute and locate the center. When is the center open? Determine what hours would be good for you if you should need help.

c. Students in your class may also help. Write the phone numbers or e-mail addresses of two or three students in your class whom you could call for help. These students may form a study group with you. When would it be a good time for all of you to meet? Set up the first meeting.

d. You can also help yourself by helping others learn the information. By working a problem with someone else, it will help you to reinforce the concepts and skills. Set aside some time to help someone else. What hours are you available? Tell someone else you are available to help. Write the person's name here.

Extra help is also there for you during class time. Remember to ask questions in class when you do not understand. Don't be afraid to raise your hand. There are probably other students who have the same question. If you work in groups, your group can also help. Use the following chart to record the times you have needed help.

Date	Topic you needed help with	Where did you find help?	Who helped you?	Were your questions answered?	Do you still have questions?

10. Have you tried creating study cards to prepare for exams?

a. Find out from your instructor what topics will be on your exam. Ask also about the format of the exam. Will it be multiple choice, true/false, problem solving, or some of each?

b. Organize your quizzes, projects, homework, and book notes from the sections to be tested.

c. Read these quizzes, projects, homework, and book notes carefully and pick out those concepts and skills that seem to be the most important. Write this information on your study card. Try to summarize in your own words and include an example of each important concept.

d. Make sure you include:

Vocabulary Words, with definitions:

Key Concepts, explained *in your own words:*

Skills, illustrated with an example or two:

e. Reviewing this study card with your instructor may be a good exam preparation activity.

11. How can you develop effective test-taking strategies?

a. A little worry before a test is good for you. Your study card will help you feel confident about what you know. Wear comfortable clothing and shoes to your exam and make yourself relax in your chair. A few deep breaths can help! Don't take stimulants such as caffeine; they only increase your anxiety!

b. As soon as you receive your test, write down on your test paper any formulas, concepts, or other information that you might forget during the exam. This is information that you would recall from your study card.

c. You may want to skim the test first to see what kind of test it is and which parts are worth the most points. This will also help you to allocate your time.

d. Read all directions and questions carefully.

e. You may want to do some easy questions first to boost your confidence.

f. Try to reason through tough problems. You may want to use a diagram or graph. Look for clues in the question. Try to estimate the answer before doing the problem. If you begin to spend too much time on one problem, you may want to mark it to come back to later.

g. Write about your test-taking strategies.

	What test-taking strategies will you use?	What test-taking strategies did you use?
Test 1		
Test 2		
Test 3		
Test 4		

12. How can you learn from your exams?

a. Errors on exams can be divided into several categories. Review your exam, identify your mistakes, and determine the category for each of your errors.

Type of Error	Meaning of Error	Question Number	Points Deducted
A. Concept	You don't understand the properties or principles required to answer the question.		
B. Application	You know the concept, but cannot apply it to the question.		
C. Skill	You know the concept and can apply it, but your skill process is incorrect.		
D. Test-taking	These errors apply to the specific way you take tests. 1. Did you change correct answers to incorrect answers? 2. Did you miscopy an answer from scrap paper? 3. Did you leave an answer blank? 4. Did you miss more questions at the beginning or in the middle or at the end of your test? 5. Did you misread the directions?		
E. Careless	Mistakes you made that you could have corrected had you reviewed your answers to the test.		
	TOTAL:		

b. You should now correct your exam and keep it for future reference.

   i. What questions do you still have after making your corrections?

   ii. Where will you go for help to get answers to these questions?

   iii. Write some strategies that you will use the next time you take a test that will help you reduce your errors.

Selected Answers

Chapter 1

Activity 1.2 Exercises: 1. a. Nine justices; **3. a.** 2, **b.** 2, **c.** 5; **4. a.** 2, **b.** 5; **5. b.** Each number is generated by multiplying the preceding number by 2 and then adding 1.

6.

COLUMN 2	COLUMN 3	COLUMN 4
5	3	8
8	5	13
13	8	21
21	13	34

Each number is generated by adding the two numbers that precede it.

Activity 1.3. Exercises: 1. b. 145, **d.** 55; **2. a.** 1960, **d.** 60; **3. b.** $83 \cdot 5 = 415$; **4. b.** $450 - 35 = 415$; **5. a.** $15 + 12 = 27$, **c.** $12 + 15 \cdot 3 - 4 = 12 + 45 - 4 = 53$; **7. b.** $\frac{20}{4} = 5$, **d.** $64 \div 4 \cdot 2 = 16 \cdot 2 = 32$, **f.** $8 \cdot 20 - 4 = 160 - 4 = 156$, **h.** $100 - (0) \div 3 = 100 - 0 = 100$; **8. b.** $80 - 27 = 53$, **d.** $50 - 40 = 10$; **9. b.** $243 + 200 = 443$, **d.** $45 \div 9 = 5$, **f.** $32 \cdot 25 = 800$, **h.** $36 + 64 = 100$, **j.** $5^2 = 25$; **10. b.** 5.551405×10^{14}; **11. b.** 45,320,000; **12.** 5.859×10^{12} miles **13. b.** $F = 120 \cdot 25 = 3000$, **d.** $F = 200 \cdot 25^2 \div 125$
$$= 200 \cdot 625 \div 125$$
$$= 125{,}000 \div 125 = 1000,$$
e. $A = (76 + 83 + 81)/3 = 240/3 = 80$

What Have I Learned? 1. b. Each of the 14 students kicks the ball to the 13 other students exactly once for a total of $14 \cdot 13 = 182$ kicks.

How Can I Practice? 1. b. Each number is generated by multiplying the preceding number by 2. **2. b.** 13 **d.** 4 **f.** 146, **h.** 3, **j.** 5, **l.** 98; **3. b.** 144, **d.** 64; **6.** 2×10^{24}; **7. b.** distributive property of multiplication over subtraction; **8. b.** false, left side $= 11$, right side $= 19$; **9. b.** $A = 49$

Activity 1.4. Exercises: 2. $1250; **4.** $7\frac{5}{12}$ hr; **6.** $4\frac{19}{24}$; **8. b.** $7\frac{5}{6}$, **d.** $5\frac{8}{35}$, **f.** $1\frac{13}{50}$

Activity 1.5. Exercises: 2. My GPA is 2.589 if I earn an F in economics. **4. b.** 60 points. **6.** The corrected GPA is 3.11.

What Have I Learned? 2. b. Each part is $\frac{1}{10}$ unit because $\frac{1}{10} + \frac{1}{10} = \frac{1}{5}$.

How Can I Practice? 2. $4\frac{9}{10}$ in. **4.** $1\frac{3}{4}$ ft. of the floor will show on each side. **6. b.** $1\frac{1}{8}$, **d.** $1\frac{19}{30}$, **f.** $\frac{1}{2}$, **h.** $\frac{5}{12}$, **j.** $\frac{11}{20}$, **l.** $\frac{7}{12}$, **n.** $1\frac{1}{2}$, **p.** 16, **r.** $8\frac{1}{18}$, **t.** $1\frac{23}{26}$, **v.** $5\frac{1}{4}$; **8. a.** 20 hr., **b.** 15.4 hr.

Activity 1.6. Exercises:
2. b. 1. and E, $\frac{12}{27} = \frac{20}{45} = 0.44\overline{4} = 44.\overline{4}\%$
2. and C, $\frac{28}{36} = \frac{21}{27} = 0.77\overline{7} = 77.\overline{7}\%$
3. and D, $\frac{45}{75} = \frac{42}{70} = 0.6 = 60\%$
4. and A, $\frac{64}{80} = \frac{60}{75} = 0.8 = 80\%$
5. and B, $\frac{35}{56} = \frac{25}{40} = 0.625 = 62.5\%$

4. b. 0.313; **6.** The relative number of women at the university is 51.2%; $\frac{2304}{4500} = 0.512 = 51.2\%$. The relative number of women is greater at the community college. **8.** 75.3%; **10. b.** 47.2%, **d.** 69.3%, **f.** 355, **h.** 32.7%, **j.** 11.3%

Activity 1.7 Exercises: 1. b. 27, **e.** 18, **h.** 240, **k.** 180; **2. c.** 570,400; **3.** $13.00; **5.** $47,500; **7. a.** General Motors: 2,983,200 cars; Ford: 1,993,200 cars; Chrysler: 1,452,000 cars **10.** 69; **11.** $47

Activity 1.8 Exercises: 2. a. $0.685, **b.** 30.6%; **4.** Acme: 33.3%; Arco: 10%; **6. a.** 400%, **b.** 80%; **8.** 200%

Activity 1.9 Exercises: 2. a. 1.08375, **b.** $21,397.56; **4. a.** $\frac{12{,}144}{11{,}952} \approx 1.016$, **b.** 1.016, **c.** 1.6%; **6. c.** $92,970; **8.** $3151.50; **9. c.** $58,950; **12.** 0.964, 3.6%; **13. b.** $0.011 = 1.1\%$; **15.** $90.97; **17.** 162.5 mg

Activity 1.10 Exercises: 2. $47,239.92; **4.** $2433; **6.** $60,000; My new salary is the same as my original salary.

Activity 1.11 Exercises: 2. approximately 2787 mi.; **4.** approximately 5.5 mi. high, which is 8.8481 km or 8848 m; **6.** approximately 4.8 qt. or 9.5 pt.; **9.** 4.8 g or 0.168 oz.

What Have I Learned? 6. a. $199 \cdot 0.90 = 179.1$ lb.; my relative will weigh 179 lb. after he loses 10% of his body weight. **b.** $179.1 \cdot 0.90 = 161.2$ lb.; $161.2 \cdot 0.90 = 145$ lb.; he must lose 10% of his body weight 3 times to reach 145 lb.

How Can I Practice? 1. a. 0.25, **d.** 0.035, **e.** 2.50;
2.

COLUMN 4
$7 \div 100 = 0.07 = 7\%$
$0.7 \div 5.0 = 0.14 = 14\%$
$0.7 \div 5.0 = 0.14 = 14\%$
$1 \div 12.5 = 0.08 = 8\%$

4. 4850 students; **6.** 89.2%; **8. a.** $168, **b.** 44%;
10. $750; **12.** 23,859 cars; **14.** 1. James; 2. Wade;
3. Bryant

Skills Check 1 1. 219; **2.** 5372; **3.** Sixteen and seven
hundred nine ten-thousandths; **4.** 3402.029; **5.** 567.05;
6. 2.599; **7.** 87.913; **8.** 28.82; **9.** 0.72108; **10.** 95.7;
11. 12%; **12.** 300%; **13.** 0.065; **14.** 3.6; **15.** $0.1\overline{3}$;
16. $\frac{3}{7} \approx 0.4286 \approx 42.86\% \approx 42.9\%$; **17.** $\frac{1}{8}$; **18.** 3.0027,
3.0207, 3.027, 3.27; **19.** 40.75; **20.** 52%; **21.** $8.67;
22. 32%; **23.** $9500 (1.055) \approx 10,023$ students;
24. 37.5%; **25.** 36.54 billion acres **26. a.** No, the
percent reductions are based on the price at the time of
the specific reduction, so it is not correct to sum up the
percent reductions. **b.** $450 before Thanksgiving, $270
in December, and $135 in January. **c.** The final price in
January would be the same as before, $135. The intermedi-
ate prices would be different; before Thanksgiving, the
price would be $250; in December, the cost would be $150.
27. The initial decay factor was 0.45. The subsequent growth
factor was 1.65. Therefore, the 2005 math scores were
$(0.45)(1.65) = 0.7425$, or approximately 74% of the 1995
scores. This is not good news. **28.** 7,121,151.5; a little over
$7 million; **29.** $1364.45; approximately $1364 in 2006.

Activity 1.12 Exercises: 3. 43-yd. gain; **6.** $-11°C$;
8. 35°C

Activity 1.13 Exercises: 4. The sign of a product will be
positive if there is an even number of factors with negative
signs; the sign of a product will be negative if there is an
odd number of factors with negative signs. **6.** -65.58;
My balance decreased by $65.58; **8.** -823.028;
10. a. $100,000, **b.** $50,000; **11. d.** $-$50.
12. I suffered a $0.20 loss per share, for a total loss of $36.

Activity 1.14 Exercises: 2. $39\frac{7}{12}$ ft. **6. c.** $-1 \cdot 6 \cdot 6 = -36$,
d. $\frac{1}{(-6)^2} = \frac{1}{36}$; **8. a.** 8.8×10^{-4} lb. = 0.00088 lb.,
b. (Answers will vary.) I would use grams because the number
is easier to say, and it may sound larger and impress my friend.

What Have I Learned? 4. a. The result is positive because
there are two negatives being multiplied, resulting in a
positive value. **c.** The sign of the product is negative because
there is an odd number of negative factors. **5. c.** A negative
number raised to an even power produces a positive result.
6. b. A negative number raised to an odd power produces a
negative result. **8. b.** Any number other than zero raised to
the zero power equals 1. **9. b.** 0.03 cm = 3.0×10^{-2} cm

How Can I Practice? 37. -6; I still need to lose 6 lb.
39. a. $-7°C$; The temperature drops 7°C. **b.** $-19°C$;
The evening temperature is expected to be $-19°C$.
d. a 9°C drop, **e.** The temperature rises 8°C; **40. a.** -9,
c. -5, **e.** 17, **g.** -9; **41. b.** $0.03937 = 3.937 \times 10^{-2}$ in.

Skills Check 2 1. 1.001; **2.** 0.019254; **3.** 1; **4.** 9;
5. $c = \sqrt{72} \approx 8.49$ m; approximately 8.5 m; **6.** 9;
7. 2.03×10^8; **8.** approximately 8.8 qt. in 8.3 L;
9. 0.00000276; **10.** approximately 9.96 oz.;
11. 16.1 sq. in.; **12.** I measured the diameter to be
$d = \frac{15}{16}$ in. Then $C = \pi\left(\frac{15}{16}\right) \approx 2.95$ in. The circumference
is approximately 2.95 in. **13.** 126 sq. in.; **14.** 8;
15. 35 hr.; **16.** $\frac{9}{1.6} = 5.625$; You can cut 5 pieces that are
1.6 yd. long. **17.** $\frac{4}{9}$; **18.** $2\frac{2}{7}$; **19.** $\frac{34}{7}$; **20.** $\frac{37}{45}$; **21.** $1\frac{7}{10}$;
22. $\frac{8}{27}$; **23.** 3; **24.** $4\frac{5}{6}$ cups; I have almost 5 cups of
mixture. **25.** $29\frac{5}{8}$ points at the end of the day;
26. $2.\overline{6}$ days; **27.** I should buy at least $4\frac{1}{4}$ lb. of hamburger;
28. 5.48 ft., 128.2 lb.; **29. a.** 52,800, **b.** 96, **c.** 80,
d. 768, **e.** 48;
30.

COLUMN 3
5×10^7
2.5×10^7
3.5×10^7
3.68×10^8
7.45×10^8
1.606×10^9
2.674×10^9

31. about 1.5 lb. of chocolate and 7 oz. of milk;
32. ≈ 0.98 in.; A 1-in. eraser is big enough to share.
33. $4\frac{1}{12}$; **34.** 9; **35.** 1; **36.** $9\frac{73}{144}$; **37.** $13\frac{7}{12}$; **38.** $24\frac{1}{6}$;
39. -16; **40.** 36; **41.** 24

Gateway Review 1. two and two hundred two ten-thousandths;
2. 14.003; **3.** hundred thousandths; **4.** 10.524;
5. 12.16; **6.** 0.30015; **7.** 0.007; **8.** 2.05; **9.** 450%;
10. 0.003; **11.** $7\frac{3}{10}$ or $\frac{73}{10}$; **12.** 60%; **13.** $1\frac{1}{8}$, 1.1, 1.01,
1.001; **14.** 27; **15.** 1; **16.** $\frac{1}{4^2} = \frac{1}{16}$; **17.** -64;
18. 16; **19.** -25 **20.** 12; **21.** 5.43×10^{-5};
22. 37,000; **23.** $\frac{1}{3}\left(\frac{25}{30} + \frac{85}{100} + \frac{60}{70}\right) = .8468 \approx 85\%$;
24. $4\frac{1}{2}$; **25.** $\frac{23}{4}$; **26.** $\frac{3}{5}$; **27.** $\frac{4}{24} + \frac{15}{24} = \frac{19}{24}$;
28. $\frac{21}{4} - \frac{15}{4} = \frac{6}{4} = \frac{3}{2} = 1\frac{1}{2}$; **29.** $\frac{\cancel{2}}{\cancel{9}} \cdot \frac{\cancel{27}}{8} = \frac{3}{4}$;
30. $\frac{\cancel{6}}{\cancel{11}} \cdot \frac{\cancel{22}}{\cancel{8}} = \frac{6}{4} = \frac{3}{2} = 1\frac{1}{2}$;
31. $\frac{\cancel{21}}{\cancel{8}} \cdot \frac{\cancel{16}}{\cancel{7}} = 6$; **32.** 20% of 80 is 16; **33.** I did better
on the second exam. **34.** There were 2500 complaints last
year. **35.** 6 students; **36.** about 19 in.; **37.** 14 in.;

38. 17.5 sq. ft.;   **39.** 3.1 sq. in.;   **40.** approximately 2093 cu. cm;   **41.** 30%;   **42.** 162;   **43.** I should be able to read 40 pages in 4 hours;   **44.** 1.364 lb.;   **45.** $\approx 0.013 \frac{\text{in.}}{\text{sec.}}$;   **46.** 10°C;   **47.** -1;   **48.** -14;   **49.** -13;   **50.** 3;   **51.** -25;   **52.** -1;   **53.** 2;   **54.** -11;   **55.** 0;   **56.** $\frac{1}{10}$;   **57.** $21 - 2 = 19$;   **58.** 80;   **59.** $192 - 9 = 183$;   **60.** $5 \cdot 4 = 20$;   **61.** The diver reaches a depth 160 ft. below the surface.   **62.** a loss of $600 in stocks;   **63.** I am overdrawn and am $63 in debt.   **64.** 4.4×10^{-27} lb.

Chapter 2

Activity 2.1   Exercises: 1. d. 1989, **f.** from 1967 to 1977, **h.** from 2005 to 2006;   **3. a.** No. Owning a negative number of DVDs doesn't make sense. **b.** Yes. It means student doesn't own any DVDs. **c.** 0, 1, 2, . . . , 1000.

Activity 2.2   Exercises: 1. d;   **3.** g;   **7.** a;   **12.** h;   **16.** f

Activity 2.3   Exercises: 1. b. Yes, $n = 0$ is reasonable. You might not purchase any notebooks at the college bookstore. **2. a.** 20, 40, 50. The output is 2 times the input. **3. b.** $-13, -10, -7, -4, -1, 2, 5, 8$;   **4. a.** $209, **b.** Gross pay is calculated by multiplying the number of hours worked by $9.50, the pay per hour.

Activity 2.4   Exercises: 2. a. His distance from me is $100 - y$;   **3. a.** The new perimeter is $3s + 15$;   **6. b.** $5(x - 6)$, **e.** $-2x - 20$;   **7. a.** -17.5, **d.** -19.5, **e.** -12, **f.** 12.5, **g.** 60

What Have I Learned? 3. a. Since input and output values are nonnegative, only the first quadrant and positive axes are necessary.

How Can I Practice? 1. c. No, negative numbers would represent distance above the surface.   **2. a.** The point having coordinates $(-3, -1500)$ is in quadrant III. **d.** The point representing the fact that there was no profit or loss in 2003 is located on the horizontal axis, between quadrant II and quadrant III.   **3. c.** The D.O. content changes the most between 11°C and 16°C, when it drops 1.6 ppm. **4. b.** From the graph, it appears to be approximately $1700. **c.** The investment would double by approximately age 13;   **6. b.** $\frac{1}{2}x + 6$;   **7. b.** 9

Activity 2.6   Exercises: 1. e. $1050, **f. i.** 5 credit hours; **2. b.** 8 in., **c.** 6 in.;   **4. d.** $590;   **5. c.** $15 million; **6. a.** $y = 52.5$, **b.** $x \approx 41.14$;   **8. a.** $y = -198.9$, **b.** $x = 23$;   **10. a.** $y = 1.8$, **b.** $x = 8.2$

Activity 2.7   Exercises: 3. b. $p = 10n - 2100$, **d.** 360 students; **4. c.** about 13 yr.;   **6. a.** approximately 181 cm;   **8.** 70 in.;   **10. a.** $x = -1$, **b.** $x = 4$, **d.** $x = -7$, **f.** $x = 0$, **h.** $x = 18$, **j.** $x = 10$; **11. b.** $y = 21.75, x = -60$,

x	y
3.5	21.75
-60	-10

e. $y = 2.8, x = 100$;

x	y
24	2.8
100	18

Activity 2.8   Exercises: 2. b. 70 mi., **c.** $3.39 \approx t$, approximately 3.4 hr.;   **5. a.** 409.2 ft., **b.** $\frac{33(p - 15)}{15} = d$; **7.** $\frac{C}{2\pi} = r$;   **9.** $\frac{P - 2l}{2} = w$;   **11.** $\frac{A - P}{Pt} = r$; **13.** $m + vt^2 = g$

Activity 2.9   Exercises: 2. a. $x = 24$;   **4.** 10.49;   **7.** 1673 consumers;   **10.** school tax $= \$3064$

What Have I Learned? 1. In both cases, you need to isolate x by performing the inverse operations indicated by $4x - 5$ in reverse order; that is add 5, and then divide by 4.

How Can I Practice? 1. b. $10 - x$, **d.** $-4x - 8$, **f.** $\frac{1}{2}x^2 - 2$;   **3. a.** $t = 2.5r + 10$;   **4. a.** $x = -3$, **c.** $x = 15.5$, **f.** $x = 4$, **h.** $x = 11.\overline{6}$;   **5. a.** Multiply the number of hours of labor by $68, and add $148 for parts to obtain the total cost of the repair. **d.** $420, **e.** No, 3.5 hr work plus parts will cost $386. **f.** 4 hr., **g.** $x = \frac{y - 148}{68}$; Solving for x allows you to answer questions such as part f more quickly. It would be especially useful if you wanted to determine the number of mechanic hours for several different amounts of money.   **6. a.** 32.8%, **b.** 2013; **8. a.** $B = 655.096 + 9.563(55) + 1.85(172) - 4.676(70)$ ≈ 1171.9 calories. He is not properly fed. **b.** $61.52 \approx A$. He is about 62 years old. **9. a.** $\frac{d}{t} = r$, **c.** $\frac{A - P}{Pt} = r$, **e.** $\frac{7}{4}(w - 3) = h$ **11. a.** $V \approx 79.5$ cu in.: $V \approx 57.7$ cu in. The second package holds 11.4 cu in. less than the original one.

Activity 2.10   Exercises: 1. $y = x - 10$;   **3.** $y = 9 + \frac{x}{6}$; **5.** $y = -4x - 15$;   **7. c.** This demonstrates the *commutative* property of multiplication.   **8. c.** No. Subtraction is *not* commutative. Reversing the order in subtraction changes the sign of the result. **9. a. i.** To obtain the output, square the input.

x	y_1
-2	4
-1	1
0	0
2	4
3	9

11.

x	$y_5 = 1 + x^2$	$y_6 = (1 + x)^2$
-1	2	0
0	1	1
2	5	9
5	26	36

y_5: Start with $x \to$ square $\to$ and $1 \to$ to obtain y; y_6: Start with $x \to$ add $1 \to$ square $\to$ to obtain y. No, They generate different outputs.

13.

x	$y_9 = 3x^2 + 1$	$y_{10} = (3x)^2 + 1$
-1	4	10
0	1	1
2	13	37
5	76	226

y_9: Start with $x \to$ square $\to$ multiply by $3 \to$ add $1 \to$ to obtain y;

y_{10}: Start with $x \to$ multiply by $3 \to$ square $\to$ add $1 \to$ to obtain y. No. They generate different outputs.

Activity 2.11 Exercises: 2. $24x - 30$; **4.** $10 - 5x$;
6. $-3p + 17$; **8.** $-12x^2 + 9x - 21$; **10.** $\frac{5}{8}x + \frac{5}{9}$;
12. $15x^2 - 12x$; **16.** $A = P + Prt$; **17. c.** $lw + 5l$;
19. a. $y - 5$, **b.** $12(y - 5) = 12y - 60$; **20. a.** $3(x + 5)$,
c. $xy(3 - 7 + 1)$, **e.** $4(1 - 3x)$, **g.** $st(4rt - 3rt + 10)$;
23. a. $5a + 8ab - 3b$, **c.** $104r - 13s^2 - 18s^3$,
e. $2x^3 + 7y^2 + 4x^2$, **g.** $3x^2y - xy^2$, **i.** $2x - 2x^2 - 5$;
25. a. $24 - x$, **c.** $5x - 75$, **e.** $9 + x$, **g.** $7x - 15$,
i. $5x - 5$, **k.** $-x^2 - 9x$; **26. c.** $398 - 26x$

Activity 2.12 Exercises: 1. c. Yes, the news is good. The value of my stock increased sixfold.
2. a. $3\left[\frac{-2n + 4}{2} - 5\right] + 6$, **b.** $-3n - 3$;
4. b. $8x - 5y + 13$, **e.** $13x - 43$

Activity 2.13 Exercises: 1. a. $C = 0.79x + 19.99$,
c. $0.79x + 19.99 = 0.59x + 29.99$, **e.** Company 2 is lower if I drive more than 50 miles. **3. d.** $y = 16.50x + 90$,
e. 288, **g.** 255, **i.** $x = \frac{y - 90}{16.5}$; This form of the equation would be useful when I am setting a salary goal and need to determine the number of hours I must work.
4. c. $0.85(12 - x)$, **f.** 7 roses and 5 carnations;
6. $x = -1$; **8.** $x = -2$; **10.** $x = 81$; **12.** $t = 3$;
14. $x = -11$; **16.** $x = 2500$; **18.** $0 = 0$; The solution is all real numbers, **22. a.** $x = \frac{y - b}{m}$, **c.** $h = \frac{A - 2\pi r^2}{2\pi r}$,
e. $y = \frac{3x - 5}{2}$, **g.** $P = \frac{A}{1 + rt}$

What Have I Learned? 1. a. By order of operations, $-x^2$ indicates to square x first, then negate. For example, $-3^2 = -(3)(3) = -9$. **b.** The negative sign can be interpreted as -1 and by the distributive property reverses the signs of the terms in parentheses. **5. b.** $2x$. The given code of instruction leads to doubling the starting value. Therefore, I would divide the resulting value by 2 to obtain the starting value.

How Can I Practice?

1.

x	$13 + 2(5x - 3)$	$(10x - 10)$	$10x + 7$
1	17	20	17
5	57	60	57
10	107	110	107

Expression a and expression c are equivalent.
2. b. $3x^2 + 15x$, **d.** $-2.4x - 2.64$, **f.** $-6x^2 - 4xy + 8x$;
3. b. $2x(3y - 4z)$, **e.** $2x(x - 3)$; **4. a.** 4, **b.** 5, **c.** -1,
d. -3, **e.** 4 and x^2 (or 4, x, x); **6. b.** $x^2y^2 + 2xy^2$;
7. b. i. add 5 to x; ii. square result; iii. subtract 15;
8. a. $6 - x$, **c.** $3x^2 + 3x$, **e.** $-4x - 10$;
9. $V = 25(2x - 3)$; **13. c.** $D = 198x - 45$ mi.;
14. c. $600 = 280 + 0.20(x - 1000)$, **d.** $x = 2600$.
I must sell \$2600 worth of furniture in order to have a gross salary of \$600. **15. a.** $S = 200 + 0.30x$,
b. $S = 350 + 0.15x$, **d.** $x = \frac{150}{0.15} = 1000$. I would have to sell about \$1000 worth of furniture per week to earn the same weekly salary from either option.
16. c. $x + (x + 10) + (x + 65) = 3x + 75$

Gateway Review

1.

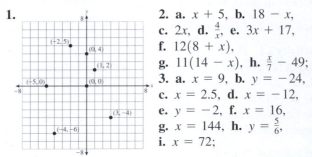

2. a. $x + 5$, **b.** $18 - x$,
c. $2x$, **d.** $\frac{4}{x}$, **e.** $3x + 17$,
f. $12(8 + x)$,
g. $11(14 - x)$, **h.** $\frac{x}{7} - 49$;
3. a. $x = 9$, **b.** $y = -24$,
c. $x = 2.5$, **d.** $x = -12$,
e. $y = -2$, **f.** $x = 16$,
g. $x = 144$, **h.** $y = \frac{5}{6}$,
i. $x = 72$;

4. a. $A = \dfrac{63 + 68 + 72 + x}{4} = \dfrac{203 + x}{4}$,
b. He must score 61. **5. a.** $x = 5$, **b.** $y = 42$,
c. $x = 6$, **d.** $x = -7.75$, **e.** $y = -66$ **f.** $x = 139.5$,
g. $x = -96$, **h.** $y = 27$; **6. a.** The input variable is the number of miles driven. **b.** The output variable is the cost of rental for a day. **c.** $y = 25 + 0.15x$;

d.

Input, x (mi)	100	200	300	400	500
Output, y (\$)	40	55	70	85	100

e.

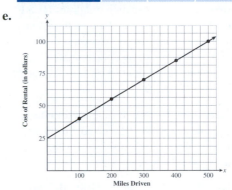

f. The cost to travel round trip from Buffalo to Syracuse is just under \$75. **g.** \$70.90, **h.** I can drive approximately 425 miles. **i.** approximately 433 mi., **j.** (Answers will vary.) **7.** My total annual home sales must be \$714,286 in order to gross \$30,000. **8. a.** $I = 2000(0.05)(I) = \$100$,
b. $I = 3000(0.06)(2) = \$360$;
9. a. $P = 2(2.8 + 3.4) = 12.4$,
b. $P = 2\left(7\frac{1}{3} + 8\frac{1}{4}\right) = 31\frac{1}{6}$;

10.

x	$(4x - 3)^2$	$4x^2 - 3$	$(4x)^2 - 3$
−1	49	1	13
0	9	−3	−3
3	81	33	141

None are equivalent.   **11. a.** $3x + 3$,
b. $-12x^2 + 12x - 18$, **c.** $4x^2 - 7x$, **d.** $-8 - 2x$,
e. $12x - 25$, **f.** $17x - 8x^2$;   **12. a.** $4(x - 3)$,
b. $x(18z + 60 - y)$, **c.** $-4(3x + 5)$;
13. a. $6x^2 - 6x + 3$, **b.** $x^2 - 3x + 7$;
14. a. $10x - 35y$, **b.** $2a + b - 3$,
c. $-10x + 20w - 10z$, **d.** $2c - 1$;   **15. a.** $x = 8.75$,
b. $x = 80$, **c.** $x = 20$, **d.** $x = 3$, **e.** $x = -13$, **f.** $x = 0$,
g. $x = 4$, **h.** $x = 7$, **i.** $x = -3$, **j.** $0 = 27$.
Since 0 does not equal 27, there is no solution.
k. $18x + 16 = 18x + 16$. All real numbers are
solutions.   **16. a.** \$280, **b.** \$196, **c.** $0.70x$,
d. $0.7(0.7x) = 0.49x$, **e.** \$196. The price is the
same **f.** The original price is \$300.
17. a. $500 - n$, **b.** $2.50n$, **c.** $4.00(500 - n)$,
d. $2.50n + 4.00(500 - n) = 1550$, **e.** $n = 300$ student
tickets and 200 adult tickets were sold.
18. a. $C = 1200 + 25x$, **b.** $R = 60x$,
c. $P = 35x - 1200$, **d.** 35 campers,
e. 52 campers, **f.** \$500;   **19. a.** $P = \frac{1}{rt}$, **b.** $t = \frac{f - v}{a}$,
c. $y = \frac{2x - 7}{3}$;   **20. a.** $C = 750 + 0.25x$, **b.** \$875,
c. A thousand booklets can be produced for \$1000.
d. $R = 0.75x$, **e.** To break even, 1500 booklets must
be sold. **f.** To make a \$500 profit, 2500 booklets must
be sold.   **21. a.** $22 - x$, **b.** $2.50x$,
c. $(22 - x)(0.30)(15) = 4.50(22 - x)$,
d. $2.50x + 4.50(22 - x) = 2.50x + 99 - 4.50x$
$$= 99 - 2x,$$
e. 7.5. I can drive 7 days. **f.** \$77.

Chapter 3

Activity 3.1   Exercises: 1. Graph 1, Generator F; Graph 2,
Generator B; Graph 3, Generator D; Graph 4, Generator C;
Graph 5, Generator E; Graph 6, Generator A;
3. a. input: selling price; output: number of units sold,
b. As the selling price increases, the number of units sold
increases slightly at first, reaches a maximum, and then
declines until none are sold.   **7. a.** The graph in part a
represents a function. It passes the vertical line test.
b. The graph in part b does *not* represent a function.
It fails the vertical line test.

Activity 3.2   Exercises: 1. d. $C(n) = 2n + 78$,
g. (Answers will vary.) The number of students who can
sign up for the trip is restricted by the capacity of the bus.
If the bus holds 45 students, then the practical domain is the
set of integers from 0 to 45. **i.** If 24 students go on the trip,
the total cost of the trip will be \$126.   **2. c.** This data set
represents a function because each input value is assigned
a single output value.   **4. a.** $\{-3, -2, -1, 0, 1, 2, 3, 4\}$,

d. The maximum value of f is 6, and it occurs when
x is 4. **g.** $f(3) = 5$;   **5. a. i.** -4, **ii.** $(2, -4)$;
7. $g(-3) = -4$;   **9.** $p(5) = 69$,   **11.** $x = 3$;
13. $w = 30$

Activity 3.3   Exercises: 2. c. The graph is horizontal during
the 1930s and 1950s. **e.** When the median age of first marriage increases, the graph rises. When the median age of
first marriage decreases, the graph falls. When the median
age of first marriage remains unchanged, the graph is horizontal. When the change in first-marriage median age is the
greatest, the graph is steepest.   **3. a.** For the years from
1900 to 1950, the average rate of change was -0.062 yr. of
age/yr. **d.** For the years from 1900 to 1940, the average
rate of change was -0.04 yr. of age/yr.   **4. a.** $\approx -0.17°$F
per hour, **c.** The temperature drops at an average rate of
$1.5°$F per hour from 6 P.M. to 10 P.M.

What Have I Learned? 1. No; the definition of a function
requires that each input value, including $x = 0$, produce
exactly one output value.   **4.** The statement $H(5) = 100$
means that for the function H, an input value of $x = 5$ corresponds to an output value of 100.   **8.** $f(1)$ represents the
output value when the input is 1. Because $f(1) = -3$, the
point $(1, -3)$ is on the graph of f.   **9.** After 10 minutes,
the ice cube weighs 4 grams.   **11.** If the rate of change of a
function is negative, then the initial output value on any interval must be larger than the final value. We can conclude that
the function is decreasing.

How Can I Practice? 1. b. Yes, for each number of credit
hours (input), there is exactly one tuition cost (output).
e. The most credit hours I can take for \$700 is 5.
4. a. $F(x) = 2.50 + 2.20x$;   **6.** $g(-4) = 6$;
8. $h(-3) = 16$;   **9.** $s(6) = 3$;   **14. c.** You leave home,
drive for 2 hours at a constant rate, and then stop for 1 hour.
Finally, you continue at a slower (but constant) speed than
before.   **16. c.** -5 lb./week, **e.** The average rate of
change indicates how quickly I am losing weight.
17. a. 9.2 gal./yr., **c.** -4.93 gal./yr., **e.** -2.48 gal./yr.

Activity 3.4   Exercises: 1. c. For all the data points to be on a
line, the rate of change between any two data points must be
the same, no matter which points are used in the calculation.
3. b. The average rate of change is not constant, so the
data is not linear.   **5. a.** 50 ft. higher, **b.** 5.2% grade;
6. b. at least 240 in. long

Activity 3.5   Exercises: 1. a. linear $\left(\text{constant average rate}\right.$
of change $= \frac{1}{2}\big)$, **c.** not linear (average rate of change not
constant);   **2. d.** Every week I hope to lose 2 pounds.
3. b. The slope of the line for this data is $-\frac{7}{10}$ beats per
minute per year of age.   **4. a. ii.** $m = \frac{-10}{3}$;   **6. a.** 50
mph, **b.** The line representing the distance traveled by car
B is steeper, so car B is going faster. Car A a rate of 50
miles per hour; car B a rate of 74 miles per hour;
7. b. No, their slopes are all different. The number of units
represented by each tick mark is different on each graph,
producing different slopes.

Activity 3.6   Exercises: 1. a. $p(n) = 13n - 2200$ dollars;
2. c. The slope is 13. The profit increases by \$13 for each
additional student attending. **d.** The vertical intercept is
$(0, -2200)$. The class project has a loss of \$2200 if no
students attend. **5. b.** 170 students are needed to break
even. **8. a.** The slope of the line is 2500, indicating that
the value of my home increased at a constant rate of \$2500
per year since 2005. **b.** The v-intercept is $(0, 125,000)$. In
2005, the value of my house was \$125,000.
c. $V(8) = 2500(8) + 125,000 = 145,000$ and
$2005 + 8 = 2013$. The market value of my house in 2013
will be \$145,000.

Activity 3.7   Exercises:

1. slope is 3;

   y-intercept is $(0, -4)$;

   x-intercept is $\left(\frac{4}{3}, 0\right)$.

3. slope is 0;

   y-intercept is $(0, 8)$;

   There is no x-intercept.

5. slope is 2;

   y-intercept is $(0, -3)$;

   x-intercept is $\left(\frac{3}{2}, 0\right)$ or $(1.5, 0)$.

7. e.

The lines all have the same
slope but different y-intercepts.
They are parallel.

9. $y = 12x + 3$;   **11. b.** The y-intercept is $(0, 5)$. The
equation is $y = 3x + 5$.

13. a.

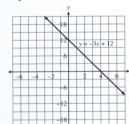

The x-intercept is $(4, 0)$;
the y-intercept is $(0, 12)$.

15. a. The slope is -0.5, indicating that the speed of the
car is decreasing at a constant rate of 0.5 mile per hour for
each foot that it travels. **b.** $v = -0.5d + 60$, **c.** 25 mph

Activity 3.8   Exercises: 1. b. 1 million more than actual
population;   **2. c.** $P(t) = 0.63t + 34.10$,
e. 40,400,000;   **4. a.** t: 0, 10; $P(t)$: 2.39, 2.36,

c. $m = \frac{2.36 - 2.39}{10 - 0} = \frac{-0.03}{10} = -0.003$ million/yr. The population of Portland is decreasing at a rate of 0.003 million
per year. **d.** The vertical intercept is $(0, 2.39)$. The
population of Portland in 1990 $(t = 0)$ was 2.39 million.
5. b. $y = -2.5x + 6$;   **6. c.** in approximately
58 years, by 2048

What Have I Learned? 4. a. The graph is a straight line. **b.** It
can be written in the form $y = mx + b$. **c.** The rate of
change is constant.

How Can I Practice? 1. b. y: 12, $m = 2$, **e.** x: 6, y: 5,
$m = -1$, **f.** x: 2, y: -14, $m = -1.5$;   **2. a.** 7.5 ft;
3. b. Yes, the data represents a linear function because the average rate of change is constant, \$28 per month.
4. d. For each second that passes during the first 5 seconds,
speed increases 11 mph. **5. b.** Yes; the three graphs represent the same linear function. They all have the same slope, 5,
and the same vertical intercept $(0, 0)$. **6. a.** The slope is
undefined. **c.** The slope is 0.5. **7. b.** The vertical
intercept is $(0, 10)$; the horizontal intercept is $\left(\frac{20}{3}, 0\right)$.
8. a. $y = 2.5x - 5$, **b.** $y = 7x + \frac{1}{2}$, **c.** $y = -4$;
11. a. $y = 0x - 2$, **b.** $y = 3x - 2$, **c.** $y = x - 2$.
The lines all have the same vertical intercept $(0, -2)$.

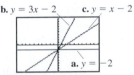

12. a. (Answers will vary.)

x	3	3	3	3
y	−4	0	2	5

c. The horizontal intercept is $(3, 0)$. There is no vertical
intercept.

Activity 3.9   Exercises: 1. c. $t = 0.15i - 361$, **f.** \$13,647;
2. $y = 3x$;

4. $y = 7x + 16$;

6. $y = 2$;

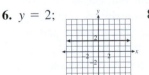

8. $y = -\frac{2}{7}x + 2$;

11. $y = 3x + 9$;   **13.** $y = 2$;
15. g. $s = 32,500 + 1625x$

Activity 3.10 Exercises:

1. h.

COLUMN 3	COLUMN 4
32	3
36.5	5.5
45.5	4.5
53	0
62	4
66.5	1.5
69.5	2.5
80	0
89	5
95	4
99.5	4.5

The error is 34.5.

2. a. 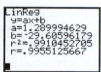 $g = 1.29t - 29.6$;

3. b. $E = 0.115t + 77.4$, **e.** A woman born in 2177 can expect to live 100 years. **4. c.** 18.045%

How Can I Practice? 1. Slope is 2; y-intercept is $(0, 1)$; x-intercept is $\left(-\frac{1}{2}, 0\right)$; **4.** Slope is $-\frac{3}{2}$;

y-intercept is $(0, -5)$; x-intercept is $\left(-\frac{10}{3}, 0\right)$;
7. Slope is -2; y-intercept is $(0, 2)$; x-intercept is $(1, 0)$;
   9. $y = 9x - 4$; **12.** $y = 0$;
16. a. $N = 4t - 160$, **b.** 88 chirps per minute;
19. a. $y = 3x + 6$, **d.** $y = 4x - 27$, **f.** $y = 2x - 13$;
20. a. $y = 2x + 6$, **c.** $y = 3$, **e.** $x = 6$;
22. a. $(-2, 0)$, **c.** $(2.6, 0)$

Activity 3.12 Exercises: 2. a. $s = 1.5n + 17.2$,
b. $s = 2.3n + 9.6$, **c.** 2019;

4. $q = 2, p = 0$;

6. $x = 6, y = 1$;

8. 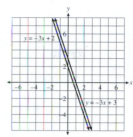 $2 = 3$; the variable drops out, leaving a false statement. Thus, there is no solution. Examining the graphs, I see that the lines are parallel.

Activity 3.13 Exercises: 1. a. $x = 0.5, y = 2.5$;
c. $x = 4, y = -1$; **2. b.** $y = 5, x = 1$;
4. a. $8x + 5y = 106$, **b.** $x + 6y = 24$, **d.** The cost of a centerpiece is \$12, and the cost of a glass is \$2.

Activity 3.14 Exercises: 1. a. $R(x) = 200x$, **b.** The slope of the cost function is 160 dollars per bundle. This represents a cost of \$160 to produce each bundle of pavers. The y-intercept is $(0, 1000)$. Fixed costs of \$1000 are incurred even when no pavers are produced. **c.** The slope of the revenue function is 200 dollars per bundle. This represents the income for each bundle sold. The y-intercept is $(0, 0)$. If no pavers are sold, there is no income. **e.** $(25, 5000)$

Activity 3.15 Exercises: 1. $l + w + d \le 61$; **4.** $x > -2$;
6. $x < 5$; **8.** $x \ge 8$; **10.** $x \ge -0.4$;
12. c. $19.95 + 0.49n < 40.95 + 0.19n$

How Can I Practice? 2. $(-45, -40)$; **4. a.** $C = 500 + 8n$,
b. $R = 19.50n$, **c.** 44 centerpieces; **6.** $x = -1, y = 275$;
8. 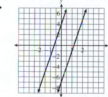 There is no solution. The lines are parallel.

10. $x \approx -6.48, y \approx -15.78$;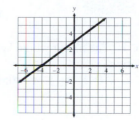

12. d. After 80 boxcars are loaded, the management will realize a profit on the purchase of the forklift.

Gateway Review 1. The slope is -2.5, the y-intercept is $(0, 4)$; **2.** The equation is $y = -2x + 2$. **3.** y-intercept is $(0, 8)$. An equation for the line is $y = \frac{-4}{3}x + 8$. **4.** row 1: c, e, b; row 2: f, a, d; **5.** y-intercept is $(0, 5)$. Equation is $y = 3x + 5$. **6.** The y-intercept is $(0, -2)$; the x-intercept is $(4, 0)$.

7. The vertical intercept is $(0, 3)$; the horizontal intercept is $(-4, 0)$.

8. $y = \frac{5}{3}x - 4$;   **9. a.** $(2, 2)$,

b.

c.
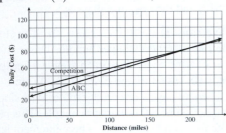

10. $a = -1; b = 6$;   **11.** $y = 2, x = 2$;
12. $q = -1.25p + 100$ **13.** $y = -0.56x + 108.8$;
14. a. ABC car rental: $C(x) = 24 + 0.30x$;
competitor: $C(x) = 34 + 0.25x$,
b.
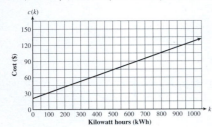

c. The lines intersect at $(200, 84)$. The daily cost would be the same at both companies when 200 miles are traveled.
d. It will be cheaper to rent from the competitor because the distance is 225 miles and the competitor's prices are cheaper if you travel over 200 miles.
15. a. $m = \frac{32 - 212}{0 - 100} = \frac{-180}{-100} = \frac{9}{5}$,   **b.** $F = 1.8C + 32$,
c. $104°$ Fahrenheit corresponds with $40°C$,
d. $170°F$ corresponds approximately to $77°C$;
16. a. \$94.97, \$110.90, \$116.21, \$126.83,
b.

c. $C(k) = 0.1062k + 20.63$,   **d.** The cost of using 875 kWh of electricity is \$113.56.   **e.** I used approximately 1218 kWh of electricity.
17. a. $P(t) = 2500t + 50,000$ dollars,   **b.** 75,100,150,
c. When $t = 40$, that is, in the year 2015, the price of the house will be \$150,000.   **18.** $V(t) = -95t + 950$;
19. a. The regression equation is approximately
$$w(h) = 6.92h - 302;$$
```
LinReg
y=ax+b
a=6.921768707
b=-301.9047619
r²=.8721857164
r=.9339088373
```

b. He would weigh about 252 lb.   **c.** The player would be about 71.1 in. tall.   **20.** The solution is $(2, 4)$.
21. The solution is $(4, 3)$;   **22. a.** $x \geq 1$,   **b.** $x \geq -6$;

23. Numerically:

x	$R(x)$	$C(x)$
0	0	2500
50	3750	5000
100	7500	7500
150	11,250	10,000
200	15,000	12,500

Graphically:
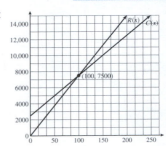

Algebraically: $x \geq 100$

Chapter 4

Activity 4.1   Exercises: 1. a. ii. Yes, it *is* a polynomial because the variable in each term contains a positive integer power of x or is a constant.   **iv.** No, it is *not* a polynomial because the variable in the second term is inside a radical sign and in the denominator.   **4.** $7x - 1$;   **6.** $8x - 7y$;
8. $2x^2 + 6$;   **10.** $4x^3 + 5x^2 + 18x - 5$;   **12.** $2.8x + 7$;
14. 12.6;   **16.** 9.723;   **18.** -75;   **21. a.** $-x^2 - x + 2$
c. $-5x^3 + 6x^2 - x$;   **23. a.** \$17,453,   **b.** \$39,315; $t = 48$ corresponds to the year 2008. If the polynomial function model continues to be valid past the year 2004, then the per capita income for U.S. residents in the year 2008 should be approximately \$39,315.

Activity 4.2   Exercises: 2. a^4;   **4.** y^9;   **6.** $-12w^7$;
8. a^{15};   **10.** $-x^{50}$;   **12.** $-5.25x^9y^2$;   **14.** $-2s^6t^8$;
16. $2xy$;   **18.** x^{-3} or $\frac{1}{x^3}$;   **20.** $11x^2$;   **22.** $\frac{81a^8}{16b^4}$;
24. $2x^2 + 6x$;   **26.** $2x^4 + 3x^3 - x^2$;   **28.** $10x^4 - 50x^3$;
30. $18t^6 - 6t^4 - 4.5t^2$;   **34. a.** $A = 5x(4x) = 20x^2$,
b. $V = 20x^2(x + 15) = 20x^3 + 300x^2$;
36. $A = (3xy^2)^2 = 9x^2y^4$;   **38.** $V = 9\pi h^3$;
40. $V = \pi r^3 + 4\pi r^2$;   **42. a.** $3x^2(6x^5 + 9x - 5)$,
c. $3x^3(2x^2 - 4x + 3)$

Activity 4.3   Exercises: 1. a. $x^2 + 8x + 7$,
e. $10 + 9c + 2c^2$,   **i.** $12w^2 + 4w - 5$,
m. $2a^2 - 5ab + 2b^2$;
2. a. $3x^2 - 2xw + 2x + 2w - 5$,
c. $3x^3 - 11x^2 - 6x + 8$,   **e.** $x^3 - 27$;
5. a. $V = 7(4)(8) = 224$ in.³,
b. $V = (7 + x)(4 + x)(8)$,   **c.** $V = 224 + 88x + 8x^2$,

d. 432 in.^{3}, 208 in.^{3}, $\frac{208}{224} \approx 92.9\%$,
e. $V = (7 - x)(4 + x)(8)$,
f. $V = 224 + 24x - 8x^2$, **g.** $x = 2$;
$V = 224 + 24(2) - 8(2)^2 = 224 + 48 - 32 = 240$ cu. in.
Check: $V = (7 - 2)(4 + 2)(8) = 5(6)(8) = 240$ cu. in.

What Have I Learned? 2. The expression $-x^2$ instructs you to square the input first, and then change the sign. The expression $(-x)^2$ instructs you to change the sign of the input and then square the result. **5.** It reverses the signs of the terms in parentheses: $-(x - y) = -1(x - y) = -x + y$.
7. Yes; in $3x^2$, only x is squared, but in $(3x)^2$, both the 3 and the x are squared: $(3x)^2 = (3x)(3x) = 9x^2$.

How Can I Practice? 2. e. $= p^{20} \cdot p^6 = p^{26}$, **f.** $12x^{10}y^8$,
i. $3y^7$, **l.** a^{12}, **q.** $-3s^5t^{12}$; **3. b.** $2x^2 + 14x$,
e. $6x^5 - 15x^3 - 3x$; **4. b.** $5xy(2x - 3y)$,
e. $8xy(y^2 + 2x - 3x^2)$;
5. b. Start with x, then add 3, square the result, subtract 12;
6. a. $7x + 10$, **c.** $4x^2 - 7x$; **7. a.** $x^2 - x - 6$,
d. $x^2 + 2xy - 8y^2$, **g.** $x^3 + 2x^2 + 9$,
i. $a^3 - 4a^2b + 4ab^2 - b^3$; **10. a.** $A = (3x)(4x) = 12x^2$,
b. $A = (3x - 5)(4x) = 12x^2 - 20x$,
c. $A = (3x - 5)(4x + 5) = 12x^2 - 5x - 25$

Activity 4.4 Exercises: 1. e. $35 = \frac{8}{3}t^2$; approximately 3.62 sec; **2. a.** $x = \pm 3$, **c.** $x^2 = -4$. There is no real number solution because a square of a real number cannot be negative.
f. $x = \pm 8$; **3. c.** $a = \pm 4$; **4. b.** $x = -1$ or $x = 3$
e. $x = 7$ or $x = -3$ **h.** $x = 8$ or $x = -16$; **5.** 13 in.; **6.** 15 in.

Activity 4.5 Exercises: 1. b. The ball will hit the water 2.5 seconds after it is tossed in the air; **3. a.** $x = 0$ or $x = 8$,
c. $x = 0$ or $x = 4$, **e.** $x = 0$ or $x = 5$; **4. c.** The rocket is on the ground $t = 0$ seconds from launch and returns to the ground $t = 30$ seconds after launch.

f.
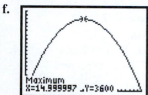
It will reach its maximum height of 3600 feet 15 seconds after launch.

5. a. $x = 0$ or $x = -10$, **c.** $y = 0$ or $y = -5$,
e. $t = 0$ or $t = 1.5$, **g.** $w = 0$ or $w = 4$

Activity 4.6 Exercises: 1. a. $x = -1$ or $x = -6$,
c. $y = -7$ or $y = -4$; **2. a.** The rocket returns to the ground in 27 seconds. **c.** The rocket is 1760 feet above the ground 5 seconds after launch and again, on the way down, 22 seconds after launch. **5. a.** At $t = 0$, the apple is 96 feet above the ground. **b.** $-16t^2 + 16t + 96 = 0$,
c. $t = 3$ or $t = -2$. Discard the result $t = -2$.
f. The maximum point of the parabola is $(0.5, 100)$. **g.** The coordinates $(0.5, 100)$ indicate that the apple at its highest point is 100 feet above the ground half a second after I toss it.

Activity 4.7 Exercises: 1. $x = -5$ or $x = 3$;
3. $x = 1$ or $x = -\frac{1}{2}$; **5.** $x \approx 10.83$ or $x \approx -0.83$;
6. b. The rocket is 2000 feet above the ground approximately 5.9 seconds and approximately 21.1 seconds after launch. **c.** The rocket is 2800 feet above the ground approximately 10.8 seconds and approximately 16.2 seconds after launch. **9. a.** There are no solutions because $\sqrt{-16}$ is not a real number. Thus, there are no x-intercepts. **b.** The graph of $y = x^2 - 2x + 5$ does not cross the x-axis so there are no x-intercepts.

What Have I Learned? 1. b. $100 - x^2 = 40$;
2. a. Divide each term by 2 and then take square roots.
c. Factor the left side of the equation and then apply the zero-product property. **3. c.** $x = -5$ or $x = 3$;
7. There will always be *one* y-intercept. The x-coordinate of the y-intercept is always zero. Therefore, $y(0) = a(0)^2 + b(0) + c = c$. This shows that the y-intercept is always $(0, c)$.

How Can I Practice? 3. a. $x = 4$ or $x = 5$,
e. $x = 3$ or $x = -1$, **g.** $x = 5$ or $x = 2$,
i. $x = 0$ or $x = -\frac{3}{2}$;
5. a. $x = -6$ or $x = 1$, **c.** $x = 9$ or $x = -2$,
e. $x = -8$ or $x = 1$, **g.** $x \approx 4.414$ or $x \approx 1.586$,
i. $x \approx 3.414$ or $x \approx 0.586$; **8. a.** 10 ft, **b.** 8 ft,
c. -80. This result is possible only if there is a hole in the ground. By the time 3 seconds pass, the ball is already on the ground. **d.** 10.24, 10.2464, 10.2496, 10.25, 10.2496, 10.2464, 10.24. Both before and after 0.625 second, the height is less than 10.25 feet. Therefore, the ball reaches its maximum height in 0.625 second.

Activity 4.8 Exercises: 2. a. The growth factor is $1 + 0.05 = 1.05$; **c.** $110, $115.50, $121.28, $127.34, $133.71, $140.39, $147.41; **3. a.** The function is increasing because the base 5 is greater than 1 and is a growth factor. **b.** The function is decreasing because the base $\frac{1}{2}$ is less than 1 and is a decay factor. **c.** The base 1.5 is a growth factor (greater than 1), so the function is increasing. **d.** The base 0.2 is a decay factor (less than 1), so the function is decreasing.

Activity 4.9 Exercises: 1. a. $s = 0.08p$,

b.

LIST PRICE, p ($)	SALES TAX s ($)
10	0.80
20	1.60
30	2.40
50	4.00
100	8.00

c.

List Price ($)

4. $y = kx$, $25 = k(5)$, $k = 5$, $y = 5x$, $y = 5(13)$, $y = 65$;
8. a. No, area varies directly as the square of the radius.
b. $k = \pi$

Activity 4.10   Exercises:
1. c.

LENGTH OF GUITAR STRING (m)	FREQUENCY (Hz)
0.2	1050
0.3	700
0.4	525
0.5	420
0.6	350

e. The frequency would be approximately 600 hertz.
3. b. The man weighs 128 pounds when he is 1000 miles above the surface of Earth.   **4. a.** $y = \frac{120}{x}$. Yes, y is a function of x. The domain is all real numbers except $x = 0$.
5. a. $x = 4$, **c.** $x = -5.5$, **d.** $x = 28$.

Activity 4.11   Exercises: 1. a. 49.9 mph;   **2. a.** $x = \pm 3$,
c. $t^2 = -9$. There is no real number solution. There is no real number whose square is a negative number. **e.** $x = 49$,
g. $x = 5$

What Have I Learned? 2. b. The formula represents an exponential function because the input variable is an exponent in the expression defining the function.   **3. c. i.** $x = 2$, a square; **ii.** The length is twice the width when the width is $\sqrt{2}$, or approximately 1.4 centimeters. The length is half the width when the width is $\sqrt{8}$, or approximately 2.8 centimeters.

How Can I Practice? 1. a. $r = \sqrt{\frac{V}{\pi h}}$, **c.** $b = \sqrt{c^2 - a^2}$;
3. a. $s(y) = 32{,}000(1.03)^y$, where s represents salary and y represents year. The practical domain covers the length of the contract, 0 to 5 years. **e.** $43{,}005$;   **4. b.** 0.95,
d. The domain is all real numbers except 0. **e.** The practical domain is all real numbers greater than 0 and less the $2 (or so).   **6.** $x = 1.5$;   **7.** $x = 0.02$;   **8.** $x = \frac{1494}{18} = 83$;
13. a. 1440, 1440, 1440, 1440, 1440, 1440, **b.** $f(x) = \frac{1440}{x}$,

c.

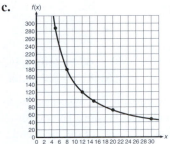

d. From the graph, $f(x) = 200$ when x is approximately 7.
e. The estimate in part d is close to the value calculated from the formula.

Gateway Review 1. $15x - 41$;   **2.** $-14x^2 + 20x - 36$;
3. $2x^3 - 9x^2 - x - 5$;   **4.** $-x^3 - 13x^2 + 16x - 2$;
5. $3x^7 + 2x^6 - 5x^5 - 20x^4$;   **6.** $9x^4 - 16x^3 - 37x^2 + 16x$;
7. $16x^{12}$;   **8.** $125x^3y^9$;   **9.** $16a^5b^5$;   **10.** $y^{11-3} = y^8$;
11. $5t^3$;   **12.** $3x^2(9x^3 - 3x - 4)$;   **13.** $4x(2x + 1)$;
14. $(5a - 13) \cdot 2 = 10a - 26$;   **15. a.** $x^2 - 10x + 24$,
b. $x^2 - 5x - 24$, **c.** $x^2 - 9$, **d.** $x^2 - 12x + 36$,
e. $2x^2 - 13x + 15$, **f.** $6x^2 - 13x - 28$,
g. $4x^2 - 25$, **h.** $x^2 + 8x + 16$, **i.** $2x^3 + 7x^2 - 9$,
j. $3x^3 + 13x^2 - 14x - 20$;
16. $A = \pi(11 + y)^2$ (factored form),
$A = \pi y^2 + 22\pi y + 121\pi$ (expanded form);
17. $A = (x + 7)(x - 2)$ (factored form),
$A = x^2 + 5x - 14$ (expanded form);
18. a. $x = \pm 2$, **b.** $x = \pm 3$, **c.** $x^2 = -4$, no solution,
d. $x = 6$ or $x = -10$, **e.** $x = 8$ or $x = -2$,
f. $x = -1$ or $x = -15$;   **19. a.** $x = 0$ or $x = -8$,
b. $x = 0$ or $x = 3$, **c.** $x = 0$ or $x = 5$,
d. $x = 0$ or $x = 5$, **e.** $x = \frac{1}{4}$ or $x = 6$;
20. a. $s = 20$ ft, **b.** The ball reaches a height of 32 feet after 1 second and then again on its way down after 2 seconds. **c.** The ball reaches the ground after 3 seconds.
21. a. $x = -6$ or $x = 2$, **b.** $x = 7$ or $x = -2$,
c. $x = 8$ or $x = -3$, **d.** $x = 3$ or $x = 7$,
e. $x = 4$ or $x = -6$, **f.** $y = -8$ or $y = 5$;
22. a. $x \approx 1.786$ or $x \approx -1.120$, **b.** $x = 1$ or $x = -2.4$,
c. $x \approx -1.345$ or $x \approx 3.345$, **d.** There is no real solution because $\sqrt{-15}$ is not a real number.
23. a. $x = 6$, **b.** $x = -7$, **c.** $x = 49$,
d. $x = 625$, **e.** $x = 33$;   **24. a.** In 5 years the tuition will be $401.47 per credit; in 10 years, $537.25.
b. approximately $20.29 per credit per year,
c. approximately $23.73 per credit per year,
d. about 12 yrs;   **25.** approximately 8840 lb;
26. a. 24 min., **b.** 30 mph;   **27. a.** approximately 0.87 sec., **b.** 1.5 sec., **c.** approximately 1.94 sec.
28. a. not linear; **b.** 4763, 6105, 7825; **c.** close, but not exact; **d.** approximately 57,022 MW

Glossary

absolute value of a number   The size or magnitude of a number. It is represented by a pair of vertical line segments enclosing the number and indicates the distance of the number from zero on the number line. The absolute value of 0, $|0|$, equals 0. The absolute value of any nonzero number is always positive.

algebraic expression   A mathematical set of instructions (containing constant numbers, variables, and the operations among them) that indicates the sequence in which to perform the computations.

angle   The figure formed by two rays (half-lines) starting from the same point.

area   The size, measured in square units, of a two-dimensional region.

average   The sum of a collection of numbers, divided by how many numbers are in the collection.

bar graph   A diagram of parallel bars depicting data.

binomial   A polynomial with exactly two terms.

break-even number   The number for which the total revenue equals the total cost.

Cartesian (rectangular) coordinate system in the plane   A system in which every point in the plane can be identified by an ordered pair of numbers, each number representing the distance of the point from a coordinate axis.

circle   A collection of points in a plane that are the same distance from some given point called its center.

circumference   The distance around a circle.

coefficient   A number (constant) that multiplies a variable.

common factor   A factor contained in each term of an algebraic expression.

commutative operation   A binary operation (one that involves two numbers) in which interchanging the order of the numbers always produces the same result. Addition and multiplication are commutative operations.

completely factored form   An algebraic expression written in factored form where none of its factors can themselves be factored any further.

constant term   A term whose value never changes. It can be a particular number such as 10 or a symbol such as c whose value, although unspecified, is fixed.

decay factor   Ratio of new (decreased) value to the original value.

degree of a monomial   The exponent on its variable. If the monomial is a constant, the degree is zero.

degree of a polynomial   The highest degree among all its terms.

delta   The name of the Greek letter, Δ. When Δ precedes a variable, as in Δx, the symbol is understood to represent a change in value of the variable.

dependent variable   Another name for the output variable of a function.

diameter   A line segment that goes through the center of a circle and connects two points on the circle; it measures twice the radius.

direct variation   A relationship defined algebraically by $y = k \cdot x$, where k is a constant, in which whenever x *increases* by a multiplicative factor (e.g., doubles), y also *increases* by the same factor.

distributive property   The property of multiplication over addition (or subtraction) that states that $a \cdot (b + c) = a \cdot b + a \cdot c$.

domain   The set (collection) of all possible input values for a function.

equation   A statement that two algebraic expressions are equal.

equivalent expressions   Two algebraic expressions are equivalent if they always produce identical outputs when given the same input value.

expand   An instruction to use the distributive property (or FOIL) to transform an algebraic expression from factored form to expanded form.

expanded form   An algebraic expression in which all parentheses have been removed using multiplication.

exponential function   A function in which the variable appears in the exponent. For example, $y = 4 \cdot 5^x$.

extrapolation   The assignment of values to a sequence of numbers beyond those actually given.

factor, a   When two or more algebraic expressions or numbers are multiplied together to form a product, those individual expressions or numbers are called the factors of that product. For example, the product $5 \cdot a \cdot (b + 2)$ contains the three factors 5, a, and $b + 2$.

factor, to   An instruction to use the distributive property to transform an algebraic expression from expanded form to factored form.

factored form   An algebraic expression written as the product of its factors.

fixed costs   Costs (such as rent, utilities, etc.) that must be paid no matter how many units are produced or sold.

formula   An algebraic statement describing the relationship among a group of variables.

function   A rule (given in tabular, graphical, or symbolic form) that relates (assigns) to any permissible input value exactly one output value.

function terminology and notation   The name of a functional relationship between input x and output y. You say that y is a function of x, and you write $y = f(x)$.

goodness-of-fit   A measure of the difference between the actual data values and the values produced by the model.

greatest common factor   The largest factor that exactly divides each term in an expression.

growth factor   Ratio of new (increased) value to the original value.

heuristic algorithm   A step-by-step procedure that can be carried out quickly but that does not necessarily yield the best solution.

histogram   A bar graph with no spaces between the bars.

horizontal axis   The horizontal number line of a rectangular coordinate system. When graphing a relationship between two variables, the input (independent) variable values are referenced on this axis.

horizontal intercept   A point where a line or curve crosses the horizontal axis.

independent variable   Another name for the input variable of a function.

input   Replacement values for a variable in an algebraic expression, table, or function. The value that is listed first in a relationship involving two variables.

integers   The set of positive and negative counting numbers and zero.

interpolation   The insertion of intermediate values in a sequence of numbers between those actually given.

inverse operation   An operation that undoes another operation. Addition and subtraction are inverses of each other as are multiplication and division. The inverse of squaring is taking the square root. Negation is its own inverse, as is taking a reciprocal.

inverse variation   A relationship defined algebraically by $y = \dfrac{k}{x}$, where k is a constant, in which whenever x *increases* by a multiplicative factor (e.g., doubles), y *decreases* by the same factor (e.g., y is halved).

like terms   Terms that contain identical variable factors (including exponents) and may differ only in their numerical coefficients.

line of best fit (to a given set of data points)   Line that will have as many of the data points as close to the line as possible.

linear function   A function that has the form $f(x) = mx + b$ and whose graph is a straight line.

linear regression equation   The equation of the line that best represents a set of plotted data points.

maximum value   The largest output value of a function.

mean   The arithmetic average of a set of data.

minimum value   The smallest output value of a function.

monomial   A single term consisting of either a number (constant), a variable, or a product of variables raised to a positive integer exponent, along with a coefficient.

numerical coefficient   A number that multiplies a variable or coefficient.

operation (arithmetic)   Addition, subtraction, multiplication, and division are binary operations (operations that involve two numbers). Exponentiation (raising a number to a power, for example, 10^4), taking roots (for example, $\sqrt{8}$), and negation (changing the sign of a number) are examples of operations on a single number.

order of operations convention   The order in which to perform each of the arithmetic operations contained in a given expression or computation.

ordered pair   Two numbers or symbols, separated by a comma and enclosed in a set of parentheses. An ordered pair can have several interpretations, depending on its context. Common interpretations are as the coordinates of a point in the plane and as an input/output pair of a function.

origin   The point at which the horizontal and vertical axes intersect.

output   Values produced by evaluating an algebraic expression, table, or function. The value that is listed second in a relationship involving two variables.

parabola   The graph of a quadratic function (second-degree polynomial function). The graph is U-shaped, opening upward or downward.

parallelogram   A four-sided figure for which each pair of opposite sides is parallel.

percent(age)   Parts per hundred.

percentage error   Describes the significance of a measurement error. It is calculated by dividing the difference between the true value and the measured (or predicted) value by the true value, then converting to a percent.

perimeter   A measure of the distance around a figure; for circles, perimeter is called *circumference*.

pie chart   A diagram of sectors of a circle depicting data.

polynomial   An algebraic expression formed by adding and/or subtracting monomials.

practical domain   The domain determined by the situation being studied.

practical range   The range determined by the situation being studied.

proportion   An equation stating that two ratios are equal.

Pythagorean theorem   The relationship among the sides of a right triangle that the sum of the squares of the lengths of the two perpendicular sides (*legs*) is equal to the square of the length of the side opposite the right angle, called the *hypotenuse*.

quadrants   The four regions of the plane separated by the vertical and horizontal axes. The quadrants are numbered from I to IV. Quadrant I is located in the upper right with the numbering continuing counterclockwise. Therefore, quadrant IV is located in the lower right.

quadratic equation   Any equation that can be written in the form $ax^2 + bx + c = 0, a \neq 0$.

quadratic formula   The formula $x = \dfrac{-b \pm \sqrt{b^2 - 4ac}}{2a}$ that represents the solutions to the quadratic equation $ax^2 + bx + c = 0$.

quadratic function   A function of the form $y = ax^2 + bx + c$.

radius   The distance from the center point of a circle to any point on the circle.

range   The set (collection) of all possible output values for a function.

rate of change   A quotient that compares the change in output values (numerator) to the corresponding change in input values (denominator). In context, a rate of change can be recognized by the word *per*, as in "miles per hour" or "people per year."

ratio   A quotient that compares two similar numerical quantities, such as "part" to "whole."

regression line A line that best approximates a set of data points as shown on a scatterplot.

rounding The process of approximating a given number by one ending at a specified place value and inserting placeholding zeros if necessary. For example, rounding 45,382 to the nearest hundred yields 45,400.

scaling The process of assigning a fixed distance between adjacent tick marks on a coordinate axis.

scatterplot A set of points in the plane whose coordinate pairs represent input/output pairs of a data set.

scientific notation A concise way for expressing very large or very small numbers as a product—(a number between 1 and 10) · (the appropriate power of 10).

signed numbers Numbers accompanied by a positive or negative sign. If no sign is indicated, the number is understood to be positive.

slope of a line A measure of the steepness of a line. It is calculated as the ratio of the rise (vertical change) to the run (horizontal change) between any two points on the line.

slope-intercept form The form $y = mx + b$ of a linear function in which m denotes the slope and b denotes the y-intercept.

solution of an equation Replacement value(s) for the variable that makes both sides of the equation equal in value.

solution of a system A point that is a solution to both equations in a system of two linear equations.

sphere A three-dimensional object for which all points are equidistant from a given point, the center.

square root function A function of the form $y = a\sqrt{x}$, where a is a constant and $x \geq 0$.

substitution method A method used to solve a system of two linear equations. Choose one equation and rewrite it to express one variable in terms of the other. In the other equation, replace the chosen variable with its expression and solve to determine the value of the second variable. Substitute the value in either original equation to determine the value of the first variable.

symbolic rule A shorthand code that indicates a sequence of operations to be performed on the input variable, x, to produce the corresponding output variable, y.

system of equations Two linear equations in the same variables considered together.

terms Parts of an algebraic expression separated by the addition, $+$, and subtraction, $-$, symbols.

trinomial A polynomial with exactly three terms.

variable A quantity that takes on specific numerical values and will often have a unit of measure (e.g., dollars, years, miles) associated with it.

verbal rule A statement that describes in words the relationship between input and output variables.

vertex   The turning point of a parabola having coordinates $\left(\frac{-b}{2a}, f\left(\frac{-b}{2a}\right)\right)$, where a and b are determined from the equation $f(x) = ax^2 + bx + c$. The vertex is the highest or lowest point of a parabola.

vertical axis   The vertical number line of a rectangular coordinate system. When graphing a relationship between two variables, the output (dependent) variable values are referenced on this axis.

vertical intercept   The point where a line or curve crosses the vertical axis.

vertical line test   A visual method of determining if a given graph is the graph of a function. A graph represents a function if and only if any vertical line that is drawn intersects the graph in at most one point.

volume   A measure of space, given in cubic units, enclosed by a three-dimensional object.

weighted average   An average in which some numbers count more heavily than others.

***x*-coordinate**   The first number of an ordered pair that indicates the horizontal directed distance of the point from the origin. A positive value indicates the point is located to the right of the origin; a negative value indicates a location to the left of the origin.

***y*-coordinate**   The second number of an ordered pair that indicates the vertical directed distance of the point from the origin. A positive value indicates the point is located above the origin; a negative value indicates a location below the origin.

zero-product rule   The algebraic principle that states if a and b are real numbers such that $a \cdot b = 0$, then either a or b, or both, must be equal to zero.

Index

Geometric Formulas

Perimeter and Area of a Triangle, and Sum of Measures of the Angles

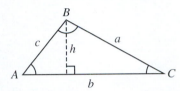

$$P = a + b + c$$
$$A = \frac{1}{2}bh$$
$$A + B + C = 180°$$

Pythagorean Theorem

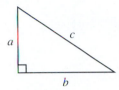

$$a^2 + b^2 = c^2$$

Perimeter and Area of a Rectangle

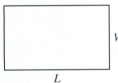

$$P = 2L + 2W$$
$$A = LW$$

Perimeter and Area of a Square

$$P = 4s$$
$$A = s^2$$

Area of a Trapezoid

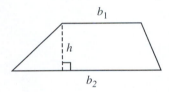

$$A = \frac{1}{2}h(b_1 + b_2)$$

Circumference and Area of a Circle

$$C = 2\pi r$$
$$A = \pi r^2$$

Volume and Surface Area of a Rectangular Solid

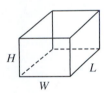

$$V = LWH$$
$$SA = 2LW + 2LH + 2WH$$

Volume and Surface Area of a Sphere

$$V = \frac{4}{3}\pi r^3$$
$$SA = 4\pi r^2$$

Volume and Surface Area of a Right Circular Cylinder

$$V = \pi r^2 h$$
$$SA = 2\pi r^2 + 2\pi rh$$

Volume and Surface Area of a Right Circular Cone

$$V = \frac{1}{3}\pi r^2 h$$
$$SA = \pi r^2 + \pi rl$$

Metric–U.S. Conversion

Length

1 meter = 3.28 feet

1 meter = 1.09 yards

1 centimeter = 0.39 inches

1 kilometer = 0.62 miles

Weight

1 gram = 0.035 ounces

1 kilogram = 2.20 pounds

1 gram = 0.0022 pounds

Volume

1 liter = 1.06 quarts

1 liter = 0.264 gallons

Temperature

Celsius to Fahrenheit

$$F = \frac{9}{5}C + 32$$
$$C = \frac{5}{9}(F - 32)$$